AF323660

ELEMENTS OF MATHEMATICAL THEORY OF EVOLUTIONARY EQUATIONS IN BANACH SPACES

WORLD SCIENTIFIC SERIES ON NONLINEAR SCIENCE

Editor: Leon O. Chua
University of California, Berkeley

Series A.   MONOGRAPHS AND TREATISES*

Volume 71:   A Practical Guide for Studying Chua's Circuits
R. Kiliç

Volume 72:   Fractional Order Systems: Modeling and Control Applications
R. Caponetto, G. Dongola, L. Fortuna & I. Petráš

Volume 73:   2-D Quadratic Maps and 3-D ODE Systems: A Rigorous Approach
E. Zeraoulia & J. C. Sprott

Volume 74:   Physarum Machines: Computers from Slime Mould
A. Adamatzky

Volume 75:   Discrete Systems with Memory
R. Alonso-Sanz

Volume 76:   A Nonlinear Dynamics Perspective of Wolfram's New Kind of Science
(Volume IV)
L. O. Chua

Volume 77:   Mathematical Mechanics: From Particle to Muscle
E. D. Cooper

Volume 78:   Qualitative and Asymptotic Analysis of Differential Equations
with Random Perturbations
A. M. Samoilenko & O. Stanzhytskyi

Volume 79:   Robust Chaos and Its Applications
Z. Elhadj & J. C. Sprott

Volume 80:   A Nonlinear Dynamics Perspective of Wolfram's New Kind of Science
(Volume V)
L. O. Chua

Volume 81:   Chaos in Nature
C. Letellier

Volume 82:   Development of Memristor Based Circuits
H. H.-C. Iu & A. L. Fitch

Volume 84:   Topology and Dynamics of Chaos:
In Celebration of Robert Gilmore's 70th Birthday
C. Letellier & R. Gilmore

Volume 85:   A Nonlinear Dynamics Perspective of Wolfram's New King of Science:
(Volume VI)
L. O. Chua

Volume 86:   Elements of Mathematical Theory of Evolutionary Equations
in Banach Spaces
A. M. Samoilenko & Y. V. Teplinsky

*To view the complete list of the published volumes in the series, please visit:
http://www.worldscientific.com/series/wssnsa

WORLD SCIENTIFIC SERIES ON
NONLINEAR SCIENCE

Series A   Vol. 86

Series Editor: Leon O. Chua

ELEMENTS OF MATHEMATICAL THEORY OF EVOLUTIONARY EQUATIONS IN BANACH SPACES

Anatoly M Samoilenko
National Academy of Sciences, Ukraine

Yuriy V Teplinsky
Kamyanets-Podilsky National University, Ukraine

World Scientific

NEW JERSEY · LONDON · SINGAPORE · BEIJING · SHANGHAI · HONG KONG · TAIPEI · CHENNAI

Published by

World Scientific Publishing Co. Pte. Ltd.

5 Toh Tuck Link, Singapore 596224

USA office: 27 Warren Street, Suite 401-402, Hackensack, NJ 07601

UK office: 57 Shelton Street, Covent Garden, London WC2H 9HE

British Library Cataloguing-in-Publication Data
A catalogue record for this book is available from the British Library.

World Scientific Series on Nonlinear Science, Series A — Vol. 86
ELEMENTS OF MATHEMATICAL THEORY OF EVOLUTIONARY
EQUATIONS IN BANACH SPACES

ISBN 978-981-4434-82-9

Printed in Singapore by World Scientific Printers.

Preface

This book is devoted to the further development of the theory of difference and differential equations in an abstract Banach space and also in the Banach space $\mathfrak{M}$ of bounded numerical sequences. It focuses on studying the problems of reduction to the canonical form in neighborhoods of the invariant sets of linear and nonlinear difference equations defined on tori, the development of the theory of invariant tori and bounded semi-invariant manifolds for such equations, studying the questions of existence and approximate construction of the periodic solutions of difference equations in infinite-dimensional spaces, and the possibility to extend solutions in degenerate cases. For the nonlinear differential equations in the space $\mathfrak{M}$, the book presents new results concerning the theory of countably point boundary-value problems.

It is well known that the systematic study of difference equations started in the second half of the 20th century is primarily due to the development of technical sciences. These equations proved to be useful while investigating the discrete dynamical systems, impulsive systems, and systems that include numerical computing devices. The difference equations were also widely applied to the numerical solution of differential equations of various types, which was associated with employing the finite-difference method. The known monographs of Ya.V. Bykov and V.L. Linenko [13], A. Halanay and D. Wexler [32], D.I. Martynyuk [54], Yu.A. Mitropol'skii, A.M. Samoilenko and D.I. Martynyuk [74], O.M. Sharkovskii, Yu.L. Maistrenko and E.Yu. Romanenko [137] have considered various questions of the theory of difference equations. Numerous results concerning the development of asymptotic methods, search for oscillatory solutions, construction of invariant manifolds, problems of reducibility and stability of solutions of difference-differential and difference equations have been obtained on the theory of nonlinear oscillations by representatives of the Kiev mathematical school. Here, we point out the works by N.N. Bogolyubov and Yu.A. Mitropol'skii [5; 6], N.N. Bogolyubov, Yu.A. Mitropol'skii and A.M. Samoilenko [7], Yu.A. Mitropol'skii and O.B. Lykova [69], Yu.A. Mitropol'skii and D.I. Martynyuk [70], Yu.A. Mitropol'skii, D.I. Martynyuk and V.I. Tynnyi [71], Yu.A. Mitropol'skii and N.O. Mykhailivs'ka [72], A.M. Samoilenko and M.I. Ronto [114] − [116], D.I. Martynyuk [55], D.I. Martynyuk and N.A. Perestyuk [57] − [59], A.M. Samoilenko and N.A. Perestyuk [111], A.M. Samoilenko, D.I. Martynyuk and N.A. Perestyuk [109; 110], G.P. Pelyukh [80] − [82], V.I. Tkachenko [165], A.M. Samoilenko and R.I. Petrishin [112; 113], A.M. Samoilenko and V.N. Laptinskii [108], I.M. Cherevko [14], V.Ya. Danilov [17], V.Ya. Danilov and D.I. Martynyuk [18], A.A. Boichuk [8], A.M. Ateiwi [1; 2], M. Kwapisz [51] and many other mathematicians. Therewith, most of the obtained results were related to the equations in finite-dimensional spaces.

In addition in the 1970s, the results of Yu.A. Mitropol'skii [68] and A.G. Ilyukhin [36], and other researchers showed that it is convenient to apply the apparatus of countable systems of ordinary differential equations, i.e., the equations in the space $\mathfrak{M}$ to the solution of many problems dealing with oscillations of systems with distributed parameters. As an example, we mention the problem of transverse vibrations of a rod loaded with an axial periodically varying force. Such equations are also involved in the problems from various sections of mathematical and theoretical physics. At that time, the results of A.N. Tikhonov [164], K.P. Persidskii [84], Yu.L. Daletskii and M.G. Krein [16], O.A. Zhautykov [176] − [178], Z.I. Khalilov [40; 41], V.H. Kharasakhal [42], L.A. Ermolaev [24], S.I. Gorshin [30], M.T. Reshetov [85] and other authors, who created the basis of the theory of countable systems of differential equations, were obtained. The development of this theory was promoted by the monographs of K.G. Valeev and O.A. Zhautykov [169] and monographs [120; 121] of the authors of this book. Since difference equations are the discrete analogs of differential ones, one can realize the urgency of development of difference equations theory in Banach spaces and particularly, in a space of bounded numerical sequences. The foundation of this scientific direction was laid in research by the authors of this book and also in the works of A.M. Samoilenko, D.I. Martynyuk and N.A. Perestyuk [110], D.I. Martynyuk, V.I. Kravets and B.Kh. Zhanbusinova [169], D.I. Martynyuk and G.V. Ver'ovkina [60], A.M. Samoilenko, V.E. Slyusarchuk and V.V. Slyusarchuk [118], A.Ya. Dorogovtsev [22],

V.Yu. Slyusarchuk [138] – [142], S.I. Trofimchuk [167; 168], M.F. Gorod-nii [29], M.G. Filippov [28], N.A. Evhuta and P.P. Zabreiko [25] – [27], M. Ronto, A. Ronto and S.I. Trofimchuk [91], A.M. Ronto [86] – [89], Yu.V. Tomilov [166], and their disciples. The questions related to estimating the periods of periodic motions of dynamical systems in Banach spaces have been studied by A. Lasota and J. Yorke [52], S. Busenberg, D. Fisher and M. Martelli [10; 11], S. Busenberg and M. Martelli [12], M. Medveď [66], as well as the authors of the above-indicated works [88; 91].

This monograph is based on the results of the authors and the results obtained in co-authorship with disciples [118; 119], [122] – [129], [138] – [158]. It comprises four chapters.

The first chapter contains the results on the reducibility of the difference equations with periodic coefficients in the space $\mathfrak{M}$ and the difference equations defined on finite-dimensional and infinite-dimensional tori, as well as the results of studying a discrete dynamical system defined in an abstract Banach space in a neighborhood of its invariant set.

In the second chapter, the theory of invariant tori for the difference equations in the space $\mathfrak{M}$, which contain deviations of a discrete argument, has been constructed. The current mathematical literature does not have the theorems analogous to the theorems of existence and continuous Fréchet-differentiability of the invariant tori of linear, quasilinear, and nonlinear systems defined on infinite-dimensional tori with respect to the angular variable and the parameter, which belongs to the space $\mathfrak{M}$. For a nonlinear degenerate difference equation, the problem of existence of a semi-invariant bounded smooth manifold has been investigated.

The third chapter is devoted to the construction of the periodic solutions of degenerate and nondegenerate quasilinear difference equations in the space $\mathfrak{M}$ in nonresonance and resonance cases, as well as the construction of such solutions of difference equations of the first and second orders in abstract Banach spaces. The problem of extension of solutions of degenerate nonlinear difference equations of the n-th order in abstract Banach spaces is investigated as well.

The fourth chapter supplements the contents of monographs of the authors [120; 121]. It deals with countably point boundary-value problems for ordinary differential equations of the normal form and the equations unsolved with respect to a derivative and defined in the space of bounded numerical sequences. A special attention is focused on the problem of reduction of the stated problems to multipoint boundary-value problem in finite-dimensional spaces. The results obtained can be easily extended to the case of difference equations in Banach spaces.

Let us mention that the methods described in this book can be successfully applied to the investigation of similar problems for differential-difference equations. As an example, one can look at papers by A.M. Samoilenko, Yu.V. Teplinsky and K.V. Pasyuk [130] – [132], Yu.V. Teplinsky [161], Yu.V. Teplinsky and K.V. Pasyuk [159] – [163] in which sufficient conditions for the existence of infinite continuous invariant tori for countable systems of differential-difference equations in general form with an infinite number of time deviations of different signs are found.

For the convenience of the readers, the book chapters are written so that they are almost independent of each other. In the text, we use the double numbering of formulas and statements (definitions, lemmas, theorems, etc.). The first number corresponds to the number of chapter, the second number denotes the number of the formula or the statement in this chapter.

The basic results of the monograph have been presented many times at scientific conferences and seminars both in Ukraine and abroad. A part of these results is being used in special courses delivered by A.M. Samoilenko at the Mechanical and Mathematical Faculty of Taras Shevchenko National University of Kyiv and by Yu.V. Teplinsky at the Physical and Mathematical Faculty of Ivan Ohienko Kam'yanets-Podilsky National University.

We hope that this monograph will be helpful to specialists who are concerned with the relevant mathematical problems.

We are grateful to Academicians of NAS of Ukraine I.O. Lukovsky and M.O. Perestyuk for reviewing the manuscript of this book.

Anatoliy Samoilenko
Yuriy Teplinsky

Contents

Preface   v

1. Reducibility problems for difference equations   1

    1.1   On analogs of the Erugin and Floquet–Lyapunov theorems for equations in the space $\mathfrak{M}$ 2

    1.2   Linear equations in the space $\mathfrak{M}$ defined on tori 9

    1.3   Nonlinear almost periodic equations defined on an infinite-dimensional torus 18

    1.4   Reduction of a discrete dynamical system in the space R^q to the canonical form in a neighborhood of its invariant set 29

    1.5   Investigation of a discrete dynamical system defined in an abstract Banach space in a neighborhood of its invariant set 42

2. Invariant tori of difference equations in the space $\mathfrak{M}$ 61

    2.1   Sufficient conditions of existence of a continuous invariant torus 62

    2.2   On the differentiability of an invariant torus with respect to the angular variable and the parameter in the coordinate-wise meaning 75

    2.3   Truncation method in studying the smoothness of invariant tori 89

    2.4   Case of linear and quasilinear systems defined on the infinite-dimensional tori 119

    2.5   On the existence of the invariant tori of nonlinear systems 137

    2.6   Differentiability of the invariant tori of nonlinear systems in the Fréchet meaning 149

2.7   Conditions of existence of the Green–Samoilenko function for a linear system defined on the set $\mathfrak{M} \times \mathcal{T}_\infty$. Reduction of the problem of construction of the invariant torus of this system to an analogous problem in the space $R^s \times \mathcal{T}_m$ . . . 166

2.8   On the existence of the smooth bounded semi-invariant manifold of a degenerate nonlinear system 172

3.   Periodic solutions of difference equations. Extension of solutions   193

3.1   On the periodic solutions of linear and quasilinear equations with periodic coefficients in the space $\mathfrak{M}$ 193

3.2   Periodic solutions of nonlinear difference equations of the first order in an abstract Banach space 216

3.3   Periodic solutions of nonlinear difference equations of the second order 229

3.4   Asymptotic periodicity of solutions of a linear equation in a complex Banach space 253

3.5   Extension "to the left" of solutions of nonlinear degenerate difference equations 260

4.   Countable-point boundary-value problems for nonlinear differential equations   289

4.1   Boundary-value problem on the semiaxis 289

4.2   Boundary-value problems on an interval 310

4.3   Reduction to a finite-dimensional multipoint case 315

4.4   Another means of the reduction. Conditions of commutativity of the limiting transitions (4.42) and (4.43) 335

4.5   Boundary-value problems for differential equations unsolvable with respect to the derivative 344

4.6   Reduction to a finite-dimensional multipoint problem . . . 364

Bibliography   385

Index   397

Chapter 1

Reducibility problems for difference equations

This chapter deals with the conditions of reducibility of difference equations in Banach spaces, i.e., the conditions which enable one to map the trajectories of solutions of these equations onto the trajectories of certain equations of a simpler structure. For the differential equations with periodic coefficients in the finite-dimensional space, the well-known results of Floquet–Lyapunov, which admit such a mapping, are available. The analogous results have been obtained in [54] in the case of finite-dimensional systems of difference equations and in [120; 121] in the case of countable systems of differential equations. Of considerable recent interest are the differential and difference equations with quasiperiodic and almost periodic coefficients defined in various normalized spaces. A lot of works is devoted to the problem of their reducibility and to the study of their solutions. A special attention was paid to the problem of studying the behavior of dynamical and discrete dynamical systems in neighborhoods of their invariant manifolds. The method of its solution has been suggested in works [102; 104; 106; 107].

Here, we present the sufficient conditions of reducibility of a periodic difference equation in the space of bounded numerical sequences in the Lyapunov meaning, apply the method of construction of iterations with accelerated convergence [7] and the truncation method of K.P. Persidskii [84] to study the reducibility of difference equations in this space, which are defined on finite-dimensional and infinite-dimensional tori, and employ the methods of integral manifolds [69] and differential topology [9; 35; 49; 135; 173] to investigate the behavior of a discontinuous dynamical system in the abstract Banach space in a neighborhood of its invariant manifold.

1

1.1 On analogs of the Erugin and Floquet–Lyapunov theorems for equations in the space $\mathfrak{M}$

We consider the homogeneous equation

$$x_{n+1} = A(n)x_n, \quad n \in Z, \tag{1.1}$$

where $x = (x^1, x^2, \dots) \in \mathfrak{M}$, $A(n) = [a_{ij}(n)]_{i,j=1}^{\infty}$ is an infinite matrix with real elements, $\mathfrak{M}$ is the space of bounded sequences of real numbers with the norm $\|x\| = \sup_i\{|x^i|, i = 1, 2, \dots\}$, and Z is the set of integers.

The norm of the matrix A is defined by the equality $\|A\| = \sup_i \sum_{j=1}^{\infty} |a_{ij}|$. It is obvious that this norm is consistent with the vector norm of the space $\mathfrak{M}$.

The set of all infinite constant matrices bounded by the norm $\|\cdot\|$ is denoted by $\boldsymbol{\Gamma}$. It is easy to verify that this set forms a linear space over the field of real numbers. In this case, if $P_k \in \boldsymbol{\Gamma}$ $(k = 1, 2, \dots, s)$, then

$$\prod_{k=1}^{s} P_k = P_1\Big(\prod_{k=2}^{s} P_k\Big). \tag{1.2}$$

We assume that, for all $n \in Z$, the matrices $A(n)$ are invertible, and $\{A(n), A^{-1}(n)\} \subset \boldsymbol{\Gamma}$.

By the solution of Eq. (1.1) on the interval $[a, b] \subset R^1$ (finite or infinite) with the initial values $x_l \in \mathfrak{M}, n = l \in [a, b]_Z = [a, b] \bigcap Z$, we mean a discrete function $x_n = x(n, x_l)$ such that it transforms (1.1) into the identity on $[a, b]_Z$, takes the values from the space $\mathfrak{M}$ on this set, and satisfies the condition $x_l = x(l, x_l)$.

The relation

$$x(n, x_l) = \begin{cases} A(n-1)A(n-2)\dots A(l)x_l & \text{for} \quad n > l, \\ A^{-1}(n)A^{-1}(n+1)\dots A^{-1}(l-1)x_l & \text{for} \quad n < l \end{cases}$$

implies that, for the arbitrary initial values of $l \in Z, x_l \in \mathfrak{M}$, there exists the unique solution $x_n = x(n, x_l)$ to Eq. (1.1) defined on the interval $(-\infty, +\infty)$.

We will formulate the following additional proposition.

Theorem 1.1. *If $\overset{(i)}{x}_n = \overset{(i)}{x}(n), i = 1, 2, 3, \dots$, are the solutions of Eq. (1.1), and, for certain $n = l \in Z$, the inequality*

$$\sum_{i=1}^{\infty} |\overset{(i)}{x}_l{}^s| < \beta_0 = const < \infty, s = 1, 2, 3, \dots, \tag{1.3}$$

holds, then, for any $c = (c_1, c_2, c_3, \dots) \in \mathfrak{M}$, the function $z(n) = \sum_{i=1}^{\infty} c_i \overset{(i)}{x}(n)$ is also a solution of this equation.

Proof. Without loss of generality, we set $l = 0$. With regard for the equalities

$$\overset{(i)_s}{x_1} = \sum_{j=1}^{\infty} a_{sj}(0)\,\overset{(i)_j}{x_0}, \quad i, s = 1, 2, 3, \ldots,$$

and condition (1.3), we obtain that the double series $\sum_{j=1}^{\infty}\sum_{i=1}^{\infty} a_{sj}(0)\,\overset{(i)_j}{x_0}$ absolutely converges for all $s = 1, 2, 3, \ldots$, and its sum is limited by the constant $\beta_1 = \beta_0 \|A(0)\|$. Then, for all natural s, the relations

$$\sum_{i=1}^{\infty} |\overset{(i)_s}{x_1}| \leq \sum_{i=1}^{\infty}\sum_{j=1}^{\infty} |a_{sj}(0)||\overset{(i)_j}{x_0}| = \sum_{j=1}^{\infty}\sum_{i=1}^{\infty} |a_{sj}(0)||\overset{(i)_j}{x_0}| < \beta_1$$

are valid.

By the method of complete mathematical induction, we verify that, for all $n \in Z^+ = \{1, 2, 3, \ldots\}$, the inequality

$$\sum_{i=1}^{\infty} |\overset{(i)_s}{x_n}| < \beta_0 \prod_{j=0}^{n-1} \|A(j)\| = \beta_n = const < \infty, \quad s \in Z^+, \qquad (1.4)$$

holds.

Using the analogous reasoning, it is easy to show that, for all $n \in Z^- = \{-1, -2, -3, \ldots\}$,

$$\sum_{i=1}^{\infty} |\overset{(i)_s}{x_n}| < \beta_0 \prod_{j=-1}^{n} \|A^{-1}(j)\| = \gamma_n = const < \infty, \quad s \in Z^+. \qquad (1.5)$$

Estimates (1.4) and (1.5) guarantee the validity of condition (1.3) for each $n \in Z$. In this case, the function $z_n = z(n)$ exists and takes the values from $\mathfrak{M}$ for all integers n.

From the inequality

$$\sum_{j=1}^{\infty}\sum_{i=1}^{\infty} |a_{sj}(n)c_i\,\overset{(i)_j}{x_n}| < \begin{cases} \displaystyle\sum_{j=1}^{\infty} \|c\||a_{sj}(n)|\beta_n \leq \|c\|\beta_n\|A(n)\|, & n \geq 0, \\[2em] \displaystyle\sum_{j=1}^{\infty} \|c\||a_{sj}(n)|\gamma_n \leq \|c\|\gamma_n\|A(n)\|, & n < 0, \end{cases}$$

it follows that the equality

$$\sum_{i=1}^{\infty}\sum_{j=1}^{\infty} c_i a_{sj}(n)\,\overset{(i)_j}{x_n} = \sum_{j=1}^{\infty}\sum_{i=1}^{\infty} a_{sj}(n)c_i\,\overset{(i)_j}{x_n}$$

is valid for all $n \in Z$ and $s \in Z^+$. This implies that

$$\sum_{i=1}^{\infty} c_i\,\overset{(i)}{x}_{n+1} = A(n)\sum_{i=1}^{\infty} c_i\,\overset{(i)}{x}_n.$$

The theorem is proved. $\qquad\qquad\square$

We will construct the matrix $\Omega(n, l)$, whose i-th column represents the solution $\overset{(i)}{x}_n = \overset{(i)}{x}(n, x_l)$ of Eq. (1.1) with the initial values $\overset{(i)}{x}_l = \underbrace{(0, 0, \ldots, 0}_{i-1}, 1, 0, 0, \ldots)$, $i = 1, 2, 3, \ldots$, $n = l$. This means that $\Omega(l, l) = E$, where E is the identity matrix. The constructed matrix will be called the matriciant of Eq. (1.1) and denoted by Ω_l^n. It is evident that the matriciant Ω_l^n is a solution of the matrix equation

$$\Omega_l^{n+1} = A(n)\Omega_l^n, \quad n \in Z, \tag{1.6}$$

which yields the relation

$$\Omega_l^n = \begin{cases} A(n-1)A(n-2)\ldots A(l), & n > l, \\ E, & n = l, \\ A^{-1}(n)A^{-1}(n+1)\ldots A^{-1}(l-1), & n < l. \end{cases} \tag{1.7}$$

The matrix $X(n)$ is called the fundamental matrix of Eq. (1.1) if, for all $n \in Z$, $X(n+1) = A(n)X(n)$ and, for any solution $x_n = x(n)$ of Eq. (1.1), there exists the constant vector $c \in \mathfrak{M}$ such that $x_n = X(n)c$. From Theorem 1.1, it follows that, for any $c \in \mathfrak{M}$, the function $x_n = \Omega_l^n c$ is a solution of Eq. (1.1). In view of equality (1.6), we come to the conclusion that the matriciant of Eq. (1.1) is its fundamental matrix.

Equation (1.1) is assumed to be reducible to an equation of the form

$$y_{n+1} = By_n, \quad n \in Z, \tag{1.8}$$

with the constant real matrix B, if there exists the infinite invertible matrix $T(n)$ with real elements, which is such that any solution $x_n = x(n)$ of Eq. (1.1) is connected with the solution $y_n = y(n)$ of Eq. (1.8) by the relation

$$x_n = T(n)y(n), \tag{1.9}$$

and $\{T(n), T^{-1}(n)\} \subset \mathbf{\Gamma} \quad \forall n \in Z$. In this case, we say that Eq. (1.8) is reduced to Eq. (1.9) with the help of the matrix $T(n)$ or this matrix reduces Eq. (1.8) to the form (1.9). Note that the matrix B in (1.8) is to be invertible and bounded along with its inverse matrix, which follows from the equality

$$B = T^{-1}(n+1)A(n)T(n), \quad n \in Z.$$

The reducibility of Eq. (1.1) is confirmed by the following statement, which is an analog of the Erugin theorem.

Theorem 1.2. *Equation (1.1) is reducible to an equation of the form (1.8) with the help of the change of variables (1.9) if and only if its some fundamental matrix $X(n)$ can be presented as*

$$X(n) = T(n)B^n, \quad n \in Z. \tag{1.10}$$

Proof. First, we will agree that, for $n \in Z^-$, the expression B^n stands for $(B^{-1})^{-n}$, and the expression B^0 means an infinite identity matrix. Let condition (1.10) be satisfied. Then the change of variables $x_n = X(n)B^{-n}y_n, n \in Z$, transforms Eq. (1.1) into the equation

$$A(n)X(n)B^{-(n+1)}y_{n+1} = A(n)X(n)B^{-n}y_n. \tag{1.11}$$

Equalities (1.10) and (1.2) yield the invertibility of the matrix $X(n)$ and the boundedness of the matrix $X^{-1}(n)$ in the norm $\forall n \in Z$. Then, from equality (1.1), we obtain Eq. (1.8).

We will deal now with proving the necessity. Let Eq. (1.1) be reducible to Eq. (1.8) by the change of variables (1.9). By Ω_0^n, we denote the matriciant of Eq. (1.8). Then its any solution $y_n = y(n)$ can be presented in the form $y_n = \Omega_0^n c$, where c is some constant vector from $\mathfrak{M}$. In this case, $\Omega_0^n = B^n, n \in Z$, which yields the equality $y_n = B^n c$. Hence, any solution $x_n = x(n)$ of Eq. (1.1) can be presented in the form $x_n = T(n)B^n c, n \in Z$. We denote $T(n)B^n$ by $X(n)$. It is easy to see that $X(n)$ is the fundamental matrix of Eq. (1.1). $\qquad \square$

Corollary 1.1. *For an arbitrary constant invertible infinite matrix B such that $\{B, B^{-1}\} \subset \mathbf{\Gamma}$, there exists the matrix $T(n)$, which reduces Eq. (1.1) to an equation of the form (1.8).*

Proof. With regard for Theorem 1.2, it is sufficient to show that the matriciant Ω_0^n of Eq. (1.1) can be presented in the form $T(n)B^n, n \in Z$, where $T(n)$ is the invertible matrix and $\{T(n), T^{-1}(n)\} \subset \mathbf{\Gamma}$. The matrix

$$T(n) = \Omega_0^n B^{-n} \tag{1.12}$$

has the required properties. $\qquad \square$

We now assume that the matrix $A(n)$ is $N-$periodic, i.e., it is periodic in n with the period $N \in Z^+ : A(n + N) = A(n), n \in Z$. Corollary 1.1 holds true for Eq. (1.1) in this case as well. The situation is complicated if the matrix $T(n)$ is required to be uniformly bounded for any $n \in Z$ in the norm $\| \cdot \|$. To this end, it is sufficient that the matrix $T(n)$ in formula (1.12) be periodic in n with a certain period. The next theorem presents the existence conditions for such a matrix.

Theorem 1.3. *The necessary and sufficient condition of the N-periodicity of a matrix $T(n)$ given by equality (1.12), in which Ω_0^n is the matriciant of Eq. (1.1) with the N-periodic matrix $A(n)$, is the existence of a real root of the N-th power of the monodromy matrix Ω_0^N such that*

$$B^N = \Omega_0^N, \quad \{B, B^{-1}\} \subset \mathbf{\Gamma}. \tag{1.13}$$

Proof.   The sufficiency of condition (1.13) follows from the equalities

$$T(n+N) = \Omega_0^n \Omega_0^N B^{-(n+N)} = \Omega_0^n B^{-n} = T(n), \quad n \in Z.$$

The assumption about the N-periodicity of matrix (1.12) yields the equality $\Omega_0^n = \Omega_0^{n+N} B^{-N}$. Setting $n = 0$ in this equality, we verify that condition (1.13) holds, i.e., this condition is necessary. $\square$

Note that the search for the N-power root of an infinite matrix is a rather complex problem. It can be connected with the problem of finding the logarithm of this matrix. So, if the matrix Ω_0^N has the real logarithm bounded in the norm, then we can set

$$B = (\Omega_0^N)^{1/N} = \exp\{\frac{1}{N} \ln \Omega_0^N\}. \tag{1.14}$$

The real branch of the logarithm exists not for any real nondegenerate matrix even in the infinite-dimensional case. But, in our case, such a branch always exists for the matrix $(\Omega_0^N)^2$. In view of relation (1.7), which ensures the equality $(\Omega_0^N)^2 = \Omega_0^{2N}$, we set

$$B = (\Omega_0^{2N})^{1/2N} = \exp\{\frac{1}{2N} \ln \Omega_0^{2N}\}$$

analogously to (1.14). Then the matrices B and $T(n)$ are real, and the latter is $2N$-periodic.

Thus, we were not able to formulate, in the infinite-dimensional case, a direct analog of the Floquet–Lyapunov theorem on reducibility of Eq. (1.1) with N-periodic matrix. We will approach the solution of this problem in a different way without finding the root of the monodromy matrix of this equation.

We consider the equation

$$\overset{(s)}{x}_{n+1} = \overset{(s)}{A}(n) \overset{(s)}{x}_n, \quad n \in Z, \tag{1.15}$$

which is obtained from (1.1) by the truncation with respect to x up to the s-th order, i.e., $\overset{(s)}{x} = (\overset{(s)}{x}^1, \overset{(s)}{x}^2, \ldots, \overset{(s)}{x}^s)$, $\overset{(s)}{A}(n) = [a_{ij}(n)]_{i,j=1}^s$.

Equation (1.15) is considered in the finite-dimensional space R^s and is reducible for any $s \in Z^+$, if the matrix $A(n)$ is N-periodic. This means that, for every $s \in Z^+$, there exists the nondegenerate matrix $\overset{(s)}{T}(n)$, which performs the reduction of Eq. (1.15) to an equation of the form

$$\overset{(s)}{y}_{n+1} = \overset{(s)}{B} \overset{(s)}{y}_n, \quad n \in Z, \tag{1.16}$$

where $\overset{(s)}{B}$ is a constant nondegenerate matrix.

The sequence of matrices $\{\overset{(k)}{C}(n)\}_{k=1}^{\infty} = \{[\overset{(k)}{c}_{ij}(n)]_{i,j=1}^{\infty}\}_{k=1}^{\infty}(n \in Z)$ is called proper one if it is uniformly bounded $\forall n \in Z$ and converges to some matrix $C(n) = [c_{ij}(n)]_{i,j=1}^{\infty}$ in the elementwise meaning as $k \to \infty$. Moreover, the series $\sum_{j=1}^{\infty} |\overset{(k)}{c}_{ij}(n)|, i = 1, 2, 3, \ldots$, uniformly converge in k $\forall n \in Z$.

In this case, if the last condition is satisfied uniformly in $i = 1, 2, 3, \ldots$, the sequence $\{\overset{(k)}{C}(n)\}_{k=1}^{\infty}$ is called strongly proper.

The analogous definition has been formulated in monograph [121] for a sequence of infinite matrices, whose elements are continuous functions of the argument $t \in [a, b] \subset R^1$.

We present the conditions, under which the elementwise convergence of the matrix sequence $\{\overset{(s)}{T}(n)\}, s \to \infty$, yields the reducibility of Eq. (1.1). Moreover, we agree to supplement the matrices of this sequence with zero elements up to infinite matrices without changing their notation.

Theorem 1.4. *Let the matrix $A(n)$ be N-periodic, and there exists $n^0 \in Z$ such that*

$$\sum_{j=m+1}^{\infty} |a_{ij}(n^0)| \le K_{n^0}(m), \quad i = 1, 2, 3, \ldots, \tag{1.17}$$

where $K_{n^0}(m) \to 0$ as $m \to \infty$, and the matrices $\overset{(s)}{A}(n)$ are not degenerate $\forall s \in Z^+, n \in Z$. If the sequence of matrices $\{\overset{(s)}{T}(n)\}_{s=1}^{\infty}$ that reduce Eq. (1.15) to an equation of the form (1.16) is strongly proper, and if the sequence of matrices $\{\overset{(s)}{T}^{-1}(n)\}_{s=1}^{\infty}$ is proper $\forall n \in Z$, then

$$\lim_{s \to \infty} \overset{(s)}{T}(n) = T(n), \quad \lim_{s \to \infty} \overset{(s)}{B} = B,$$

in the elementwise meaning, and Eq. (1.1) is reducible to Eq. (1.8) with the help of the $2N$-periodic matrix $T(n)$.

Proof. Obviously, it is sufficient to prove the statement of the theorem for each interval $[a, +\infty)_Z$, to which n belongs. Without loss of generality, we set $a = 0$, $n^0 \in Z^+$ and show that

$$\lim_{s \to \infty} \overset{(s)}{x}(n, \overset{(s)}{x}_0) = x(n, x_0) \tag{1.18}$$

coordinatewise, where $\overset{(s)}{x}_n = \overset{(s)}{x}(n, \overset{(s)}{x}_0)$ and $x_n = x(n, x_0)$ are the solutions of Eqs. (1.15) and (1.1), respectively, $n \in [0, +\infty)_Z, x_0 = (x_0^1, x_0^2, \dots) \in \mathfrak{M}, \overset{(s)}{x}_0 = (x_0^1, x_0^2, \dots, x_0^s) \in R^s$.

We will prove equality (1.18) for the first coordinates of the vectors $\overset{(s)}{x}_n$ and x_n. For $n = 1$, we have

$$\lim_{s \to \infty} \overset{(s)1}{x}_1 = \lim_{s \to \infty} \sum_{j=1}^{s} a_{1j}(0) x_0^j = x_1^1,$$

and $\| \overset{(s)}{x}_1 \| \leq \|A(0)\| \|x_0\|$ uniformly in $s \in Z^+$.

In an analogous way, we obtain the equality

$$\lim_{s \to \infty} \overset{(s)1}{x}_2 = \lim_{s \to \infty} \sum_{j=1}^{\infty} a_{1j}(1) \overset{(s)j}{x}_1, \tag{1.19}$$

where we set $\overset{(s)j}{x}_1 = 0$ for $j > s$.

Since the series $\sum_{j=1}^{\infty} a_{1j}(1) \overset{(s)j}{x}_1$ is majorized by the numerical convergent series

$$\|A(0)\| \|x_0\| \sum_{j=1}^{\infty} \|a_{1j}(1)\|,$$

it uniformly converges in $s \in Z^+$. Then it follows from (1.19) that $\lim_{s \to \infty} \overset{(s)1}{x}_2 = x_2^1$, and $\| \overset{(s)}{x}_2 \| \leq \|A(0)\| \|A(1)\| \|x_0\|$ uniformly in $s \in Z^+$.

By the method of complete mathematical induction, we verify the validity of equality (1.18) for all $n \in Z^+$. Moreover,

$$\| \overset{(s)}{x}_n \| \leq \|x_0\| \prod_{k=0}^{n-1} \|A(k)\|$$

uniformly in $s \in Z^+$. However, for all $s \in Z$, $\overset{(s)}{y}_n = \overset{(s)}{T}{}^{-1}(n) \overset{(s)}{x}_n$, where $\overset{(s)}{y}_n$ is the solution of Eq. (1.16). Denoting $\overset{(s)}{T}{}^{-1}(n)$ by $[t_{ij}^{(s)}]_{i,j=1}^{s}$, we obtain

$$\overset{(s)p}{y}_n = \sum_{i=1}^{\infty} t_{pi}^{(s)}(n) \overset{(s)i}{x}_n \quad (p \in Z^+, \quad t_{pi}^{(s)}(n) \overset{(s)i}{x}_n = 0 \quad \text{for} \quad i > s, p > s).$$

$$\tag{1.20}$$

The last series converges uniformly in $s \in Z$ since it is majorized by the numerical series

$$\left(\prod_{k=0}^{n-1} \|A(k)\| \right) \|x_0\| \sum_{i=1}^{\infty} |t_{pi}^{(s)}(n)|, \quad n \in Z^+,$$

and the sequence of matrices $\{\overset{(s)}{T}{}^{-1}(n)\}_{s=1}^{\infty}$ is proper.

This implies that, as $s \to \infty$, the sequence $\{\overset{(s)}{y}{}_n\}$ coordinatewise converges to the vector $y_n = T^{-1}(n)x_n$ for any $n \in Z^+$, where the matrix $T^{-1}(n)$ is inverse to the matrix $T(n) = \lim_{s\to\infty} \overset{(s)}{T}(n)$.

It remains to show that the matrix sequence $\{\overset{(s)}{B}\}_{s=1}^{\infty}$ converges elementwise to the matrix $B \in \mathbf{\Gamma}$. Indeed, for all $n \in Z^+$, $\overset{(s)}{B} = \overset{(s)}{T}{}^{-1}(n+1)\,\overset{(s)}{A}(n)\,\overset{(s)}{T}(n)$. Since $\overset{(s)}{B}$ are constant matrices for all $s \in Z^+$, we set

$$\overset{(s)}{B} = \overset{(s)}{T}{}^{-1}(n^0+1)\,\overset{(s)}{A}(n^0)\,\overset{(s)}{T}(n^0). \tag{1.21}$$

Condition (1.17) guarantees the strong validity of the matrix sequence $\{\overset{(s)}{A}(n^0)\}_{s=1}^{\infty}$. Since the product of a proper sequence by the strongly proper one is a proper sequence [97], it follows from (1.21) that the sequence $\{\overset{(s)}{B}\}_{s=1}^{\infty}$ is proper. This means that it converges elementwise to some matrix B. The invertibility and boundedness of the matrix B can be easily proved.

It is evident that the matrix $T(n)$ is $2N$-periodic if the matrices $\overset{(s)}{T}(n)$ have this property for all $s \in Z^+$. $\qquad\square$

1.2 Linear equations in the space $\mathfrak{M}$ defined on tori

We consider the system of equations

$$x_{n+1} = Ax_n + P(\varphi_n)x_n, \quad \Delta\varphi_n = \omega, \quad n \in Z, \tag{1.22}$$

where

$$\varphi = (\varphi^1, \varphi^2, \ldots, \varphi^m), \quad \omega = (\omega_1, \omega_2, \ldots, \omega_m) \in R^m,$$
$$x = (x^1, x^2, x^3, \ldots) \in \mathfrak{M},$$

$A = diag\{a_1, a_2, a_3, \ldots\}$ is a diagonal infinite matrix with real elements, $P(\varphi) = [p_{ij}(\varphi)]_{i,j=1}^{\infty}$ is the infinite matrix 2π-periodic in $\varphi^i (i = 1, 2, 3, \ldots, m)$, and $\Delta\varphi_n = \varphi_{n+1} - \varphi_n$.

We denote the solution of the equation $\Delta\varphi_n = \omega$, $n \in Z$, by $\varphi_n = \varphi(n, \varphi_0)$, where $\varphi(n, \varphi_0)$ is a discrete function such that $\varphi(0, \varphi_0) = \varphi_0 \in \mathcal{T}_m$, $\mathcal{T}_m$ is an m-dimensional torus in the space R^m.

By the solution of the system of equations (1.22) on the interval $[a, b] \subset R^1$ (finite or infinite) with the initial values of $n = l \in [a, b]_Z, \varphi_0 \in R^m, x_l \in \mathfrak{M}$, we mean a discrete function $x_n = x(n, \varphi_0, x_l)$ such that it transforms the equation

$$x_{n+1} = Ax_n + P(\varphi(n, \varphi_0))x_n, \quad n \in Z,$$

into the identity on $[a, b]_Z$, takes the values from the space $\mathfrak{M}$ on this set, and satisfies the condition $x_l = x(l, \varphi_0, x_l)$.

We introduce the following notation: $k = (k_1, k_2, \ldots, k_m)$ is an integer-valued vector, E is the identity matrix,

$$|k| = \sum_{i=1}^{m} |k_i|, \quad \langle k, \omega \rangle = \sum_{i=1}^{m} k_i \omega_i, \quad \|P(\varphi)\| = \sup_i \sum_{j=1}^{\infty} |p_{ij}(\varphi)|,$$

$$\|P(\varphi)\|_0 = \sup_i \sum_{j=1}^{\infty} \max_{\varphi} |p_{ij}(\varphi)|, \quad |Im\varphi| = \sup_i \{|Im\varphi^i|\}.$$

Theorem 1.5. *Let the elements of the matrix A be different from zero, $\inf_{i \neq j} |a_i - a_j| = r_0 > 0$, and let the matrix $P(\varphi)$ be analytic in the domain $|Im\varphi| \leq \rho_0 > 0$ and real for real φ. Assume that there exist the positive constants ε and d such that, for all integer-valued vectors k, the inequality*

$$\left| \sin \frac{\langle k, \omega \rangle}{2} \right| \geq \varepsilon |k|^{-d}, \quad |k| \neq 0,$$

is valid. Then there exists the positive constant M_0 such that, for $\|P(\varphi)\|_0 \leq M_0$, the system of equations (1.22) is reduced by the change of variables $x_n = \Phi(\varphi_n)y_n$ to a system of the form

$$y_{n+1} = A_0 y_n, \quad \Delta\varphi_n = \omega, \quad n \in Z, \tag{1.23}$$

where A_0 is a constant diagonal matrix with real elements, $\Phi(\varphi)$ is a matrix, which is 2π-periodic in $\varphi^i (i = 1, 2, 3, \ldots, m)$, invertible and analytic in the domain $|Im\varphi| \leq \rho_0/2$, and real for real φ.

Proof. Proof is performed by the method of construction of iterations with accelerated convergence analogously to that of Theorem 5.1 from [74]. Let $x_n = x(n)$ be a solution of the system of equation (1.22) on the segment $[a, b]$. By the change of variables

$$x_n = (E + U_1(\varphi_n))y_n^{(1)} = \Phi_1(\varphi_n)y_n^{(1)}, \tag{1.24}$$

the system of equations (1.22) is reduced to a system of the form

$$y_{n+1}^{(1)} = A_1 y_n^{(1)} + P_1(\varphi_n)y_n^{(1)}, \quad \Delta\varphi_n = \omega, \tag{1.25}$$

where

$$P_1(\varphi_n) = (E + U_1(\varphi_n + \omega))^{-1}(P(\varphi_n)U_1(\varphi_n) - U_1(\varphi_n + \omega)\bar{D}_1),$$

$$A_1 = A + \bar{D}_1, \quad \bar{D}_1 = \frac{1}{(2\pi)^m}\int_0^{2\pi}\cdots\int_0^{2\pi} D_1(\varphi)d\varphi^1\ldots d\varphi^m,$$

$$D_1(\varphi) = diag\{p_{11}(\varphi), p_{22}(\varphi), p_{33}(\varphi), \ldots\},$$

and $U_1(\varphi)$ is a solution, which is 2π-periodic in $\varphi^i(i = 1, 2, 3, \ldots, m)$ in the domain $|Im\varphi| \le \rho_0 - 2\delta_0 = \rho_1 > 0$, of the matrix equation

$$U_1(\varphi + \omega)A = AU_1(\varphi) + P(\varphi) - \bar{D}_1.$$

The estimate $\|U_1(\varphi)\|_0 \le M_0^{\gamma-1}/8$ for $M_0 < 1, 1 < \gamma = const < 2$ ensures the invertibility of the infinite matrix $E + U_1(\varphi_n + \omega)$:

$$(E + U_1(\varphi_n + \omega))^{-1} = E + \sum_{\alpha=1}^{\infty}(-1)^\alpha(U_1(\varphi_n + \omega))^\alpha.$$

It is easy to see that the vector $y_n^{(1)}$ belongs to $\mathfrak{M}$ on the segment $[a, b]_Z$.

For sufficiently small M_0, we can apply the same transformation to the system of equations (1.25) and system (1.22).

It is easy to show that, with the appropriate choice of the constants M_0, δ_0, the process of transformation of system (1.22) by changes of the form (1.24) can go on infinitely.

After p sequential changes, the transformation

$$x_n = (E + U_1(\varphi_n))\ldots(E + U_p(\varphi_n))y_n^{(p)} = \Phi_p(\varphi_n)y_n^{(p)}$$

reduces the system of equations (1.22) to a system of the form

$$y_{n+1}^{(p)} = A_p y_n^{(p)} + P_p(\varphi_n)y_n^{(p)}, \quad \Delta\varphi_n = \omega,$$

where $A_p = A + \bar{D}_1 + \bar{D}_2 + \cdots + \bar{D}_p$ is a diagonal infinite matrix,

$$P_p(\varphi_n) = (E + U_p(\varphi_n + \omega))^{-1}(P_{p-1}(\varphi_n)U_p(\varphi_n) - U_p(\varphi_n + \omega)\bar{D}_p),$$

$U_p(\varphi)$ is a solution, which is 2π-periodic in the domain

$$|Im\varphi| \le \rho_{p-1} - 2\delta_0^{p-1} = \rho_p > 0,$$

of the matrix equation

$$U_p(\varphi + \omega)A_{p-1} = A_{p-1}U_p(\varphi) + P_{p-1}(\varphi) - \bar{D}_p,$$

$$\bar{D}_p = \frac{1}{(2\pi)^m}\int_0^{2\pi}\cdots\int_0^{2\pi} D_p(\varphi)d\varphi^1\ldots d\varphi^m, \quad P_{p-1}(\varphi) = [p_{ij}^{p-1}(\varphi)]_{i,j=1}^{\infty},$$

$$D_p(\varphi) = diag\{p_{11}^{(p-1)}(\varphi), p_{22}^{(p-1)}(\varphi), p_{33}^{(p-1)}(\varphi), \dots\}.$$

It easy to verify that, for $n \in [a,b]_Z$, the sequence $y_n^{(p)}$ converges in the norm $\|\cdot\|$ to the function y_n, which defines the solution $y_n = y(n)$ of the system of equations (1.23), where $A_0 = \lim_{p\to\infty}(A + \bar{D}_1 + \bar{D}_2 + \cdots + \bar{D}_p)$. The inequality

$$\|\Phi_{s+k_0}(\varphi) - \Phi_s(\varphi)\|_0$$

$$\leq \frac{1}{8}\prod_{\alpha=1}^{\infty}(1 + \frac{M_{\alpha-1}^{\gamma-1}}{8})(\sum_{\alpha=s}^{s+k_0-1} M_\alpha^{\gamma-1}) < c\sum_{\alpha=s}^{\infty} M_\alpha^{\gamma-1},$$

where $M_\alpha = M_{\alpha-1}^\gamma, c = const > 0$ ensures that the sequence $\{\Phi_p(\varphi)\}$ converges uniformly in φ to the matrix $\Phi(\varphi)$ in the norm $\|\cdot\|_0$. The invertibility of the matrix $\Phi(\varphi)$ follows from the inequality $\|\Phi_p(\varphi) - E\| \leq l_0 < 1, p = 1, 2, 3, \dots$. $\qquad\square$

Corollary 1.2. *If, under conditions of Theorem 1.5, the function $y_n = y(n)$ is a solution of the system of equations (1.23) on the interval $[a,b]$, then $x_n = \Phi(\varphi_n)y_n$ is a solution of the system of equations (1.22) on the same interval.*

Proof. To prove this fact, it is sufficient to verify the validity of the equality

$$\Phi(\varphi_n + \omega)A_0 - A\Phi(\varphi_n) - P(\varphi_n)\Phi(\varphi_n) = 0 \tag{1.26}$$

for all $n \in Z$.

We consider the operator

$$L_p(\varphi_n) = \Phi_p(\varphi_n + \omega)A_p - A\Phi_p(\varphi_n)$$
$$- P(\varphi_n)\Phi_p(\varphi_n) + \Phi_p(\varphi_n + \omega)P_p(\varphi_n) \tag{1.27}$$

and, by employing the method of complete mathematical induction, show that $L_p(\varphi_n) = 0$ identically with respect to φ_n for all natural p and integer n.

We set $p = 1$. Then

$$L_1(\varphi_n) = (E + U_1(\varphi_n + \omega))(A + \bar{D}_1)$$
$$+ (E + U_1(\varphi_n + \omega))(E + U_1(\varphi_n + \omega))^{-1}$$
$$\times (P(\varphi_n)U_1(\varphi_n) - U_1(\varphi_n + \omega)\bar{D}_1)$$
$$- A(E + U_1(\varphi_n)) - P(\varphi_n)(E + U_1(\varphi_n)) =$$
$$= U_1(\varphi_n + \omega)A - AU_1(\varphi_n) - P(\varphi_n) + \bar{D}_1 = 0.$$

Let now $L_p(\varphi_n) = 0$ for $p \le k$. We will show that $L_{k+1}(\varphi_n) = 0$. This follows from the equality

$$
\begin{aligned}
L_{k+1}(\varphi_n) &= \Phi_k(\varphi_n + \omega)A_k - A\Phi_k(\varphi_n) - P(\varphi_n)\Phi_k(\varphi_n) \\
&\quad + \Phi_k(\varphi_n + \omega)P_k(\varphi_n) + (\Phi_k(\varphi_n + \omega)A_k \\
&\quad + \Phi_k(\varphi_n + \omega)P_k(\varphi_n) - A\Phi_k(\varphi_n) - P(\varphi_n)\Phi_k(\varphi_n)U_{k+1}(\varphi_n) \\
&\quad + \Phi_k(\varphi_n + \omega)(-P_k(\varphi_n) - A_k U_{k+1}(\varphi_n) + \bar{D}_{k+1} + U_{k+1}(\varphi_n + \omega)A_k) = \\
&= L_k(\varphi_n) + L_k(\varphi_n)U_{k+1}(\varphi_n) + \Phi_k(\varphi_n + \omega)0 = 0.
\end{aligned}
$$

Passing in (1.27) to the limit as $p \to \infty$, we obtain (1.26), that completes the proof of Corollary 1.2. $\qquad\square$

It is easy to verify that, for a sufficiently small M_0, the elements $\lambda_i, i = 1, 2, 3, \ldots$, of the diagonal matrix A_0 are separated. This implies that this matrix is not bounded in the norm. However, it has the inverse diagonal matrix A_0^{-1} bounded in the norm. Therefore, for any vector $y_l \in \mathfrak{M}$, there exists the solution $y_n = y(n, \varphi_0, y_l)$, $y_l = y(l, \varphi_0, y_l)$ of the system of equations (1.23) on the interval $(-\infty, l]_Z$. The extension of the solution to the right depends on the initial value of y_l. There exist the solutions of the system (1.23) defined on the interval $(-\infty, \infty)_Z$. Since only a finite number of elements of the matrix A_0 can belong to the segment $[-1, 1]$, the example of a solution is the solution with the initial value of

$$
y_l = \{ \frac{1}{|\lambda_1|}; \frac{1}{|\lambda_2|^2}; \frac{1}{|\lambda_3|^3}; \ldots \} \in \mathfrak{M}.
$$

We now consider the case where the matrix A is bounded in the norm. In this case, its elements cannot be separated. Hence, we cannot employ the method of proving Theorem 1.5 to solve the problem of reducibility of the system of equations (1.22) to a system of the form (1.23). In this case, we use the method of truncation of the system of equations (1.22).

Along with (1.22), we consider the system of equations

$$
\overset{(s)}{x}_{n+1} = \overset{(s)}{A}\,\overset{(s)}{x}_n + \overset{(s)}{P}(\varphi_n)\overset{(s)}{x}_n, \quad \Delta\varphi_n = \omega, \quad n \in Z, \tag{1.28}
$$

where

$$
\overset{(s)}{x} = (\overset{(s)1}{x}, \overset{(s)2}{x}, \ldots, \overset{(s)s}{x}), \quad \overset{(s)}{A} = diag\{a_1, a_2, \ldots, a_s\}, \quad \overset{(s)}{P}(\varphi) = [p_{ij}(\varphi)]^s_{i,j=1}.
$$

This system is finite-dimensional and follows from (1.22) by the truncation with respect to x up to the s-th order.

We assume that all elements of the matrix A are different, there exists the bounded inverse matrix for the matrix $A + P(\varphi)$, and all conditions of Theorem 1.5 are satisfied for system (1.22) except for the separation of the elements of the matrix A. In this case according to Theorem 5.1 from [60], there exist the constant $M_0(s) > 0$ for any $s = 1, 2, \ldots$ and the nondegenerate matrix $\overset{(s)}{\Phi}(\varphi)$ such that the change of variables

$$\overset{(s)}{x}_n = \overset{(s)}{\Phi}(\varphi_n) \overset{(s)}{y}_n \tag{1.29}$$

reduces the system of equations (1.28) to a system of the form

$$\overset{(s)}{y}_{n+1} = \overset{(s)}{A_0} \overset{(s)}{y}_n, \quad \Delta\varphi_n = \omega, \quad n \in Z, \tag{1.30}$$

on the interval $(-\infty, \infty)$.

By $\overset{(s)}{\Phi}{}^{-1}(\varphi)$, we denote the matrix inverse to $\overset{(s)}{\Phi}(\varphi)$, $s = 1, 2, 3, \ldots$, and formulate the following proposition.

Theorem 1.6. *Let the matrix $P(\varphi)$ be such that*

$$\sum_{j=n+1}^{\infty} \sup_{\varphi} |p_{ij}(\varphi)| \leq K(n), \quad i = 1, 2, 3, \ldots, \tag{1.31}$$

where $K(n) \to 0$ as $n \to \infty$. Assume that the sequences of matrices $\{\overset{(s)}{\Phi}(\varphi)\}_{s=1}^{\infty}$ and $\{\overset{(s)}{\Phi}{}^{-1}(\varphi)\}_{s=1}^{\infty} \ \forall \varphi \in \mathcal{T}_m$ are strongly proper and proper, respectively. Then, for

$$K(0) \leq \inf_{s=1,2,\ldots} M_0(s),$$

the system of equations (1.22) is reduced to a system of the form (1.23) on the interval $(-\infty, +\infty)$ with the help of the matrix $\Phi(\varphi)$, which can be presented as the elementwise limit

$$\Phi = \lim_{s \to \infty} \overset{(s)}{\Phi}(\varphi).$$

Proof. Obviously, it is sufficient to prove the statement of the theorem for each interval $[a, +\infty)_Z$ of changing n. Without loss of generality, we set $a = 0$ and show that

$$\lim_{s \to \infty} \overset{(s)}{x}(n, \varphi_0, \overset{(s)}{x}_0) = x(n, \varphi_0, x_0) \tag{1.32}$$

coordinatewise, where $\overset{(s)}{x}_n = \overset{(s)}{x}(n, \varphi_0, \overset{(s)}{x}_0)$ and $x_n = x(n, \varphi_0, x_0)$ are the solutions of the system of equations (1.28) and (1.22), respectively,

$$n \in [0, +\infty)_Z, \quad x_0 = (x_0^1, x_0^2, \ldots) \in \mathfrak{M}, \quad \overset{(s)}{x}_0 = (x_0^1, x_0^2, \ldots, x_0^s) \in R^s.$$

We will prove relation (1.32) for the first coordinates of the vectors $\overset{(s)}{x}_n$ and x_n. For $n = 1$, we have

$$\lim_{s\to\infty} \overset{(s)1}{x}_1 = \lim_{s\to\infty} a_1 x_0^1 + \lim_{s\to\infty} \sum_{j=1}^{s} p_{1j}(\varphi) x_0^j = x_1^1,$$

and $\| \overset{(s)}{x}_1 \| \leq (\|A\| + K(0)) \|x_0\|$ uniformly in $s = 1, 2, 3, \cdots$.

Analogously, we obtain the equality

$$\lim_{s\to\infty} \overset{(s)1}{x}_2 = a_1 x_1^1 + \lim_{s\to\infty} \sum_{j=1}^{\infty} p_{1j}(\varphi_1) \overset{(s)j}{x}_1, \tag{1.33}$$

where, for $j > s$, we set $\overset{(s)j}{x}_1 = 0$. Since

$$\sum_{j=1}^{\infty} p_{1j}(\varphi_1) \overset{(s)j}{x}_1 \leq (\|A\| + K(0)) \|x_0\| \sum_{j=1}^{\infty} |p_{1j}(\varphi_1)|,$$

the series on the left-hand side of the last inequality converges uniformly in s under condition (1.31). Then relation (1.33) yields

$$\lim_{s\to\infty} \overset{(s)1}{x}_2 = a_1 x_1^1 + \sum_{j=1}^{\infty} p_{1j}(\varphi_1) x_1^j = x_2^1,$$

and

$$\| \overset{(s)}{x}_2 \| \leq (\|A\| + K(0)) \|x_1\|$$

uniformly in $s = 1, 2, 3, \cdots$.

By the method of complete mathematical induction, we verify the validity of equality (1.32) for all $n \in [0, +\infty)_Z$.

By $\overset{(s)}{\Psi}(\varphi) = [\psi_{ij}^{(s)}(\varphi)]_{i,j=1}^s$, we denote the matrix $\overset{(s)}{\Phi}{}^{-1}(\varphi), s = 1, 2, 3, \ldots$. It follows from (1.29) that $\overset{(s)}{y}_n = \overset{(s)}{\Psi}(\varphi_n) \overset{(s)}{x}_n$ for all natural s. Then

$$\overset{(s)p}{y}_n = \sum_{i=1}^{\infty} \psi_{pi}^{(s)}(\varphi_n) \overset{(s)i}{x}_n, \quad s = 1, 2, 3, \ldots;$$

$$\psi_{pi}^{(s)}(\varphi_n) \overset{(s)i}{x}_n = 0, \quad i > s, p > s.$$

However, the series on the right-hand side of the last equality converges uniformly in $s = 1, 2, 3, \ldots$, since it is majorized by the numerical series

$$(\|A\| + K(0))^n \|x_0\| \sum_{i=1}^{\infty} |\psi_{p,i}^{(s)}(\varphi_n)|,$$

and the sequence of matrices $\{\overset{(s)}{\Psi}(\varphi)\}_{s=1}^{\infty}$ is proper.

This implies that, as $s \to \infty$, the sequence $\{\overset{(s)}{y}_n\}_{s=1}^{\infty}$ converges coordinatewise to the vector $y_n = \Psi(\varphi_n)x_n \; \forall n \in [0, +\infty)_Z$, where the matrix

$$\Psi(\varphi) = \lim_{s \to \infty} \overset{(s)}{\Psi}(\varphi)$$

is inverse to the matrix

$$\Phi(\varphi) = \lim_{s \to \infty} \overset{(s)}{\Phi}(\varphi),$$

and the last two limiting transitions are taken in the elementwise meaning.

It remains to show that the sequence of matrices $\{\overset{(s)}{A}_0\}_{s=1}^{\infty}$ is elementwise convergent to some diagonal matrix A_0.

Indeed, for $n \in [0, +\infty)_Z$, the matrices

$$\overset{(s)}{A}_0 = \overset{(s)}{\Phi}{}^{-1}(\varphi_n + \omega) \, \overset{(s)}{A} \, \overset{(s)}{\Phi}(\varphi_n) + \overset{(s)}{\Phi}{}^{-1}(\varphi_n + \omega) \, \overset{(s)}{P}(\varphi_n) \, \overset{(s)}{\Phi}(\varphi_n)$$

are diagonal for all $s = 1, 2, 3, \dots$.

It is evident that the matrix sequences $\{\overset{(s)}{\Phi}{}^{-1}(\varphi_n + \omega) \, \overset{(s)}{A}\}_{s=1}^{\infty}$ and $\{\overset{(s)}{P}(\varphi_n)\}_{s=1}^{\infty}$ are proper and strongly proper, respectively. Since the sum of proper sequences and the product of a proper sequence and a strongly proper one are proper sequences themselves, the sequence of matrices $\{\overset{(s)}{A}_0\}_{s=1}^{\infty}$ is proper, which means that it is elementwise convergent. $\square$

We complete the sub by considering the case where the initial system of equations is defined on an infinite-dimensional torus $\mathcal{T}_\infty$.

We write down the system of equations

$$x_{n+1} = Ax_n + P(\varphi_n)x_n, \quad \Delta\varphi_n = \omega, \quad n \in Z, \tag{1.34}$$

where

$$x = (x^1, x^2, \dots, x^p) \in R^p, \quad \varphi = (\varphi^1, \varphi^2, \varphi^3, \dots) \in \mathcal{T}_\infty,$$

$$\omega = (\omega_1, \omega_2, \dots) \in \mathfrak{M}, \quad A = diag(a_1, \dots, a_p), \quad P(\varphi) = [p_{ij}(\varphi)]_{i,j=1}^{p}.$$

We set

$$\overset{(m)}{\varphi} = (\varphi^1, \varphi^2, \dots, 0, 0, \dots), \quad \overset{(m)}{\omega} = (\omega_1, \omega_2, \dots, \omega_m, 0, 0, \cdots)$$

and consider, along with (1.34), the system of equations

$$\overset{(m)}{x}_{n+1} = A \, \overset{(m)}{x}_n + P(\overset{(m)}{\varphi}_n) \, \overset{(m)}{x}_n, \quad \Delta \overset{(m)}{\varphi}_n = \overset{(m)}{\omega}, \quad n \in Z, \tag{1.35}$$

where $\overset{(m)}{x}_n = (\overset{(m)}{x}{}^1_n, \overset{(m)}{x}{}^2_n, \ldots, \overset{(m)}{x}{}^p_n)$.

Let us assume that, for any natural m, there exist the positive numbers $\delta(m)$ and $d(m)$ such that

$$|\sin \frac{\langle \overset{(m)}{k}, \overset{(m)}{\omega} \rangle}{2}| \geq \delta(m)|\overset{(m)}{k}|^{-d(m)} \tag{1.36}$$

for all nonzero integer-valued vectors $\overset{(m)}{k} = (\overset{(m)}{k}_1, \ldots, \overset{(m)}{k}_m, 0, 0, \ldots)$.

Then, under the conditions of Theorem 1.5 for any natural m, there exist the constant $M_o(m)$ and the matrix $\Phi_m(\overset{(m)}{\varphi})$ such that, for $\|P(\overset{(m)}{\varphi})\|_0 \leq M_0(m)$, the change of variables

$$\overset{(m)}{x}_n = \Phi_m(\overset{(m)}{\varphi}_n)\, \overset{(m)}{y}_n$$

reduces the system of equations (1.35) to a system of the form

$$\overset{(m)}{y}_{n+1} = A_m \overset{(m)}{y}_n, \quad \Delta \overset{(m)}{\varphi}_n = \overset{(m)}{\omega}, \quad n \in Z,$$

on the interval $(-\infty, +\infty)$.

It is said that a mapping $f : \mathfrak{M} \to \mathfrak{B}$, where $\mathfrak{B}$ is an arbitrary linear space with norm $\| \cdot \|_*$ satisfies the sharpened Cauchy–Lipschitz conditions with respect to $\varphi \in \mathfrak{M}$, if, for any points $\varphi = (\varphi^1, \ldots, \varphi^m, \varphi^{m+1}, \varphi^{m+2}, \ldots)$ and $\bar{\varphi} = (\varphi^1, \ldots, \varphi^m, \bar{\varphi}^{m+1}, \bar{\varphi}^{m+2}, \ldots)$ from the set $\mathfrak{M}$, whose first m corresponding coordinates coincide pairwise, the inequality

$$\|f(\varphi) - f(\bar{\varphi})\|_* \leq \alpha(m)\|\varphi - \bar{\varphi}\|$$

holds, where $\alpha(m) \to 0$ as $m \to \infty$. The set of all such mappings will be denoted by $L_{Lip}(\varphi)$.

The following proposition is presented without proof, since it is analogous to that of Theorem 1.6.

Theorem 1.7. *Let $P(\varphi) \in L_{Lip}(\varphi,)$ let, for all natural m, the matrices $P(\overset{(m)}{\varphi})$ be analytic in the domain $|Im \overset{(m)}{\varphi}| \leq \rho_0$ $(\rho_0 > 0)$, and let condition (1.36) be satisfied. Then if $\|P(\varphi)\| \leq \inf_m M_0(m)$, and if the matrix sequences*

$$\{\Phi_m(\overset{(m)}{\varphi})\}^\infty_{m=1}, \quad \{\Phi_m^{-1}(\overset{(m)}{\varphi})\}^\infty_{m=1}$$

are convergent in the norm $\| \cdot \|$ as $m \to \infty$, then the system of equations (1.34) is reducible to system (1.23) on the interval $(-\infty, +\infty)$ with the help of the matrix $\Phi(\varphi) = \lim_{m\to\infty} \Phi_m(\overset{(m)}{\varphi})$.

If the variables x and φ in the system of equations (1.22) belong to the space $\mathfrak{M}$, then there exist various ways of their truncation. We will formulate one of the possible results as a corollary of Theorems 1.6 and 1.7.

We assume that, in this case, system (1.22) satisfies the conditions of Theorem 1.7 and relation (1.31). Then, for any natural numbers s and m, there exist the constants $M_0(s,m) > 0$ such that, for $\| \overset{(s)}{P}(\overset{(m)}{\varphi}) \| \leq M_0(s,m)$, the systems of equations corresponding to system (1.22), which are truncated with respect to the variable x to the s-th and the variable φ to the m-th orders, are reducible on the entire axis with the help of the matrices denoted by $\overset{(s)}{\Phi}_m(\overset{(m)}{\varphi})$.

Theorem 1.8. *Let, under the above assumptions for any natural number s,*

$$\{\overset{(s)}{\Phi}_m(\overset{(m)}{\varphi})\}_{m=1}^{\infty} \to \overset{(s)}{\Phi}(\varphi), \quad \{\overset{(s)}{\Phi}_m^{-1}(\overset{(m)}{\varphi})\}_{m=1}^{\infty} \to \overset{(s)}{\Phi}{}^{-1}(\varphi)$$

as $m \to \infty$ in the norm $\| \cdot \|$, and let the sequences of matrices $\{\overset{(s)}{\Phi}(\varphi)\}_{s=1}^{\infty}$ and $\{\overset{(s)}{\Phi}{}^{-1}(\varphi)\}_{s=1}^{\infty}$ be strongly proper and proper, respectively. Then, for

$$\|P(\varphi)\|_0 \leq \inf_{s,m=1,2,\ldots} M_0(s,m),$$

the system of equations (1.22) is reducible on the interval $(-\infty, +\infty)$. Moreover, the matrix $\Phi(\varphi)$ realizing the corresponding change of variables can be presented in the form of a double limit

$$\Phi(\varphi) = \lim_{s\to\infty} \lim_{m\to\infty} \overset{(s)}{\Phi}_m(\overset{(m)}{\varphi}),$$

where the external and internal limiting transitions are taken in the coordinatewise meaning and in the norm, respectively.

1.3 Nonlinear almost periodic equations defined on an infinite-dimensional torus

1^0. First, we consider the difference equation

$$x(t+1) = x(t) + \omega + P(x(t), t) + \lambda, \tag{1.37}$$

where $\{x, \omega, \lambda\} \subset R^n$, $P(x,t)$ is an n-dimensional real vector-function, which is 2π-periodic in x_i $(i = 1, 2, \ldots, n)$ and almost periodic in t with the basis of frequencies α, where $\alpha = (\alpha_1, \alpha_2, \ldots, \alpha_m, \ldots)$ is an infinite sequence of real rationally noncommensurable numbers. We assume that $x = (x_1, x_2, \ldots, x_n)$ are the angular coordinates on an m-dimensional torus $\mathcal{T}_m$, with Eq. (1.37) being defined on it.

We set the problem: to find a change of variables $x = y + U_0(y, t)$, which reduces Eq. (1.37) to the equation

$$y(t+1) = y(t) + \omega. \tag{1.38}$$

We denote sequences of integers and complex numbers by $r = (r_1, r_2, r_3, \ldots)$ and $(x, \varphi) = (x_1, x_2, \ldots, x_n, \varphi_1, \varphi_2, \varphi_3, \ldots)$, respectively, and set

$$r^m = (r_1, r_2, \ldots, r_m); \quad |r^m| = \max_{1 \leq i \leq m} \{|r_i|\}, \quad \|r^m\| = \sum_{i=1}^{m} |r_i|;$$

$$(x, \varphi^m) = (x_1, x_2, \ldots, x_n, \varphi_1, \varphi_2, \ldots, \varphi_m); \quad |\varphi^m| = \max_{1 \leq i \leq m} \{|\varphi_i|\};$$

$$\langle r^m, \varphi^m \rangle = \sum_{i=1}^{m} r_i \varphi_i; \quad |Imx| = \max_i \{|Imx_i|\}; \quad |Im\varphi| = \sup_i \{|Im\varphi_i|\};$$

$$\Pi(\rho) = \{(x, \varphi)| \quad |Imx| < \rho, \quad |Im\varphi| < \rho\};$$
$$\Pi(\rho, m) = \{(x, \varphi^m)| \quad |Imx| < \rho, \quad |Im\varphi^m| < \rho\};$$
$$\Pi_0 = \{(x, \varphi)| \quad |Imx| = 0, \quad |Im\varphi| = 0\};$$
$$\Pi(\rho)L = \{(x, \varphi, \lambda)| \quad (x, \varphi) \in \Pi(\rho), \quad \|\lambda\| < L\}.$$

We introduce the following norms: for the function

$$F(x, \varphi) = (F^1(x, \varphi), \ldots, F^n(x, \varphi))$$

defined in the domain $\Pi(\rho)$, we set

$$\|F(x, \varphi)\|_\rho = \sum_{i=1}^{n} \sup_{\Pi(\rho)} |F^i(x, \varphi)|;$$

for the n-dimensional function $F(x, \varphi, \lambda)$, defined in the domain $\Pi(\rho)L$, we set

$$\|F(x, \varphi, \lambda)\|_{\rho, L} = \sum_{i=1}^{n} \sup_{\Pi(\rho)L} |F^i(x, \varphi, \lambda)|;$$

and, for the $n \times n$-dimensional matrix $R(x, \varphi, \lambda) = [R_{ij}(x, \varphi, \lambda)]_{i,j=1}^{n}$ defined in the domain $\Pi(\rho)L$, we set

$$\|R(x, \varphi, \lambda)\|_{\rho, L} = \max_{1 \leq i \leq n} \sum_{j=1}^{n} \sup_{\Pi(\rho)L} |R_{ij}(x, \varphi, \lambda)|.$$

Definition 1.1. The n-dimensional vector-function $F(x,\varphi), (x,\varphi) \in \Pi_0$ belongs to the class $A(\Theta), \Theta > 0$, if there exist the positive monotonically nonde-creasing sequences of integers $n(k)$ and real numbers $g(k)$, as well as the sequence of complex n-dimensional vector-functions $F_k(x,\varphi) = F_k(x,\varphi^{n(k)})$, which satisfy the following conditions:

$$1) \quad \lim_{k\to\infty} n(k) = \infty, \quad \overline{\lim_{k\to\infty}} \frac{\ln n(k)}{g(k)} = b < \infty,$$

$$g(k+1) - g(k) \geq \ln \frac{k+1}{k}, \quad k \geq 1;$$

2) $F_1(x,\varphi) = 0$, and, for each $k \geq 2$, the function $F_k(x,\varphi)$ is 2π-periodic in each variable, real for real (x,φ), and analytic in the domain $\Pi(\rho_k, n(k))$, where

$$\rho_k = c_0 \sum_{i=k}^{\infty} \exp\{-ag(i)\}, \quad k \geq 1, \quad \overline{\lim_{k\to\infty}} \frac{\ln k}{g(k)} < a, \quad c_0 > 0;$$

3) $\lim_{k\to\infty} F_k(x,\varphi) = F(x,\varphi)$ for $(x,\varphi) \in \Pi_0$;

4) there exists the constant $C_F > 0$ such that, for any $k > 1$, the inequality

$$\|F_k(x,\varphi) - F_{k-1}(x,\varphi)\|_{\rho_k} \leq C_F \exp\{-\Theta n(k)g(k)\}$$

holds.

Definition 1.2. The n-dimensional vector-function $P(x,t), x \in R^n, t \in R^1$ belongs to a class $A(\Theta, \alpha)$, where $\alpha = (\alpha_1, \alpha_2, \dots)$ is the infinite sequence of real rationally noncommensurable numbers, if there exists the function $F(x,\varphi)$ from $A(\Theta)$ such that $P(x,t) = F(x,\alpha t), x \in R^n, t \in R^1$.

Definition 1.3. The n-dimensional vector-function

$$\Upsilon(x,\varphi^m,\lambda), \quad \{x,\varphi^m,\lambda\} \subset C^n$$

belongs to a class

$$H(k,\rho,L,c), \quad \rho > 0, L > 0, c > 0, \quad k = 1,2,3,\dots,$$

if it is defined and analytic in the domain

$$\Pi(\rho,m)L = \{(x,\varphi^m,\lambda)|(x,\varphi^m) \in \Pi(\rho,m), \|\lambda\| < L\},$$

2π-periodic in x_i, φ_j, and real for real x, φ^m, and λ and satisfies the estimate

$$\|\Upsilon(x,\varphi^m,\lambda)\|_{\rho,L} = \sum_{i=1}^{n} \sup_{\Pi(\rho,m)L} |\Upsilon^i(x,\varphi^m,\lambda)| \leq c \exp\{-\Theta n(k)g(k)\}.$$

In what follows, we consider that the sequences $n(k)$ and $g(k)$, apart from conditions contained in Definition 3.1, satisfy the equalities

$$\lim_{k\to\infty}\frac{n(k+1)}{n(k)} = \lim_{k\to\infty}\frac{g(k+1)}{g(k)} = 1.$$

The following inductive lemma is valid.

Lemma 1.1. *The system of difference equations*

$$x(t+1) = x(t) + \omega + \Upsilon(x(t), \varphi^{n(k)}, \lambda) + \lambda, \quad \Delta\varphi_t^{n(k)} = \alpha^{n(k)}, \qquad (1.39)$$

where the vector-function $\Upsilon(x, \varphi^{n(k)}, \lambda)$ *belongs to the class* $H(k, \sigma_k, L_k, c_1)$,

$$\sigma_k = \rho_k/\sigma_0, \quad \sigma_0 = 2p_1(1+\sigma_1),$$

$$p_1 = \prod_{i=1}^{\infty}(1 + c_1 c_3 \exp\{-\Theta_1 n(i)g(i)\}),$$

$$\sigma_1 = \sum_{i=1}^{\infty} c_1 c_3 \exp\{-\Theta_1 n(i)g(i)\},$$

$$L_k = c_1 c_2 \exp\{-\Theta_2 n(k)g(k)\},$$

$\Theta, \Theta_1, \Theta_2$ *are the positive constants,* $\Theta > 2(a+b)$, $\Theta - \Theta_1 < \Theta_2 < \Theta/2 < \Theta_1 < \Theta - a - b$, *and* c_1, c_2, *and* c_3 *are positive constants, is reduced by the change of variables*

$$x = y + U(y, \varphi^{n(k)}, \lambda), \quad \lambda = \lambda(\mu) \qquad (1.40)$$

to the system

$$y(t+1) = y(t) + \omega + R(y, \varphi^{n(k)}, \mu) + \mu, \quad \Delta\varphi_t^{n(k)} = \alpha^{n(k)},$$

if the condition

$$\left|cosec\frac{\langle \omega, s\rangle + \langle \alpha^m, r^m\rangle}{2}\right| \le c(m)(\|s\| + \|r^m\|)^{n+m},$$

where $s = (s_1, \ldots, s_n)$ *and* $r^m = (r_1, \ldots, r_m)$ *are any nonzero integer-valued vectors, is satisfied.*

Proof. After change (1.40), the system of equations (1.39) turns into the system

$$y(t+1) + U(y(t+1), \varphi_{t+1}^{n(k)}, \lambda(\mu)) = y(t) + U(y(t), \varphi_t^{n(k)}, \lambda(\mu))$$

$$+ \omega + \Upsilon(y(t) + U(y(t), \varphi_t^{n(k)}, \lambda(\mu)), \varphi_t^{n(k)}, \lambda(\mu)) + \lambda(\mu).$$

We present it in the form

$$y(t+1) - y(t) - \omega = U(y(t), \varphi_t^{n(k)}, \lambda(\mu)) + U(y(t) + \omega, \varphi_{t+1}^{n(k)}, \lambda(\mu))$$
$$- U(y(t+1) - y(t) - \omega + y(t) + \omega, \varphi_{t+1}^{n(k)}, \lambda(\mu)) - U(y(t) + \omega, \varphi_{t+1}^{n(k)}, \lambda(\mu))$$
$$+ S_{N(k)}\Upsilon(y(t), \varphi_t^{n(k)}, \lambda(\mu)) - S_{N(k)}\Upsilon(y(t), \varphi_t^{n(k)}, \lambda(\mu))$$
$$+ \bar{\Upsilon}(\lambda(\mu)) - \bar{\Upsilon}(\lambda(\mu)) + \Upsilon(y(t) + U(y(t), \varphi_t^{n(k)}, \lambda(\mu)), \varphi_t^{n(k)}, \lambda(\mu)), \quad (1.41)$$

where

$$S_{N(k)}\Upsilon = \sum_{\|s\| + \|r^{n(k)}\| \le N(k)} \Upsilon_{s, r^{n(k)}}(\lambda) \exp\{i\langle y, s\rangle + i\langle \varphi^{n(k)}, r^{n(k)}\rangle\},$$

$$\Upsilon_{s, r^{n(k)}}(\lambda) = \frac{1}{(2\pi)^{n+n(k)}} \int_0^{2\pi} \cdots \int_0^{2\pi} \Upsilon \exp\{-i\langle y, s\rangle$$
$$+ i\langle \varphi^{n(k)}, r^{n(k)}\rangle\} dy_1 \cdots dy_n d\varphi_1 \cdots d\varphi_{n(k)},$$

$$\bar{\Upsilon}(\lambda) = \frac{1}{(2\pi)^{n+n(k)}} \int_0^{2\pi} \cdots \int_0^{2\pi} \Upsilon\, dy_1 \cdots dy_n d\varphi_1 \cdots d\varphi_{n(k)},$$

$$N(k) = \Theta_4 \frac{2\sigma_0}{c_0} n(k) g(k) \exp\{\alpha g(k)\}.$$

We choose the functions $U(y, \varphi^{n(k)}, \lambda)$ and $\lambda(\mu)$ as the solutions of the equations

$$\mu = \lambda(\mu) + \bar{\Upsilon}(\lambda(\mu)),$$

$$U(y + \omega, \varphi^{n(k)} + \alpha^{n(k)}, \lambda(\mu))$$
$$= U(y, \varphi^{n(k)}, \lambda(\mu)) - \bar{\Upsilon}(\lambda(\mu)) + S_{N(k)}\Upsilon(y, \varphi^{n(k)}, \lambda(\mu)). \quad (1.42)$$

By virtue of such a choice, relations (1.41) can be written in the form

$$y(t+1) - y(t) - \omega = U(y(t) + \omega, \varphi_{t+1}^{n(k)}, \lambda(\mu))$$
$$- U(y(t+1) - y(t) - \omega + y(t) + \omega, \varphi_{t+1}^{n(k)}, \lambda(\mu))$$
$$+ \Upsilon(y(t) + U(y(t), \varphi_t^{n(k)}, \lambda(\mu)), \varphi_t^{n(k)}, \lambda(\mu))$$
$$- S_{N(k)}\Upsilon(y(t), \varphi_t^{n(k)}, \lambda(\mu)) + \mu. \quad (1.43)$$

Upon solving (1.43) with respect to $y(t+1) - y(t) - \omega$, we obtain the system of equations

$$y(t+1) = y(t) + \omega + R(y(t), \varphi_t^{n(k)}, \mu) + \mu, \quad \Delta\varphi_t^{n(k)} = \alpha^{n(k)},$$

where

$$R(y, \varphi^{n(k)}, \mu) = [(E + \frac{\partial U}{\partial y})^{-1} - E]\mu$$

$$+ (E + \frac{\partial U}{\partial y})^{-1}[\Upsilon(y(t) + U(y(t), \varphi_t^{n(k)}, \lambda(\mu)), \varphi_t^{n(k)}, \lambda(\mu))$$

$$- S_{N(k)} \Upsilon(y(t), \varphi_t^{n(k)}, \lambda(\mu))].$$

According to [57], [74], Eq. (1.42) has the solution

$$U(\psi^{m(k)}, \lambda(\mu)) = \sum_{0 < \|r^{m(k)}\| \leq N(k)} \frac{\Upsilon_{r^{m(k)}}(\lambda(\mu))}{e^{i\langle \beta^{m(k)}, r^{m(k)} \rangle} - 1} \exp\{i\langle \psi^{m(k)}, r^{m(k)} \rangle\},$$

where $m(k) = n + n(k)$, $\psi^{m(k)} = (y, \varphi^{n(k)})$, $\beta^{m(k)} = (\omega, \alpha^{n(k)})$.

In this case, the following estimates hold:

$$\|\lambda(0)\| \leq c_1 \exp\{-\Theta n(k)g(k)\},$$

$$\|\lambda(\mu)\|_{L_{k+1}} \leq L_k - c_1 c_4 \exp\{-\Theta_3 n(k)g(k)\},$$

$$\|\frac{\partial \lambda(\mu)}{\partial \mu}\|_{L_{k+1}} \leq 1 + \frac{c_5}{c_4}n^2 \exp\{-(\Theta - \Theta_3)n(k)g(k)\},$$

$$\|R\|_{\sigma_{k+1}, L_{k+1}} \leq \frac{c_1}{c_2} \exp\{-\Theta n(k+1)g(k+1)\},$$

$$\|U\|_{\sigma_{k+1}, L_{k+1}} \leq c_1 c_3 \exp\{-\Theta_1 n(k+1)g(k+1)\},$$

$$\sup_{\Pi(\sigma_{k+1}, n(k))L_{k+1}} \|ImU\| \leq \frac{\delta_k}{2\sigma_0} = \frac{c_0}{2\sigma_0} \exp\{-\alpha g(k)\},$$

$$\|\frac{\partial U}{\partial y}\|_{\sigma_{k+1}, L_{k+1}} \leq c_1 c_3 \exp\{-\Theta_1 n(k+1)g(k+1)\} \leq \frac{1}{2},$$

$$\|\frac{\partial U}{\partial \varphi^{n(k)}}\|_{\sigma_{k+1}, L_{k+1}} \leq c_1 c_3 \exp\{-\Theta_1 n(k+1)g(k+1)\},$$

$$\|\frac{\partial U}{\partial \mu}\|_{\sigma_{k+1}, L_{k+1}} \leq \frac{c_3 c_5}{c_4}n^2 \exp\{-(\Theta_1 - \Theta_3)n(k+1)g(k+1)\},$$

where $\Theta_1 < \Theta_3 < \Theta_2$ and $1 \leq c_4 \leq c_2$.

It is easy to verify that the function $R(y, \varphi^{n(k)}, \mu)$ is analytic, 2π-periodic in y_j, φ_j, and real for real $(y, \varphi^{n(k)}, \mu)$ in the domain $\Pi(\sigma_{k+1}, n(k))L_{k+1}$. $\qquad\square$

Using the inductive lemma, we will prove the theorem of reducibility of Eq. (1.37).

Theorem 1.9. *Let Eq. (1.37) satisfy the following conditions:*

1) $P(x,t) \in A(\Theta, \alpha), \quad \Theta > 2(a+b),$

$$\lim_{k \to \infty} \frac{n(k+1)}{n(k)} = \lim_{k \to \infty} \frac{g(k+1)}{g(k)} = 1;$$

2) for any integer $m > 0$ and arbitrary nonzero integer-valued vectors $s = (s_1, \ldots, s_n), r^m = (r_1, \ldots, r_m),$ *the inequality*

$$\left| cosec \frac{\langle \omega, s \rangle + \langle \alpha^m, r^m \rangle}{2} \right| \le c(m)(\| s \| + \| r^m \|)^{n+m},$$

$$c(m) = \gamma d_1^m m^{d_2}, \quad \gamma > 1, d_1 > 1, d_2 > 0$$

holds. Then there exists the constant $M_0 > 0$ such that, for $C_F \le M_0$, there exist the vector $\lambda = \lambda_0$ and the change $x = y + U_0(t,y), \quad U_0(t,y) \in A(\Theta_1, \alpha)$, which reduces Eq. (1.37) to Eq. (1.38).

Proof.   We consider the sequence of systems

$$x(t+1) = x(t) + \omega + F_k(x(t), \varphi^{n(k)}) + \lambda, \quad \Delta\varphi_t^{n(k)} = \alpha^{n(k)} \qquad (1.44)$$

and prove that, for any $k \ge 1$, it is possible construct the change of variables

$$x = V_k(x^{(k)}, \varphi^{n(k-1)}, \lambda^{(k)}), \quad \lambda = \Phi_k(\lambda^{(k)}), \qquad (1.45)$$

which transforms system (1.37) into the system of equations

$$x^{(k)}(t+1) = x^{(k)}(t) + \omega + \Upsilon_k(x^{(k)}, \varphi^{n(k)}, \lambda^{(k)}) + \lambda^{(k)}, \quad \Delta\varphi_t^{n(k)} = \alpha^{n(k)},$$
$$(1.46)$$

and

$$V_k - x^{(k)} \in H_1(k, \sigma_k, L_k), \quad \Upsilon_k \in H(k, \sigma_k, L_k, c_1).$$

Setting

$$x = V_1(x^{(1)}) = x^{(1)}, \quad \lambda = \Phi_1(\lambda^{(1)}) = \lambda^{(1)}$$

in the system of equations (1.44) for $k = 1$, we transform it into the system

$$x^{(1)}(t+1) = x^{(1)}(t) + \omega + \Upsilon_1(x^{(1)}, \varphi^{n(1)}, \lambda^{(1)}) + \lambda^{(1)}, \quad \Delta\varphi_t^{n(1)} = \alpha^{n(1)},$$

where $\Upsilon_1 = 0$.

Under assumption that the change of variables (1.45) transforms the system of equations (1.37) into system (1.46), we notice that the last system satisfies all conditions of the inductive lemma. Moreover, at the appropriate choice of constants c_1, c_2, and c_3, we can construct the change of variables

$$x^{(k)} = x^{(k+1)} + U_{k+1}(x^{(k+1)}, \varphi^{n(k)}, \lambda^{(k)}(\lambda^{(k+1)})), \quad \lambda^{(k)} = \lambda^{(k)}(\lambda^{(k+1)}),$$

which reduces system (1.46) to the system

$$x^{(k+1)}(t+1) = x^{(k+1)}(t) + \omega + R_{k+1}(x^{(k+1)}(t), \varphi^{n(k)}, \lambda^{(k+1)}) + \lambda^{(k+1)},$$

$$\Delta\varphi_t^{n(k)} = \alpha^{n(k)}.$$

In the system of equations

$$x(t+1) = x(t) + \omega + F_k(x(t), \varphi^{n(k)})$$
$$+ \lambda + (F_{k+1}(x(t), \varphi^{n(k+1)}) - F_k(x(t), \varphi^{n(k)})),$$
$$\Delta\varphi_t^{n(k+1)} = \alpha^{n(k+1)},$$

let us make the change of variables

$$x = V_{k+1}(x^{(k+1)}, \varphi^{n(k)}, \lambda^{(k+1)}), \quad \lambda = \Phi_{k+1}(\lambda^{(k+1)}),$$

where

$$V_{k+1}(x^{(k+1)}, \varphi^{n(k)}, \lambda^{(k+1)}) = V_k(x^{(k+1)} + U_{k+1}, \varphi^{n(k-1)}, \lambda^{(k)}(\lambda^{(k+1)})),$$

$$\Phi_{k+1}(\lambda^{(k+1)}) = \Phi_k(\lambda^{(k)}(\lambda^{(k+1)})).$$

We obtain the system

$$x^{(k+1)}(t+1) = x^{(k+1)}(t) + \omega + \Upsilon_{k+1}(x^{(k+1)}(t), \varphi^{n(k+1)}, \lambda^{(k+1)}) + \lambda^{(k+1)},$$

$$\Delta\varphi_t^{n(k+1)} = \alpha^{n(k+1)},$$

and

$$V_{k+1} \in H(k+1, \sigma_{k+1}, L_{k+1}, c_1).$$

Analogously to the substantiation of the modified method of accelerated convergence [28], we can prove the convergence of the sequences $\Phi_k(0)$ and $\tilde{V}_k(x, \varphi) = V_k(x, \varphi^{n(k-1)}, 0)$ as $k \to \infty$: $\lim_{k\to\infty} \Phi_k(0) = \lambda_0$, $\lim_{k\to\infty} \tilde{V}_k(x, \varphi) = V(x, \varphi)$, $U_0(x, t) = V(x, \alpha t) - x$.

Making the change of variables $x = V_k(y, \alpha^{n(k-1)}t, 0)$, $\lambda = \Phi_k(0)$ in the equation

$$x(t+1) = x(t) + \omega + F_k(x, \alpha^{n(k)}t) + \lambda + (F(x(t), \alpha t) - F_k(x(t), \alpha^{n(k)}t)),$$

we obtain the equation

$$y(t+1) = y(t) + \omega + \Upsilon_k(y, \alpha^{n(k-1)}t, 0)$$
$$+ (E + \frac{\partial V_k}{\partial y})^{-1}(F(V_k, \alpha t) - F_k(V_k, \alpha^{n(k)}t)).$$

Since

$$\|\Upsilon_k(y, \alpha^{n(k)}t, 0) + (E + \frac{\partial \tilde{V}_k}{\partial y})^{-1}(F(\tilde{V}_k, \varphi) - F_k(\tilde{V}_k, \varphi^{n(k)}))\|_0 \to 0$$

as $k \to \infty$, Eq. (1.37) is reduced to an equation of the form (1.38) by the change $x = y + U_0(t, y)$. This completes the proof of the theorem. $\qquad\square$

2^0. We now consider the equation of the form (1.37)

$$x(t+1) = x(t) + \omega + P(x(t), t) + \lambda, \qquad (1.47)$$

where $\{x, \omega, \lambda\} \subset \mathfrak{M}$ is an infinite-dimensional vector-function $P(x, t)$, which is real, 2π-periodic in $x_i, i = 1, 2, \ldots$, and almost periodic in t with the basis of frequencies α. In this case, we consider Eq. (1.47) to be defined on an infinite-dimensional torus.

For the sequences $x = (x_1, x_2, \ldots)$ and $\varphi = (\varphi_1, \varphi_2, \ldots)$ of complex numbers, we introduce the notation:

$$x^n = (x_1, x_2, \ldots, x_n, 0, 0, \ldots);$$
$$|x| = \sup_{i=1,2,\ldots} \{|x_i|\}; \quad |Imx| = \sup_i \{|Imx_i|\};$$
$$\Pi(\rho) = \{(x, \varphi) |\ |Imx| < \rho, \quad |Im\varphi| < \rho\};$$
$$\Pi(\rho, n_1, n_2) = \{(x^{n_1}, \varphi^{n_2}) |\ |Imx^{n_1}| < \rho, \quad |Im\varphi^{n_2}| < \rho\};$$
$$\Pi_0 = \{(x, \varphi) |\ |Imx| = 0, \quad |Im\varphi| = 0\};$$
$$\Pi(\rho)L = \{(x, \varphi, \lambda) |\ (x, \varphi) \in \Pi(\rho), \quad \|\lambda\| < L\};$$
$$\Pi(\rho, n_1, n_2)L = \{(x^{n_1}, \varphi^{n_2}, \lambda) | (x^{n_1}, \varphi^{n_2}) \in \Pi(\rho, n_1, n_2), \quad \|\lambda\| < L\}.$$

For the infinite-dimensional vector-function

$$F(x, \varphi) = (F^{(1)}(x, \varphi), \ldots, F^{(n)}(x, \varphi), \ldots)$$

we introduce the following norms:

$$\|F(x, \varphi)\|_\rho = \sup_{\Pi(\rho)} \sum_{i=1}^{\infty} |F^{(i)}(x, \varphi)|, \quad \|F(x, \varphi)\|_0 = \sup_{\Pi_0} \sum_{i=1}^{\infty} |F^{(i)}(x, \varphi)|.$$

Definition 1.4. The infinite-dimensional vector-function $F(x, \varphi), (x, \varphi) \in \Pi_0$ belongs to the class $A_1(\Theta), \Theta > 0$, if it is possible to point out the positive monotonically nondecreasing sequences of integers $n_1(k)$ and $n_2(k)$ and real numbers $g(k)$, as well as the sequence of vector-functions

$$F_k(x, \varphi) = F_k^{(n_1(k))}(x^{n_1(k)}, \varphi^{n_2(k)})$$
$$= (F_k^{(1)}(x^{n_1(k)}, \varphi^{n_2(k)}), \ldots, F_k^{(n_1(k))}(x^{n_1(k)}, \varphi^{n_2(k)}), 0, 0, \ldots),$$

which satisfy the conditions:

$$1) \quad \lim_{k \to \infty} n(k) = \infty, \quad \overline{\lim_{k \to \infty}} \frac{\overline{\ln n(k)}}{g(k)} = b < \infty,$$

$$g(k+1) - g(k) \geq \ln \frac{k+1}{k}, \quad n(k) \geq n_1(k) + n_2(k);$$

2) $F_1(x, \varphi) = 0$, and, $\forall k \geq 2$, the function $F_k(x, \varphi)$ is 2π-periodic in each variable, real for the real (x, φ), and analytic in the domain $\Pi(\rho_k, n_1(k), n_2(k))$, where

$$\rho_k = c_0 \sum_{i=k}^{\infty} \exp\{-ag(i)\}, \quad \overline{\lim_{k \to \infty}} \frac{\ln k}{g(k)} < a, \quad c_0 > 0;$$

3) $\lim_{k \to \infty} F_k(x, \varphi) = F(x, \varphi)$ for $(x, \varphi) \in \Pi_0$;

4) there exists the constant $C_F > 0$ such that, for each $k > 1$, the inequality

$$\|F_k(x, \varphi) - F_{k-1}(x, \varphi)\|_{\rho_k} \leq C_F \exp\{-\Theta n(k)g(k)\}$$

is satisfied.

Definition 1.5. The infinite-dimensional vector-function $P(x, t), x \in \mathfrak{M}, t \in R^1$ belongs to the class $A_1(\Theta, \alpha)$, if there exists the function $F(x, \varphi)$ from $A_1(\Theta)$ such that $P(x, t) = F(x, \alpha t), \forall x \in \mathfrak{M}, t \in R^1$.

Definition 1.6. The n_1-dimensional vector-function $\Upsilon(x^{n_1}, \varphi^{n_2}, \lambda^{n_1})$ belongs to the class $H(k, \rho, L, c), \rho > 0, L > 0, c > 0$, if it is defined and analytic in the domain $\Pi(\rho, n_1, n_2)L, 2\pi$-periodic in x_i and φ_j, real for real $(x^{n_1}, \varphi^{n_2}, \lambda^{n_1})$, and satisfies the estimate

$$\|\Upsilon\|_{\rho, L} = \sum_{i=1}^{n_1} \sup_{\Pi(\rho, n_1, n_2)L} \|\Upsilon^i\| \leq c \exp\{-\Theta n(k)g(k)\}.$$

The following proposition is an analog of Lemma 1.1.

Lemma 1.2. *The system of equations*

$$x^{n_1(k)}(t+1) = x^{n_1(k)}(t) + \omega^{n_1(k)} + \Upsilon^{n_1(k)}(x^{n_1(k)}, \varphi^{n_2(k)}, \lambda^{n_1(k)}) + \lambda^{n_1(k)},$$

$$\Delta \varphi_p^{n_2(k)} = \alpha^{n_2(k)},$$

where the $n_1(k)$-dimensional vector-function

$$\Upsilon^{n_1(k)}(x^{n_1(k)}(t), \varphi^{n_2(k)}, \lambda^{n_1(k)})$$

is analytic in the domain $\Pi(\sigma_{k_1}, n_1(k), n_2(k))L_k, 2\pi$-periodic in the variables x_i and φ_j, real for real $(x^{n_1(k)}, \varphi^{n_2(k)}, \lambda^{n_1(k)})$, and satisfies the estimate

$$\|\Upsilon\|_{\sigma_k, L_k} \leq c_1 \exp\{-\Theta n(k)g(k)\},$$

$$\sigma_k = \frac{\rho_k}{\sigma_0}, \quad \sigma_0 = 2p_1(1 + q_1),$$

$$p_1 = \prod_{i=1}^{\infty}(1 + c_1 c_3 \exp\{-\Theta_1 n(i)g(i)\},$$

$$q_1 = \sum_{i=1}^{\infty} c_1 c_3 \exp\{-\Theta_1 n(i)g(i)\},$$

$$L_k = c_1 c_2 \exp\{-\Theta_2 n(k)g(k)\},$$

$\Theta, \Theta_1,$ *and* Θ_2 *satisfy the conditions*

$$\Theta > 2(a + b), \quad \Theta - \Theta_1 < \Theta_2 < \Theta/2 < \Theta - a - b;$$

$$c_1 > 0, \quad c_2 > 0, \quad c_3 > 0,$$

is reducible to the system

$$y^{n_1(k)}(t + 1) = y^{n_1(k)}(t) + \omega^{n_1(k)}$$

$$+ R(y^{n_1(k)}(t), \varphi_t^{n_2(k)} \mu^{n_1(k)}, \mu^{n_1(k)}) + \mu^{n_1(k)},$$

$$\Delta\varphi_t^{n_2(k)} = \alpha^{n_2(k)}$$

by the change of variables

$$x^{n_1(k)} = y^{n_1(k)} + U(y^{n_1(k)}, \varphi_t^{n_2(k)}, \lambda^{n_1(k)}),$$

$$\lambda^{n_1(k)} = \lambda^{n_1(k)}(\mu^{n_1(k)}),$$

if the condition

$$\left|\cosec\frac{\langle\omega^n, \bar{s}\rangle + \langle\alpha^m, \bar{r}^m\rangle}{2}\right| \le c(m)(\|\bar{s}\| + \|\bar{r}^m\|)^{n+m}$$

is satisfied, where $\bar{s} = (s_1, \ldots, s_n)$ *and* $\bar{r}^m = (r_1, \ldots, r_m)$ *are arbitrary nonzero integer-valued vectors,* $c(m) = \gamma d_1^m m^{d_2}, \quad \gamma > 1, d_1 > 1, d_2 > 0.$

Theorem 1.10. *Let Eq. (1.47) satisfy the conditions:*

1) $P(x, t) \in A_1(\Theta, \alpha), \quad \Theta > 2(a + b),$

$$\lim_{k\to\infty} \frac{n(k + 1)}{n(k)} = \lim_{k\to\infty} \frac{g(k + 1)}{g(k)} = 1;$$

2) *for all* $n_1 > 0, n_2 > 0m$ *and any nonzero integer-valued vectors* $\bar{s}^{n_1} = (s_1, \ldots, s_{n_1})$ *and* $\bar{r}^{n_2} = (r_1, \ldots, r_{n_2}),$ *the inequality*

$$\left|\cosec\frac{\langle\omega^{n_1}, \bar{s}^{n_1}\rangle + \langle\alpha^{n_2}, \bar{r}^{n_2}\rangle}{2}\right| \le c(n)(\|\bar{s}^{n_1}\| + \|\bar{r}^{n_2}\|)^n,$$

$$n = n_1 + n_2; \quad c(n) = \gamma d_1^n n^{d_2}, \quad \gamma > 1, d_1 > 1, d_2 > 0,$$

is valid. Then there exists the constant $M_0 > 0$ *such that, for* $C_F \le M_0$*, there exist the vector* λ *and the change of variables* $x = y + U_0(t, y), \quad U_0(t, y) \in A_1(\Theta, \alpha),$ *which reduces Eq. (1.47) to the equation* $y(t + 1) = y(t) + \omega.$

The complete proofs of Lemma 1.2 and Theorem 1.10 are presented in [110]. They are quite analogous to proofs of Lemma 1.1 and Theorem 1.9, respectively.

1.4 Reduction of a discrete dynamical system in the space R^q to the canonical form in a neighborhood of its invariant set

We consider the dynamical system $x = x(n, x_0)$, $x(0) = x_0$, $x_0 \in R^q$, $n \in Z$ defined in the space R^q by the difference equation

$$x(n+1) - x(n) = X(x(n)), \tag{1.48}$$

where $n = 0, \pm 1, \ldots$ is the discrete time, $x = \{x^1, x^2, \ldots, x^q\} \in R^q$, $X(x)$ is the function r times continuously differentiable on the set R^q with respect to $x^i (i = 1, 2, \ldots, q)$.

Let Eq. (1.48) have the invariant surface

$$\mathcal{M} : x = f(\varphi), \quad \varphi \in \mathcal{T}_m, \tag{1.49}$$

where $\mathcal{T}_m$ is an m-dimensional torus, the function $f(\varphi)$ belongs to the space $C^r(\mathcal{T}_m)$ of functions r-times continuously differentiable on this torus, and the surface $\mathcal{M}$ is filled with quasiperiodic trajectories

$$x(n, f(\varphi)) = f(\omega n + \varphi), \quad n \in Z, \quad \varphi \in \mathcal{T}_m. \tag{1.50}$$

Here, $\omega = (\omega_1, \ldots, \omega_m)$ is the frequency basis of the quasiperiodic function $f(\omega t)$, and φ is an arbitrary point from $\mathcal{T}_m$.

We assume that

$$\mathrm{rank}\frac{\partial f(\varphi)}{\partial \varphi} = m, \quad \varphi \in \mathcal{T}_m, \tag{1.51}$$

and the matrix $\frac{\partial f(\varphi)}{\partial \varphi}$ can be supplemented up to the 2π-periodic basis in R^q, i.e., there exists the matrix $B(\varphi) \in C^r(\mathcal{T}_m)$ such that

$$\det[\frac{\partial f(\varphi)}{\partial \varphi}, B(\varphi)] \neq 0, \quad \varphi \in \mathcal{T}_m. \tag{1.52}$$

Under the above assumptions, we pass to the study of the behavior of trajectories of Eq. (1.48), which start in a small neighborhood of the manifold $\mathcal{M}$. First of all, we note that the invariance of the manifold $\mathcal{M}$ and the quasiperiodicity of trajectories on it require the validity of the identity

$$f(\varphi + \omega) = f(\varphi) + X(f(\varphi)), \quad \varphi \in \mathcal{T}_m. \tag{1.53}$$

Assumptions (1.51) and (1.52) guarantee the introduction of the local coordinates $(\varphi, h) = (\varphi^1, ..., \varphi^m, h^1, ..., h^{q-m})$ in a neighborhood of the manifold $\mathcal{M}$ by the formula

$$x = f(\varphi) + B(\varphi)h \tag{1.54}$$

and mapping (1.48) in a neighborhood of $\mathcal{M}$ in the form

$$\varphi(n+1) - \varphi(n) = \omega + A(\varphi(n), h(n))h(n),$$

$$h(n+1) - h(n) = P(\varphi(n), h(n))h(n) \tag{1.55}$$

with the matrices $A(\varphi, h)$ and $P(\varphi, h)$ of appropriate sizes, which are 2π-periodic in φ and sufficiently smooth with respect to φ, h in the set

$$\{(\varphi, h)| \quad \|h\| \leq \delta, \quad \varphi \in \mathcal{T}_m\}. \tag{1.56}$$

Here, $n \in Z$, and δ is a sufficiently small positive number. To establish this fact, we perform the change of variables (1.54) in (1.48) and obtain the system of equations

$$f(\varphi + \omega + A(\varphi, h)h) + B(\varphi + \omega + A(\varphi, h)h)[h + P(\varphi, h)h]$$
$$- [f(\varphi) + B(\varphi)h] = X(f(\varphi) + B(\varphi)h)$$

for the matrices $A = A(\varphi, h)$ and $P = P(\varphi, h)$ or, taking identity (1.53) into account,

$$f(\varphi + \omega + Ah) - f(\varphi + \omega) + Bf(\varphi + \omega + Ah)(h + Ph) - Bf(\varphi)h$$
$$= X(f(\varphi) + B(\varphi)h) - X(f(\varphi)).$$

We present the last equation in the form

$$\frac{\partial f(\varphi + \omega)}{\partial \varphi} Ah + B(\varphi + \omega + Ah)Ph = X(f(\varphi) + B(\varphi)h) - X(f(\varphi))$$

$$- \{[f(\varphi + \omega + Ah) - f(\varphi + \omega) - \frac{\partial f(\varphi + \omega)}{\partial \varphi} Ah]$$

$$+ [B(\varphi + \omega + Ah) - B(\varphi + \omega)]h + [B(\varphi + \omega) - B(\varphi)]h\}.$$

This yields the equality

$$\frac{\partial f(\varphi + \omega)}{\partial \varphi} A + B(\varphi + \omega + Ah)P = \int_0^1 \frac{\partial X(f(\varphi) + tB(\varphi)h)}{\partial x} dt B(\varphi)$$

$$- \{\int_0^1 [\frac{\partial f(\varphi + \omega + tAh)}{\partial \varphi} - \frac{\partial f(\varphi + \omega)}{\partial \varphi}]dt A$$

$$+ [B(\varphi + \omega + Ah) - B(\varphi + \omega)] + [B(\varphi + \omega) - B(\varphi)]\}. \tag{1.57}$$

According to assumption (1.52) for the fixed $M > 0$, there exists $\delta = \delta(M) > 0$ such that

$$\det[\frac{\partial f(\varphi + \omega)}{\partial \varphi}, B(\varphi + \omega + Ah)] \neq 0 \tag{1.58}$$

for all φ, h from set (1.56) and for an arbitrary matrix A, which satisfies the condition

$$\|A\| \leq M, \tag{1.59}$$

where the norm of the matrix A is consistent with the norm of the vector h.

From a condition (1.58) follows that Eq. (1.57) has the solution of the form

$$A = L_1(\varphi, Ah)Q(\varphi, h, A), \quad P = L_2(\varphi, Ah)Q(\varphi, h, A) \tag{1.60}$$

for all φ, h from set (1.56) and for A, which satisfies estimate (1.59), where $Q = Q(\varphi, h, A)$ is a matrix function, which is defined by the right-hand side of Eq. (1.57); $L_1(\varphi, Ah)$ and $L_2(\varphi, Ah)$ are the blocks of the matrix inverse to the matrix

$$[\frac{\partial f(\varphi + \omega)}{\partial \varphi}, B(\varphi + \omega + Ah)].$$

The matrix Q can be presented in the form

$$Q = B(\varphi + \omega) - B(\varphi) + \frac{\partial X(f(\varphi))}{\partial x}B(\varphi) + Q_1(\varphi, h, A). \tag{1.61}$$

In the equation (1.61) $Q_1 = Q_1(\varphi, h, A)$ is the matrix defined in the domain $\|h\| \leq \delta, \varphi \in \mathcal{T}_m, \|A\| \leq M, r - 1$ times continuously differentiable with respect to its variables, and satisfies the condition $Q_1(\varphi, 0, A) = 0$.

The matrices $L_1(\varphi, Ah)$ and $L_2(\varphi, Ah)$ have properties analogous to those of the matrix Q; in this case, $L_1(\varphi, 0)$ and $L_2(\varphi, 0)$ are the blocks of the matrix inverse to the matrix

$$[\frac{\partial f(\varphi + \omega)}{\partial \varphi}, B(\varphi + \omega)].$$

By $C^r_{Lip}(\mathcal{T}_m \times \mathcal{K}_\mu)$, we denote the space of functions of the variable (φ, h), which are defined in the set $\mathcal{T}_m \times \mathcal{K}_\mu$, $\mathcal{K}_\mu = \{h | \|h\| \leq \mu\}$, and have the continuous partial derivatives up to the r-th order inclusive in this set, and these derivatives satisfy the Lipschitz condition with respect to (φ, h).

In the space $C_{Lip}(\mathcal{T}_m \times \mathcal{K}_\mu)$, we define the subset $C(M, K)$ of the matrix functions $A = A(\varphi, \mu)$, which satisfy the conditions

$$\|A(\varphi, h)\| \leq M, \quad \|A(\varphi', h') - A(\varphi, h)\| \leq K(\|\varphi' - \varphi\| + \|h' - h\|)$$

for any (φ', h') and (φ, h) from $\mathcal{T}_m \times \mathcal{K}_\mu$.

We will show that the first equality in (1.60) under the appropriate choice of the constants M, K, and μ has a solution in the set $C(M, K)$. For this purpose, we define the operator $S : A \to SA = L_1(\varphi, Ah)Q(\varphi, h, A)$ on the set $C(M, K)$. For $r \geq 2$, this operator transfer $C(M, K)$ into a subset of the space $C_{Lip}(\mathcal{T}_m \times \mathcal{K}_\mu)$. In addition, for $SA = (SA)(\varphi, h)$, the estimates

$$\|(SA)(\varphi, h)\| \leq c_1(1 + \mu + \mu M^2), \tag{1.62}$$

$$\|(SA)(\varphi', h') - (SA)(\varphi, h)\|$$
$$\leq c_2(1 + \mu M K + M^2)(\|\varphi' - \varphi\| + \|h' - h\|) \tag{1.63}$$

$\forall\{(\varphi, h), (\varphi', h')\} \subset \mathcal{T}_m \times \mathcal{K}_\mu$ hold, where c_1 and c_2 are positive constants indepen-dent of M, K, and $\mu \leq \delta/K$. By the appropriate choice of M, K, and μ, it is possible to attain these inequalities (1.62) and (1.63) yield the membership of SA to the set $C(M, K)$.

For the pair of matrix functions $A = A(\varphi, h)$ and $A_1 = A_1(\varphi, h)$ from the set $C(M, K)$, we have the estimate

$$\|SA - SA_1\| \leq c_3\mu(1 + M)\|A - A_1\|,$$

where c_3 is a constant independent of M, K, and μ. At sufficiently small μ, this estimate implies that S is the contraction operator on $C(M, K)$. According to the principle of contracting mappings, there exists the unique solution of the equation $A = SA$ on the set $C(M, K)$.

The last equation coincides with the first equality in (1.60), and its solution defines the unique matrix $A = A(\varphi, h)$ on the right-hand side of system (1.55) in $C_{Lip}(\mathcal{T}_m \times \mathcal{K}_\mu)$ for sufficiently small $\mu > 0$. The theorem of implicit function ensures the membership of $A(\varphi, h)$ to the space $C_{Lip}^{r-1}(\mathcal{T}_m \times \mathcal{K}_\mu)$.

With the help of the obtained matrix $A(\varphi, h)$, the second equality in (1.60) defines the matrix $P = P(\varphi, h)$ on the right-hand side of system (1.55), which also belongs to the space $C_{Lip}^{r-1}(\mathcal{T}_m \times \mathcal{K}_\mu)$.

Thus, the dynamical system (1.48) in a small neighborhood of the manifold $\mathcal{M}$ is reduced to the system of equations (1.55) with the matrices A and P, which belong to $C_{Lip}^{r-1}(\mathcal{T}_m \times \mathcal{K}_\mu)$. The problem consists in the determination of the conditions, under which there exists the change of variables $\varphi \to \psi$, which reduces system (1.48) to the quasiperiodic system

$$\psi(n + 1) - \psi(n) = \omega,$$

$$h(n + 1) - h(n) = R(\psi(n), h(n))h(n). \tag{1.64}$$

Theorem 1.11. *Let the above conditions be satisfied, and let the matrix $P(\varphi, 0)$ satisfy the inequality*

$$\|E + P(\varphi, 0)\| \le d < 1, \quad \varphi \in \mathcal{T}_m. \tag{1.65}$$

Then it is possible to point out $\mu > 0$ and a matrix $U(\varphi, h) \in C_{Lip}^{r-2}(\mathcal{T}_m \times \mathcal{K}_\mu), 2 \le r < \infty$, such that the change of variables

$$\varphi = \psi + U(\psi, h)h \tag{1.66}$$

reduces the system of equations (1.55) to system (1.64) with the matrix

$$R(\psi, h) = P(\psi + U(\psi, h)h, h). \tag{1.67}$$

Proof. We define the required transformation $\varphi \to \psi$ by the formula

$$\psi = \varphi + V(\varphi, h)h, \tag{1.68}$$

where $V = V(\varphi, h)$ is a matrix function from $C(\mathcal{T}_m \times \mathcal{K}_\mu)$. With regard for (1.55), (1.64), and change (1.68), the matrix V can be determined from the relation

$$V(\varphi(n) + \omega + A(\varphi(n), h(n))h(n), P_1(\varphi(n), h(n))h(n))P_1(\varphi(n), h(n))h(n)$$
$$- V(\varphi(n), h(n))h(n) + A(\varphi(n), h(n))h(n) = 0,$$

which implies that V satisfies the equation

$$V(\varphi, h) = V(\varphi + \omega + A(\varphi, h)h, P_1(\varphi, h)h)P_1(\varphi, h) + A(\varphi, h), \tag{1.69}$$

where $P_1(\varphi, h) = E + P(\varphi, h)$. We set

$$\varphi_1(\varphi, h) = \varphi + \omega + A(\varphi, h)h, \quad h_1(\varphi, h) = P_1(\varphi, h)h. \tag{1.70}$$

Eq. (1.70) lead expression (1.68) to the form

$$V(\varphi, h) = V(\varphi_1(\varphi, h), h_1(\varphi, h))P_1(\varphi, h) + A(\varphi, h). \tag{1.71}$$

At a sufficiently small $\mu > 0$, condition (1.65) yields the inequality

$$\|P_1(\varphi, h)\| \le d_1 = const < 1 \tag{1.72}$$

for all $(\varphi, h) \in \mathcal{T}_m \times \mathcal{K}_\mu$. The inequality (1.72) allows one to construct the successive approximations to the solution of Eq. (1.71):

$$V_1(\varphi, h) = A(\varphi, h),$$

$$V_{i+1}(\varphi, h) = V_i(\varphi_1(\varphi, h), h_1(\varphi, h))P_1(\varphi, h) + A(\varphi, h), \quad i \ge 1. \tag{1.73}$$

They satisfy the estimates

$$\|V_1(\varphi, h)\| \le \|A(\varphi, h)\| \le \max_{\mathcal{T}_m \times \mathcal{K}_\mu} \|A(\varphi, h)\| = M_1,$$

$$\|V_{i+1}(\varphi, h)\| \leq M_1 \sum_{\nu=0}^{i} d_1^{\nu}, \quad i \geq 1,$$

are valid. It is easy to verify that sequence (1.73) uniformly converges with respect to $(\varphi, h) \in \mathcal{T}_m \times \mathcal{K}_\mu$, and its limiting function

$$V(\varphi, h) = \lim_{i \to \infty} V_i(\varphi, h)$$

is a solution of Eq. (1.69), which belongs to the space $C(\mathcal{T}_m \times \mathcal{K}_\mu)$.

We now clarify the question on the smoothness of the function $V(\varphi, h)$. For this purpose, we consider the function $W = W(\varphi, h, \mu) = V(\varphi, \mu h)$ for $(\varphi, h) \in \mathcal{T}_m \times \mathcal{K}_\mu$ and for a sufficiently small $\mu > 0$. This function satisfies the equation

$$W(\varphi, h, \mu) = W(\varphi_1(\varphi, \mu h), h_1(\varphi, \mu h), \mu) P_1(\varphi, \mu h) + A(\varphi, \mu h) \qquad (1.74)$$

and is the limit of successive approximations

$$W_1(\varphi, h, \mu) = A(\varphi, \mu h), \quad W_{i+1}(\varphi, h, \mu) = W_i(\varphi_1(\varphi, \mu h), h_1(\varphi, \mu h), \mu)$$
$$\times P_1(\varphi, \mu h) + A(\varphi, \mu h), \quad i \geq 1. \qquad (1.75)$$

By differentiating expressions (1.75), we obtain the equalities

$$\frac{\partial W_1(\varphi, h, \mu)}{\partial \varphi^\nu} = \frac{\partial A(\varphi, \mu h)}{\partial \varphi^\nu}, \quad \frac{\partial W_1}{\partial h^\nu} = \mu \frac{\partial A(\varphi, \mu h)}{\partial (\mu h)^\nu},$$

$$\frac{\partial W_{i+1}(\varphi, h, \mu)}{\partial \varphi^\nu} = [\sum_{j=1}^{m} \frac{\partial W_i(\varphi_1(\varphi, \mu h), h_1(\varphi, \mu h), \mu)}{\partial \varphi_1^j} \frac{\partial \varphi_1^j(\varphi, \mu h)}{\partial \varphi^\nu}$$

$$+ \sum_{j=1}^{q-m} \frac{\partial W_i(\varphi_1(\varphi, \mu h), h_1(\varphi, \mu h), \mu)}{\partial h_1^j} \frac{\partial h_1^j(\varphi, \mu h)}{\partial \varphi^\nu}] P_1(\varphi, \mu h)$$

$$+ W_i(\varphi_1(\varphi, \mu h), h_1(\varphi, \mu h), \mu) \frac{\partial P_1(\varphi, \mu h)}{\partial \varphi^\nu} + \frac{\partial A(\varphi, \mu h)}{\partial \varphi^\nu},$$

$$\frac{\partial W_{i+1}(\varphi, h, \mu)}{\partial h^\nu} = \mu[\sum_{j=1}^{m} \frac{\partial W_i(\varphi_1(\varphi, \mu h), h_1(\varphi, \mu h), \mu)}{\partial \varphi_1^j} \frac{\partial \varphi_1^j(\varphi, \mu h)}{\partial (\mu h)^\nu}$$

$$+ \sum_{j=1}^{q-m} \frac{\partial W_i(\varphi_1(\varphi, \mu h), h_1(\varphi, \mu h), \mu)}{\partial h_1^j} \frac{\partial h_1^j(\varphi, \mu h)}{\partial (\mu h)^\nu}] P_1(\varphi, \mu h)$$

$$+ \mu W_i(\varphi_1(\varphi, \mu h), h_1(\varphi, \mu h), \mu) \frac{\partial P_1(\varphi, \mu h)}{\partial (\mu h)^\nu} + \mu \frac{\partial A(\varphi, \mu h)}{\partial (\mu h)^\nu}, \quad i \geq 1, \qquad (1.76)$$

where φ_1^j and h_1^j are the j-th coordinates of the vectors $\varphi_1(\varphi, \mu h)$ and $h_1(\varphi, \mu h)$. For $\mu = 0$, equalities (1.76) take the form

$$\frac{\partial W_1(\varphi, h, 0)}{\partial \varphi^\nu} = \frac{\partial A(\varphi, 0)}{\partial \varphi^\nu}, \quad \frac{\partial W_1(\varphi, h, 0)}{\partial h^\nu} = 0,$$

$$\frac{\partial W_{i+1}(\varphi, h, 0)}{\partial \varphi^\nu} = \frac{\partial W_i(\varphi + \omega, h, 0)}{\partial \varphi^\nu} P_1(\varphi, 0) + W_1(\varphi + \omega, h, 0)$$
$$\times \frac{\partial P_1(\varphi, 0)}{\partial \varphi^\nu} + \frac{\partial A(\varphi, 0)}{\partial \varphi^\nu}, \quad \frac{\partial W_{i+1}(\varphi, h, 0)}{\partial h^\nu} = 0, \quad i \geq 1, \quad (1.77)$$

and lead to the estimate

$$\max_{\mathcal{T}_m \times \mathcal{K}_\mu} \|W'_{i+1}\| \leq d_1 \max_{\mathcal{T}_m \times \mathcal{K}_\mu} \|W'_i\| + M_1, \quad i \geq 1, \qquad (1.78)$$

where W'_i are the matrices of derivatives of the iteration W_i, and M_1 is some positive constant.

Inequality (1.78) yields the estimate

$$\max_{\mathcal{T}_m \times \mathcal{K}_\mu} \|W'_{i+1}\| \leq \frac{M_1}{1 - d_1}, \quad i \geq 1. \qquad (1.79)$$

For $\mu \neq 0$, formulas (1.76) have the form of the matrix equalities

$$W'_{i+1}(\varphi, h, \mu) = W'_i(\varphi_1(\varphi, \mu h), h_1(\varphi, \mu h), \mu) P_2(\varphi, \mu, h)$$
$$+ W_i(\varphi_1(\varphi, \mu h), h_1(\varphi, \mu h), \mu) P_1^I(\varphi, h, \mu) + A_1^I(\varphi, h, \mu), \quad i = 0, 1, 2, ...,$$
$$(1.80)$$

where W'_0 and W_0 are the zero matrices; W'_i is the matrix of derivatives of the iteration W_i; P'_1, A'_1, and P_2 are the matrix functions of the variables φ, h, and μ continuously dependent on these variables for $\varphi \in \mathcal{T}_m, h \in \mathcal{K}_\mu, \mu \in [0, \mu_0], \mu_0$ is a sufficiently small positive number.

Formulas (1.77) are a particular case of formulas (1.80) and coincide with them for $\mu = 0$. Hence, for the matrices P_2, P'_1 and A'_1 for $\mu = 0$, we have the expressions

$$P_2(\varphi, h, 0) = \text{diag}\{P_1(\varphi, 0), ..., P_1(\varphi, 0), 0, ..., 0\},$$

$$P'_1(\varphi, h, 0) = \{\frac{\partial P_1(\varphi, 0)}{\partial \varphi^1}, ..., \frac{\partial P_1(\varphi, 0)}{\partial \varphi^m}, 0, ..., 0\},$$

$$A'_1(\varphi, h, 0) = \{\frac{\partial A(\varphi, 0)}{\partial \varphi^1}, ..., \frac{\partial A(\varphi, 0)}{\partial \varphi^m}, 0, ..., 0\}.$$

From the first expression, we obtain the estimate

$$\|P_2(\varphi, h, 0)\| = \|P_1(\varphi, 0)\| \leq d_1, \tag{1.81}$$

which guarantees the following estimate for the norm of the matrix $P_2(\varphi, h, \mu)$:

$$\|P_2(\varphi, h, \mu)\| \leq d_1(\mu). \tag{1.82}$$

Here, $d_1(\mu) \to d_1$ as $\mu \to 0$.

Selecting a sufficiently small $\mu_0 > 0$, we obtain the inequality

$$d_1(\mu) \leq d_2 \leq const < 1 \tag{1.83}$$

for all $\mu \in [0, \mu_0]$. Then relations (1.79) – (1.83) yield the estimate

$$\max_{\mathcal{T}_m \times \mathcal{K}_\mu} \|W'_{i+1}(\varphi, h, \mu)\| \leq d_2 \max_{\mathcal{T}_m \times \mathcal{K}_\mu} \|W'_i(\varphi, h, \mu)\| + M_2, \quad i = 0, 1, 2, ...,$$

and we have

$$\max_{\mathcal{T}_m \times \mathcal{K}_\mu} \|W'_{i+1}(\varphi, h, \mu)\| \leq \frac{M_2}{1 - d_2}, \quad i = 0, 1, 2, ..., \tag{1.84}$$

where M_2 is some positive constant.

Inequality (1.84) means that

$$\max_{\nu, j} \left\{ \left\| \frac{\partial W_i(\varphi, h, \mu)}{\partial \varphi^\nu} \right\|; \left\| \frac{\partial W_i(\varphi, h, \mu)}{\partial h^j} \right\| \right\} \leq \frac{M_2}{1 - d_2}$$

for all $i = 1, 2, ...$ But then (see (1.74))

$$\max_{\nu, j} \left\{ \left\| \frac{\partial V_i(\varphi, h)}{\partial \varphi^\nu} \right\|; \left\| \frac{\partial V_i(\varphi, h)}{\partial h^j} \right\| \right\} \leq \frac{M_2}{\mu_0(1 - d_2)}.$$

Hence, the sequence of the first derivatives of approximations (1.73) is uniformly bounded. In an analogous way, we can establish the uniform boundedness of the sequence of arbitrary derivatives of approximations (1.73) to the order $r - 1$ inclusive. It is the sufficient condition for $V = V(\varphi, h)$ to belong to the space $C_{Lip}^{r-2}(\mathcal{T}_m \times \mathcal{K}_\mu)$.

To complete the proof of the theorem, we have to solve relation (1.68) for φ in the form

$$\varphi = \psi + U(\psi, h)h \tag{1.85}$$

with the function $U = U(\psi, h)$ from the space $C_{Lip}^{r-2}(\mathcal{T}_m \times \mathcal{K}_\mu)$. Substituting (1.85) into (1.68), we obtain the equation

$$U = -V(\psi + Uh, h) \tag{1.86}$$

for the matrix U. Equation (1.86) has the form of the first equation in (1.60). Therefore, the reasoning made while proving the solvability of the first equation in (1.60) is also valid while dealing with Eq. (1.86). This yields the solvability of Eq. (1.86) in the space $C_{Lip}^{r-2}(\mathcal{T}_m \times \mathcal{K}_\mu)$ nd, hence, the solvability of Eq. (1.68) in the form (1.85).

It remains to substitute φ in the matrix $P(\varphi, h)$, which defines the right-hand side of the second equation of system (1.55), by its value of (1.85) in order to obtain expression (1.67) for R. The theorem is proved. $\qquad\square$

The next theorem characterizes the behavior of trajectories of the discrete dynamical system (1.48), which start at a small neighborhood of the manifold $\mathcal{M}$.

Theorem 1.12. *Let the conditions of Theorem 1.11 be satisfied. Then there exists a sufficiently small $\delta > 0$ such that, for each y_0, which satisfy the inequality $\rho(y_0, \mathcal{M}) \leq \inf_{x \in \mathcal{M}} \|y_0 - x\| \leq \delta$, it is posssible to find the values of φ_0 and ψ_0 from the space $\mathcal{T}_m$ such that*

$$\|x(n, y_0) - f(\omega n + \psi_0)\| \leq \bar{K}_1 d_3^n \|y_0 - f(\varphi_0)\| \tag{1.87}$$

for all $n = 0, 1, \ldots$ and for some positive $\bar{K}_1$ and d_3, where $d_3 = d_3(\delta) \to d_2$ and $\|\varphi_0 - \psi_0\| \to 0$ as $\delta \to 0$.

The proof of Theorem 1.12 is analogous to that of Theorem 4 in [102].

Corollary 1.3. *If the conditions of Theorem 1.11 are satisfied, then the quasiperiodic solution*

$$x = x(n, f(\varphi)) = f(\omega n + \varphi)$$

of system (1.48) is Lyapunov-stable for any $\varphi \in \mathcal{T}_m$.

The proof of Corollary 1.3 is analogous to that of Corollary 1 in [102].

Corollary 1.4. *Let the conditions of Theorem 1.11 be satisfied, and let $\langle k, \omega \rangle \neq 0 \mod 2\pi$ for each integer-valued vector $k = (k_1, \ldots, k_m) \neq 0$. Then for an arbitrary function $F = F(x)$ continuous in a neighborhood of $\mathcal{M}$ and any solution $x = x(n, y_0)$ of system (1.48), for which $\rho(y_0, \mathcal{M}) \leq \delta$, the relation*

$$\lim_{n \to \infty} \frac{1}{n} \sum_{\nu=0}^{n-1} F(x(\nu, y_0)) = F_0 = (2\pi)^{-m} \int_0^{2\pi} \ldots \int_0^{2\pi} F(f(\varphi)) d\varphi_1 \ldots d\varphi_m$$

$$\tag{1.88}$$

is valid.

Proof.   We present the function F as the sum

$$F(x) = P(x, \varepsilon) + R(x, \varepsilon),$$

where $P(x, \varepsilon)$ is a polynomial, which approximates F in a neighborhood of $\mathcal{M}$ to within an arbitrary fixed $\varepsilon > 0$, i.e., $|R(x, \varepsilon)| \leq \varepsilon \; \forall x \in U_\delta(\mathcal{M}) \equiv \{x | \rho(x, \mathcal{M}) \leq \varepsilon\}$. This yields the estimate

$$\frac{1}{n} | \sum_{\nu=0}^{n-1} [F(x(\nu, y_0)) - P(x(\nu, y_0), \varepsilon)]| \leq \varepsilon \tag{1.89}$$

for any $n = 1, 2, \ldots.$ From inequality (1.87), we have the estimate

$$\frac{1}{n} | \sum_{\nu=0}^{n-1} [P(x(\nu, y_0)) - P(f(\omega\nu + \psi_0))]| \leq \frac{K(\varepsilon)\bar{K}_1}{n}$$

$$\times \sum_{\nu=0}^{n-1} d_3^\nu \|y_0 - f(\varphi_0)\| \leq \frac{1}{n} \frac{K_2(\varepsilon)}{1 - d_3}, \tag{1.90}$$

where $K(\varepsilon)$ is the Lipschitz constant of the polynomial P for $x \in U_\delta$. We present the function $P(f(\varphi), \varepsilon)$ as the sum

$$P(f(\varphi), \varepsilon) = Q(\varphi, \varepsilon) + R_1(\varphi, \varepsilon),$$

where $Q(\varphi, \varepsilon)$ is a trigonometric polynomial, which approximates $P(f(\varphi), \varepsilon)$ to within ε, $|R_1(\varphi, \varepsilon)| \leq \varepsilon \quad \forall \varphi \in \mathcal{T}_m$. Therefore, the estimate

$$\frac{1}{n} | \sum_{\nu=0}^{n-1} [P(f(\omega\nu + \psi_0)) - Q(\omega\nu + \psi_0)]| \leq \varepsilon \tag{1.91}$$

holds for any $n = 1, 2, \ldots.$ According to the definition

$$Q(\varphi, \varepsilon) = \sum_{\|k\| \leq N} Q_k e^{i\langle k, \varphi\rangle},$$

where $N = N(\varepsilon)$ is a sufficiently large integer, and $Q_k = Q_k(\varepsilon)$ are the Fourier coefficients of the function $Q(\varphi, \varepsilon)$. This yields the equalities

$$\frac{1}{n} \sum_{\nu=0}^{n-1} Q(\omega\nu + \psi_0, \varepsilon) = Q_0 + \frac{1}{n} \sum_{\nu=0}^{n-1} \sum_{1 \leq \|k\| \leq N} Q_k e^{i\langle k, \omega\rangle\nu} e^{i\langle k, \psi_0\rangle}$$

$$= Q_0 + \frac{1}{n} \sum_{1 \leq \|k\| \leq N} Q_k [\sum_{\nu=0}^{n-1} e^{i\langle k, \omega\rangle\nu}] e^{i\langle k, \psi_0\rangle}$$

and the estimates

$$|\frac{1}{n}\sum_{\nu=0}^{n-1}Q(\omega\nu+\psi_0,\varepsilon)-Q_0| \le \frac{1}{n}\sum_{1\le\|k\|\le N}|Q_k||\sum_{\nu=0}^{n-1}e^{i\langle k,\omega\rangle\nu}|$$

$$\le \max_{1\le\|k\|\le N}\frac{1}{n}|\sum_{\nu=0}^{n-1}e^{i\langle k,\omega\rangle\nu}|\sum_{1\le\|k\|\le N}|Q_k| = M(\varepsilon)\max_{1\le\|k\|\le N}\frac{1}{n}|\sum_{\nu=0}^{n-1}e^{i\langle k,\omega\rangle\nu}|.$$

$$(1.92)$$

For the last sum of inequality (1.92), the estimate

$$|\sum_{\nu=0}^{n-1}e^{i\langle k,\omega\rangle\nu}| = [(\sum_{\nu=0}^{n-1}\cos\nu\langle k,\omega\rangle)^2 + (\sum_{\nu=0}^{n-1}\sin\nu\langle k,\omega\rangle)^2]^{\frac{1}{2}}$$

$$= |\sin\frac{n\langle k,\omega\rangle}{2}||\text{cosec}\,\frac{\langle k,\omega\rangle}{2}| \le |\text{cosec}\,\frac{\langle k,\omega\rangle}{2}|, \quad \langle k,\omega\rangle \ne 0 \bmod 2\pi,$$

is valid. Due to it, we have

$$|\frac{1}{n}\sum_{\nu=0}^{n-1}Q(\omega\nu+\psi_0,\varepsilon)-Q_0| \le \frac{1}{n}M(\varepsilon)\max_{1\le\|k\|\le N}|\text{cosec}\,\frac{\langle k,\omega\rangle}{2}| \qquad (1.93)$$

and the inequality

$$|F_0 - Q_0| \le |F_0 - P_0| + |P_0 - Q_0| \le 2\varepsilon \qquad (1.94)$$

for the average values of $F_0, Q_0,$ and P_0 of the functions $F(f(\varphi)), Q(\varphi,\varepsilon),$ and $P(f(\varphi),\varepsilon)$.

By combining inequalities (1.89) – (1.94), we obtain the estimate

$$|\frac{1}{n}\sum_{\nu=0}^{n-1}F(x(\nu,y_0))-F_0| \le \frac{1}{n}|\sum_{\nu=0}^{n-1}[F(x(\nu,y_0))-P(x(\nu,y_0),\varepsilon)]|$$

$$+ \frac{1}{n}|\sum_{\nu=0}^{n-1}[P(x(\nu,y_0))-P(f(\omega\nu+\psi_0),\varepsilon)]| + \frac{1}{n}|\sum_{\nu=0}^{n-1}[P(f(\omega\nu$$

$$+\psi_0),\varepsilon)-Q(\omega\nu+\psi_0),\varepsilon)]| + \frac{1}{n}|\sum_{\nu=0}^{n-1}Q(\omega\nu+\psi_0),\varepsilon)-Q_0|$$

$$+ |Q_0 - F_0| \le 4\varepsilon + \frac{1}{n}M_1(\varepsilon), \quad (1.95)$$

where

$$M_1(\varepsilon) = \frac{K_2(\varepsilon)}{1-d_3} + M(\varepsilon)\max_{1\le\|k\|\le N}|\text{cosec}\,\frac{\langle k,\omega\rangle}{2}|.$$

We choose $n_0 = n_0(\varepsilon)$ so large that the inequality $M_1(\varepsilon)/n \leq \varepsilon$ is satisfied for all $n \geq n_0$. Then relation (1.95) takes the form

$$|\frac{1}{n}\sum_{\nu=0}^{n-1} F(x(\nu, y_0)) - F_0| \leq 5\varepsilon \quad \forall n \geq n_0,$$

which yields the limiting relation (1.88). The corollary is proved. $\qquad\square$

Let us consider the perturbed system of equations

$$x(n+1) - x(n) = X(x(n)) + \varepsilon Y(x(n)), \tag{1.96}$$

where $Y = Y(x) \in C^r(R^q)$, and ε is a small positive parameter. Under the change of variables (1.54), system of equations (1.96) in a neighborhood of the manifold $\mathcal{M}$ turns into the system

$$\varphi(n+1) - \varphi(n) = \omega + \varepsilon a(\varphi(n)) + A(\varphi(n), h(n), \varepsilon)h(n),$$

$$h(n+1) - h(n) = P(\varphi(n), h(n), \varepsilon)h(n) + \varepsilon b(\varphi(n)), \tag{1.97}$$

where $a = a(\varphi)$, $A = A(\varphi, h, \varepsilon)$, $P = P(\varphi, h, \varepsilon)$, and $b = b(\varphi)$ are functions from the space $C^{r-1}_{Lip}(\mathcal{T}_m \times \mathcal{K}_\mu)$ for all sufficiently small $\varepsilon > 0$.

To system (1.97), we apply the perturbation theory of the invariant toroidal manifold of a discrete dynamical system [74], [32] and the above-mentioned technique of reduction of system (1.55) to the form (1.64) and arrive at the following proposition.

Theorem 1.13. *Let the conditions of Theorem 1.11 be satisfied. Then, for sufficiently small positive values of μ and ε_0, there exists the change of variables*

$$\varphi = \psi + U(\psi, z, \varepsilon)h, \quad h = u(\varphi, \varepsilon) + z,$$

which reduces the system of equations (1.97) to the system

$$\psi(n+1) - \psi(n) = \omega + F(\psi(n), \varepsilon),$$

$$z(n+1) - z(n) = R(\psi(n), z(n), \varepsilon)z(n),$$

$u(\varphi, \varepsilon), U(\varphi, h, \varepsilon), F(\varphi, \varepsilon)$, and $R(\varphi, h, \varepsilon)$ belong to the space $C^{r-2}_{Lip}(\mathcal{T}_m \times \mathcal{K}_\mu)$ for every $\varepsilon \in [0, \varepsilon_0]$, and

$$\lim_{\varepsilon \to 0}(\|u(\varphi, \varepsilon)\|_{r-2, Lip} + \|U(\varphi, h, \varepsilon) - U(\varphi, h)\|_{r-2, Lip}) = 0.$$

For the function $u = u(\varphi, h) \in C^r_{Lip}(\mathcal{T}_m \times \mathcal{K}_\mu)$, we denote the quantity $\|u\|_r + K_r$ by $\|\cdot\|_{r,Lip}$, where $\|u\|_r = \max_{0 \le \rho \le r} \|D^\rho u\|$, D^ρ is an arbitrary derivative of the function u with respect to (φ, h) of the order ρ, and K_r is the Lipschitz constant for the r-th derivatives of the function u.

We now present a result [134], which concerns the reducibility of Eq. (1.48) to the canonical form in a neighborhood of the invariant manifold $\mathcal{M}$ filled with trajectories having the property

$$x(n, f(\varphi)) = f(\varphi_n(\varphi)), \quad \varphi \in \mathcal{T}_m, n \in Z,$$

where the function $\varphi_n(\varphi) \; \forall \varphi \in \mathcal{T}_m, n \in Z$ satisfies the equation

$$\varphi_{n+1}(\varphi) - \varphi_n(\varphi) = a(\varphi_n(\varphi)),$$

and $a(\varphi) \in C^r(\mathcal{T}_m)$.

Analogously to the proof of Theorem 1.11, it can be shown that if condition (1.51) is satisfied, there exists a change of variables of the form (1.54), which allows one in a neighborhood of the set $\mathcal{M}$ to pass from the Euclidean coordinates x to the local coordinates $(\varphi, h) \in \mathcal{T}_m \times \mathcal{K}_\delta$, with respect to which Eq. (1.48) takes the form

$$\varphi(n+1) - \varphi(n) = a(\varphi(n)) + A(\varphi(n), h(n))h(n),$$

$$h(n+1) - h(n) = P(\varphi(n), h(n))h(n). \tag{1.98}$$

In view of this, the following proposition is proved.

Theorem 1.14. *Let* $r \ge 2$, *and let there exist the positive constants* $d_0, p_0 < 1$, *and* $\gamma_0 < 1$ *such that*

$$\left\|[E + \frac{\partial a(\varphi)}{\partial \varphi}]^{-1}\right\| \le d_0 \quad \forall \varphi \in \mathcal{T}_m,$$

$$\|E + P(\varphi, 0)\| \le p_0 \quad \forall \varphi \in \mathcal{T}_m,$$

$$d_0 p_0 (1 + \max_{\varphi \in \mathcal{T}_m} \left\|\frac{\partial a(\varphi)}{\partial \varphi}\right\|) \le \gamma_0.$$

Then there exist $\mu > 0$ *and a matrix* $U(\varphi, h)$ *from the space* $C_{Lip}(\mathcal{T}_m, \mathcal{K}_\mu)$ *such that the change of variables* $\varphi = \psi + U(\psi, h)h$ *reduces the system of equations (1.98) to the system*

$$\psi(n+1) - \psi(n) = a(\psi(n)),$$

$$h(n+1) - h(n) = R(\psi(n), h(n))h(n),$$

where the matrix $R(\psi, h) = P(\psi + U(\psi, h)h, h)$.

Note that Theorems 1.11 and 1.14 can be extended to the case where the equation of the form (1.48) defines a discrete dynamical system in the space $\mathfrak{M}$ of bounded numerical sequences, i.e., where $x = (x^1, x^2, \dots) \in \mathfrak{M}$, the function $X(x(n)) = \{X_1(x(n)), X_2(x(n)), \dots\}$ maps $\mathfrak{M} \to \mathfrak{M}\ \forall n \in Z$, and the invariant set $\mathcal{M}$ is defined by an equation of the form (1.49), in which $f(\varphi) = \{f_1(\varphi), f_2(\varphi), \dots\} : \mathcal{T}_m \to \mathfrak{M}$. For this purpose, we can employ the method of truncation of this equation with respect to x. It is necessary to attain the fulfillment of conditions of Theorem 1.11 or 1.14 for the sequence of relevant truncated equations defined in the spaces R^q of growing dimension, i.e., when $q \to \infty$. For each such equation, there exists a change of variables of the form (1.66), which reduces it to the canonical form in a neighborhood of the corresponding invariant set in the finite-dimensional space. The problem consists in finding the sufficient conditions, under which the required change of variables for the input equation can be obtained from the sequence of changes of the form (1.66) for the truncated equations due to the limiting transition as $q \to \infty$ at least in the coordinatewise meaning . This problem is significantly complicated if the function $f(\varphi)$ generating the invariant set $\mathcal{M}$ maps the infinite-dimensional torus $\mathcal{T}_\infty$ into the set $\mathfrak{M}$. The first studies in the mentioned direction have been performed in [170], [171], [174].

1.5   Investigation of a discrete dynamical system defined in an abstract Banach space in a neighborhood of its invariant set

In the infinite-dimensional Banach space E, we define a dynamical system by the difference equation

$$x(n+1) = X(x(n)), \quad n \in Z, \tag{1.99}$$

where $X : E \to E$ is a C^r-mapping.

Let this equation have an invariant manifold $\mathcal{M}$ of the class C^r $(r \geq 2)$, which is defined by the equation

$$x = f(\varphi), \quad \varphi \in \mathcal{T}_m,$$

where $f : \mathcal{T}_m \to E$ is a C^r-mapping, and $\mathcal{T}_m$ is an m-dimensional torus filled with quasiperiodic trajectories

$$x(n, f(\varphi)) = f(n\omega + \varphi), \quad n \in Z, \quad \varphi \in \mathcal{T}_m. \tag{1.100}$$

Here, $\omega = (\omega_1, \dots, \omega_m)$ is the frequency basis of the quasiperiodic function $f(t\omega)$ (see (1.49) and (1.50)).

We will investigate the behavior of trajectories of the dynamical system (1.99) in a sufficiently small neighborhood of the manifold $\mathcal{M}$.

1^0. Construction of the mapping which splits system (1.99) in a neighborhood of the manifold $\mathcal{M}$. Let df_φ be the derivative of the smooth mapping $f : \mathcal{T}_m \to E$ at the point φ, and let $T(\mathcal{T}_m)_\varphi$ and $TM_{f(\varphi)}$ be the tangent spaces to the manifolds $\mathcal{T}_m$ and $\mathcal{M}$ at the points φ and $f(\varphi)$, respectively [67], [35]. Then $df_\varphi : T(\mathcal{T}_m)_\varphi \to TM_{f(\varphi)}$ is a linear mapping. Since the manifold $\mathcal{T}_m$ is an m-dimensional torus, the space $T(\mathcal{T}_m)_{f(\varphi)}$ is also m-dimensional for any $\varphi \in \mathcal{T}_m$. We require that, for the dimension $\dim TM_{f(\varphi)}$ of the space $TM_{f(\varphi)}$, the analogous property

$$\dim TM_{f(\varphi)} = m, \quad \varphi \in \mathcal{T}_m. \tag{1.101}$$

be true. Since the space $TM_{f(\varphi)}$ is finite-dimensional, the infinite-dimensional Banach space E can be presented in the form of the direct sum of the subspace $TM_{f(\varphi)}$ and some subspace E_φ [45], [46]:

$$E = TM_{f(\varphi)} + E_\varphi.$$

By $P(\varphi)$, we denote the projector on E_φ in parallel to $TM_{f(\varphi)}$ [39]. Then $I - P(\varphi)$ is the projector on $TM_{f(\varphi)}$ in parallel to E_φ (I is the identity operator which acts in E).

Since the mapping f is a C^r-mapping, the derivative df is a C^{r-1}-mapping. With regard for (1.101), the projectors $P(\varphi)$ and $I - P(\varphi)$ are also assumed C^{r-1}-mappings.

We now define a C^{r-1} mapping F, which puts the vector $x \in E$ to the correspondence to each pair (φ, h) of elements $\varphi \in \mathcal{T}_m$ and $h \in E_\varphi$:

$$x = f(\varphi) + P(\varphi)h. \tag{1.102}$$

This mapping has the inverse mapping. The exact meaning of this assertion is clarified by the following proposition.

Lemma 1.3. *Let $V_\delta(\mathcal{T}_m) = \{(\varphi, h)|\ \varphi \in \mathcal{T}_m, h \in E_\varphi, \|h\| < \delta\}$ and*

$$U_\delta(M) = \{x \in E|\ \inf_{y \in \mathcal{M}} \|x - y\| < \delta\}.$$

Then there exists a number $\varepsilon > 0$, such that

$$FV_\varepsilon(\mathcal{T}_m) \supset U_\mu(\mathcal{M})$$

for some number $\mu > 0$, and the mapping $F : V_\varepsilon(\mathcal{T}_m) \to FV_\varepsilon(\mathcal{T}_m)$ is a C^{r-1}- diffeomorphism.

Proof.   Since $r \geq 2$ and F is C^{r-1}-mapping (because f and P_2 are C^r- and C^{r-1}-mappings respectively), there exists the derivative $dF_{(\varphi,0)}$: $T(\mathcal{T}_m)_\varphi \times E_\varphi \to E$ of the mapping F at the point $(\varphi, 0)$ for each $\varphi \in \mathcal{T}_m$. This derivative has the form

$$dF_{(\varphi,0)}(\psi, h) = (df_\varphi)\psi + P(\varphi)h,$$

where $\psi \in T(\mathcal{T}_m)_\varphi$ and $h \in E_\varphi$. From the definition of $P(\varphi)$, it follows that the image of the operator

$$dF_{(\varphi,0)} : T(\mathcal{T}_m)_\varphi \times E_\varphi \to E$$

coincides with the Banach space E, and the kernel $\ker dF_{(\varphi,0)}$ of this operator consists of the alone zero element (the equality $(df_\varphi)\psi + P(\varphi)h = 0$ yields $(df_\varphi)\psi \in TM_{f(\varphi)}$ and $P(\varphi)h \in E_\varphi$; therefore, $(df_\varphi)\psi = 0$, $P(\varphi)h = 0$, hence, $\psi = 0$ and $h = 0$, with regard for (1.101) and the fact that $P(\varphi)$ is the identity operator on E_φ). Then, by the Banach theorem [43], the operator $dF_{(\varphi,0)}$ has the continuous inverse operator

$$(dF_{(\varphi,0)})^{-1} : E \to T(\mathcal{T}_m)_\varphi \times E_\varphi.$$

According to the theorem of the existence of an implicit function, there exists a neighborhood $U_\gamma(f(\varphi)) = \{y \in E | \|y - f(\varphi)\| < \gamma\}$ of a point $f(\varphi)$ of the manifold $\mathcal{M}$ such that Eq. (1.102) for (φ, h) has the unique solution (φ, h) for each $x \in U_\gamma(f(\varphi))$, and the relevant mapping

$$F : \{(\varphi, h) | \varphi \in \mathcal{T}_m, h \in E_\varphi,$$

$$x = f(\varphi) + P(\varphi)h, x \in U_\gamma(f(\varphi))\} \to U_\gamma(f(\varphi))$$

is a C^{r-1}-diffeomorphism.

The compactness of the torus $\mathcal{T}_m$ and the continuity of the derivative $dF_{(\varphi,0)}$ on it yield

$$\sup_{\varphi \in \mathcal{T}_m} \|(dF_{(\varphi,0)})^{-1}\| < \infty.$$

Therefore, the above consideration is valid for every point $\varphi \in \mathcal{T}_m$ with the conservation of the number γ. This allows one [35] to extend the local properties of the mapping F as a C^{r-1}-diffeomorphism to all points of the manifold $\mathcal{M}$.

Lemma 5.1 is proved. $\square$

Lemma 1.4. *In a sufficiently small neighborhood of the manifold $\mathcal{M}$, the dynamical system (1.99) for the new variables $(\varphi, h) \in \mathcal{T}_m \times E_\varphi$ takes the form*

$$(\varphi(n+1), h(n+1)) = F^{-1} X F(\varphi(n), h(n)), \tag{1.103}$$

where $(\varphi(n), h(n)) \in \mathcal{T}_m \times E_{\varphi(n)}$ and $(\varphi(n+1), h(n+1)) \in \mathcal{T}_m \times E_{\varphi(n+1)}$.

Proof.   According to Lemma 1.3 and the invariance of the manifold $\mathcal{M}$ with respect to the C^r-mapping $X : E \to E$, there exist neighborhoods U and V of the manifold $\mathcal{M}$ such that the mapping $F^{-1} : U \to V$ is a C^{r-1}-diffeomorphism, and $XU \subset V$. Hence, if $x(n) \in U$, then $x(n+1) = X(x(n)) \in V$. In this case,

$$F^{-1}x(n+1) = F^{-1}XFF^{-1}x(n). \tag{1.104}$$

According to the same lemma, there exist the elements $(\varphi(n), h(n)) \in \mathcal{T}_m \times E_{\varphi(n)}$ and $(\varphi(n+1), h(n+1)) \in \mathcal{T}_m \times E_{\varphi(n+1)}$ such that

$$F^{-1}x(n) = (\varphi(n), h(n)),$$
$$F^{-1}x(n+1) = (\varphi(n+1), h(n+1)).$$

Equality (1.104) and two last equalities yield relation (1.103).

Lemma 1.4 is proved. $\qquad\qquad\qquad\qquad\qquad\qquad\qquad\qquad\qquad\square$

2^0. **The representation of the mapping $F^{-1}XF$ in a neighborhood of the manifold $\mathcal{T}_m \times \{0\}$.** From Lemma 1.3 and the invariance of the manifold $\mathcal{M}$ with respect to the mapping X, it follows that the mapping $F^{-1}XF$ is defined in a sufficiently small neighborhood of the manifold $\mathcal{T}_m \times \{0\} \subset \mathcal{T}_m \times E$. If $\|h\| < \varepsilon$ and ε is sufficiently small number, the image of the element $(\varphi, h) \in \mathcal{T}_m \times E_\varphi$ is the element $(\varphi_1(\varphi, h), h_1(\varphi, h))$ of the set $\mathcal{T}_m \times E_{\varphi_1(\varphi, h)}$ at the mapping $F^{-1}XF$. Hence,

$$(\varphi_1(\varphi, h), h_1(\varphi, h)) = F^{-1}XF(\varphi, h).$$

Since the mapping $F^{-1}XF$ is a C^{r-1}-mapping, the mappings φ_1 and h_1 are also C^{r-1}-mappings.

We present $\varphi_1(\varphi, h)$ and $h_1(\varphi, h)$ via F, X, φ, and h and define the mappings $Q : \mathcal{T}_m \times E \to \mathcal{T}_m$ and $R : \mathcal{T}_m \times E \to E$ by the equalities

$$Q(\varphi, h) = \varphi,$$
$$R(\varphi, h) = h, \quad \varphi \in \mathcal{T}_m, \ h \in E.$$

Then

$$Q(\varphi_1(\varphi, h), h_1(\varphi, h)) = \varphi_1(\varphi, h),$$
$$R(\varphi_1(\varphi, h), h_1(\varphi, h)) = h_1(\varphi, h)$$

for all $(\varphi, h) \in \mathcal{T}_m \times E_\varphi$. Hence,

$$\varphi_1(\varphi, h) = QF^{-1}XF(\varphi, h), \tag{1.105}$$

$$h_1(\varphi, h) = RF^{-1}XF(\varphi, h). \tag{1.106}$$

According to the Taylor formula,

$$\varphi_1(\varphi, h) = \varphi_1(\varphi, 0) + (d_h\varphi_1)_{(\varphi,0)}h + o(h),$$
$$h_1(\varphi, h) = h_1(\varphi, 0) + (d_h h_1)_{(\varphi,0)}h + o(h)$$

as $h \to 0$. In view of (1.99), (1.100), and (1.102), we obtain

$$F(\varphi, 0) = f(\varphi), \tag{1.107}$$

$$X(f(\varphi)) = f(\varphi + \omega) \tag{1.108}$$

for all $\varphi \in \mathcal{T}_m$. Therefore,

$$F^{-1}f(\varphi + \omega) = (\varphi + \omega, 0), \tag{1.109}$$

and relations (1.105) and (1.106) $\forall \varphi \in \mathcal{T}_m$ yield the equalities

$$\varphi_1(\varphi, 0) = \varphi + \omega,$$
$$h_1(\varphi, 0) = 0.$$

Thus,

$$\varphi_1(\varphi, h) = \varphi + \omega + (d_h\varphi_1)_{(\varphi,0)}h + o(h), \tag{1.110}$$

$$h_1(\varphi, h) = (d_h h_1)_{(\varphi,0)}h + o(h) \tag{1.111}$$

as $h \to 0$ for all $\varphi \in \mathcal{T}_m$.

Using relations (1.107) – (1.109) and (1.102) and the chain rule [67], we will determine the derivatives $(d_h\varphi_1)_{(\varphi,0)}$, $(d_h h_1)_{(\varphi,0)}$.

For them, we obtain the following representations:

$$(d_h\varphi_1)_{(\varphi,0)} = (d_h Q F^{-1} X F)_{(\varphi,0)}$$
$$= (dQ)_{F^{-1}XF(\varphi,0)}(dF^{-1})_{XF(\varphi,0)}(dX)_{F(\varphi,0)}(d_h F)_{(\varphi,0)}$$
$$= (dQ)_{(\varphi+\omega,0)}(dF^{-1})_{f(\varphi+\omega)}(dX)_{f(\varphi)}P(\varphi), \quad \varphi \in \mathcal{T}_m;$$

$$(d_h h_1)_{(\varphi,0)} = (dR)_{(\varphi+\omega,0)}(dF^{-1})_{f(\varphi+\omega)}(dX)_{f(\varphi)}P(\varphi), \quad \varphi \in \mathcal{T}_m.$$

We now show that

$$(dQ)_{(\varphi+\omega,0)}(dF^{-1})_{f(\varphi+\omega)} = (df)^{-1}_{\varphi+\omega}(I - P(\varphi + \omega)), \tag{1.112}$$

$$(dR)_{(\varphi+\omega,0)}(dF^{-1})_{f(\varphi+\omega)} = P(\varphi + \omega). \tag{1.113}$$

Indeed, since

$$F(\varphi + \omega, h) = f(\varphi + \omega) + P(\varphi + \omega)h$$

for all $(\varphi, h) \in \mathcal{T}_m \times E_{\varphi+\omega}$, we have

$$(dF)_{(\varphi+\omega,0)}(\psi, h) = (df)_{\varphi+\omega}\psi + P(\varphi + \omega)h$$

for all $(\psi, h) \in T(\mathcal{T}_m)_{\varphi+\omega} \times E_{\varphi+\omega}$. Therefore,

$$(dF^{-1})_{f(\varphi+\omega)}x = (df)^{-1}_{\varphi+\omega}(I - P(\varphi + \omega))x, P(\varphi + \omega)x) \qquad (1.114)$$

for all $x \in E$ (the mapping $(df)^{-1}_{\varphi+\omega}$ acts from the tangent space $T\mathcal{M}_{f(\varphi+\omega)}$ onto the tangent space $T(\mathcal{T}_m)_{\varphi+\omega}$). If A is the identity mapping of the manifold $\mathcal{N}$, then, as is known, $(dA)_x$ is the identity mapping of the tangent space $T\mathcal{N}_x$ [67]. Taking this result into account, we come to the conclusion that

$$(dQ)_{(\varphi+\omega,0)}(\psi, h) = \psi,$$
$$(dR)_{(\varphi+\omega,0)}(\psi, h) = h$$

for all $(\psi, h) \in T(\mathcal{T}_m)_{\varphi+\omega} \times E_{\varphi+\omega}$. Therefore, relation (1.114) yields equalities (1.112) and (1.113).

Thus, with regard for relations (1.110) and (1.111), we verify the validity of the following proposition on the representation of the mapping $F^{-1}XF$.

Theorem 1.15. *There exists the number $\varepsilon > 0$ such that, $\forall(\varphi, h) \in \mathcal{T}_m \times E_\varphi$, for which $\|h\| < \varepsilon$,*

$$F^{-1}XF(\varphi, h)$$
$$= (\varphi + \omega + (df)^{-1}_{\varphi+\omega}(I - P(\varphi + \omega))(dX)_{f(\varphi)}P(\varphi)h + \alpha(\varphi, h),$$
$$P(\varphi + \omega)(dX)_{f(\varphi)}P(\varphi)h + \beta(\varphi, h)),$$

where $\alpha(\varphi, h)$ and $\beta(\varphi, h)$ are C^{r-1}-mappings satisfying the relation

$$\sup_{(\varphi,h)\in\mathcal{T}_m\times E_\varphi, \|h\|<\varepsilon} (\|\alpha(\varphi, h)\| + \|\beta(\varphi, h)\|) = o(\varepsilon) \qquad (1.115)$$

as $\varepsilon \to 0$.

Remark 1.1. Relation (1.115) follows from (1.110), (1.111), the compactness of the set $\mathcal{T}_m$, and the continuity of the C^{r-1}-mappings

$$\alpha(\varphi, h) = \varphi_1(\varphi, h) - \varphi - \omega - (d_h\varphi_1)_{(\varphi,0)}h,$$
$$\beta(\varphi, h) = h_1(\varphi, h) - (d_h h_1)_{(\varphi,0)}h$$

in some neighborhood of the set $\mathcal{T}_m \times \{0\}$.

The mapping $F^{-1}XF$ can be also represented in a different way.

Theorem 1.16. *There exists the number $\varepsilon > 0$ such that, on the set $G_\varepsilon = \{(\varphi, h) \in \mathcal{T}_m \times E_\varphi | \|h\| < \varepsilon\}$, the mapping $F^{-1}XF$ can be represented as*

$$F^{-1}XF(\varphi, h) = (\varphi + \omega + A(\varphi, h)h, B(\varphi, h)h),$$

where $A(\varphi, h)$ and $B(\varphi, h)$ are linear bounded (for fixed $(\varphi, h) \in G_\varepsilon$) mappings, which act from E_φ in R^m and E, respectively, and are C^{r-1}-mappings.

Proof.   Since, for a sufficiently small ε, the mapping $F^{-1}XF$ is defined on G_ε,

$$F^{-1}XF(\varphi, h) = (\varphi_1(\varphi, h), h_1(\varphi, h)),$$
$$\varphi_1(\varphi, h) = QF^{-1}XF(\varphi, h),$$
$$h_1(\varphi, h) = RF^{-1}XF(\varphi, h),$$
$$\varphi_1(\varphi, h) = \varphi + \omega + (\varphi_1(\varphi, h) - \varphi_1(\varphi, 0)),$$
$$h_1(\varphi, h) = h_1(\varphi, h) - h_1(\varphi, 0)$$

(see the proof of Theorem 1.15) and

$$\varphi_1(\varphi, h) - \varphi_1(\varphi, 0) = \left(\int_0^1 (d_h\varphi_1)_{(\varphi, th)} dt\right)h,$$

$$h_1(\varphi, h) - h_1(\varphi, 0) = \left(\int_0^1 (d_h h_1)_{(\varphi, th)} dt\right)h,$$

the mappings $A(\varphi, h)$ and $B(\varphi, h)$ are defined by the equalities

$$A(\varphi, h) = \int_0^1 (d_h\varphi_1)_{(\varphi, th)} dt, \tag{1.116}$$

$$B(\varphi, h) = \int_0^1 (d_h h_1)_{(\varphi, th)} dt. \tag{1.117}$$

The linearity and the boundedness of these mappings for fixed $(\varphi, h) \in G_\varepsilon$ follow from equalities (1.116) and (1.117) and the membership of the mappings φ_1 and h_1 to the class C^{r-1}. These mappings are C^{r-1}-mappings, since the mappings

$$(d_h\varphi_1)_{(\varphi, th)} = (d_h QF^{-1}XF)_{(\varphi, th)}$$
$$= (dQ)_{F^{-1}XF(\varphi, th)}(dF^{-1})_{XF(\varphi, th)}(dX)_{F(\varphi, th)}(dF)_{(\varphi, th)}$$
$$= (dQ)_{F^{-1}X(f(\varphi)+P(\varphi)th)}(dF^{-1})_{X(f(\varphi)+P(\varphi)th)}(dX)_{f(\varphi)+P(\varphi)th}P(\varphi),$$

$$(d_h h_1)_{(\varphi,th)} = (d_h R F^{-1} X F)_{(\varphi,th)}$$

$$= (dR)_{F^{-1}X(f(\varphi)+P(\varphi)th)}(dF^{-1})_{X(f(\varphi)+P(\varphi)th)}(dX)_{f(\varphi)+P(\varphi)th}P(\varphi)$$

have the analogous properties for all $t \in [0,1]$, which completes the proof of Theorem 1.16. $\qquad\square$

3^0. Theorems on the splitting of system (1.99) in a neighborhood of the manifold $\mathcal{M}$. Lemma 1.4 and Theorem 1.15 imply that the following proposition is valid.

Theorem 1.17. *In a sufficiently small neighborhood of the manifold $\mathcal{M}$, system (1.99) with respect to the variables $(\varphi, h) \in \mathcal{T}_m \times E_\varphi$ is represented in the form*

$$\varphi(n+1) = \varphi(n) + \omega + \Phi(\varphi(n))h(n) + \Phi_1(\varphi(n), h(n)),$$

$$h(n+1) = H(\varphi(n))h(n) + H_1(\varphi(n), h(n)), \quad n \in Z, \qquad (1.118)$$

where $\Phi(\varphi) : E_\varphi \to T(\mathcal{T}_m)_{\varphi+\omega}$ and $H(\varphi) : E_\varphi \to E_{\varphi+\omega}$ ($\varphi \in \mathcal{T}_m$) are linear mappings defined by the equalities

$$\Phi(\varphi) = (df)^{-1}_{\varphi+\omega}(I - P(\varphi + \omega))(dX)_{f(\varphi)}P(\varphi),$$
$$H(\varphi) = P(\varphi + \omega)(dX)_{f(\varphi)}P(\varphi),$$

whereas $\Phi_1(\varphi, h)$ and $H_1(\varphi, h)$ are, in a general case, nonlinear mappings, which satisfy the relation

$$\sup_{(\varphi,h)\in\mathcal{T}_m\times E_\varphi,\ \|h\|<\varepsilon} (\|\Phi_1(\varphi, h)\| + \|H_1(\varphi, h)\|) = o(\varepsilon) \qquad (1.119)$$

as $\varepsilon \to 0$, and the mappings Φ, Φ_1, H, and H_1 are C^{r-1}-mappings.

The following analogous proposition follows from Lemma 1.4 and Theorem 1.16.

Theorem 1.18. *In a sufficiently small neighborhood of the manifold $\mathcal{M}$, system (1.99) with respect to the variables $(\varphi, h) \in \mathcal{T}_m \times E_\varphi$ is represented in the form*

$$\varphi(n+1) = \varphi(n) + \omega + A(\varphi(n), h(n))h(n),$$

$$h(n+1) = B(\varphi(n), h(n))h(n), \quad n \in Z, \qquad (1.120)$$

where $A(\varphi, h)$ and $B(\varphi, h)$ are linear bounded mappings (for fixed $\varphi \in \mathcal{T}_m$, $h \in E_\varphi$, and a sufficiently small h in the norm), which act from E_φ in R^m and E, respectively, and are C^{r-1}-mappings.

Remark 1.2. It is evident that

$$A(\varphi, h)h = \Phi(\varphi)h + \Phi_1(\varphi, h),$$
$$B(\varphi, h)h = H(\varphi)h + H_1(\varphi, h)$$

for all $\varphi \in \mathcal{T}_m$ and $h \in \{x \in E_\varphi | \|x\| < \varepsilon\}$, where ε is a sufficiently small number.

4^0. Auxiliary operator equation. We now study the operator equation, which allows us to simplify Eqs. (1.118) and (1.120).

Consider the equation

$$Y(\varphi, h) = A(\varphi, h) + Y(\varphi + \omega + A(\varphi, h)h, B(\varphi, h)h)B(\varphi, h), \qquad (1.121)$$

where $A(\varphi, h)$ and $B(\varphi, h)$ are C^{r-1}-mappings, which are linear for fixed $(\varphi, h) \in \{(\psi, \delta) \in \mathcal{T}_m \times E_\varphi | \|\delta\| < \varepsilon\}$ (ε is a sufficiently small number) act from E_φ into R^m and E, respectively, and are considered in Theorems 1.16 and 1.18, and $Y(\varphi, h)$ is the required solution of the equation, which is a linear mapping acting from E_φ into R^m for each fixed $(\varphi, h) \in G_\varepsilon$.

Lemma 1.5. *If*

$$\sup_{\varphi \in \mathcal{T}_m} \|P(\varphi + \omega)(dX)_{f(\varphi)}P(\varphi)\| < 1, \qquad (1.122)$$

then there exists the number $\varepsilon_0 > 0$ such that Eq. (1.121) has the unique solution $V(\varphi, h)$, which is defined on G_{ε_0}, belongs to the class C^0 which for each fixed $(\varphi, h) \in G_{\varepsilon_0}$, and is a linear mapping that acts from E_φ into R^m. This solution can be presented as

$$V(\varphi, h) = A(\varphi_0, h_0)$$
$$+ \sum_{n=1}^{\infty} A(\varphi_n, h_n)B(\varphi_{n-1}, h_{n-1})B(\varphi_{n-2}, h_{n-2})\dots B(\varphi_1, h_1)B(\varphi_0, h_0),$$

$$(1.123)$$

where

$$\varphi_n = \varphi_{n-1} + \omega + A(\varphi_{n-1}, h_{n-1})h_{n-1}, \qquad (1.124)$$

$$h_n = B(\varphi_{n-1}, h_{n-1})h_{n-1}, \quad n \geq 1, \qquad (1.125)$$

and

$$\varphi_0 = \varphi, \quad h_0 = h.$$

For sufficiently small ε and $\sup_{\varphi \in \mathcal{T}_m} \sum_{k=0}^{r-1} \|(d^k B)_{(\varphi, 0)}\|$, the solution $V(\varphi, h)$ is a C^{r-1}-mapping.

Proof. We note that, according to Theorem 1.17 and the proof of Theorem 1.16,

$$B(\varphi, h) = B(\varphi, 0) + \int_0^1 [(d_h h_1)_{(\varphi, th)} - (d_h h_1)_{(\varphi, 0)}] dt,$$

where

$$B(\varphi, 0) = P(\varphi + \omega)(dX)_{f(\varphi)} P(\varphi),$$

the derivative $(d_h h_1)_{(\varphi, h)}$ is uniformly continuous on the set $\mathcal{T}_m \times \{0\}$ by virtue of the compact-ness of the set $\mathcal{T}_m$. Hence,

$$\lim_{h \to 0} \sup_{\varphi \in \mathcal{T}_m} \left\| \int_0^1 [(d_h h_1)_{(\varphi, th)} - (d_h h_1)_{(\varphi, 0)}] dt \right\| = 0.$$

Therefore, according to (1.122), there exists the number $\varepsilon_0 > 0$ such that the operator-functions $A(\varphi, h)$ and $B(\varphi, h)$ are continuous on the closure $\overline{G_{\varepsilon_0}}$ of the set G_{ε_0}, and

$$q = \sup_{(\varphi, h) \in \overline{G_{\varepsilon_0}}} \| B(\varphi, h) \| < 1. \tag{1.126}$$

Taking relation (1.125) into account, we obtain

$$\| h_n \| \le q^n \| h \|$$

for all $n \ge 1$ if $\| h \| \le \varepsilon_0$. Therefore, by virtue of (1.124) and (1.125), $(\varphi_n, h_n) \in \overline{G_{\varepsilon_0}}$ for all $n \ge 1$ if $(\varphi, h) \in \overline{G_{\varepsilon_0}}$. Then it follows from (1.126) that the operator series, via which operator-function (1.123) is defined, is majorized by the numerical series

$$M + Mq + Mq^2 + \ldots + Mq^n + \ldots,$$

where

$$M = \sup_{(\varphi, h) \in \overline{G_{\varepsilon_0}}} \| A(\varphi, h) \| < \infty. \tag{1.127}$$

The finiteness of the quantity $\sup_{(\varphi, h) \in \overline{G_{\varepsilon_0}}} \| A(\varphi, h) \|$ follows from the following facts:

1) Theorem 1.17 and the proof of Theorem 1.16 ensure the validity of the relation

$$A(\varphi, h) = A(\varphi, 0) + \int_0^1 [(d_h \varphi_1)_{(\varphi, th)} - (d_h \varphi_1)_{(\varphi, 0)}] dt,$$

where

$$A(\varphi, 0) = (df)^{-1}_{\varphi + \omega}(I - P(\varphi + \omega))(dX)_{f(\varphi)} P(\varphi);$$

2) by virtue of the continuity of the operator-function $A(\varphi, 0)$ on the compact set $\mathcal{T}_m$, the inequality

$$\sup_{\varphi \in \mathcal{T}_m} \|A(\varphi, 0)\| < \infty$$

is valid;

3) since the derivative $(d_h \varphi_1)_{(\varphi, h)}$ is uniformly continuous on the compact set $\mathcal{T}_m \times \{0\}$, we have

$$\lim_{h \to 0} \sup_{\varphi \in \mathcal{T}_m} \| \int_0^1 [(d_h \varphi_1)_{(\varphi, th)} - (d_h \varphi_1)_{(\varphi, 0)}]dt \| = 0.$$

Thus, the sum of the operator series (see (1.123)), which defines the mapping $V(\varphi, h)$, is continuous on G_{ε_0}. By the usual substitution of this mapping into Eq. (1.121) with regard for (1.124) and (1.125), we verify that this mapping is a solution of the equation under study.

From (1.126), it follows that the existence of solutions of Eq. (1.121) can be established by the principle of contracting mappings. Therefore, the solution of Eq. (1.121) represented in the form (1.123) is unique.

Now, we give the scheme of solving the problem of the membership of the solution $V(\varphi, h)$ to the class C^{r-1}.

Consider the Banach space $\mathcal{L}$ of functions $Z(\varphi, h)$, which are $r - 1$ times continuously differentiable on G_{ε_0} (ε_0 is a sufficiently small positive number) and take the values in the space $L(E_\varphi, R^m)$ of linear continuous operators acting from E_φ into R^m, with the norm

$$\|z\|_{\mathcal{L}} = \sup_{(\varphi, h) \in G_{\varepsilon_0}} (\|Z(\varphi, h)\|_{L_1} + \|(dZ)_{(\varphi, h)}\|_{L_2} + \dots$$
$$+ \|(d^{r-1}Z)_{(\varphi, h)}\|_{L_r}),$$

where

$$L_1 = L(E_\varphi, R^m), L_2 = L(E_\varphi, L_1), ..., L_r = L(E_\varphi, L_{r-1}).$$

Further, we will consider the linear continuous operator

$$\mathcal{D} : \mathcal{L} \to \mathcal{L},$$

defined by the equality

$$(\mathcal{D}Z)(\varphi, h) = A(\varphi, h) + Z(\varphi_1(\varphi, h), h_1(\varphi, h))B(\varphi, h),$$

where

$$\varphi_1(\varphi, h) = \varphi + \omega + A(\varphi, h)h \quad \text{and} \quad h_1(\varphi, h) = B(\varphi, h)h.$$

Consider the relations

$$(d^k \mathcal{D} Z)_{(\varphi,h)} = (d^k A)_{(\varphi,h)} + \sum_{l=0}^{k} C_k^l (d^l Z(\varphi_1, h_1))_{(\varphi,h)} (d^{k-1-l} B)_{(\varphi,h)}$$

$(Z \in \mathcal{L}, k = \overline{1, r-1})$, which follow from the rule of differentiation of operator functions and the Leibniz formula. With regard for the boundedness of the quantities

$$\alpha_l = \sup_{(\varphi,h) \in G_{\varepsilon_0}} \left(\|(d^l A)_{(\varphi,h)}\| + \|(d^l B)_{(\varphi,h)}\| \right.$$
$$\left. + \|(d^l \varphi_1)_{(\varphi,h)}\| + \|(d^l h_1)_{(\varphi,h)}\| \right), \quad l = \overline{1, r-1},$$

which can be established by a reasoning analogous to that used above in the proof of Lemma 1.5, we can conclude that

$$\|\mathcal{D} Z_1 - \mathcal{D} Z_2\|_{\mathcal{L}} \leq q \|Z_1 - Z_2\|_{\mathcal{L}} \tag{1.128}$$

for all $Z_i \in \mathcal{L}$ $(i = \overline{1,2})$, and the coefficient q is presented in the form

$$q = Q(\alpha_1, \alpha_2, ... \alpha_{r-1}) \sup_{\varphi \in \mathcal{T}_m} \sum_{k=1}^{r-1} \|(d^k B)_{(\varphi,0)}\|, \tag{1.129}$$

where $Q(\alpha_1, \alpha_2, ... \alpha_{r-1})$ is some continuous function nonnegative on R^{r-1}.

From (1.128) and (1.129), it follows that the mapping $\mathcal{D} : \mathcal{L} \to \mathcal{L}$ is a contraction mapping if the value of

$$\beta = \sup_{\varphi \in \mathcal{T}_m} \sum_{k=1}^{r-1} \|(d^k B)_{(\varphi,0)}\|$$

is sufficiently small. If this condition is satisfied, the mapping $\mathcal{D}$ has the single fixed point $V(\varphi, h) \in \mathcal{L}$, which is evidently a solution of Eq. (1.121). This completes the substantiation of Lemma 1.5. $\qquad\square$

5^0. C^{r-1}-**mapping** S **simplifying system (1.120) in a neighborhood of the manifold** $\mathcal{T}_m \times \{0\}$. For a sufficiently small $\varepsilon_0 > 0$, let us consider the mapping S, which transforms every point $(\varphi, h) \in (\mathcal{T}_m \times E_\varphi) \cap G_{\varepsilon_0}$ into some point $(\psi, \delta) \in (\mathcal{T}_m \times E_\psi) \cap G_{\varepsilon_1}$, where ε_1 is some positive number dependent on ε_0. We set this mapping with the help of the equalities

$$\psi = \varphi + V(\varphi, h)h, \quad \delta = P(\varphi + V(\varphi, h)h)h, \tag{1.130}$$

where $V(\varphi, h)$ is the mapping from Lemma 5.3, and $P(\psi)$ is the projector considered in it. 1^0.

Lemma 1.6. *For sufficiently small $\varepsilon_0 > 0$ and*

$$\sup_{\varphi \in \mathcal{T}_m} \sum_{k=0}^{r-1} \|(d^k B)_{(\varphi, 0)}\|,$$

the mapping $S : G_{\varepsilon_0} \to SG_{\varepsilon_0}$ is a C^{r-1}-diffeomorphism. In this case, the inverse mapping S^{-1} can be presented in the form

$$\varphi = \psi + U_1(\psi, \delta)\delta, \quad h = \delta + U_2(\psi, \delta)\delta, \tag{1.131}$$

where $U_1(\psi, \delta)$ and $U_2(\psi, \delta)$ are mappings, which are linear and continuous for fixed $(\psi, \delta) \in G_{\varepsilon_0}$ act from E_φ into R^m and E, respectively, and are C^{r-1}-mappings.

Proof.   Note that the mapping $S : G_{\varepsilon_0} \to SG_{\varepsilon_0}$ for a sufficiently small ε_0 is a C^{r-1}-mapping by Lemma 5.3 taking into account that the projector $P(\varphi + V(\varphi, h)h)$ as the composition of two C^{r-1}-mappings is also a C^{r-1}-mapping. We will show that the mapping S has the continuous inverse mapping.

For this purpose, we present firstly the second equality of system (1.130) in the form

$$\delta = h + W(\varphi, h)h, \tag{1.132}$$

where

$$W(\varphi, h) = P(\varphi + V(\varphi, h)h) - P(\varphi). \tag{1.133}$$

The linear mapping $W(\varphi, h)$ $(\varphi, h) \in G_{\varepsilon_0}$ is a C^{r-1}-mapping.

In view of (1.123), (1.126), and (1.127), we can assume that

$$\sup_{(\varphi, h) \in G_{\varepsilon_0}} \|V(\varphi, h)\| < \infty. \tag{1.134}$$

Then, according to the compactness of the set $\mathcal{T}_m$, continuity of $P(\varphi)$ on $\mathcal{T}_m$, and relation (1.133), we have

$$\sup_{\varphi \in \mathcal{T}_m} \|P(\varphi)\| < \infty \tag{1.135}$$

and

$$\sup_{(\varphi, h) \in G_{\varepsilon_0}} \|W(\varphi, h)\| = o(\varepsilon_0) \quad \text{as} \quad \varepsilon_0 \to 0. \tag{1.136}$$

We now apply the operator $P(\varphi)$ to the both sides of equality (1.132). Taking into account that $h \in E_\varphi$, we obtain

$$P(\varphi)\delta = (P(\varphi) + P(\varphi)W(\varphi, h))h. \tag{1.137}$$

Without loss of generality, we can assume by virtue of (1.135) and (1.136) that the number ε_0 is chosen so small that

$$\sup_{(\varphi,h)\in G_{\varepsilon_0}} \|P(\varphi)W(\varphi,h)\| < 1. \tag{1.138}$$

The last relation guarantees the existence of the continuous inverse operator

$$(P(\varphi) + P(\varphi)W(\varphi,h))^{-1}$$

for the linear continuous operator

$$(P(\varphi) + P(\varphi)W(\varphi,h)) : E_\varphi \to E_\varphi$$

(the operator $P(\varphi)$ plays the role of the identity operator in the space E_φ) for all fixed $(\varphi,h) \in G_{\varepsilon_0}$. Therefore, relation (1.137) yields

$$h = (P(\varphi) + P(\varphi)W(\varphi,h))^{-1}P(\varphi)\delta. \tag{1.139}$$

Note that the mapping $(P(\varphi) + P(\varphi)W(\varphi,h))^{-1}P(\varphi)$ is a C^{r-1}-mapping, and, according to (1.138),

$$\sup_{(\varphi,h)\in G_{\varepsilon_0}} \|(P(\varphi) + P(\varphi)W(\varphi,h))^{-1}P(\varphi)\| < \infty. \tag{1.140}$$

With regard for (1.132) and (1.139), we present the system of relations (1.130) as

$$\psi = \varphi + V_1(\varphi,h)\delta, \quad \delta = h + W_1(\varphi,h)\delta, \tag{1.141}$$

where the mappings

$$V_1(\varphi,h) = V(\varphi,h)(P(\varphi) + P(\varphi)W(\varphi,h))^{-1}P(\varphi),$$
$$W_1(\varphi,h) = W(\varphi,h)(P(\varphi) + P(\varphi)W(\varphi,h))^{-1}P(\varphi)$$

linear for fixed $(\varphi,h) \in G_{\varepsilon_0}$ are C^{r-1}-mappings. For them, we have

$$\sup_{(\varphi,h)\in G_{\varepsilon_0}} (\|V_1(\varphi,h)\| + \|W_1(\varphi,h)\|) < \infty \tag{1.142}$$

by virtue of (1.134), (1.135) and (1.140).

After the preliminary work, we will show that the mapping S has the continuous inverse mapping. Let us find the derivative $(dS)_{(\varphi,0)}$. From (1.130) and (1.132), it follows that

$$(dS)_{(\varphi,0)}(\eta,\xi) = (\eta + V(\varphi,0)\xi,\xi)$$

for all $\varphi \in \mathcal{T}_m$ and $(\eta,\xi) \in (T\mathcal{T}_m)_\varphi \times E_\varphi$. This implies that the derivative $(dS)_{(\varphi,0)}$ has the continuous inverse mapping $(dS)^{-1}_{(\varphi,0)}$, and

$$(dS)^{-1}_{(\varphi,0)}(\eta,\xi) = (\eta - V(\varphi,0)\xi,\xi) \tag{1.143}$$

for all $\varphi \in \mathcal{T}_m$ and $(\eta, \xi) \in (T\mathcal{T}_m)_\varphi \times E_\varphi$. Moreover,

$$\sup_{\varphi \in \mathcal{T}_m} \|(dS)^{-1}_{(\varphi,0)}\| < \infty,$$

which follows from (1.134). Then the theorem of implicit function, compactness of the set $\{(dS)^{-1}_{(\varphi,0)} : \varphi \in \mathcal{T}_m\}$ (by virtue of the continuous dependence of the derivative $(dS)^{-1}_{(\varphi,0)}$ on φ on the compact set $\mathcal{T}_m$), and relations (1.134), (1.135) and (1.136) imply that, for a sufficiently small $\varepsilon_0 > 0$, the mapping $S : G_{\varepsilon_0} \to SG_{\varepsilon_0}$ has the continuous inverse mapping S^{-1}. Since the mapping S is a C^{r-1}-mapping, the mapping S^{-1} has the analogous property.

Hence, for a sufficiently small $\varepsilon_0 > 0$, the mapping $S : G_{\varepsilon_0} \to SG_{\varepsilon_0}$ is a C^{r-1}-diffeomorphism.

Then system (1.130) can be solved for φ and h for all $(\psi, \delta) \in SG_{\varepsilon_0}$. Let $\varphi = \varphi(\psi, \delta)$, and $h = h(\psi, \delta)$ be the solutions of this system. Then the membership of S^{-1} to the class C^{r-1} implies that the mappings $\varphi(\psi, \delta)$ and $h(\psi, \delta)$ are C^{r-1}-mappings. Using these mappings and system (1.141), we verify that the mappings $U_1(\psi, \delta)$ and $U_2(\psi, \delta)$ from (1.131) can be presented as

$$U_1(\psi, \delta) = V_1(\varphi(\psi, \delta), h(\psi, \delta)), \tag{1.144}$$

$$U_2(\psi, \delta) = W_1(\varphi(\psi, \delta), h(\psi, \delta)) \tag{1.145}$$

and have the properties indicated in the assertion of Lemma 1.6. This completes the proof. $\qquad\square$

We now use Lemmas 1.5 and 1.6 to simplify the system of difference equations (1.120).

Theorem 1.19. *For sufficiently small $\varepsilon_0 > 0$ and*

$$\sup_{\varphi \in \mathcal{T}_m} \sum_{k=0}^{r-1} \|(d^k B)_{(\varphi,0)}\|,$$

system (1.120) in the neighborhood G_{ε_0} of the manifold $\mathcal{T}_m \times \{0\}$ is reduced with the help of change (1.130) to the system

$$\psi(n+1) = \psi(n) + \omega,$$

$$\delta(n+1) = C(\psi(n), \delta(n))\delta(n), \quad n \in Z, \tag{1.146}$$

where $C(\psi, \delta)$ is a continuous mapping, which is linear for each $(\psi, \delta) \in G_{\varepsilon_0}$, acts from E_ψ into $E_{\psi+\omega}$, is a C^{r-1}-mapping and satisfies the relation

$$\sup_{\varphi \in \mathcal{T}_m} \|C(\psi, \delta) - P(\psi + \omega)(dX)_{f(\psi)}P(\psi)\| = o(\delta) \quad as \quad \delta \to 0. \tag{1.147}$$

Proof.   Lemma 1.5 and its proof imply that the solution $(\varphi(n), h(n))$ of system (1.120) satisfies the relation

$$V(\varphi(n), h(n))h(n) = A(\varphi(n), h(n))h(n) + V(\varphi(n) + \omega$$

$$+A(\varphi(n), h(n))h(n), B(\varphi(n), h(n))h(n))B(\varphi(n), h(n))h(n)$$

for all $n \in Z$, for which $\|h(n)\| < \varepsilon_0$. In view of the equalities

$$\psi(n) = \varphi(n) + V(\varphi(n), h(n))h(n),$$

$$\psi(n + 1) = \varphi(n + 1)$$

$$+V(\varphi(n)+\omega+A(\varphi(n), h(n))h(n), B(\varphi(n), h(n))h(n))B(\varphi(n), h(n))h(n),$$

which follow from (1.130), it is easy to verify that the first equation of system (1.120) turns into the first equation of system (1.146).

Using successively relations (1.130), (1.120), and (1.131) and the equality $\psi(n+1) = \psi(n) + \omega$, we obtain

$$\delta(n + 1) = P(\psi(n + 1))h(n + 1) = P(\psi(n) + \omega)B(\varphi(n), h(n))h(n)$$

$$= P(\psi(n) + \omega)B(\psi(n) + U_1(\psi(n), \delta(n))\delta(n), \delta(n) + U_2(\psi(n), \delta(n))\delta(n))$$

$$\times (I + U_2(\psi(n), \delta(n)))\delta(n).$$

Therefore, change (1.130) reduces system (1.120) in a neighborhood G_{ε_0} of the manifold $\mathcal{T}_m \times \{0\}$ to system (1.144), where

$$C(\psi, \delta) = P(\psi + \omega)B(\psi + U_1(\psi, \delta)\delta, \delta + U_2(\psi, \delta)\delta)(I + U_2(\psi, \delta)). \quad (1.148)$$

The last relation and Lemma 1.6 imply that the mapping $C(\psi, \delta)$ for the fixed $(\psi, \delta) \in \mathcal{T}_m \times \{0\}$ (δ is sufficiently small) is linear and continuous and acts from E_ψ into $E_{\psi+\omega}$. Since$P$, B, U_1, and U_2 are C^{r-1}-mappings, equality (1.148) endows the mapping $C(\psi, \delta)$ with the analogous property.

We now prove the validity of relation (1.147). The proof of Lemma 1.6 and relations (1.142) $-$ (1.144) yield the estimate

$$\sup_{(\psi,\delta)\in G_{\varepsilon_0}} (\|U_1(\psi, \delta)\| + \|U_2(\psi, \delta)\|) < \infty.$$

In view of the uniform continuity $B(\psi, 0)$ in ψ on the compact set $\mathcal{T}_m$, this estimate leads to

$$\sup_{(\psi,\delta)\in G_\varepsilon} \|B(\psi + U_1(\psi, \delta)\delta, \delta + U_2(\psi, \delta)\delta) - B(\psi, 0)\| = o(\varepsilon)$$

as $\varepsilon \to 0$. With regard for relation (1.148), the equalities $P^2(\psi + \omega) = P(\psi + \omega)$ and $B(\psi, 0) = P(\psi + \omega)(dX)_{f(\psi)}P(\psi)$ (see Theorem 1.17 and Remark 1.2), and the relation

$$\sup_{(\psi,\delta)\in G_\varepsilon} \|U_2(\psi, \delta)\| = o(\varepsilon) \quad \text{as} \quad \varepsilon \to 0,$$

which follows from (1.136), (1.145), and definition of $W_1(\varphi, h)$ (see the proof of Lemma 1.6), we verify the validity of relation (1.147). The proof of Theorem 1.19 is completed. $\square$

Remark 1.3. Since $P(\psi)\delta = \delta$ if $\delta \in E_\psi$, we have $C(\psi, \delta) = C(\psi, P(\psi)\delta)$ for all $(\psi, \delta) \in G_{\varepsilon_0}$. Therefore, system (1.146) can be presented in the form

$$\psi(n + 1) = \psi(n) + \omega,$$

$$\delta(n + 1) = C(\psi(n), P(\psi(n))\delta(n))\delta(n), \quad n \in Z. \tag{1.149}$$

Relations (1.148), (1.144), (1.145), (1.133), (1.123), (1.124), (1.125), (1.116) and (1.117) and the 2π-periodicity of the function $f(\psi) = f(\psi_1, \psi_2, ..., \psi_n)$ in each variable ψ_k $(k = \overline{1, n})$ imply that the operator-function

$$C(\psi, P(\psi)\delta) = C(\psi_1, ..., \psi_n, P(\psi_2, ..., \psi_n)\delta)$$

is also 2π-periodic in each variable ψ_k $(k = \overline{1, n})$ for all $\delta \in E$ sufficiently small in the norm.

6^0. Asymptotic behavior of trajectories of system (1.99) in a neighborhood of the manifold $\mathcal{M}$. The restrictions on the mappings X and f, which allowed us to pass from system (1.99) to system (1.146) give a possibility to study the asymptotic behavior of trajectories of the dynamical system (1.99) in a neighborhood of the manifold $\mathcal{M}$.

Theorem 1.20. *Let relation (1.122) be satisfied, and let the value of*

$$\sup_{\varphi \in \mathcal{T}_m} \sum_{k=0}^{r-1} \|(d^k B)_{(\varphi, 0)}\|$$

be sufficiently small. Then the manifold $\mathcal{M}$ is an attractor of system (1.99).

Proof. Theorem 1.19 and relation (1.122) ensure the existence of a number $\varepsilon_0 > 0$, for which

$$q = \sup_{(\psi, \delta) \in G_{\varepsilon_0}} \|C(\psi, \delta)\| < 1. \tag{1.150}$$

There also exists the number $\varepsilon_1 > 0$ such that, in a neighborhood

$$U_{\varepsilon_1}(\mathcal{M}) = \{x \in E| \inf_{y \in \mathcal{M}} \|x - y\| < \varepsilon_1\}$$

of the manifold $\mathcal{M}$, the mapping

$$SF^{-1} : U_{\varepsilon_1}(\mathcal{M}) \to SF^{-1}U_{\varepsilon_1}(\mathcal{M}) \quad (SF^{-1}U_{\varepsilon_1}(\mathcal{M}) \subset G_{\varepsilon_0})$$

is defined and is a C^{r-1}-diffeomorphism according to Lemmas 1.3 and 1.6. This mapping gives possibility to reduce the study of the behavior of trajectories of system (1.99) in the neighborhood $U_{\varepsilon_1}(\mathcal{M})$ to that of the behavior of trajectories of system (1.146) in the neighborhood $SF^{-1}U_{\varepsilon_1}(\mathcal{M}) \subset G_{\varepsilon_0}$. Therefore, to prove the statement of the theorem, it is sufficient to show that the set $\mathcal{T}_m \times \{0\}$ is an attractor [53] of system (1.146).

We now consider the mapping

$$g(\psi, \delta) = (\psi + \omega, C(\psi, \delta)\delta),$$

which is generated by the right-hand side of system (1.146) and is defined on G_{ε_0}. From (1.150), it follows that $g^n G_{\varepsilon_0} \subset g^{n-1} G_{\varepsilon_0}$ and $g^n G_{\varepsilon_0} \subset G_{\varepsilon_n}$, where $\varepsilon_n = q^n \varepsilon_0$ and $n \in N$, N being the set of natural numbers. Since also $\mathcal{T}_m \times \{0\} \subset g^n G_{\varepsilon_0}$ for $n \in N$, which follows from the definition of the mapping g, we have

$$\bigcap_{n \geq 1} g^n G_{\varepsilon_0} = \mathcal{T}_m \times \{0\}.$$

Therefore, $\mathcal{T}_m \times \{0\}$ is an attractor of system (1.146). $\qquad \square$

Theorem 1.21. *Let the conditions of Theorem 1.20 be satisfied. Then it is possible to indicate a sufficiently small positive number δ such that each solution $x(n)$ of the difference equation (1.99), for which $x(0) \in U_\delta(\mathcal{M})$, is Lyapunov-stable.*

Proof. Let ε_1 be the number considered in the proof of Theorem 1.20. As δ, we take the number such that the relation

$$x(n) \in U_{\varepsilon_1}(\mathcal{M}) \quad \forall n \geq 0. \tag{1.151}$$

holds. This is feasible, because $\mathcal{M}$ is an attractor. Since the mapping SF^{-1} is defined on $U_{\varepsilon_1}(\mathcal{M})$, $SF^{-1}x(n)$ is the solution of system (1.146) by virtue of (1.151). Moreover, since the mapping $SF^{-1} : U_{\varepsilon_1}(\mathcal{M}) \to SF^{-1}U_{\varepsilon_1}(\mathcal{M})$ is a diffeomorphism, it is sufficient to prove the stability of the solution $SF^{-1}x(n)$ of system (1.146) in order to prove that of the solution $x(n)$ of system (1.99).

We now present the solution $SF^{-1}x(n)$ in the form $(\psi(n), \delta(n))$.

Since $(\psi(n), \delta(n)) \in G_{\varepsilon_0}$ for all $n \geq 0$ by virtue of the proof of Theorem 1.20, $\|\delta(0)\| < \varepsilon_0$. We consider an arbitrary positive number $\varepsilon < \varepsilon_1 - \|\delta(0)\|$ and show that there exists the number $\gamma \in (0, \varepsilon/2)$ such that, for any solution $(\psi_1(n), \delta_1(n))$ of system (1.146) such that

$$\|\psi(0) - \psi_1(0)\| + \|\delta(0) - \delta_1(0)\| < \gamma, \tag{1.152}$$

the relation

$$\|\psi(n) - \psi_1(n)\| + \|\delta(n) - \delta_1(n)\| < \varepsilon, \quad n \geq 1 \qquad (1.153)$$

is valid. From the first equation of system (1.146), it follows that

$$\|\psi(n) - \psi_1(n)\| = \|\psi(0) - \psi_1(0)\| < \frac{\varepsilon}{2}, \quad n \geq 1. \qquad (1.154)$$

Let n_0 be a natural number such that

$$\varepsilon_0 q^{n_0} < \frac{\varepsilon}{4}, \qquad (1.155)$$

where q is the same number as that in relation (1.150). We note that

$$\|\delta(n)\| \leq q^n \varepsilon_0,$$
$$\|\delta_1(n)\| \leq q^n \varepsilon_0$$

for all $n \geq 1$, which follows from (1.150) and the second equation of system (1.146), and

$$\|\delta(n) - \delta_1(n)\| \leq \|\delta(n)\| + \|\delta_1(n)\|, \quad n \geq 1.$$

Therefore, relations (1.154) and (1.155) yield the inequality

$$\|\psi(n) - \psi_1(n)\| + \|\delta(n) - \delta_1(n)\| < \varepsilon \qquad (1.156)$$

for all $n \geq n_0$. It follows from the continuous dependence of $C(\psi, \delta)$ on ψ and δ (see (1.146)) that it is possible to select a small number $\gamma > 0$ such that relation (1.152) guarantees the validity of the inequality

$$\|\delta(n) - \delta_1(n)\| < \frac{\varepsilon}{2}$$

for all $n \in \{1, 2, ..., n_0 - 1\}$. Then, according to (1.154),

$$\|\psi(n) - \psi_1(n)\| + \|\delta(n) - \delta_1(n)\| < \varepsilon \qquad (1.157)$$

for all $n \in \{1, 2, ..., n_0 - 1\}$.

Thus, inequalities (1.156) and (1.157) yield inequality (1.153) if estimate (1.152) with a sufficiently small γ holds. The arbitrariness of the choice of $\varepsilon \in (0, \varepsilon_0 - \|\delta(0)\|)$ yields the stability of the solution $SF^{-1}x(n)$ of system (1.146). The theorem is proved. $\qquad \square$

Chapter 2

Invariant tori of difference equations in the space $\mathfrak{M}$

It is well known [105] that the first profound results on the invariant toroidal manifolds of systems of nonlinear mechanics have been obtained in the works by N.M. Krylov and N.N. Bogolyubov [47] and N.N. Bogolyubov [4]. Later on, these results were developed in works by Yu.A. Mitropol'skii that resulted in the creation of the method of integral manifolds of nonlinear mechanics [69]. The significant contribution to the perturbation theory of invariant toroidal manifolds was made by Ya. Kurzweil [50], S.P. Diliberto [19] – [21], J.K. Hale [33; 34], I. Kupka [48], and J.H. Kyner [44]. In the 1960–1970s, J. Moser and R.J. Sacker published a series of works [75] – [79] and [97; 98], which practically completed the creation of this theory.

In 1970, a new method of construction and investigation of the invariant toroidal manifolds of ordinary differential equations defined on m-dimensional tori was suggested [103]. Now, this method is called the Green–Samoilenko function method of solving the problem of the invariant tori of linear expansions of the dynamical systems on tori. In what follows, we will use the abbreviated notation GSF for this function. For three past decades, a great deal of scientific works was devoted to this method and its applications (e.g., [17; 18; 31; 55; 60; 73; 74; 105; 111; 174]). In works [3], [120] – [124], the mentioned method was applied to studying the invariant tori of the countable systems of differential equations defined on tori.

During the past years, only several scientific works apart from those of this books's authors were published, in particular [23], [56], [60], [170] – [172], [175], where this method was employed in investigating the invariant tori of countable systems of difference-differential and difference equations. However, all these works did not solve completely the problem of constructing the theory of invariant tori for systems of the indicated type.

In this chapter, we will create the basis of the theory of invariant toroidal manifolds for linear and nonlinear difference equations in the spaces of bounded numerical sequences, which are defined on finite-dimensional and infinite-dimensional tori and contain independent deviations of a discrete argument. The basic problem consists in investigating the conditions of existence and properties of smoothness of these manifolds.

2.1 Sufficient conditions of existence of a continuous invariant torus

We will consider the system of equations

$$\varphi_{n+1} = \varphi_n + a(\varphi_n, \mu), \quad x_{n+1} = P(\varphi_{n+p}, \mu)x_n + c(\varphi_{n+g+1}, \mu), \qquad (2.1)$$

in which $\varphi = (\varphi^1, \varphi^2, \ldots, \varphi^m) \in R^m$, $x = (x^1, x^2, x^3, \ldots) \in \mathfrak{M}$, where the functions

$$a(\varphi, \mu) = \{a_1(\varphi, \mu), a_2(\varphi, \mu), \ldots, a_m(\varphi, \mu)\},$$

$$c(\varphi, \mu) = \{c_1(\varphi, \mu), c_2(\varphi, \mu), \ldots\}$$

and the infinite matrix $P(\varphi, \mu) = [p_{ij}(\varphi, \mu)]_{i,j=1}^{\infty}$ are real and 2π-periodic in $\varphi^i (i = \overline{1,m})$; $n \in Z$; p and g are the integer-valued parameters, which determine a deviation of the argument; $\mu \in \sigma = (\mu_1, \mu_2) \subset R^1$ is a real parameter. Interpreting φ^i as angular coordinates, we consider that the system of equations (2.1) is defined on an m-dimensional torus $\mathcal{T}_m$.

Below, we consider also that the mapping $\Phi(\varphi, \mu) = \varphi + a(\varphi, \mu) : R^m \to R^m$ is invertible for every $\mu \in \sigma$,

$$\|a(\varphi, \mu)\| \leq A^0, \|c(\varphi, \mu)\| \leq C^0, \|P(\varphi, \mu)\| = \sup_i \sum_{j=1}^{\infty} |p_{ij}(\varphi, \mu)| \leq P^0,$$

and A^0, P^0, C^0 are positive constants independent of $\varphi \in \mathcal{T}_m$, $\mu \in \sigma$.

By $\varphi_n(\varphi, \mu)$, we denote the solution of the first equation in (2.1) such that $\varphi_0(\varphi, \mu) = \varphi \in \mathcal{T}_m$ for every $\mu \in \sigma$.

Definition 2.1. By the invariant torus $\mathcal{T}(p, g, \mu)$ of the system of equation (2.1), we call the set of points $x \in \mathfrak{M}$,

$$x = u(p, g, \mu, \varphi) = (u_1(p, g, \mu, \varphi), u_2(p, g, \mu, \varphi), \ldots), \quad \varphi \in \mathcal{T}_m,$$

if the function $u(p, g, \mu, \varphi)$ is defined for any $\{p, g\} \subset Z, \varphi \in R^m, \mu \in \sigma$, 2π-periodic in $\varphi^i (i = 1, 2, 3, \ldots, m)$, bounded in the norm and for any $\varphi \in \mathcal{T}_m$, $\mu \in \sigma$ satisfies the equality

$$u(p, g, \mu, \varphi_{n+1}(\varphi, \mu)) = P(\varphi_{n+p}(\varphi, \mu), \mu)u(p, g, \mu, \varphi_n(\varphi, \mu))$$

$$+ c(\varphi_{n+g+1}(\varphi, \mu), \mu).$$

We assume that the homogeneous equation

$$x_{n+1} = P(\varphi_{n+p}(\varphi,\mu),\mu)x_n, \quad n \in Z, \tag{2.2}$$

has GSF. This means that there exist the matriciant $\Omega_l^n(p,\varphi,\mu)$ of Eq. (2.2) and the infinite matrix $C(\varphi,\mu)$, which is 2π-periodic in $\varphi^i (i = 1,2,3,\ldots,m)$ and bounded in the norm, such that the function

$$G_0(l,p,\mu,\varphi) = \begin{cases} \Omega_l^0(p,\varphi,\mu)C(\varphi_{l+p}(\varphi,\mu),\mu), & \text{if} \quad l \le 0; \\ \Omega_l^0(p,\varphi,\mu)[C(\varphi_{l+p}(\varphi,\mu),\mu) - E], & \text{if} \quad l > 0 \end{cases}$$

satisfies the inequality $\|G_0(l,p,\mu,\varphi)\| \le M\lambda^{|l|}$ for all $\{p,l\} \subset Z$, $\varphi \in \mathcal{T}_m$, $\mu \in \sigma$, where M and $\lambda < 1$ are positive constants independent of p,l,φ,μ, and E is the infinite identity matrix.

It is easy to see that, in order that the GSF of Eq. (2.2) for arbitrary $p \in Z$ exist, it is sufficient that it exist for $p = 0$. We also note that if the inverse matrix $P^{-1}(\varphi,\mu)$ bounded in the norm exists, the matriciant of Eq. (2.2) is presented in the form

$$\Omega_l^n(\varphi,p,\mu) = \begin{cases} \displaystyle\prod_{i=n+p-1}^{l+p} P(\varphi_i(\varphi,\mu),\mu) & \text{for} \quad n > l; \\[6pt] E & \text{for} \quad n = l; \\[6pt] \displaystyle\prod_{i=n+p}^{l+p-1} P^{-1}(\varphi_i(\varphi,\mu),\mu) & \text{for} \quad n < l. \end{cases}$$

In this case, the matrix $\Omega_l^n(\varphi,p,\mu)$ is invertible, and $(\Omega_l^n(\varphi,p,\mu))^{-1} = \Omega_n^l(\varphi,p,\mu)$.

If Eq. (2.2) has GSF, it is easy to verify that, $\forall\{p,g\} \subset Z$, $\mu \in \sigma$, the system of equations (2.1) has an invariant torus generated by the function

$$u(p,g,\mu,\varphi) = \sum_{l=-\infty}^{\infty} G_0(l,p,\mu,\varphi)c(\varphi_{l+g}(\varphi,\mu),\mu). \tag{2.3}$$

This torus is called continuous or smooth in φ, μ, if its generating function $u(p,g,\mu,\varphi)$ possesses the corresponding property. First, we consider the system of equations of the form (2.1), which is independent of the parameter μ:

$$\varphi_{n+1} = \varphi_n + a(\varphi_n), \quad x_{n+1} = P(\varphi_{n+p})x_n + c(\varphi_{n+g+1}). \tag{2.4}$$

By $C_{Lip}^0(\mathcal{T}_m)$, we denote the set of Lipschitz mappings $f(\varphi)$ defined on $\mathcal{T}_m$. The positive constant K, which ensures the inequality $\|f(\varphi) - f(\bar{\varphi})\| \le K\|\varphi - \bar{\varphi}\|$, will be called the coefficient, with which $f(\varphi)$ enters this set.

Theorem 2.1. *Let the GSF of the equation $x_{n+1} = P(\varphi_{n+p}(\varphi))x_n$, $n \in Z$, exist for $p = 0$, and let the following conditions be satisfied:*

1) for any $\varphi \in \mathcal{T}_m$ and $p = 0$, this equation has the unique solution $x_n = 0$ bounded on the set Z;

2) $\{a(\varphi), P(\varphi), c(\varphi), \Phi^{-1}(\varphi)\} \subset C^0_{Lip}(\mathcal{T}_m)$ with the coefficients $\alpha, \beta, \gamma, \xi$, respectively.

In this case, the invariant torus of the system of equations (2.4) is generated by a function of the form (2.3), which satisfies the Hölder condition

$$\|u(p, g, \varphi) - u(p, g, \bar{\varphi})\| \leq \Delta\|\varphi - \bar{\varphi}\|^{\frac{\nu}{2(\nu+1)}},$$

where $\Delta > 0$ is a constant independent of $\{\varphi, \bar{\varphi}\} \subset \mathcal{T}_m$, ν is any positive real number, which satisfies the inequality $\nu/(\nu + 1) < -\log_{1+\alpha}\lambda$ for $\xi \leq 1$ and the inequality $\nu/(\nu + 1) < \min\{-\log_{1+\alpha}\lambda, -\log_{\xi}\lambda\}$ for $\xi > 1$.

Proof.   Under the conditions of the formulated theorem $\forall\{p, g, l\} \subset Z, \varphi \in \mathcal{T}_m$, the equality

$$\bar{G} = \sum_{k=-\infty}^{\infty} G_0(k, p, \varphi)\bar{P}G_{k-1}(l, p, \bar{\varphi}), \tag{2.5}$$

is valid, where by $\bar{G}$ and $\bar{P}$ stand for the differences $G_0(l, p, \varphi) - G_0(l, p, \bar{\varphi})$ and $P(\varphi_{k+p-1}(\varphi)) - P(\varphi_{k+p-1}(\bar{\varphi}))$, respectively.

It is easy to obtain the estimates

$$\|P(\varphi) - P(\bar{\varphi})\| \leq K\|\varphi - \bar{\varphi}\|^{\frac{\nu}{\nu+1}}, \quad K = (2P^0\beta^\nu)^{\frac{1}{\nu+1}};$$

$$\|c(\varphi) - c(\bar{\varphi})\| \leq K_1\|\varphi - \bar{\varphi}\|^{\frac{\nu}{\nu+1}}, \quad K_1 = (2C^0\gamma^\nu)^{\frac{1}{\nu+1}}, \tag{2.6}$$

which are valid for any positive real number ν.

By the inductive reasoning, we can verify the validity of the inequalities

$$\|\varphi_n(\varphi) - \varphi_n(\bar{\varphi})\| \leq \xi^{-n}\|\varphi - \bar{\varphi}\|, \quad n < 0, n \in Z,$$

$$\|\varphi_n(\varphi) - \varphi_n(\bar{\varphi})\| \leq (1 + \alpha)^n\|\varphi - \bar{\varphi}\|, \quad n \geq 0, n \in Z, \tag{2.7}$$

for all $\{\varphi, \bar{\varphi}\} \in \mathcal{T}_m$.

Relations (2.5) and (2.6) ensure the validity of the estimate

$$\|\bar{G}\| \leq M^2 K(I_1 + I_2), \tag{2.8}$$

where

$$I_1 = \sum_{k=-\infty}^{-p} \lambda^{|k|+|k-1-l|}\|\varphi_{k+p-1}(\varphi) - \varphi_{k+p-1}(\bar{\varphi})\|^{\frac{\nu}{\nu+1}},$$

$$I_2 = \sum_{k=-p+1}^{\infty} \lambda^{|k|+|k-1-l|} \|\varphi_{k+p-1}(\varphi) - \varphi_{k+p-1}(\bar{\varphi})\|^{\frac{\nu}{\nu+1}}.$$

With regard for (2.7), we obtain the inequality

$$I_1 \le \|\varphi - \bar{\varphi}\|^{\frac{\nu}{\nu+1}} \sum_{k=-\infty}^{-p} \lambda^{|k|} \xi^{-\frac{\nu(k+p-1)}{\nu+1}}.$$

Then we consider two cases.

Case p > 0. For $\xi \le 1$ $\forall \nu > 0$, we have $\lambda \xi^{\frac{\nu}{\nu+1}} < 1$ and $I_1 \le K_p^1 \|\varphi - \bar{\varphi}\|^{\frac{\nu}{\nu+1}}$, where $K_p^1 = \lambda^p \xi^{\frac{\nu}{\nu+1}} / (1 - \lambda \xi^{\frac{\nu}{\nu+1}})$. If $\xi > 1$, then I_1 satisfies the same estimate, but ν is chosen from the condition

$$\frac{\nu}{\nu+1} < -\log_\xi \lambda. \tag{2.9}$$

Case p $\le$ 0. For $\xi \le 1$ $\forall \nu > 0$, we have $I_1 \le \Theta_1 \|\varphi - \bar{\varphi}\|^{\frac{\nu}{\nu+1}}$ with $\Theta_1 = S_p^1 + K_p^2$, where, in turn,

$$S_p^1 = \frac{\lambda^{-p} \xi^{\frac{\nu}{\nu+1}} [(\lambda^{-1} \xi^{\frac{\nu}{\nu+1}})^{-p} - 1]}{\lambda^{-1} \xi^{\frac{\nu}{\nu+1}} - 1}, \qquad K_p^2 = \frac{\xi^{\frac{\nu(-p+1)}{\nu+1}}}{1 - \lambda \xi^{\frac{\nu}{\nu+1}}}.$$

If $\xi > 1$, then I_1 satisfies the same estimate, but ν should be consistent with condition (2.9).

Analogously, we write down the estimate for I_2:

$$I_2 \le \|\varphi - \bar{\varphi}\|^{\frac{\nu}{\nu+1}} \sum_{k=-p+1}^{\infty} \lambda^{|k|} (1+\alpha)^{\frac{\nu(k+p-1)}{\nu+1}}.$$

Again, we should consider two cases.

Case p > 0. In this case, the estimate $I_2 \le \Theta_2 \|\varphi - \bar{\varphi}\|^{\frac{\nu}{\nu+1}}$ holds. Here, $\Theta_2 = S_p^2 + \bar{K}_p^1$,

$$S_p^2 = \frac{\lambda^{p-1} [(\lambda^{-1}(1+\alpha)^{\frac{\nu}{\nu+1}})^{p-1} - 1]}{\lambda^{-1}(1+\alpha)^{\frac{\nu}{\nu+1}} - 1}, \qquad \bar{K}_p^1 = \frac{(1+\alpha)^{\frac{\nu(p-1)}{\nu+1}}}{1 - \lambda(1+\alpha)^{\frac{\nu}{\nu+1}}},$$

and $\nu > 0$ satisfies the condition

$$\frac{\nu}{\nu+1} < -\log_{1+\alpha} \lambda. \tag{2.10}$$

Case p $\le$ 0. For I_2, the inequality $I_2 \le \bar{K}_p^2 \|\varphi - \bar{\varphi}\|^{\frac{\nu}{\nu+1}}$ holds. Here,

$$\bar{K}_p^2 = \frac{\lambda^{-p+1}}{1 - \lambda(1+\alpha)^{\frac{\nu}{\nu+1}}},$$

and $\nu > 0$ is chosen from condition (2.10).

In view of (2.8), we have the estimate

$$\|\bar{G}\| \le K^*\|\varphi - \bar{\varphi}\|^{\frac{\nu}{\nu+1}}, \tag{2.11}$$

where

$$K^* = \begin{cases} M^2 K(K_p^1 + \Theta_2), & \text{if } p > 0; \\ M^2 K(\Theta_1 + \bar{K}_p^2), & \text{if } p \le 0. \end{cases}$$

In the cases where $\xi \le 1$ and $\xi > 1$, estimate (2.11) is valid for all $\nu > 0$, which satisfy, respectively, condition (2.10) and the condition

$$\frac{\nu}{\nu+1} < \min\{-\log_{1+\alpha}\lambda, -\log_\xi \lambda\}. \tag{2.12}$$

By $\bar{u}$, we denote the difference $u(p, g, \varphi) - u(p, g, \bar{\varphi})$ and write down the inequality

$$\|\bar{u}\| \le \sum_{l=-\infty}^{\infty} \{C^0\|\bar{G}\| + M\lambda^{|l|}\|c(\varphi_{l+g}(\varphi)) - c(\varphi_{l+g}(\bar{\varphi}))\|\}.$$

By analogy with (2.6), it is easy to obtain the estimate

$$\|\bar{G}\| \le (2MK^*)^{\frac{1}{2}}\|\varphi - \bar{\varphi}\|^{\frac{\nu}{2(\nu+1)}}\lambda^{\frac{|l|}{2}},$$

where $\nu > 0$ is chosen in view of ξ, as was indicated above. Then

$$\|\bar{u}\| \le M^0\|\varphi - \bar{\varphi}\|^{\frac{\nu}{2(\nu+1)}} + M\sum_{l=-\infty}^{\infty}\lambda^{|l|}\|c(\varphi_{l+g}(\varphi)) - c(\varphi_{l+g}(\bar{\varphi}))\|,$$

where

$$M^0 = C^0(2MK^*)^{\frac{1}{2}}\frac{1+\sqrt{\lambda}}{1-\sqrt{\lambda}}.$$

Taking relations (2.6) and (2.7) into account and denoting the expression

$$(2C^0)^{\frac{1}{2}} \times (2C^0\gamma^\nu)^{\frac{1}{2(\nu+1)}}$$

by η, we write down the inequality

$$\|c(\varphi_{l+g}(\varphi)) - c(\varphi_{l+g}(\bar{\varphi}))\| \le \begin{cases} \eta\xi^{-\frac{(l+g)\nu}{2(\nu+1)}}\|\varphi - \bar{\varphi}\|^{\frac{\nu}{2(\nu+1)}}, & \text{if } l+g < 0; \\ \eta(1+\alpha)^{\frac{(l+g)\nu}{2(\nu+1)}}\|\varphi - \bar{\varphi}\|^{\frac{\nu}{2(\nu+1)}}, & \text{if } l+g \ge 0, \end{cases}$$

which leads, in turn, to the estimate

$$\sum_{l=-\infty}^{\infty}\lambda^{|l|}\|c(\varphi_{l+g}(\varphi)) - c(\varphi_{l+g}(\bar{\varphi}))\| \le \eta\{\sum_{l=-\infty}^{-g-1}\lambda^{|l|}\xi^{-\frac{(l+g)\nu}{2(\nu+1)}}$$

$$\times \|\varphi - \bar{\varphi}\|^{\frac{\nu}{2(\nu+1)}} + \sum_{l=-g}^{\infty}\lambda^{|l|}(1+\alpha)^{\frac{(l+g)\nu}{2(\nu+1)}}\|\varphi - \bar{\varphi}\|^{\frac{\nu}{2(\nu+1)}}\}.$$

By I_3 and I_4, we denote, respectively, the first and second sums on the right-hand side of the last inequality. For each of them, we again consider two cases.

Case g > 0. It is easy to verify that $I_3 \leq K_g^1 \|\varphi - \bar{\varphi}\|^{\frac{\nu}{2(\nu+1)}}$, where

$$K_g^1 = \frac{\lambda^{g+1}\xi^{\frac{\nu}{2(\nu+1)}}}{1 - \lambda\xi^{\frac{\nu}{2(\nu+1)}}}$$

and $\nu > 0$ is any real number for $\xi \leq 1$ and satisfies the condition

$$\frac{\nu}{\nu+1} < -2\log_\xi \lambda \tag{2.13}$$

for $\xi > 1$.

Case g $\leq$ 0. By denoting $\Theta_1^0 = S_g^1 + K_g^2$,

$$S_g^1 = \frac{\lambda^{-g-1}\xi^{\frac{\nu}{2(\nu+1)}}\left[(\lambda^{-1}\xi^{\frac{\nu}{2(\nu+1)}})^{-g-1} - 1\right]}{\lambda^{-1}\xi^{\frac{\nu}{2(\nu+1)}} - 1}, \qquad K_g^2 = \frac{\xi^{-\frac{g\nu}{2(\nu+1)}}}{1 - \lambda\xi^{\frac{\nu}{2(\nu+1)}}},$$

we obtain the estimate $I_3 \leq \Theta_1^0 \|\varphi - \bar{\varphi}\|^{\frac{\nu}{2(\nu+1)}}$. In this case, $\nu > 0$ is any real number for $\xi \leq 1$ and satisfies condition (2.13) for $\xi > 1$.

We now consider cases $g > 0$ and $g \leq 0$ for sum I_4.

Case g > 0. The estimate $I_4 \leq \Theta_2^0 \|\varphi - \bar{\varphi}\|^{\frac{\nu}{2(\nu+1)}}$ is valid, where

$$\Theta_2^0 = S_g^2 + \bar{K}_g^1, \qquad S_g^2 = \frac{\lambda^g\left[(\lambda^{-1}(1+\alpha)^{\frac{\nu}{2(\nu+1)}})^g - 1\right]}{\lambda^{-1}(1+\alpha)^{\frac{\nu}{2(\nu+1)}} - 1},$$

$$\bar{K}_g^1 = \frac{(1+\alpha)^{\frac{g\nu}{2(\nu+1)}}}{1 - \lambda(1+\alpha)^{\frac{\nu}{2(\nu+1)}}},$$

and $\nu > 0$ satisfies the condition

$$\frac{\nu}{\nu+1} < -2\log_{1+\alpha} \lambda. \tag{2.14}$$

Case g $\leq$ 0. The inequality $I_4 \leq \bar{K}_g^2 \|\varphi - \bar{\varphi}\|^{\frac{\nu}{2(\nu+1)}}$ holds. Here,

$$\bar{K}_g^2 = \frac{\lambda^{-g}}{1 - \lambda(1+\alpha)^{\frac{\nu}{2(\nu+1)}}},$$

and $\nu > 0$ satisfies condition (2.14).

Now, it is easy to verify the validity of the inequality

$$\sum_{l=-\infty}^{\infty} \lambda^{|l|} \|c(\varphi_{l+g}(\varphi)) - c(\varphi_{l+g}(\bar{\varphi}))\| \leq \bar{K}^* \|\varphi - \bar{\varphi}\|^{\frac{\nu}{2(\nu+1)}}, \tag{2.15}$$

where

$$\bar{K}^* = \begin{cases} \eta(K_g^1 + \Theta_2^0), & \text{if} \quad g > 0; \\ \eta(\bar{K}_g^2 + \Theta_1^0), & \text{if} \quad g \leq 0. \end{cases}$$

In the cases where $\xi \le 1$ and $\xi > 1$, estimate (2.15) is valid for all $\nu > 0$, which satisfy, respectively, condition (2.14) and the inequality

$$\frac{\nu}{\nu + 1} < \min\{-2\log_{1+\alpha}\lambda, -2\log_{\xi}\lambda\}.$$

Now, we can write down the estimate

$$\|\bar{u}\| \le (M^0 + M\bar{K}^*)\|\varphi - \bar{\varphi}\|^{\frac{\nu}{2(\nu+1)}}.$$

For $\xi \le 1$ and $\xi > 1$, it is valid $\forall \nu > 0$, which satisfy, respectively, condition (2.10) and inequality (2.12). Setting $\Delta = M^0 + M\bar{K}^*$, we complete the proof of Theorem 2.1. $\qquad\square$

Example 2.1. To find the value of Δ in a special case where $p > 0, g > 0, \xi \le 1$.

Having constructed the function

$$f(s,r) = \frac{\lambda^s \xi^r}{1 - \lambda\xi^r} + \frac{(1+\alpha)^{(s-1)r}}{1 - \lambda(1+\alpha)^r} + \frac{\lambda^{s-1}[(\lambda^{-1}(1+\alpha)^r)^{s-1} - 1]}{\lambda^{-1}(1+\alpha)^r - 1},$$

it is easy to obtain the equality

$$\Delta = C^0[2M^3(2P^0\beta^\nu)^{\frac{1}{\nu+1}}f(p, \frac{\nu}{\nu+1})]^{\frac{1}{2}}\frac{1+\sqrt{\lambda}}{1-\sqrt{\lambda}}$$
$$+ M(2C^0)^{\frac{1}{2}}(2C^0\gamma^\nu)^{\frac{1}{2(\nu+1)}}f(g+1, \frac{\nu}{2(\nu+1)}),$$

in which $\nu > 0$ is any real number satisfying condition (2.10).

Now, we will come back to the system of equations (2.1). By $\omega(z)$, we denote some continuous scalar function nondecreasing on the segment $[0; \mu_2 - \mu_1]$ and such that $\omega(0) = 0$.

Theorem 2.2. *For $p = 0$, let the GSF of Eq. (2.2) exist, and let, for all $\{\varphi, \bar{\varphi}\} \subset \mathcal{T}_m$ and $\{\mu, \bar{\mu}\} \subset \sigma$:*

1)

$$\|a(\varphi, \mu) - a(\bar{\varphi}, \bar{\mu})\| \le \alpha_1\|\varphi - \bar{\varphi}\| + \alpha_2\omega(|\mu - \bar{\mu}|),$$

$$\|P(\varphi, \mu) - P(\bar{\varphi}, \bar{\mu})\| \le \beta_1\|\varphi - \bar{\varphi}\| + \beta_2\omega(|\mu - \bar{\mu}|),$$

$$\|c(\varphi, \mu) - c(\bar{\varphi}, \bar{\mu})\| \le \gamma_1\|\varphi - \bar{\varphi}\| + \gamma_2\omega(|\mu - \bar{\mu}|),$$

$$\|\Phi^{-1}(\varphi, \mu) - \Phi^{-1}(\bar{\varphi}, \bar{\mu})\| \le \xi_1\|\varphi - \bar{\varphi}\| + \xi_2\omega(|\mu - \bar{\mu}|),$$

where $\alpha_i, \beta_i, \gamma_i,$ and $\xi_i (i = 1, 2)$ are positive constants independent of $\varphi, \mu, \bar{\varphi},$ and $\bar{\mu}$;

2) for $p = 0$, Eq. (2.2) has the unique solution $x_n = 0$ bounded on Z;

3) $\xi_1 > 1$, $\lambda < \min\{\frac{1}{1+\alpha_1}; \frac{1}{\xi_1}\}$.

Then the function $u(p, g, \mu, \varphi)$ generating the invariant torus of the system of equations (2.1) is continuous in the totality of variables φ, μ, and the inequality

$$\|u(p, g, \mu, \varphi) - u(p, g, \bar{\mu}, \bar{\varphi})\|$$

$$\leq M_*\{\|\varphi - \bar{\varphi}\| + \omega(|\mu - \bar{\mu}|)\}^{\frac{1}{2}} + \|\varphi - \bar{\varphi}\| + \omega(|\mu - \bar{\mu}|), \quad (2.16)$$

where M_* is a positive constant independent of $\varphi, \mu, \bar{\varphi}$, and $\bar{\mu}$, holds.

Proof. With the help of the inductive reasoning, it is easy to verify the validity of the following inequalities:

$$\|\varphi_n(\varphi, \mu) - \varphi_n(\bar{\varphi}, \bar{\mu})\| \leq (1 + \alpha_1)^n\{\|\varphi - \bar{\varphi}\| + \frac{\alpha_2}{\alpha_1}\omega(|\mu - \bar{\mu}|)\}, n \geq 0,$$

$$\|\varphi_n(\varphi, \mu) - \varphi_n(\bar{\varphi}, \bar{\mu})\| \leq \xi_1^{-n}\{\|\varphi - \bar{\varphi}\| + \frac{\xi_2}{\xi_1 - 1}\omega(|\mu - \bar{\mu}|)\}, n < 0. \quad (2.17)$$

For convenience, we introduce the notation

$$\omega(|\mu - \bar{\mu}|) = \omega, \quad G_0(l, p, \mu, \varphi) - G_0(l, p, \bar{\mu}, \bar{\varphi}) = \tilde{G},$$

$$c(\varphi_{l+g}(\varphi, \mu), \mu) - c(\varphi_{l+g}(\bar{\varphi}, \bar{\mu}), \bar{\mu}) = \tilde{c},$$

$$u(p, g, \mu, \varphi) - u(p, g, \bar{\mu}, \bar{\varphi}) = \tilde{u},$$

$$\frac{1}{1 - \lambda\xi_1} + \frac{\lambda}{\xi_1 - \lambda} = \lambda_\xi, \quad \frac{1}{1 + \alpha_1 - \lambda} + \frac{1}{1 - \lambda(1 + \alpha_1)} = \lambda_\alpha.$$

Writing down the equality analogous to (2.5), we obtain the estimate of the form (2.8):

$$\|\tilde{G}\| \leq M^2(I_1 + I_2), \quad\quad\quad (2.18)$$

where

$$I_1 = \sum_{k=-\infty}^{-p} \lambda^{|k|}\{\beta_1\xi_1^{-(k+p-1)}\|\varphi - \bar{\varphi}\| + [\frac{\beta_1\xi_2\xi_1^{-(k+p-1)}}{\xi_1 - 1} + \beta_2]\omega\},$$

$$I_2 = \sum_{k=-p+1}^{+\infty} \lambda^{|k|}\{\beta_1(1 + \alpha_1)^{k+p-1}\|\varphi - \bar{\varphi}\| + [\frac{\beta_1(\alpha_2)(1 + \alpha_1)^{k+p-1}}{\alpha_1} + \beta_2]\omega\}.$$

For I_1, we have

$$I_1 \leq \beta_1 \xi_1^{1-p} \|\varphi - \bar{\varphi}\| \sum_{k=-\infty}^{+\infty} \lambda^{|k|} \xi_1^{-k} + \frac{\beta_2(1+\lambda)\omega}{1-\lambda}$$

$$+ \frac{\beta_1 \xi_2 \xi_1^{1-p}}{\xi_1 - 1} \omega \sum_{k=-\infty}^{+\infty} \lambda^{|k|} \xi_1^{-k}.$$

Under condition 3 of Theorem 2.2, this yields the inequality

$$I_1 \leq K^1 \|\varphi - \bar{\varphi}\| + \bar{K}^1 \omega,$$

where

$$K^1 = \beta_1 \xi_1^{1-p} \lambda_\xi, \quad \bar{K}^1 = \beta_2 \frac{1+\lambda}{1-\lambda} + \frac{\beta_1 \xi_2 \xi_1^{1-p}}{\xi_1 - 1} \lambda_\xi.$$

Analogously, we obtain the estimate for I_2,

$$I_2 \leq K^2 \|\varphi - \bar{\varphi}\| + \bar{K}^2 \omega,$$

where

$$K^2 = \beta_1 (1+\alpha_1)^{p-1} \lambda_\alpha, \quad \bar{K}^2 = \beta_2 \frac{1+\lambda}{1-\lambda} + \frac{\beta_1 \alpha_2 (1+\alpha_1)^{p-1}}{\alpha_1} \lambda_\alpha.$$

Using (2.18), we obtain the inequality

$$\|\tilde{G}\| \leq \psi_1 \|\varphi - \bar{\varphi}\| + \psi_2 \omega, \tag{2.19}$$

where $\psi_1 = M^2(K^1 + K^2), \psi_2 = M^2(\bar{K}^1 + \bar{K}^2)$.

In view of (2.19), it is easy to see that

$$\|\tilde{G}\| \leq (2M)^{\frac{1}{2}} \lambda^{\frac{|l|}{2}} (\psi_1 \|\varphi - \bar{\varphi}\| + \psi_2 \omega)^{\frac{1}{2}},$$

which yields the estimate

$$\|\tilde{u}\| \leq M^0 \{\psi_1 \|\varphi - \bar{\varphi}\| + \psi_2 \omega\}^{\frac{1}{2}} + M \sum_{l=-\infty}^{+\infty} \lambda^{|l|} \|\tilde{c}\|, \tag{2.20}$$

where $M^0 = C^0 (2M)^{\frac{1}{2}} \frac{1+\sqrt{\lambda}}{1-\sqrt{\lambda}}$.

The relation

$$\|\tilde{c}\| \leq \begin{cases} \gamma_1 \xi_1^{-(l+g)} (\|\varphi - \bar{\varphi}\| + \frac{\xi_2 \omega}{\xi_1 - 1}) + \gamma_2 \omega & \text{at} \quad l+g < 0; \\[2ex] \gamma_1 (1+\alpha_1)^{l+g} (\|\varphi - \bar{\varphi}\| + \frac{\alpha_2}{\alpha_1} \omega) + \gamma_2 \omega & \text{at} \quad l+g \geq 0 \end{cases}$$

yields the estimate

$$M \sum_{l=-\infty}^{+\infty} \lambda^{|l|} \bar{c} \leq M(I_3 + I_4), \tag{2.21}$$

in which

$$I_3 = \sum_{l=-\infty}^{-g-1} \lambda^{|l|}\{\gamma_1 \xi_1^{-(l+g)}(\|\varphi - \bar{\varphi}\| + \frac{\xi_2 \omega}{\xi_1 - 1}) + \gamma_2 \omega\},$$

$$I_4 = \sum_{l=-g}^{+\infty} \lambda^{|l|}\{\gamma_1 (1 + \alpha_1)^{l+g}(\|\varphi - \bar{\varphi}\| + \frac{\alpha_2}{\alpha_1}\omega) + \gamma_2 \omega\}.$$

For I_3 and I_4, we obtain the estimates

$$I_3 \le K^3\|\varphi - \bar{\varphi}\| + \bar{K}^3 \omega, \quad I_4 \le K^4\|\varphi - \bar{\varphi}\| + \bar{K}^4 \omega, \tag{2.22}$$

where

$$K^3 = \gamma_1 \xi_1^{-g}\lambda_\xi, \quad \bar{K}^3 = \gamma_2 \frac{1+\lambda}{1-\lambda} + \frac{\gamma_1 \xi_2 \xi_1^{-g}}{\xi_1 - 1}\lambda_\xi,$$

$$K^4 = \gamma_1 (1 + \alpha_1)^g \lambda_\alpha, \quad \bar{K}^4 = \gamma_2 \frac{1+\lambda}{1-\lambda} + \frac{\gamma_1 \alpha_2 (1 + \alpha_1)^g}{\alpha_1}\lambda_\alpha.$$

Taking relations $(2.20) - (2.22)$ into account and denoting the expressions $M(K^3 + K^4)$ and $M(\bar{K}^3 + \bar{K}^4)$ by η_1 and η_2, respectively, we obtain the inequality

$$\|\tilde{u}\| \le M^0\{\psi_1\|\varphi - \bar{\varphi}\| + \psi_2 \omega(|\mu - \bar{\mu}|)\}^{\frac{1}{2}} + \eta_1\|\varphi - \bar{\varphi}\| + \eta_2 \omega(|\mu - \bar{\mu}|),$$

which leads to estimate (2.16) if we denote the expression

$$\max\{M^0(\max\{\psi_1, \psi_2\})^{\frac{1}{2}}, \eta_1, \eta_2\}$$

by M_*. The theorem is proved. $\qquad\square$

Remark 2.1. Since a certain moment in the process, where $\|\varphi - \bar{\varphi}\| \to 0$ and $|\mu - \bar{\mu}| \to 0$, estimate (2.16) leads to the inequality

$$\|u(p, g, \mu, \varphi) - u(p, g, \bar{\mu}, \bar{\varphi})\| \le 2M_*\{\|\varphi - \bar{\varphi}\| + \omega(|\mu - \bar{\mu}|)\}^{\frac{1}{2}}. \tag{2.23}$$

We now prove the proposition, which allows us to omit condition 3 in the statement of Theorem 2.2.

Corollary 2.1. *Let all conditions of Theorem 2.2 except the third one be satisfied. Then the function $u(p, g, \mu, \varphi)$, which is continuous in the variables φ and μ, satisfies the inequality*

$$\|u(p, g, \mu, \varphi) - u(p, g, \bar{\mu}, \bar{\varphi})\| \le M^*\{\|\varphi - \bar{\varphi}\| + \omega(|\mu - \bar{\mu}|)\}^{\frac{\nu}{2(\nu+1)}}, \tag{2.24}$$

since a certain moment as $\|\varphi - \bar{\varphi}\| \to 0$ and $|\mu - \bar{\mu}| \to 0$, where M^* is a positive constant independent of $\{\varphi, \bar{\varphi}\} \subset \mathcal{T}_m$ and $\{\mu, \bar{\mu}\} \subset \sigma$, and ν is any positive real number, which satisfies the condition

$$\frac{\nu}{\nu + 1} < \min\{-\log_{\xi_1} \lambda; -\log_{(1+\alpha_1)} \lambda\}, \quad \xi_1 > 1. \tag{2.25}$$

Proof. We set $\xi_1 > 1$ and denote $P(\varphi_n(\varphi,\mu),\mu) - P(\varphi_n(\bar\varphi,\bar\mu),\bar\mu)$ by $\tilde P$. In view of (2.17), we come to the estimate

$$\|\tilde P\| \le \begin{cases} P_1\xi_1^{-\frac{n\nu}{\nu+1}}(\|\varphi-\bar\varphi\|+\omega)^{\frac{\nu}{\nu+1}} & \text{for} \quad n < 0; \\ P_2(1+\alpha_1)^{\frac{n\nu}{\nu+1}}(\|\varphi-\bar\varphi\|+\omega)^{\frac{\nu}{\nu+1}} & \text{for} \quad n \ge 0, \end{cases} \tag{2.26}$$

where ν is any positive real number, and, by P_1 and P_2, we denote the expressions

$$(2P^0)^{\frac{1}{\nu+1}}(\max\{\beta_1; \frac{\beta_1\xi_2}{\xi_1-1} + \beta_2\})^{\frac{\nu}{\nu+1}},$$

$$(2P^0)^{\frac{1}{\nu+1}}(\max\{\beta_1; \frac{\beta_1\alpha_2}{\alpha_1} + \beta_2\})^{\frac{\nu}{\nu+1}},$$

respectively.

Relation (2.26) yields the inequality

$$\|\tilde G\| \le M^2(I_1^0 + I_2^0), \tag{2.27}$$

where

$$I_1^0 = \sum_{k=-\infty}^{-p} \lambda^{|k|} P_1 \xi_1^{-\frac{\nu(k+p-1)}{\nu+1}}(\|\varphi-\bar\varphi\|+\omega)^{\frac{\nu}{\nu+1}},$$

$$I_2^0 = \sum_{k=-p+1}^{+\infty} \lambda^{|k|} P_2(1+\alpha_1)^{\frac{\nu(k+p-1)}{\nu+1}}(\|\varphi-\bar\varphi\|+\omega)^{\frac{\nu}{\nu+1}}.$$

According to condition (2.25), $\lambda\xi_1^{\frac{\nu}{\nu+1}} < 1$ and $\lambda(1+\alpha_1)^{\frac{\nu}{\nu+1}} < 1$, which allows us to write down the inequalities

$$I_i^0 \le \bar P_i(\|\varphi-\bar\varphi\|+\omega)^{\frac{\nu}{\nu+1}} \quad (i=1,2), \tag{2.28}$$

where

$$\bar P_1 = P_1\lambda^{(1)}\xi_1^{-\frac{\nu(p-1)}{\nu+1}}, \quad \bar P_2 = P_2\lambda^{(2)}(1+\alpha_1)^{\frac{\nu(p-1)}{\nu+1}},$$

$$\lambda^{(1)} = (1-\lambda\xi_1^{\frac{\nu}{\nu+1}})^{-1} + \lambda\xi_1^{-\frac{\nu}{\nu+1}}(1-\lambda\xi_1^{-\frac{\nu}{\nu+1}})^{-1},$$

$$\lambda^{(2)} = (1-\lambda(1+\alpha_1)^{-\frac{\nu}{\nu+1}})^{-1} + \lambda(1+\alpha_1)^{\frac{\nu}{\nu+1}}(1-\lambda(1+\alpha_1)^{-\frac{\nu}{\nu+1}})^{-1}.$$

It is easy to see that relations (2.27) and (2.28) yield the continuity of the GSF of Eq. (2.2) by the totality of variables φ and μ since

$$\|\tilde G\| \le M^2(\bar P_1 + \bar P_2)\{\|\varphi-\bar\varphi\|+\omega\}^{\frac{\nu}{\nu+1}}.$$

From the last estimate, it is easy to pass to the following one:

$$\|\tilde{G}\| \leq [2M^3(\bar{P}_1 + \bar{P}_2)]^{\frac{1}{2}} \lambda^{\frac{|l|}{2}} \{\|\varphi - \bar{\varphi}\| + \omega\}^{\frac{\nu}{2(\nu+1)}}. \qquad (2.29)$$

Finally, the relation

$$\|\tilde{c}\| \leq \begin{cases} C_1 \xi_1^{-\frac{\nu(l+g)}{\nu+1}} (\|\varphi - \bar{\varphi}\| + \omega)^{\frac{\nu}{\nu+1}} & \text{for} \quad l+g < 0; \\ C_2(1+\alpha_1)^{\frac{\nu(l+g)}{\nu+1}} (\|\varphi - \bar{\varphi}\| + \omega)^{\frac{\nu}{\nu+1}} & \text{for} \quad l+g \geq 0, \end{cases}$$

where

$$C_1 = (2C^0)^{\frac{1}{\nu+1}} (\max\{\gamma_1; \frac{\gamma_1 \xi_2}{\xi_1 - 1} + \gamma_2\})^{\frac{\nu}{\nu+1}},$$

$$C_2 = (2C^0)^{\frac{1}{\nu+1}} (\max\{\gamma_1; \frac{\gamma_1 \alpha_2}{\alpha_1} + \gamma_2\})^{\frac{\nu}{\nu+1}},$$

holds. This yields

$$M \sum_{l=-\infty}^{+\infty} \lambda^{|l|} \|\tilde{c}\| \leq M(\|\varphi - \bar{\varphi}\| + \omega)^{\frac{\nu}{\nu+1}} \{C_1 \xi_1^{-\frac{\nu g}{\nu+1}} \sum_{l=-\infty}^{-g-1} \lambda^{|l|} \xi_1^{-\frac{\nu l}{\nu+1}}$$

$$+ C_2(1+\alpha_1)^{\frac{\nu g}{\nu+1}} \sum_{l=-g}^{+\infty} \lambda^{|l|} (1+\alpha_1)^{\frac{\nu l}{\nu+1}}\} \leq C_3(\|\varphi - \bar{\varphi}\| + \omega)^{\frac{\nu}{\nu+1}}, \qquad (2.30)$$

where

$$C_3 = M\{C_1 \xi_1^{-\frac{\nu g}{\nu+1}} \lambda^{(1)} + C_2(1+\alpha_1)^{\frac{\nu g}{\nu+1}} \lambda^{(2)}\}.$$

From relations (2.29) and (2.30), we obtain the inequality

$$\|\tilde{u}\| \leq \sum_{l=-\infty}^{+\infty} \{C^0\|\tilde{G}\| + M\lambda^{|l|}\|\tilde{c}\|\} \leq C_4(\|\varphi - \bar{\varphi}\| + \omega(|\mu - \bar{\mu}|))^{\frac{\nu}{2(\nu+1)}}$$

$$+ C_3(\|\varphi - \bar{\varphi}\| + \omega(|\mu - \bar{\mu}|))^{\frac{\nu}{\nu+1}},$$

where

$$C_4 = C^0[2M^3(\bar{P}_1 + \bar{P}_2)]^{\frac{1}{2}} \frac{1+\sqrt{\lambda}}{1-\sqrt{\lambda}}.$$

Denoting $2\max\{C_3, C_4\}$ by M^*, we arrive at estimate (2.24) since a certain moment in the process of $\|\varphi - \bar{\varphi}\| \to 0$, $\|\mu - \bar{\mu}\| \to 0$, which completes the proof of Corollary 2.1. $\qquad \square$

Note that there is no real number $\nu > 0$ such that this estimate would yield inequality (2.23).

If the function $a(\varphi, \mu)$ is independent of μ, i.e., $a(\varphi, \mu) = a(\varphi)$, the conditions of continuity of the function $u(p, g, \mu, \varphi)$ in the parameter μ are essentially simplified.

Corollary 2.2. *Let $a(\varphi, \mu)$ be independent of μ, let there exist the GSF of the equation $x_{n+1} = P(\varphi_{n+p}(\varphi), \mu)x_n$, $n \in Z$ for $p = 0$, and let requirement 2 of Theorem 2.2 be satisfied. Then the inequalities*

$$\sup_{\varphi} \|P(\varphi, \mu) - P(\varphi, \bar{\mu})\| \leq \omega_1(|\mu - \bar{\mu}|),$$

$$\sup_{\varphi} \|c(\varphi, \mu) - c(\varphi, \bar{\mu})\| \leq \omega_2(|\mu - \bar{\mu}|),$$

in which the functions $\omega_1(z)$ and $\omega_2(z)$ have the properties of the function $\omega(z)$, guarantee the continuity of the function $u(p, g, \mu, \varphi)$ in the parameter μ.

Proof.   Indeed, the estimate

$$\|\tilde{G}\| \leq \sum_{k=-\infty}^{+\infty} M^2 \lambda^{|k|} \|P(\varphi_{k+p-1}(\varphi), \mu) - P(\varphi_{k+p-1}(\varphi), \bar{\mu})\|$$

$$\leq M^2 \frac{1+\lambda}{1-\lambda} \omega_1(|\mu - \bar{\mu}|)$$

leads to the inequality

$$\|\tilde{G}\| \leq (M^3 \frac{1+\lambda}{1-\lambda})^{\frac{1}{2}} \lambda^{\frac{|l|}{2}} \omega_1^{\frac{1}{2}}(|\mu - \bar{\mu}|).$$

Analogously,

$$M \sum_{l=-\infty}^{+\infty} \lambda^{|l|} \|c(\varphi_{l+g}(\varphi), \mu) - c(\varphi_{l+g}(\varphi), \bar{\mu})\| \leq M \frac{1+\lambda}{1-\lambda} \omega_2(|\mu - \bar{\mu}|).$$

Then

$$\|u(p, g, \mu, \varphi) - u(p, g, \bar{\mu}, \varphi)\|$$

$$\leq \sum_{l=-\infty}^{+\infty} C^0 (M^3 \frac{1+\lambda}{1-\lambda})^{\frac{1}{2}} \lambda^{\frac{|l|}{2}} \omega_1^{\frac{1}{2}}(|\mu - \bar{\mu}|) + M \frac{1+\lambda}{1-\lambda} \omega_2(|\mu - \bar{\mu}|)$$

$$\leq C^0 \frac{1+\sqrt{\lambda}}{1-\sqrt{\lambda}} (M^3 \frac{1+\lambda}{1-\lambda})^{\frac{1}{2}} \omega_1^{\frac{1}{2}}(|\mu - \bar{\mu}|) + M \frac{1+\lambda}{1-\lambda} \omega_2(|\mu - \bar{\mu}|).$$

The last estimate completes the proof. $\square$

Note that the results obtained in this subsection are preserved in the case where the torus, on which the system of equations (2.1) is considered, is infinite-dimensional, i.e., $\varphi = \{\varphi^1, \varphi^2, \varphi^3, \dots\}$.

2.2 On the differentiability of an invariant torus with respect to the angular variable and the parameter in the coordinatewise meaning

First, we consider the case where the matriciant $\Omega_l^n(p,\varphi,\mu)$ of Eq. (2.2) exists and satisfies the inequality

$$\|\Omega_l^n(0,\varphi,\mu)\| \le M\lambda^{l-n}, \quad l > n, \tag{2.31}$$

where $M > 0$ and $0 < \lambda < 1$ are constants independent of φ and μ. It is evident that, in this case, Eq. (2.2) has GSF of the form

$$G_0(l,p,\mu,\varphi) = \begin{cases} 0, & \text{if } l \le 0; \\ -\Omega_l^0(p,\varphi,\mu), & \text{if } l > 0, \end{cases}$$

and the invariant torus $\mathcal{T}(p,g,\mu)$ of the system of equations (2.1) is defined by the function

$$u(p,g,\mu,\varphi) = -\sum_{l=1}^{\infty} \Omega_l^0(p,\varphi,\mu)c(\varphi_{l+g}(\varphi,\mu),\mu).$$

This torus is covered with a family of trajectories of the system of equations (2.1) defined by the equality

$$x_n = -\sum_{l=n+1}^{\infty} \Omega_l^n(p,\varphi,\mu)c(\varphi_{l+g}(\varphi,\mu),\mu).$$

By $C^1(\varphi)$, $C^1(\mu)$, and $C^1(\varphi,\mu)$, we denote the sets of vector-functions and matrices, which are 2π-periodic in $\varphi^i(i = \overline{1,m})$ and depend on $\varphi \in R^m$ and $\mu \in \sigma$. Moreover, their coordinates and elements are continuously differentiable with respect to $\varphi^i(i = \overline{1,m})$ and $\mu \in \sigma$. The equality $i = \overline{1,m}$ means that the index i takes the values of $1, 2, ..., m$.

We introduce the notation

$$\|P(\varphi,\mu)\|_{\varphi} = \sup_{i} \sum_{j=1}^{\infty} \sup_{\varphi \in \mathcal{T}_m} |p_{ij}(\varphi,\mu)|,$$

$$\|P(\varphi,\mu)\|_{\varphi\mu} = \sup_{i} \sum_{j=1}^{\infty} \sup_{\varphi \in \mathcal{T}_m, \mu \in \sigma} |p_{ij}(\varphi,\mu)|$$

and formulate the following proposition.

Lemma 2.1. *Let, $\forall \mu \in \sigma, \varphi \in \mathcal{T}_m$, there exist the matrix, which is bounded in the norm $\|\cdot\|$ and inverse to $P(\varphi,\mu)$, and let the following conditions hold:*

1) the matriciant of Eq. (2.2) satisfies inequality (2.31), in which the norm $\| \cdot \|$ is replaced by the norm $\| \cdot \|_\varphi$;

2) $\{a(\varphi, \mu), P(\varphi, \mu), c(\varphi, \mu)\} \subset C^1(\varphi)$, and $\|\frac{\partial P(\varphi,\mu)}{\partial \varphi^i}\|_\varphi \leq P^*$, $\|\frac{\partial a(\varphi,\mu)}{\partial \varphi^i}\| \leq A^*$, $\|\frac{\partial c(\varphi,\mu)}{\partial \varphi^i}\| \leq C_*$, where P^*, A^*, and C_* are positive constants independent of $i = \overline{1, m}$ and φ;

3) $\lambda(1 + mA^*) < 1$.

Then, for any $p \geq 0, g \geq -1, \mu \in \sigma$, the function $u(p, g, \mu, \varphi) \in C^1(\varphi)$.

Proof.   First, employing the method of complete mathematical induction, we show that $\varphi_n(\varphi, \mu) \in C^1(\varphi)$ for all $n \geq 0$, and

$$\|\frac{\partial \varphi_n(\varphi, \mu)}{\partial \varphi^i}\| \leq N_n^*,$$

where N_n^* is a positive constant dependent on n. Indeed, for $n = 0$, the statement is proper, and $N_0^* = 1$. We assume that it is valid for $n \leq k \in Z$. We have $\varphi_{k+1}(\varphi, \mu) = \varphi_k(\varphi, \mu) + a(\varphi_k(\varphi, \mu), \mu)$, which yields

$$\frac{\partial \varphi_{k+1}(\varphi, \mu)}{\partial \varphi^i} = \frac{\partial \varphi_k(\varphi, \mu)}{\partial \varphi^i} + \sum_{j=1}^{m} \frac{\partial a(\varphi_k(\varphi, \mu), \mu)}{\partial \varphi_k^j(\varphi, \mu)} \frac{\partial \varphi_k^j(\varphi, \mu)}{\partial \varphi^i}.$$

The last equality leads to the estimate

$$\|\frac{\partial \varphi_{k+1}(\varphi, \mu)}{\partial \varphi^i}\| \leq N_k^* + mA^* N_k^* = N_{k+1}^*.$$

The chain of inequalities

$$\|\frac{\partial \varphi_n(\varphi, \mu)}{\partial \varphi^i}\| \leq \|\frac{\partial \varphi_{n-1}(\varphi, \mu)}{\partial \varphi^i}\| + \sum_{j=1}^{m} \|\frac{\partial a(\varphi_{n-1}(\varphi, \mu), \mu)}{\partial \varphi_{n-1}^j(\varphi, \mu)}\| \| \frac{\partial \varphi_{n-1}^j(\varphi, \mu)}{\partial \varphi^i}|$$

$$\leq \|\frac{\partial \varphi_{n-1}(\varphi, \mu)}{\partial \varphi^i}\|(1 + mA^*) \leq \|\frac{\partial \varphi_{n-2}(\varphi, \mu)}{\partial \varphi^i}\|(1 + mA^*)^2 \leq \ldots$$

$$\leq \|\frac{\partial \varphi}{\partial \varphi^i}\|(1 + mA^*)^n = (1 + mA^*)^n,$$

which are valid for all integers $n \geq 0$, shows that $N_n^* = (1 + mA^*)^n$.

Condition (2.31) states the fact that Eq. (2.2) has only the unique solution bounded on the set Z . This allows us to write down the equality

$$\bar{\Omega} = \sum_{k=1}^{\infty} \Omega_k^0(p, \varphi, \mu) \bar{P} G_{k-1}(l, p, \mu, \bar{\varphi}),$$

in which $\bar{\Omega}$ and $\bar{P}$ stand for the differences

$$\Omega_l^0(p, \varphi, \mu) - \Omega_l^0(p, \bar{\varphi}, \mu) \text{ and } P(\varphi_{k+p-1}(\varphi, \mu), \mu) - P(\varphi_{k+p-1}(\bar{\varphi}, \mu), \mu)$$

respectively. Moreover,

$$G_{k-1}(l,p,\mu,\bar{\varphi}) = \begin{cases} 0, & \text{if } \ l \leq k-1; \\ -\Omega_l^{k-1}(p,\bar{\varphi},\mu), & \text{if } \ l > k-1. \end{cases}$$

Then, for $l > 0$,

$$\bar{\Omega} = -\sum_{k=1}^{l} \Omega_k^0(p,\varphi,\mu)\bar{P}\Omega_l^{k-1}(p,\bar{\varphi},\mu).$$

Assuming that φ and $\bar{\varphi}$ differ only by the i-th coordinates, we divide the last equality by $\varphi^i - \bar{\varphi}^i$. We obtain

$$\frac{\bar{\Omega}}{\varphi^i - \bar{\varphi}^i} = -\sum_{k=1}^{l} \Omega_k^0(p,\varphi,\mu)\frac{\bar{P}}{\varphi^i - \bar{\varphi}^i}\Omega_l^{k-1}(p,\bar{\varphi},\mu). \tag{2.32}$$

We now set $\Omega_s^\gamma(p,\varphi,\mu) = [\omega_{sij}^\gamma(\varphi)]_{i,j=1}^\infty$ and write down the element of the matrix in Eq. (2.32) under the sign of sum. This element belongs to the r-th column and the s-th row:

$$W_{sr}(k) = \sum_{q=1}^{\infty} \omega_{ksq}^0(\varphi)$$

$$\times \sum_{j=1}^{\infty} \frac{p_{qj}(\varphi_{k+p-1}(\varphi,\mu),\mu) - p_{qj}(\varphi_{k+p-1}(\bar{\varphi},\mu),\mu)}{\varphi^i - \bar{\varphi}^i}\omega_{ljr}^{k-1}(\bar{\varphi}).$$

The series

$$S = \sum_{j=1}^{\infty} \sup_{\varphi} \left| \frac{\partial p_{qj}(\varphi_{k+p-1}(\varphi,\mu),\mu)}{\partial \varphi^i} \right| |\omega_{ljr}^{k-1}(\bar{\varphi})|$$

converges uniformly in $\bar{\varphi}^i$. Indeed, since

$$\frac{\partial p_{qj}(\varphi_{k+p-1}(\varphi,\mu),\mu)}{\partial \varphi^i}$$

$$= \sum_{s=1}^{m} \frac{\partial p_{qj}(\varphi_{k+p-1}(\varphi,\mu),\mu)}{\partial \varphi_{k+p-1}^s(\varphi,\mu)} \frac{\partial \varphi_{k+p-1}^s(\varphi,\mu)}{\partial \varphi^i}$$

and $k + p - 1 \geq 0$, we have

$$S \leq \sum_{j=1}^{\infty} MN_{k+p-1}^* \sum_{s=1}^{m} \sup_{\varphi} \left| \frac{\partial p_{qj}(\varphi_{k+p-1}(\varphi,\mu),\mu)}{\partial \varphi_{k+p-1}^s(\varphi,\mu)} \right| \leq mMP^*N_{k+p-1}^*.$$

Then the series $W_{sr}(k)$ also converges uniformly in $\bar{\varphi}^i$:

$$|W_{sr}(k)| \leq \sum_{q=1}^{\infty} |\omega_{ksq}^0(\varphi)|S \leq mM^2P^*N_{k+p-1}^*\lambda^k.$$

It is easy to verify that, under the conditions of Lemma 2.1, $a(\varphi, \mu)$ and $P(\varphi, \mu)$ satisfy the Lipschitz condition with respect to φ. For example,

$$\|P(\varphi, \mu) - P(\bar{\varphi}, \mu)\| \leq \sup_i \sum_{j=1}^{\infty} \sum_{k=1}^{m} \sup_{\varphi} \left| \frac{\partial p_{ij}(\varphi, \mu)}{\partial \varphi^k} \right| \|\varphi^k - \bar{\varphi}^k\|$$

$$\leq \|\varphi - \bar{\varphi}\| \sum_{k=1}^{m} \sup_i \sum_{j=1}^{\infty} \sup_{\varphi} \left| \frac{\partial p_{ij}(\varphi, \mu)}{\partial \varphi^k} \right| \leq mP^* \|\varphi - \bar{\varphi}\|.$$

This yields the continuity of the matrix $\Omega_l^0(p, \varphi, \mu)$ $(l > 0)$ in φ in the norm $\|\cdot\|$.

The last result means that we can pass in (2.32) elementwise to the limit as $\bar{\varphi}^i \to \varphi^i$ and obtain the equality

$$\frac{\partial \Omega_l^0(p, \varphi, \mu)}{\partial \varphi^i} = - \sum_{k=1}^{l} \Omega_k^0(p, \varphi, \mu) \frac{\partial P(\varphi_{k+p-1}(\varphi, \mu), \mu)}{\partial \varphi^i} \Omega_l^{k-1}(p, \varphi, \mu).$$

It is easy to see that the formula for the s-th coordinate of the function $u(p, g, \mu, \varphi)$ has the form

$$u_s = - \sum_{l=1}^{\infty} \sum_{j=1}^{\infty} \omega_{lsj}^0(\varphi) c_j(\varphi_{l+g}(\varphi, \mu), \mu).$$

We will show that the equality

$$\frac{\partial u_s}{\partial \varphi^i} = - \sum_{l=1}^{\infty} \sum_{j=1}^{\infty} \frac{\partial}{\partial \varphi^i} (\omega_{lsj}^0(\varphi) c_j(\varphi_{l+g}(\varphi, \mu), \mu)), \quad i = \overline{1, m}, \tag{2.33}$$

is valid. Since

$$\frac{\partial}{\partial \varphi^i} (\omega_{lsj}^0(\varphi) c_j(\varphi_{l+g}(\varphi, \mu), \mu)) = \frac{\partial \omega_{lsj}^0(\varphi)}{\partial \varphi^i} c_j(\varphi_{l+g}(\varphi, \mu), \mu)$$

$$+ \omega_{lsj}^0(\varphi) \sum_{r=1}^{m} \frac{\partial c_j(\varphi_{l+g}(\varphi, \mu), \mu)}{\partial \varphi_{l+g}^r(\varphi, \mu)} \frac{\partial \varphi_{l+g}^r(\varphi, \mu)}{\partial \varphi^i},$$

we have

$$I(l, \varphi, s) = \sum_{j=1}^{\infty} \left| \frac{\partial}{\partial \varphi^i} (\omega_{lsj}^0(\varphi) c_j(\varphi_{l+g}(\varphi, \mu), \mu)) \right|$$

$$\leq C^0 \sum_{j=1}^{\infty} \sup_{\varphi} \left| \frac{\partial}{\partial \varphi^i} (\omega_{lsj}^0(\varphi)) \right| + \sum_{j=1}^{\infty} \sup_{\varphi} |\omega_{lsj}^0(\varphi)| m C_* (1 + mA^*)^{l+g}$$

$$\leq C^0 M^2 m P^* \sum_{k=1}^{l} \lambda^{l+1} (1 + mA^*)^{k+p-1} + M m C_* \lambda^l (1 + mA^*)^{l+g},$$

which means the uniform convergence of the internal series in (2.33) in φ and $s = 1, 2, 3, \ldots$. Now, we will show that the external series also converges uniformly in φ and $s = 1, 2, 3, \ldots$, i.e., equality (2.33) is valid.

Indeed,

$$\sum_{l=1}^{\infty} I(l, \varphi, s) \le K_1 \sum_{l=1}^{\infty} \sum_{k=1}^{l} \lambda^{l+1} (1 + mA^*)^k + K_2 \sum_{l=1}^{\infty} (\lambda(1 + mA^*))^l,$$

where

$$K_1 = C^0 P^* M^2 m (1 + mA^*)^{p-1}, \quad K_2 = MmC_*(1 + mA^*)^g.$$

Finally, the relations

$$\sum_{l=1}^{\infty} (\lambda(1 + mA^*))^l = \frac{\lambda(1 + mA^*)}{1 - \lambda(1 + mA^*)},$$

$$\sum_{l=1}^{\infty} \sum_{k=1}^{l} \lambda^{l+1}(1 + mA^*)^k \le \sum_{l=1}^{\infty} \sum_{k=1}^{l} (\lambda(1 + mA^*))^l = \sum_{l=1}^{\infty} l(\lambda(1 + mA^*))^l$$

guarantee the uniform convergence of the series $\sum_{l=1}^{\infty} I(l, \varphi, s)$ in φ i $s = 1, 2, 3, \ldots$, because the series $\sum_{l=1}^{\infty} l(\lambda(1 + mA^*))^l$ is convergent. The last result proves equality (2.33), as well as Lemma 2.1, since the membership $u(p, g, \mu, \varphi) \in C^1(\varphi)$ becomes now obvious. $\qquad\square$

In the case where the function $a(\varphi, \mu)$ is independent of φ, the mapping $\Phi(\varphi, \mu)$ is invertible on $R^m \; \forall \mu \in \sigma$, and the inverse mapping $\Phi^{-1}(\varphi, \mu)$ is differentiable with respect to $\varphi^i, i = \overline{1, m}$. This allows us to formulate the following proposition.

Corollary 2.3. *Let, $\forall \mu \in \sigma, \varphi \in \mathcal{T}_m$, there exist the matrix bounded in the norm $\| \cdot \|$ and inverse to $P(\varphi, \mu)$, $\{P(\varphi, \mu), c(\varphi, \mu)\} \subset C^1(\varphi)$, $a(\varphi, \mu) = a(\mu)$, let condition 1 of Lemma 2.1 be satisfied, and let the estimates*

$$\left\| \frac{\partial c(\varphi, \mu)}{\partial \varphi^i} \right\| \le C_*, \quad \left\| \frac{\partial P(\varphi, \mu)}{\partial \varphi^i} \right\|_\varphi \le P^*,$$

where P^* and C_* are positive constants independent of φ, and $i = \overline{1, m}$, hold.

Then the function $u(p, g, \mu, \varphi) \in C^1(\varphi)$ for any $\{p, g\} \subset Z, \mu \in \sigma$.

The proof of Corollary 2.3 is analogous to that of Lemma 2.1 and is not presented here.

We will study the smoothness of the invariant torus of the system of equations (2.1) in the parameter μ.

Theorem 2.3. *Let, $\forall \mu \in \sigma, \varphi \in \mathcal{T}_m$, there exist the matrix bounded in the norm $\| \cdot \|$ and inverse to $P(\varphi, \mu)$, and let the following conditions hold:*

1) the matriciant of Eq. (2.2) satisfies inequality (2.31), in which the norm $\| \cdot \|$ is replaced by the norm $\| \cdot \|_{\varphi \mu}$;

2) $\{a(\varphi, \mu), P(\varphi, \mu), c(\varphi, \mu)\} \subset C^1(\varphi, \mu)$, and

$$\max\{\|\frac{\partial P(\varphi, \mu)}{\partial \varphi^i}\|_{\varphi\mu}, \|\frac{\partial P(\varphi, \mu)}{\partial \mu}\|_{\varphi\mu}\} \leq P^*,$$

$$\max\{\|\frac{\partial a(\varphi, \mu)}{\partial \varphi^i}\|, \|\frac{\partial a(\varphi, \mu)}{\partial \mu}\|\} \leq A^*,$$

$$\max\{\|\frac{\partial c(\varphi, \mu)}{\partial \varphi^i}\|, \|\frac{\partial c(\varphi, \mu)}{\partial \mu}\|\} \leq C_*,$$

where P^*, A^*, and C_* are positive constants independent of $i = \overline{1, m}$, μ, and φ;

3) $\lambda(1 + mA^*) < 1$.

Then, for any $p \geq 0, g \geq -1$, the function $u(p, g, \mu, \varphi) \in C^1(\varphi, \mu)$.

Proof. Proof is performed analogously to that of Lemma 2.1. By the inductive reasoning, it is easy to verify that $\varphi_n(\varphi, \mu) \in C^1(\varphi, \mu)$ for all $n \geq 0$, and $\|\partial \varphi_n(\varphi, \mu)/\partial \mu\| \leq N_n$, where $N_n = [(1 + mA^*)^n - 1]/m$.

Analogously to (2.32) for $l > 0$, we obtain the equality

$$\frac{\tilde{\Omega}}{\mu - \bar{\mu}} = -\sum_{k=1}^{l} \Omega_k^0(p, \varphi, \mu) \frac{\tilde{P}}{\mu - \bar{\mu}} \Omega_l^{k-1}(p, \varphi, \bar{\mu}), \qquad (2.34)$$

where, by $\tilde{\Omega}$ and $\tilde{P}$, we denoted the differences $\Omega_l^0(p, \varphi, \mu) - \Omega_l^0(p, \varphi, \bar{\mu})$ and $P(\varphi_{k+p-1}(\varphi, \mu), \mu) - P(\varphi_{k+p-1}(\varphi, \bar{\mu}), \bar{\mu})$, respectively, $\{\mu, \bar{\mu}\} \subset \sigma$. It is easy to make sure that the conditions of Theorem 2.3 allow us to pass elementwise in equality (2.34) to the limit as $\bar{\mu} \to \mu$ and to obtain the expression for the derivative $\partial \Omega_l^0(p, \varphi, \mu)/\partial \mu$.

It remains to prove that the equality

$$\frac{\partial u_s}{\partial \mu} = -\sum_{l=1}^{\infty} \sum_{j=1}^{\infty} \frac{\partial}{\partial \mu} (\omega_{lsj}^0(\varphi) c_j(\varphi_{l+g}(\varphi, \mu), \mu))$$

is valid.

This proof can be performed analogously to that of equality (2.33). $\square$

Corollary 2.4. *Let, $\forall \mu \in \sigma, \varphi \in \mathcal{T}_m$, there exist the matrix bounded in the norm $\| \cdot \|$ and inverse to $P(\varphi, \mu)$, let condition 1 of Theorem 2.3 be satisfied,*

$a(\varphi, \mu) = \omega$, where ω is a constant vector, and let $\{P(\varphi, \mu), c(\varphi, \mu)\} \subset C^1(\varphi, \mu)$. Moreover, let

$$\max\{\|\frac{\partial P(\varphi, \mu)}{\partial \mu}\|_\mu, \|\frac{\partial P(\varphi, \mu)}{\partial \varphi^i}\|_\varphi, \|\frac{\partial c(\varphi, \mu)}{\partial \mu}\|, \|\frac{\partial c(\varphi, \mu)}{\partial \varphi^i}\|\} \le C,$$

where C is a positive constant independent of φ, μ and $i = \overline{1, m}$.

Then the function $u(p, g, \mu, \varphi) \in C^1(\varphi, \mu)$ for any $\{p, g\} \subset Z$.

Now, we will consider the general case.

Lemma 2.2. *Let,* $\forall \mu \in \sigma, \varphi \in \mathcal{T}_m$, *the following conditions be valid:*

1) for $p = 0$, *Eq.* (2.2) *has the unique solution* $x_n = 0$, *which is bounded on the set* Z, *and GSF such that* $\|G_0(l, 0, \mu, \varphi)\|_\varphi \le M\lambda^{|l|}$;

2) $\{a(\varphi, \mu), P(\varphi, \mu), c(\varphi, \mu), \Phi^{-1}(\varphi, \mu)\} \subset C^1(\varphi)$, *and*

$$\|\frac{\partial P(\varphi, \mu)}{\partial \varphi^i}\|_\varphi \le P^*, \|\frac{\partial a(\varphi, \mu)}{\partial \varphi^i}\| \le A^*,$$

$$\|\frac{\partial c(\varphi, \mu)}{\partial \varphi^i}\| \le C_*, \|\frac{\partial \Phi^{-1}(\varphi, \mu)}{\partial \varphi^i}\| \le \Phi_*,$$

where P^*, A^*, C_*, *and* Φ_* *are positive constants independent of* $i = \overline{1, m}$ *and* φ;

3) $\lambda < \min\{\frac{1}{m\Phi_*}; \frac{1}{1+mA^*}\}$.

Then the function $u(p, g, \mu, \varphi) \in C^1(\varphi)$ *for any* $\{p, g\} \subset Z, \mu \in \sigma$.

Proof. Analogously to (2.32), we write down the equality

$$\frac{\bar{G}}{\varphi^i - \bar{\varphi}^i} = \sum_{k=-\infty}^{\infty} G_0(k, p, \mu, \varphi) \frac{\bar{P}}{\varphi^i - \bar{\varphi}^i} G_{k-1}(l, p, \mu, \bar{\varphi}), \qquad (2.35)$$

in which, by $\bar{G}$, we denote the difference $G_0(l, p, \mu, \varphi) - G_0(l, p, \mu, \bar{\varphi})$.

We set $G_\gamma(s, p, \mu, \varphi) = [g_{sij}^\gamma(\varphi, \mu)]_{i,j=1}^\infty$. For the matrix in equality (2.35) standing under the sign of sum, we present the element, which belongs to the r-th column and s-th row, in the form

$$W_{sr}^k = \sum_{q=1}^\infty g_{ksq}^0(\varphi, \mu)$$

$$\times \sum_{j=1}^\infty \frac{p_{qj}(\varphi_{k+p-1}(\varphi, \mu), \mu) - p_{qj}(\varphi_{k+p-1}(\bar{\varphi}, \mu), \mu)}{\varphi^i - \bar{\varphi}^i} g_{ljr}^{k-1}(\bar{\varphi}, \mu).$$

It is easy to verify that, $\forall n \in Z$, there exists the derivative $\frac{\partial \varphi_n(\varphi,\mu)}{\partial \varphi^i}$ $(i = \overline{1,m})$, and $\|\frac{\partial \varphi_n(\varphi,\mu)}{\partial \varphi^i}\| \leq N_n^*$, where

$$N_n^* = \begin{cases} (1 + mA^*)^n, & \text{if} \quad n \geq 0, i = \overline{1,m}; \\ m^{-n-1}\Phi_*^{-n}, & \text{if} \quad n < 0, i = \overline{1,m}. \end{cases}$$

This leads to the uniform convergence of the series

$$S = \sum_{j=1}^{\infty} \sup_{\varphi} \left| \frac{\partial p_{qj}(\varphi_{k+p-1}(\varphi,\mu),\mu)}{\partial \varphi^i} \right| \|g_{ljr}^{k-1}(\bar{\varphi})\|$$

in $\bar{\varphi}^i$, and $S \leq N_{k+p-1}^* mMP^*$. Then the series W_{sr}^k also converges uniformly in $\bar{\varphi}^i$, and $|W_{sr}^k| \leq N_{k+p-1}^* mM^2 P^* \lambda^{|k|}$.

It is evident that, in order that the series $\sum_{k=-\infty}^{\infty} W_{sr}^k$ be uniformly convergent in $\bar{\varphi}^i$, it is sufficient that the series $S_1 = \sum_{k=-\infty}^{\infty} N_{k+p-1}^* \lambda^{|k|}$ be convergent.

We write down the inequality $S_1 \leq S_1^0 + S_2^0$, where

$$S_1^0 = \sum_{k=-\infty}^{-p} \lambda^{|k|} m^{-k-p} \Phi_*^{-k-p+1}$$

and

$$S_2^0 = \sum_{k=-p+1}^{\infty} \lambda^{|k|} (1 + mA^*)^{k+p-1}.$$

For $p \geq 0$, we have

$$S_1^0 = \frac{\lambda^p \Phi_*}{1 - \lambda m \Phi_*},$$

since $\lambda m \Phi_* < 1$, and

$$S_2^0 = \xi(p) + \frac{\lambda(1 + mA^*)^p}{1 - \lambda(1 + mA^*)},$$

since $\lambda(1 + mA^*) < 1$. By $\xi(p)$, we denoted the sum of a finite number of terms in the geometrical progression $\sum_{k=-p+1}^{0} \lambda^{|k|} (1 + mA^*)^{k+p-1}$.

For $p < 0$:

$$S_2^0 = \frac{\lambda^{-p+1}}{1 - \lambda(1 + mA^*)} \quad \text{and} \quad S_1^0 = \eta(p) + \frac{m^{-p}\Phi_*^{-p+1}}{1 - \lambda m \Phi_*},$$

where $\eta(p) = \sum_{k=1}^{-p} \lambda^k m^{-k-p} \Phi_*^{-k-p+1}$.

Hence,

$$S_1 \leq \begin{cases} \dfrac{\lambda^p \Phi_*}{1 - \lambda m \Phi_*} + \xi(p) + \dfrac{\lambda(1 + mA^*)^p}{1 - \lambda(1 + mA^*)}, & \text{if } p \geq 0; \\[3mm] \dfrac{m^{-p}\Phi_*^{-p+1}}{1 - \lambda m \Phi_*} + \eta(p) + \dfrac{\lambda^{-p+1}}{1 - \lambda(1 + mA^*)}, & \text{if } p < 0, \end{cases}$$

i.e., the series S_1 is convergent. This means that we can elementwise pass to the limit in (2.35) as $\bar{\varphi}^i \to \varphi^i$.

We will show that, for the s-th coordinate of the function $u(p, g, \mu, \varphi)$, the equality

$$\frac{\partial u_s}{\partial \varphi^i} = - \sum_{l=-\infty}^{\infty} \sum_{j=1}^{\infty} \frac{\partial}{\partial \varphi^i}(g_{lsj}^0(\varphi, \mu)c_j(\varphi_{l+g}(\varphi, \mu), \mu)), \quad i = \overline{1, m}, \quad (2.36)$$

is valid. Indeed, it is easy to verify that the internal series in (2.36) converges uniformly in φ and $s = 1, 2, 3, \ldots$, and, $\forall i \in \{1, 2, \ldots, m\}$,

$$\sum_{j=1}^{\infty} |\frac{\partial}{\partial \varphi^i}(g_{lsj}^0(\varphi, \mu)c_j(\varphi_{l+g}(\varphi, \mu), \mu))|$$

$$\leq C^0 \|\frac{\partial G_0(l, p, \mu, \varphi)}{\partial \varphi^i}\|_\varphi + mMC_* N_{l+g}^* \lambda^{|l|};$$

$$\|\frac{\partial G_0(l, p, \mu, \varphi)}{\partial \varphi^i}\|_\varphi$$

$$\leq \sum_{k=-\infty}^{\infty} M\lambda^{|k|} M\lambda^{|l-k+1|} m \|\frac{\partial P(\varphi, \mu)}{\partial \varphi^i}\|_\varphi N_{k+p-1}^* \leq M^2 mP^* S_1.$$

It remains to prove that the series

$$\sum_{l=-\infty}^{\infty} \{C^0 M^2 mP^* \sum_{k=-\infty}^{\infty} \lambda^{|k|+|l-k+1|} N_{k+p-1}^* + mMC_* N_{l+g}^* \lambda^{|l|}\} \quad (2.37)$$

converges $\forall \{p, g\} \subset Z$. To prove the convergence of the series $\sum_{l=-\infty}^{\infty} N_{l+g}^* \lambda^{|l|}$ is not difficult.

The equalities

$$S_2 = \sum_{k=-\infty}^{\infty} \sum_{l=-\infty}^{\infty} \lambda^{|k|+|l-k+1|} N_{k+p-1}^*$$

$$= \sum_{k=-\infty}^{\infty} \lambda^{|k|} N_{k+p-1}^* \{ \sum_{l=-\infty}^{k-1} \lambda^{-l+k-1l} + \sum_{l=k}^{\infty} \lambda^{l-k+1l} \} = \frac{1+\lambda}{1-\lambda} S_1$$

ensure the convergence of the double series S_2. Since all terms of the last series are positive, it is absolutely convergent. In this case, the double series

$$\sum_{l=-\infty}^{\infty} \sum_{k=-\infty}^{\infty} \lambda^{|k|+|l-k+1|} N^*_{k+p-1}$$

converges to the same sum. Hence, series (2.37) converges, which completes the proof of Lemma 2.2. $\square$

Theorem 2.4. *Let, $\forall \mu \in \sigma, \varphi \in \mathcal{T}_m$, the following conditions be satisfied:*

1) for $p = 0$, Eq. (2.2) has the unique solution $x_n = 0$, which is bounded on the set Z, and GSF such that $\|G_0(l, 0, \mu, \varphi)\|_{\varphi\mu} \le M\lambda^{|l|}$;

2) $\{a(\varphi, \mu), P(\varphi, \mu), c(\varphi, \mu), \Phi^{-1}(\varphi, \mu)\} \subset C^1(\varphi, \mu)$, and

$$\max\{\|\frac{\partial P(\varphi, \mu)}{\partial \varphi^i}\|_{\varphi\mu}, \|\frac{\partial P(\varphi, \mu)}{\partial \mu}\|_{\varphi\mu}\} \le P^*;$$

$$\max\{\|\frac{\partial a(\varphi, \mu)}{\partial \varphi^i}\|, \|\frac{\partial a(\varphi, \mu)}{\partial \mu}\|\} \le A^*; \max\{\|\frac{\partial c(\varphi, \mu)}{\partial \varphi^i}\|, \|\frac{\partial c(\varphi, \mu)}{\partial \mu}\|\} \le C_*;$$

$$\max\{\|\frac{\partial \Phi^{-1}(\varphi, \mu)}{\partial \varphi^i}\|, \|\frac{\partial \Phi^{-1}(\varphi, \mu)}{\partial \mu}\|\} \le \Phi_*,$$

where P^*, A^*, C_*, and Φ_* are positive constants independent of $i = \overline{1, m}$, φ, and μ;

3) $\lambda < \min\{\frac{1}{m\Phi_*}; \frac{1}{1+mA^*}\}$.

Then the function $u(p, g, \mu, \varphi) \in C^1(\varphi, \mu)$ for any $\{p, g\} \subset Z$.

Proof.   It is clear that, under the conditions of Theorem 2.4, the inequalities

$$\|a(\varphi, \mu) - a(\bar{\varphi}, \bar{\mu})\| \le mA^*\|\varphi - \bar{\varphi}\| + A^*|\mu - \bar{\mu}|;$$

$$\|P(\varphi, \mu) - P(\bar{\varphi}, \bar{\mu})\| \le mP^*\|\varphi - \bar{\varphi}\| + P^*|\mu - \bar{\mu}|;$$

$$\|\Phi^{-1}(\varphi, \mu) - \Phi^{-1}(\bar{\varphi}, \bar{\mu})\| \le m\Phi_*\|\varphi - \bar{\varphi}\| + \Phi_*|\mu - \bar{\mu}|,$$

which guarantee the continuity of GSF of Eq. (2.2) in the totality of variables φ and μ, hold $\forall\{\mu, \bar{\mu}\} \subset \sigma, \{\varphi, \bar{\varphi}\} \subset \mathcal{T}_m$.

It is easy to verify that the relation

$$\|\frac{\partial \varphi_n(\varphi, \mu)}{\partial \mu}\| \le N_n^0 = \begin{cases} \dfrac{1}{m}(1 + mA^*), & \text{if } n \ge 0; \\[2ex] \dfrac{\Phi_*((m\Phi_*)^{-n} - 1)}{m\Phi_* - 1}, & \text{if } n < 0, m\Phi_* \ne 1; \\[2ex] -n\Phi_* & \text{if } n < 0, m\Phi_* = 1 \end{cases}$$

holds $\forall \mu \in \sigma, \varphi \in \mathcal{T}_m$. We will prove it, for example, for $n < 0, m\Phi_* \neq 1$. In this case, the recurrence inequality

$$\left\| \frac{\partial \varphi_n(\varphi, \mu)}{\partial \mu} \right\| \leq m\Phi_* \left\| \frac{\partial \varphi_{n+1}(\varphi, \mu)}{\partial \mu} \right\| + \Phi_*$$

yields the required estimate:

$$\left\| \frac{\partial \varphi_n(\varphi, \mu)}{\partial \mu} \right\| \leq (m\Phi_*)^2 \left\| \frac{\partial \varphi_{n+2}(\varphi, \mu)}{\partial \mu} \right\| + m\Phi_*^2 + \Phi_* \leq \cdots$$

$$\leq (m\Phi_*)^{-n-1} \left\| \frac{\partial \varphi_{-1}(\varphi, \mu)}{\partial \mu} \right\| + \Phi_* + m\Phi_*^2 + m^2\Phi_*^3 + \cdots + m^{-n-2}(\Phi_*)^{-n-1}$$

$$\leq \sum_{i=0}^{-n-1} m^i \Phi_*^{i+1} = \frac{\Phi_*((m\Phi_*)^{-n} - 1)}{m\Phi_* - 1}.$$

Analogously to (2.35), we write down the equality

$$\frac{\tilde{G}}{\mu - \bar{\mu}} = \sum_{k=-\infty}^{\infty} G_0(k, p, \mu, \varphi) \frac{\tilde{P}}{\mu - \bar{\mu}} G_{k-1}(l, p, \bar{\mu}, \varphi), \qquad (2.38)$$

in which $\tilde{G}$ and $\tilde{P}$ stand for the differences $G_0(l, p, \mu, \varphi) - G_0(l, p, \bar{\mu}, \varphi)$ and $P(\varphi_{k+p-1}(\varphi, \mu), \mu) - P(\varphi_{k+p-1}(\varphi, \bar{\mu}), \bar{\mu})$, respectively.

The series

$$S_* = \sum_{j=1}^{\infty} \frac{p_{qj}(\varphi_{k+p-1}(\varphi, \mu), \mu) - p_{qj}(\varphi_{k+p-1}(\varphi, \bar{\mu}), \bar{\mu})}{\mu - \bar{\mu}} \omega_{ljr}^{k-1}(\varphi, \bar{\mu})$$

converges uniformly in $\bar{\mu} \in \sigma$. Indeed,

$$|S_*| \leq \sum_{j=1}^{\infty} \{ \sum_{r=1}^{m} \sup_{\varphi, \mu} \left| \frac{\partial p_{qj}(\varphi_{k+p-1}(\varphi, \mu), \mu)}{\partial \varphi_{k+p-1}^r(\varphi, \mu)} \right| |N_{k+p-1}^0$$

$$+ \sup_{\varphi, \mu} \left| \frac{\partial p_{qj}(\varphi, \mu)}{\partial \mu} \right| \} |g_{ljr}^{k-1}(\varphi, \bar{\mu})| \leq P^* M(mN_{k+p-1}^0 + 1).$$

Then the series $W_{sr}^0(k) = \sum_{q=1}^{\infty} g_{ksq}^0(\varphi, \mu) S_*$ also uniformly converges in $\bar{\mu} \in \sigma$, and $|W_{sr}^0(k)| \leq M^2 P^* \lambda^{|k|}(mN_{k+p-1}^0 + 1)$.

It is evident that, in order that the series $\sum_{k=-\infty}^{\infty} W_{sr}^0(k)$ be uniformly convergent in $\bar{\mu}$, it is sufficient that the series $S_1^* = \sum_{k=-\infty}^{\infty} N_{k+p-1}^0 \lambda^{|k|}$ be convergent.

Let $m\Phi_* \neq 1$. In this case, $S_1^* \leq Z_1 + Z_2$, where

$$Z_1 = \sum_{k=-\infty}^{-p} \lambda^{|k|} \frac{\Phi_*((m\Phi_*)^{-k-p+1} - 1)}{m\Phi_* - 1};$$

$$Z_2 = \sum_{k=-p+1}^{\infty} \frac{1}{m} \lambda^{|k|} (1 + mA^*)^{k+p-1}.$$

As was shown above,

$$Z_2 = \begin{cases} \dfrac{1}{m}[\xi(p) + \dfrac{\lambda(1+mA^*)^p}{1 - \lambda(1+mA^*)}], & \text{if } p \geq 0; \\[4mm] \dfrac{\lambda^{-p+1}}{m(1 - \lambda(1+mA^*))}, & \text{if } p < 0. \end{cases} \tag{2.39}$$

We denote the fraction $\Phi_*(m\Phi_*)^{-p+1}/|m\Phi_* - 1|$ by $\delta(p)$ and write down the relation

$$Z_1 \leq \delta(p) \sum_{k=-\infty}^{-p} \lambda^{|k|}(m\Phi_*)^{-k} + \frac{\Phi_*}{|m\Phi_* - 1|} \sum_{k=-\infty}^{-p} \lambda^{|k|}.$$

It is obvious that, for $p \geq 0$,

$$Z_1 \leq \delta(p) \frac{(\lambda m\Phi_*)^p}{1 - \lambda m\Phi_*} + \frac{\Phi_*}{|m\Phi_* - 1|} \frac{\lambda^p}{1 - \lambda} = \varepsilon_1(p).$$

For $p < 0$, we have

$$Z_1 \leq \delta(p)\{ \sum_{k=-\infty}^{0} (\lambda m\Phi_*)^{-k}$$

$$+ \sum_{k=1}^{-p} \lambda^k (m\Phi_*)^{-k}\} + \frac{\Phi_*}{|m\Phi_* - 1|}\{ \sum_{k=-\infty}^{0} \lambda^{-k} + \sum_{k=1}^{-p} \lambda^k\}$$

$$= \delta(p)\{\frac{1}{1 - \lambda m\Phi_*} + \delta_1(p)\} + \frac{\Phi_*}{|m\Phi_* - 1|}\{\frac{1}{1 - \lambda} + \delta_2(p)\} = \varepsilon_2(p),$$

where $\delta_1(p)$ and $\delta_2(p)$ stand for the sums $\sum_{k=1}^{-p} \lambda^k(m\Phi_*)^{-k}$ and $\sum_{k=1}^{-p} \lambda^k$, respectively. Thus, we obtain the following relation:

$$S_1^* \leq \begin{cases} \varepsilon_1(p) + \dfrac{1}{m}[\xi(p) + \dfrac{\lambda(1+mA^*)^p}{1 - \lambda(1+mA^*)}], & \text{if } p \geq 0; \\[4mm] \varepsilon_2(p) + \dfrac{\lambda^{-p+1}}{m(1 - \lambda(1+mA^*))}, & \text{if } p < 0. \end{cases}$$

But if $m\Phi_* = 1$, equality (2.39) is preserved, and we have

$$Z_1 = \sum_{k=-\infty}^{-p} \lambda^{|k|}(-k - p + 1)\Phi_* \leq \Phi_* \sum_{k=-\infty}^{-p} \lambda^{|k|}|k| + |p - 1|\Phi_* \sum_{k=-\infty}^{-p} \lambda^{|k|}.$$

In this case,

$$\varepsilon_1(p) = \Phi_* \gamma(p) + |p - 1|\Phi_* \frac{\lambda^p}{1 - \lambda},$$

where $\gamma(p)$ is the sum of the convergent series $\sum_{k=-\infty}^{-p} \lambda^{-k}(-k)(p \geq 0)$, and

$$\varepsilon_2(p) = \Phi_*\{\delta_3 + \delta_4(p)\} + |p - 1|\Phi_*\{\frac{1}{1-\lambda} + \delta_2(p)\},$$

where we denote the sum of the convergent series $\sum_{k=-\infty}^{0} \lambda^{-k}(-k)$ and the finite sum $\sum_{k=1}^{-p} \lambda^k k \, (p < 0)$ by δ_3 and $\delta_4(p)$, respectively. Hence, the series S_1^* converges, which gives the opportunity to pass elementwise to the limit in equality (2.38) as $\bar{\mu} \to \mu$ and to obtain the derivative of the GSF of Eq. (2.2) with respect to μ. In this case, the estimate

$$\|\frac{\partial G_0(l, p, \mu, \varphi)}{\partial \mu}\|_{\varphi\mu} \leq M^2 P^*\{m \sum_{k=-\infty}^{\infty} \lambda^{|k|} N_{k+p-1}^0 + \sum_{k=-\infty}^{\infty} \lambda^{|k|}\}$$

$$\leq M^2 P^*\{m S_1^* + \frac{1+\lambda}{1-\lambda}\}$$

holds. It remains to verify that the equality

$$\frac{\partial u_s}{\partial \mu} = -\sum_{l=-\infty}^{\infty} \sum_{j=1}^{\infty} \frac{\partial}{\partial \mu}(g_{lsj}^0(\varphi, \mu)c_j(\varphi_{l+g}(\varphi, \mu), \mu)), \quad i = \overline{1, m}, \qquad (2.40)$$

is valid. The relations

$$\sum_{j=1}^{\infty} |\frac{\partial}{\partial \mu}(g_{lsj}^0(\varphi, \mu)c_j(\varphi_{l+g}(\varphi, \mu), \mu))| \leq \sum_{j=1}^{\infty} \{\sup_{\varphi,\mu} |\frac{\partial g_{lsj}^0(\varphi, \mu)}{\partial \mu}|$$

$$\times |c_j(\varphi_{l+g}(\varphi, \mu), \mu))| + \sup_{\varphi,\mu} |g_{lsj}^0(\varphi, \mu)|[\sum_{r=1}^{m} |\frac{\partial c_j(\varphi_{l+g}(\varphi, \mu), \mu)}{\partial \varphi_{l+g}^r(\varphi, \mu)}||\frac{\partial \varphi_{l+g}^r(\varphi, \mu)}{\partial \mu}|$$

$$+ |\frac{\partial c_j(\varphi, \mu)}{\partial \mu}|_{\varphi=\varphi_{l+g}(\varphi,\mu)}|]\} \leq C^0\|\frac{\partial G_0(l, p, \mu, \varphi)}{\partial \mu}\|_{\varphi\mu} + M\lambda^{|l|}(mC_* N_{l+g}^0 + C_*)$$

$$\leq C^0 M^2 P^*\{m S_1^* + \frac{1+\lambda}{1-\lambda}\} + M\lambda^{|l|} C_*(m N_{l+g}^0 + 1)$$

testify to the convergence, which is uniform in $\mu \in \sigma$, of the internal series on the right-hand side of (2.40). Finally, we will verify that the series

$$\sum_{l=-\infty}^{\infty} \{C^0\|\frac{\partial G_0(l, p, \mu, \varphi)}{\partial \mu}\|_{\varphi\mu} + M\lambda^{|l|} C_*(m N_{l+g}^0 + 1)\}$$

converges for any integers p and g. The convergence of the series $\sum_{l=-\infty}^{\infty} \lambda^{|l|}(m N_{l+g}^0 + 1)$ is obvious.

It remains to prove the convergence of the series

$$Z_3 = \sum_{l=-\infty}^{\infty} \sum_{k=-\infty}^{\infty} \lambda^{|k|+|l-k+1|} N_{k+p-1}^0$$

and

$$Z_4 = \sum_{l=-\infty}^{\infty} \sum_{k=-\infty}^{\infty} \lambda^{|k|+|l-k+1|}.$$

It follows from the absolute convergence of the double series

$$Z_5 = \sum_{k=-\infty}^{\infty} \lambda^{|k|} N_{k+p-1}^0 \sum_{l=-\infty}^{\infty} \lambda^{|l-k+1|}$$

and

$$Z_6 = \sum_{k=-\infty}^{\infty} \lambda^{|k|} \sum_{l=-\infty}^{\infty} \lambda^{|l-k+1|}.$$

Moreover, $Z_3 = Z_5$, $Z_4 = Z_6$. Theorem 2.4 is proved. $\qquad\square$

We now present the example of a system of equations of the form (2.4), which has a smooth invariant torus in the case where the matrix $\Omega_l^0(p, \varphi)$ satisfies the inequality $\|\Omega_l^0(0, \varphi)\|_\varphi \leq M\lambda^{-l}\ \forall l \leq 0$, where $M > 0$ and $0 < \lambda < 1$ are constant numbers. By constructing the relevant GSF and by performing a reasoning analogous to the proof of Lemma 2.1, we obtain that, in this case, the analog of Corollary 2.3 holds, if condition 1 of Lemma 2.1 in its statement is replaced by the inequality indicated above; moreover, the condition of invertibility of the matrix $P(\varphi)$ can be omitted.

Example 2.2. On a two-dimensional torus, we set a system of equations of the form (2.4), in which $\varphi = \{\varphi^1, \varphi^2\}$, $\varphi_{n+1}^1 = \varphi_n^1 + 1$, $\varphi_{n+1}^2 = \varphi_n^2 + 2$, the matrix

$$P(\varphi) = P(\varphi^1, \varphi^2) = [p_{ij}(\varphi^1, \varphi^2)]_{i,j=1}^{\infty},$$

where

$$p_{i,j} = \frac{1}{i+2}\left(\frac{1}{2}\right)^{j-1} \sin(\varphi^1 + \varphi^2), \quad c_i(\varphi^1, \varphi^2) = \cos(\varphi^1 + \varphi^2)$$

$(i = 1, 2, \dots)$, $p = 2, g = 1$. It is easy to see that $\|P(\varphi)\| \leq 2/3 < 1$ and $\|c\| \leq 1$. Then, for all $l < 0$,

$$\|\Omega_l^0(p, \varphi)\| = \|\prod_{i=-1}^{l} P(\varphi_{i+2}(\varphi))\| \leq \left(\frac{2}{3}\right)^{-l}.$$

In this case, the invariant torus of the input system of equations exists and is generated by the function

$$u(2, 1, \varphi) = c(\varphi_1(\varphi)) + \sum_{l=-\infty}^{-1} \prod_{i=-1}^{l} P(\varphi_{i+2}(\varphi))c(\varphi_{l+1}(\varphi)). \qquad (2.41)$$

In addition, since $\|\frac{\partial P(\varphi)}{\partial \varphi^i}\|_\varphi \leq \frac{2}{3}$ and $\|\frac{\partial c(\varphi)}{\partial \varphi^i}\| \leq 1$, the function $u(2, 1, \varphi) \in C^1(\varphi)$. Note that the last inclusion can be verified directly, taking equality (2.41) into account. It is clear that the result will not change if we choose any integers instead of $p = 2$ and $g = 1$.

2.3 Truncation method in studying the smoothness of invariant tori

1^0. **Coordinatewise differentiability of the first order.** We will consider the truncated system of equations

$$\varphi_{n+1} = \varphi_n + a(\varphi_n, \mu),$$

$$\overset{(s)}{x}_{n+1} = \overset{(s)}{P}(\varphi_{n+p}, \mu)\overset{(s)}{x}_n + \overset{(s)}{c}(\varphi_{n+g+1}, \mu), \quad n \in Z, \qquad (2.42)$$

which corresponds to system (2.1). Here,

$$\overset{(s)}{x} = (x^1, x^2, \ldots, x^s), \quad \overset{(s)}{c}(\varphi, \mu) = \{c_1(\varphi, \mu), c_2(\varphi, \mu), \ldots, c_s(\varphi, \mu)\},$$

$$\overset{(s)}{P}(\varphi) = [p_{ij}(\varphi)]_{i,j=1}^s.$$

By $\overset{(s)}{\Omega}_l^n(p, \varphi, \mu)$, we denote the matriciant of the equation

$$\overset{(s)}{x}_{n+1} = \overset{(s)}{P}(\varphi_{n+p}(\varphi, \mu), \mu)\overset{(s)}{x}_n, \quad n \in Z.$$

Statement 2.1. *Let the following conditions be satisfied $\forall \varphi \in \mathcal{T}_m, \mu \in \sigma$:*

1) the matrix $\overset{(s)}{P}(\varphi, \mu)$ is nondegenerate;

2) $\forall l > n$, $\|\overset{(s)}{\Omega}_l^n(0, \varphi, \mu)\| \le M\lambda^{l-n}$, where the constants $M > 0$, $0 < \lambda < 1$ are independent of φ and μ;

3) $\{a(\varphi, \mu), \overset{(s)}{P}(\varphi, \mu), \overset{(s)}{c}(\varphi, \mu)\} \subset C^1(\varphi, \mu)$, and

$$\max\{\|\frac{\partial \overset{(s)}{P}(\varphi, \mu)}{\partial \varphi^i}\|, \|\frac{\partial \overset{(s)}{P}(\varphi, \mu)}{\partial \mu}\|\} \le P^*, \max\{\|\frac{\partial a(\varphi, \mu)}{\partial \varphi^i}\|, \|\frac{\partial a(\varphi, \mu)}{\partial \mu}\|\} \le A^*,$$

$$\max\{\|\frac{\partial \overset{(s)}{c}(\varphi, \mu)}{\partial \varphi^i}\|, \|\frac{\partial \overset{(s)}{c}(\varphi, \mu)}{\partial \mu}\|\} \le C_*,$$

where P^*, A^*, and C_* are positive constants independent of $i = \overline{1, m}, \mu$, and φ;

4) $\lambda(1 + mA^*) < 1$.

Then, for all integers $p \ge 0$ and $g \ge -1$, the system of equations (2.42) has the invariant torus $\mathcal{T}^{(s)}(p, g, \mu)$, whose generating function

$$\overset{(s)}{u}(p, g, \mu, \varphi) = -\sum_{l=1}^{\infty} \overset{(s)}{\Omega}_l^0(p, \varphi, \mu)\overset{(s)}{c}(\varphi_{l+g}(\varphi, \mu), \mu)$$

belongs to $C^1(\varphi, \mu)$.

Proof of this statement is a simple consequence of Theorem 2.3 from the previous subsection.

In what follows, we agree to supplement, if necessary, a finite matrix to an infinite matrix by zeros, by preserving its notation.

We recall that the sequence of matrices $\{\overset{(n)}{C}(\varphi) = [\overset{(n)}{c}_{ij}(\varphi)]_{i,j=1}^\infty\}_{n=1}^\infty$ is called proper on the set $\mathcal{T}_m$, if, for every $\varphi \in \mathcal{T}_m$, it is bounded, elementwise converges to some matrix $C(\varphi)$, and the series $\sum_{j=1}^\infty |\overset{(n)}{c}_{ij}(\varphi)|$ $(i = 1, 2, \ldots)$ converge uniformly in n.

Lemma 2.3. *Let conditions 1 and 2 of Statement 2.1 be satisfied for all natural s, let M and λ be independent of s, and let $\| \overset{(s)}{P}^{-1}(\varphi, \mu)\| \leq P_1$, where the constant P_1 is independent of $s \in Z^+ = \{1, 2, 3, \ldots\}, \mu \in \sigma, \varphi \in \mathcal{T}_m$.*

If the sequence of matrices $\{\overset{(s)}{P}^{-1}(\varphi, \mu)\}_{s=1}^\infty$ supplemented with zeros to infinite matrices is proper, then

$$\lim_{s \to \infty} \overset{(s)}{u}(p, g, \mu, \varphi) = u(p, g, \mu, \varphi)$$

in the coordinatewise meaning, where $u(p, g, \mu, \varphi)$ is the function generating the invariant torus $\mathcal{T}(p, g, \mu)$ of the system of equations (2.1).

Proof. Let $\lim_{s \to \infty} \overset{(s)}{P}^{-1}(\varphi, \mu) = \Delta(\varphi, \mu)$ elementwise. The validity of the sequence $\{\overset{(s)}{P}^{-1}(\varphi, \mu)\}_{s=1}^\infty$ allows us to show that

$$\Delta(\varphi, \mu)P(\varphi, \mu) = E \quad \text{and} \quad P(\varphi, \mu)\Delta(\varphi, \mu) = E,$$

where E is the identity matrix. Hence, the matrix $P(\varphi, \mu)$ is invertible, and $P^{-1}(\varphi, \mu) = \Delta(\varphi, \mu)$. It is evident that $\|P^{-1}(\varphi, \mu)\| \leq P_1$.

Let $x_n = x_n(p, \varphi, \mu, x_l)$ be a solution of the equation

$$x_{n+1} = P(\varphi_{n+p}(\varphi, \mu), \mu)x_n,$$

and let $\overset{(s)}{x}_n = \overset{(s)}{x}_n(p, \varphi, \mu, \overset{(s)}{x}_l)$ be a solution of the relevant truncated equation. For $n = l - 1$ for the r-th coordinates, we have

$$\lim_{s \to \infty} \overset{(s)}{x}_{l-1}^r = \lim_{s \to \infty} \sum_{i=1}^\infty \overset{(s)}{p}_{ri}^{-1}(\varphi_{i+p-1}(\varphi, \mu), \mu) \overset{(s)}{x}_l^i, \quad r \in Z^+,$$

where $\overset{(s)}{x}{}^i_l = 0$ for $i > s$, $\overset{(s)}{x}{}^i_l = x^i_l$ for $i \leq s$, and $\overset{(s)}{P}{}^{-1}(\varphi, \mu) = [\overset{(s)}{p}{}^{-1}_{ri}(\varphi, \mu)]^s_{r,i=1}$. The last series converges uniformly in $s \in Z^+$. Therefore,

$$\lim_{s \to \infty} \overset{(s)}{x}{}^r_{l-1} = \sum_{i=1}^{\infty} p^{-1}_{ri}(\varphi_{i+p-1}(\varphi, \mu), \mu) x^i_l = x^r_{l-1}.$$

In this case,

$$P^{-1}(\varphi, \mu) = [p^{-1}_{ri}(\varphi, \mu)]^\infty_{r,i=1}, \quad \| \overset{(s)}{x}{}_{l-1} \| \leq \|x_l\| P_1$$

for all $s \in Z^+$.

Analogously, we obtain the equalities

$$\lim_{s \to \infty} \overset{(s)}{x}{}^r_{l-2} = \sum_{i=1}^{\infty} p^{-1}_{ri}(\varphi_{i+p-2}(\varphi, \mu), \mu) x^i_{l-1} = x^r_{l-2}, \quad r \in Z^+.$$

Using the method of complete mathematical induction, we verify that

$$\lim_{s \to \infty} \overset{(s)}{x}{}_n = x_n$$

in the coordinatewise meaning for all integer $n < l$. In this case,

$$\lim_{s \to \infty} \overset{(s)}{\Omega}{}^0_l(p, \varphi, \mu) = \Omega^0_l(p, \varphi, \mu)$$

elementwise, and $\|\Omega^0_l(p, \varphi, \mu)\| \leq M\lambda^l$ for all integer $l > 0$.

Taking into account that

$$u(p, g, \mu, \varphi) = -\sum_{l=1}^{\infty} \Omega^0_l(p, \varphi, \mu) c(\varphi_{l+g}(\varphi, \mu), \mu)$$

and using condition 2 of Statement 2.1, we complete the proof of the Lemma. $\qquad\square$

We say that the function $f(\varphi, \mu)$ together with its derivatives satisfy the Cauchy–Lipschitz condition with a coefficient $\xi > 0$ with respect to φ, μ on the Cartesian product $\mathcal{T}_m \times \sigma$ if

$$\|f(\varphi, \mu) - f(\bar\varphi, \bar\mu)\| \leq \xi(\|\varphi - \bar\varphi\| + |\mu - \bar\mu|),$$

$$\left\| \frac{\partial f(\varphi, \mu)}{\partial \varphi^i} - \frac{\partial f(\bar\varphi, \bar\mu)}{\partial \varphi^i} \right\| \leq \xi(\|\varphi - \bar\varphi\| + |\mu - \bar\mu|),$$

$$\left\| \frac{\partial f(\varphi, \mu)}{\partial \mu} - \frac{\partial f(\bar\varphi, \bar\mu)}{\partial \mu} \right\| \leq \xi(\|\varphi - \bar\varphi\| + |\mu - \bar\mu|),$$

and the constant ξ is independent of $\{\varphi, \bar{\varphi}\} \subset \mathcal{T}_m, \{\mu, \bar{\mu}\} \subset \sigma$ and $i = \overline{1, m}$. In the set $C^1(\varphi, \mu)$, we define a subset $C^1_{Lip}(\varphi, \mu)$, whose elements together with their derivatives satisfy the Cauchy–Lipschitz condition with respect to φ, μ on $\mathcal{T}_m \times \sigma$.

Lemma 2.4. *If* $a(\varphi, \mu) \in C^1_{Lip}(\varphi, \mu)$ *with a coefficient* $\alpha_0 > 0$, *then,* $\forall n \geq 0, \{\varphi, \bar{\varphi}\} \subset \mathcal{T}_m, \{\mu, \bar{\mu}\} \subset \sigma, i = \overline{1, m}$, *the following estimates hold:*

$$\left\| \frac{\partial \varphi_n(\varphi, \mu)}{\partial \mu} \right\| \leq (1 + mA^*)^n,$$

$$\|\varphi_n(\varphi, \mu) - \varphi_n(\bar{\varphi}, \bar{\mu})\| \leq (1 + \alpha_0)^n (\|\varphi - \bar{\varphi}\| + |\mu - \bar{\mu}|), \tag{2.43}$$

$$\left\| \frac{\partial \varphi_n(\varphi, \mu)}{\partial \mu} - \frac{\partial \varphi_n(\bar{\varphi}, \bar{\mu})}{\partial \mu} \right\| \leq 4m(1 + mA^*)^n(1 + \alpha_0)^n(\|\varphi - \bar{\varphi}\| + |\mu - \bar{\mu}|), \tag{2.44}$$

$$\left\| \frac{\partial \varphi_n(\varphi, \mu)}{\partial \varphi^i} - \frac{\partial \varphi_n(\bar{\varphi}, \bar{\mu})}{\partial \varphi^i} \right\| \leq 2m(1 + mA^*)^n(1 + \alpha_0)^n(\|\varphi - \bar{\varphi}\| + |\mu - \bar{\mu}|). \tag{2.45}$$

Proof. The first of the above inequalities is proved in the previous section. By using the method of complete mathematical induction, we will substantiate only inequalities (2.43) and (2.44), since the proof of inequality (2.45) is quite analogous.

It is obvious that the estimate

$$\|\varphi_n(\varphi, \mu) - \varphi_n(\bar{\varphi}, \bar{\mu})\| \leq \|\varphi - \bar{\varphi}\|(1 + \alpha_0)^n + \alpha_0 \sum_{i=0}^{n-1}(1 + \alpha_0)^i |\mu - \bar{\mu}| \tag{2.46}$$

yields (2.43). For $n = 0$, we have $\|\varphi_n(\varphi, \mu) - \varphi_n(\bar{\varphi}, \bar{\mu})\| = \|\varphi - \bar{\varphi}\|$, and inequality (2.46) is proper. Assume that it is proper for all n such that $0 < n \leq k \in Z$. Then

$$\|\varphi_{k+1}(\varphi, \mu) - \varphi_{k+1}(\bar{\varphi}, \bar{\mu})\| \leq \|\varphi_k(\varphi, \mu) - \varphi_k(\bar{\varphi}, \bar{\mu})\| + \|a(\varphi_k(\varphi, \mu), \mu)$$
$$- a(\varphi_k(\bar{\varphi}, \bar{\mu}), \bar{\mu})\| \leq \|\varphi_k(\varphi, \mu) - \varphi_k(\bar{\varphi}, \bar{\mu})\| + \alpha_0(\|\varphi_k(\varphi, \mu) - \varphi_k(\bar{\varphi}, \bar{\mu})\|$$
$$+ |\mu - \bar{\mu}|) \leq (1 + \alpha_0)\|\varphi_k(\varphi, \mu) - \varphi_k(\bar{\varphi}, \bar{\mu})\| + \alpha_0|\mu - \bar{\mu}|$$
$$\leq \|\varphi - \bar{\varphi}\|(1 + \alpha_0)^{k+1} + \{\alpha_0(1 + \alpha_0)\sum_{i=0}^{k-1}(1 + \alpha_0)^i + \alpha_0\}|\mu - \bar{\mu}|$$
$$\leq \|\varphi - \bar{\varphi}\|(1 + \alpha_0)^{k+1} + \alpha_0 \sum_{i=0}^{k}(1 + \alpha_0)^i |\mu - \bar{\mu}|$$
$$= (1 + \alpha_0)^{k+1}(\|\varphi - \bar{\varphi}\| + |\mu - \bar{\mu}|),$$

i.e., estimate (2.46) holds for all $n \geq 0$.

Analogously, the estimate

$$\left\|\frac{\partial \varphi_n(\varphi, \mu)}{\partial \mu} - \frac{\partial \varphi_n(\bar{\varphi}, \bar{\mu})}{\partial \mu}\right\|$$

$$\leq 4m\alpha_0(1 + mA^*)^n \sum_{i=0}^{n-1}(1 + \alpha_0)^i(\|\varphi - \bar{\varphi}\| + |\mu - \bar{\mu}|) \quad (2.47)$$

yields inequality (2.44).

For $n = 0$, we have

$$\left\|\frac{\partial \varphi_n(\varphi, \mu)}{\partial \mu} - \frac{\partial \varphi_n(\bar{\varphi}, \bar{\mu})}{\partial \mu}\right\| = 0,$$

and inequality (2.47) is satisfied. Assume that it holds for all n such that $0 < n \leq k \in Z$. Then

$$\left\|\frac{\partial \varphi_{k+1}(\varphi, \mu)}{\partial \mu} - \frac{\partial \varphi_{k+1}(\bar{\varphi}, \bar{\mu})}{\partial \mu}\right\| \leq \left\|\frac{\partial \varphi_k(\varphi, \mu)}{\partial \mu} - \frac{\partial \varphi_k(\bar{\varphi}, \bar{\mu})}{\partial \mu}\right\|$$

$$+ \sum_{\nu=1}^{m}\left\{\left\|\frac{\partial a(\varphi_k(\varphi, \mu), \mu)}{\partial \varphi_k^\nu(\varphi, \mu)}\right\|\left\|\frac{\partial \varphi_k(\varphi, \mu)}{\partial \mu} - \frac{\partial \varphi_k(\bar{\varphi}, \bar{\mu})}{\partial \mu}\right\|\right.$$

$$+ \left\|\frac{\partial a(\varphi_k(\varphi, \mu), \mu)}{\partial \varphi_k^\nu(\varphi, \mu)} - \frac{\partial a(\varphi_k(\bar{\varphi}, \bar{\mu}), \bar{\mu})}{\partial \varphi_k^\nu(\bar{\varphi}, \bar{\mu})}\right\|\left\|\frac{\partial \varphi_k(\bar{\varphi}, \bar{\mu})}{\partial \mu}\right\|\right\}$$

$$+ \left\|\frac{\partial a(\varphi, \mu)}{\partial \mu}\Big|_{\varphi=\varphi_k(\varphi, \mu)} - \frac{\partial a(\varphi, \bar{\mu})}{\partial \mu}\Big|_{\varphi=\varphi_k(\bar{\varphi}, \bar{\mu})}\right\|$$

$$\leq \left\|\frac{\partial \varphi_k(\varphi, \mu)}{\partial \mu} - \frac{\partial \varphi_k(\bar{\varphi}, \bar{\mu})}{\partial \mu}\right\|(1 + A^*) + m\alpha_0(\|\varphi_k(\varphi, \mu) - \varphi_k(\bar{\varphi}, \bar{\mu})\|$$

$$+ |\mu - \bar{\mu}|)(1 + mA^*)^k + \alpha_0(\|\varphi_k(\varphi, \mu) - \varphi_k(\bar{\varphi}, \bar{\mu})\| + |\mu - \bar{\mu}|)$$

$$\leq 4m\alpha_0(1 + mA^*)^{k+1}\sum_{i=0}^{k-1}(1 + \alpha_0)^i(\|\varphi - \bar{\varphi}\| + |\mu - \bar{\mu}|) + 4m\alpha_0(1 + \alpha_0)^k$$

$$\times (1 + mA^*)^k(\|\varphi - \bar{\varphi}\| + |\mu - \bar{\mu}|)$$

$$\leq 4m\alpha_0(1 + mA^*)^{k+1}\Big[\sum_{i=0}^{k-1}(1 + \alpha_0)^i + (1 + \alpha_0)^k\Big]$$

$$\times (\|\varphi - \bar{\varphi}\| + |\mu - \bar{\mu}|) \leq 4m(1 + mA^*)^{k+1}(1 + \alpha_0)^{k+1}(\|\varphi - \bar{\varphi}\| + |\mu - \bar{\mu}|),$$

i.e., estimate (2.47) holds for all $n \in Z_0^+$, which completes the proof of Lemma 2.4. $\qquad\square$

We now state the following important proposition.

Theorem 2.5. *Let the conditions of Lemma 2.3 be satisfied; let*

$$\{a(\varphi,\mu), P(\varphi,\mu), c(\varphi,\mu)\} \subset C^1_{Lip}(\varphi,\mu)$$

so that $a(\varphi,\mu)$ has a coefficient α_0, and let the inequalities

$$\max\{\|\frac{\partial P(\varphi,\mu)}{\partial\varphi^i}\|, \|\frac{\partial P(\varphi,\mu)}{\partial\mu}\|\} \leq P^*,$$

$$\max\{\|\frac{\partial a(\varphi,\mu)}{\partial\varphi^i}\|, \|\frac{\partial a(\varphi,\mu)}{\partial\mu}\|\} \leq A^*,$$

$$\max\{\|\frac{\partial c(\varphi,\mu)}{\partial\varphi^i}\|, \|\frac{\partial c(\varphi,\mu)}{\partial\mu}\|\} \leq C_*, \tag{2.48}$$

where the positive constants A^*, P^*, C_* are independent of $\varphi \in \mathcal{T}_m, \mu \in \sigma, i = \overline{1,m}$, hold.

If $\lambda(1 + mA^*)(1 + \alpha_0) < 1$ in this case, then, for all integers $p \geq 0, g \geq -1$, the system of equations (2.1) has an invariant torus, whose generating function $u(p,g,\mu,\varphi) \in C^1_{Lip}(\varphi,\mu)$.

Prior to the proof of this theorem, we formulate three auxiliary propositions.

Lemma 2.5. *For all $\{l,s\} \subset Z^+$ under conditions of Theorem 2.5, the estimates*

$$\|\frac{\partial \overset{(s)}{\Omega^0_l}(p,\varphi,\mu)}{\partial\varphi^i}\| \leq K_1\lambda^l(1 + mA^*)^l,$$

$$\|\frac{\partial \overset{(s)}{\Omega^0_l}(p,\varphi,\mu)}{\partial\mu}\| \leq K_2\lambda^l(1 + mA^*)^l,$$

$$\| \overset{(s)}{\Omega^0_l}(p,\varphi,\mu) - \overset{(s)}{\Omega^0_l}(p,\bar\varphi,\bar\mu)\| \leq K_3\lambda^l(1 + mA^*)^l(\|\varphi - \bar\varphi\| + |\mu - \bar\mu|),$$

where the positive constants $K_1, K_2,$ and K_3 are independent of $\{\varphi, \bar\varphi\} \subset \mathcal{T}_m, \{\mu, \bar\mu\} \subset \sigma, s \in Z^+, i = \overline{1,m}$, hold.

Proof. First, we write down the equalities

$$\tilde\Omega = -\sum_{k=1}^{l} \overset{(s)}{\Omega^0_k}(p,\varphi,\mu)\tilde P \overset{(s)}{\Omega^{k-1}_l}(p,\bar\varphi,\bar\mu),$$

$$\frac{\partial \overset{(s)}{\Omega^0_l}(p,\varphi,\mu)}{\partial\varphi^i} = -\sum_{k=1}^{l} \overset{(s)}{\Omega^0_k}(p,\varphi,\mu)\frac{\partial \overset{(s)}{P}(\varphi_{k+p-1}(\varphi,\mu),\mu)}{\partial\varphi^i} \overset{(s)}{\Omega^{k-1}_l}(p,\varphi,\mu),$$

$$\frac{\partial \overset{(s)}{\Omega}{}_l^0(p,\varphi,\mu)}{\partial \mu} = -\sum_{k=1}^{l} \overset{(s)}{\Omega}{}_k^0(p,\varphi,\mu)\frac{\partial \overset{(s)}{P}(\varphi_{k+p-1}(\varphi,\mu),\mu)}{\partial \mu}\,\overset{(s)}{\Omega}{}_l^{k-1}(p,\varphi,\mu),$$

where

$$\tilde{\Omega} = \overset{(s)}{\Omega}{}_l^0(p,\varphi,\mu) - \overset{(s)}{\Omega}{}_l^0(p,\bar{\varphi},\bar{\mu}),$$

$$\tilde{P} = \overset{(s)}{P}(\varphi_{k+p-1}(\varphi,\mu),\mu) - \overset{(s)}{P}(\varphi_{k+p-1}(\bar{\varphi},\bar{\mu}),\bar{\mu}), \quad k+p-1 \geq 0.$$

This yields

$$\left\|\frac{\partial \overset{(s)}{\Omega}{}_l^0(p,\varphi,\mu)}{\partial \varphi^i}\right\|$$

$$\leq \sum_{k=1}^{l} M^2\lambda^{l+1}\sum_{\nu=1}^{m}\left\|\frac{\partial P(\varphi_{k+p-1}(\varphi,\mu),\mu)}{\partial \varphi_{k+p-1}^{\nu}(\varphi,\mu)}\right\|\left\|\frac{\partial \varphi_{k+p-1}(\varphi,\mu)}{\partial \varphi^i}\right\|$$

$$\leq M^2\lambda^l\sum_{k=1}^{l}\sum_{\nu=1}^{m}P^*(1+mA^*)^{k+p-1}$$

$$\leq M^2 P^*(1+mA^*)^p\lambda^l\frac{(1+mA^*)^l - 1}{A^*}$$

$$\leq K_1\lambda^l(1+mA^*)^l, \quad K_1 = \frac{1}{A^*}M^2 P^*(1+mA^*)^p.$$

Analogously,

$$\left\|\frac{\partial \overset{(s)}{\Omega}{}_l^0(p,\varphi,\mu)}{\partial \mu}\right\|$$

$$\leq \sum_{k=1}^{l} M^2\lambda^l\Bigl\{\sum_{\nu=1}^{m}\left\|\frac{\partial P(\varphi_{k+p-1}(\varphi,\mu),\mu)}{\partial \varphi_{k+p-1}^{\nu}(\varphi,\mu)}\right\|\left\|\frac{\partial \varphi_{k+p-1}(\varphi,\mu)}{\partial \mu}\right\|$$

$$+\left\|\frac{\partial P(\varphi,\mu)}{\partial \mu}\right|_{\varphi=\varphi_{k+p-1}(\varphi,\mu)}\right\|\Bigr\} \leq K_1\lambda^l(1+mA^*)^l + M^2\lambda^l P^*$$

$$\leq K_2\lambda^l(1+mA^*)^l, \quad K_2 = K_1 + M^2 P^*.$$

Finally, we have

$$\|\tilde{\Omega}\| \leq \sum_{k=1}^{l} M^2\lambda^{l+1}\|\tilde{P}\|$$

$$\leq M^2\lambda^l\sum_{k=1}^{l}\beta_0(\|\varphi_{k+p-1}(\varphi,\mu) - \varphi_{k+p-1}(\bar{\varphi},\bar{\mu})\| + |\mu - \bar{\mu}|)$$

$$\leq \beta_0 M^2 \lambda^l \sum_{k=1}^{l} \{(1+\alpha_0)^{k+p-1}(\|\varphi - \bar{\varphi}\| + |\mu - \bar{\mu}|) + |\mu - \bar{\mu}|\}$$

$$\leq K_3 \lambda^l (1+\alpha_0)^l (\|\varphi - \bar{\varphi}\| + |\mu - \bar{\mu}|),$$

where $K_3 = 2\beta_0 M^2 (1+\alpha_0)^p / \alpha_0$, β_0 is a coefficient, with which the matrix $P(\varphi, \mu)$ enters the set $C^1_{Lip}(\varphi, \mu)$. $\qquad \square$

Lemma 2.6. *Under conditions of Theorem 2.5 for all $n \geq 0$, the estimates*

$$\left\| \frac{\partial \overset{(s)}{P}(\varphi_n(\varphi, \mu), \mu)}{\partial \varphi^i} - \frac{\partial \overset{(s)}{P}(\varphi_n(\bar{\varphi}, \bar{\mu}), \bar{\mu})}{\partial \varphi^i} \right\|$$

$$\leq K_4 (1+\alpha_0)^n (1+mA^*)^n (\|\varphi - \bar{\varphi}\| + |\mu - \bar{\mu}|),$$

$$\left\| \frac{\partial \overset{(s)}{P}(\varphi_n(\varphi, \mu), \mu)}{\partial \mu} - \frac{\partial \overset{(s)}{P}(\varphi_n(\bar{\varphi}, \bar{\mu}), \bar{\mu})}{\partial \mu} \right\|$$

$$\leq K_5 (1+\alpha_0)^n (1+mA^*)^n (\|\varphi - \bar{\varphi}\| + |\mu - \bar{\mu}|),$$

$$\left\| \frac{\partial \overset{(s)}{c}(\varphi_n(\varphi, \mu), \mu)}{\partial \varphi^i} - \frac{\partial \overset{(s)}{c}(\varphi_n(\bar{\varphi}, \bar{\mu}), \bar{\mu})}{\partial \varphi^i} \right\|$$

$$\leq K_6 (1+\alpha_0)^n (1+mA^*)^n (\|\varphi - \bar{\varphi}\| + |\mu - \bar{\mu}|),$$

$$\left\| \frac{\partial \overset{(s)}{c}(\varphi_n(\varphi, \mu), \mu)}{\partial \mu} - \frac{\partial \overset{(s)}{c}(\varphi_n(\bar{\varphi}, \bar{\mu}), \bar{\mu})}{\partial \mu} \right\|$$

$$\leq K_7 (1+\alpha_0)^n (1+mA^*)^n (\|\varphi - \bar{\varphi}\| + |\mu - \bar{\mu}|),$$

where the positive constants K_4–K_7 are independent of $\{\varphi, \bar{\varphi}\} \subset \mathcal{T}_m$, $\{\mu, \bar{\mu}\} \subset \sigma, s \in Z^+, i = \overline{1, m}$, hold.

Proof. The following chain of inequalities is satisfied:

$$\left\| \frac{\partial \overset{(s)}{P}(\varphi_n(\varphi, \mu), \mu)}{\partial \varphi^i} - \frac{\partial \overset{(s)}{P}(\varphi_n(\bar{\varphi}, \bar{\mu}), \bar{\mu})}{\partial \varphi^i} \right\|$$

$$\leq \sum_{\nu=1}^{m} \left\| \frac{\partial \overset{(s)}{P}(\varphi_n(\varphi, \mu), \mu)}{\partial \varphi_n^\nu(\varphi, \mu)} \frac{\partial \varphi_n^\nu(\varphi, \mu)}{\partial \varphi^i} - \frac{\partial \overset{(s)}{P}(\varphi_n(\bar{\varphi}, \bar{\mu}), \bar{\mu})}{\partial \varphi_n^\nu(\bar{\varphi}, \bar{\mu})} \frac{\partial \varphi_n^\nu(\bar{\varphi}, \bar{\mu})}{\partial \varphi^i} \right\|$$

$$\leq \sum_{\nu=1}^{m} \left\{ \left\| \frac{\partial P(\varphi_n(\varphi, \mu), \mu)}{\partial \varphi_n^\nu(\varphi, \mu)} \right\| \left\| \frac{\partial \varphi_n(\varphi, \mu)}{\partial \varphi^i} - \frac{\partial \varphi_n(\bar{\varphi}, \bar{\mu})}{\partial \varphi^i} \right\| \right.$$

$$+ \left\| \frac{\partial P(\varphi_n(\varphi,\mu),\mu)}{\partial \varphi_n^\nu(\varphi,\mu)} - \frac{\partial P(\varphi_n(\bar\varphi,\bar\mu),\bar\mu)}{\partial \varphi_n^\nu(\bar\varphi,\bar\mu)} \right\| \left\| \frac{\partial \varphi_n(\bar\varphi,\bar\mu)}{\partial \varphi^i} \right\| \}$$

$$\leq \sum_{\nu=1}^{m} \{ P^* 2m(1+mA^*)^n (1+\alpha_0)^n (\|\varphi-\bar\varphi\|+|\mu-\bar\mu|)$$

$$+ \beta_0[(1+\alpha_0)^n(\|\varphi-\bar\varphi\|+|\mu-\bar\mu|)+|\mu-\bar\mu|](1+mA^*)^n \}$$

$$\leq K_4(1+mA^*)^n(1+\alpha_0)^n(\|\varphi-\bar\varphi\|+|\mu-\bar\mu|).$$

Here, $K_4 = 2m(mP^* + \beta_0)$.

Analogously,

$$\left\| \frac{\partial \overset{(s)}{P}(\varphi_n(\varphi,\mu),\mu)}{\partial \mu} - \frac{\partial \overset{(s)}{P}(\varphi_n(\bar\varphi,\bar\mu),\bar\mu)}{\partial \mu} \right\|$$

$$\leq \sum_{\nu=1}^{m} \left\| \frac{\partial \overset{(s)}{P}(\varphi_n(\varphi,\mu),\mu)}{\partial \varphi_n^\nu(\varphi,\mu)} \frac{\partial \varphi_n^\nu(\varphi,\mu)}{\partial \mu} - \frac{\partial \overset{(s)}{P}(\varphi_n(\bar\varphi,\bar\mu),\bar\mu)}{\partial \varphi_n^\nu(\bar\varphi,\bar\mu)} \frac{\partial \varphi_n^\nu(\bar\varphi,\bar\mu)}{\partial \mu} \right\|$$

$$+ \left\| \frac{\partial \overset{(s)}{P}(\varphi,\mu)}{\partial \mu}\Big|_{\varphi=\varphi_n(\varphi,\mu)} - \frac{\partial \overset{(s)}{P}(\varphi,\bar\mu)}{\partial \mu}\Big|_{\varphi=\varphi_n(\bar\varphi,\bar\mu)} \right\|$$

$$\leq 2K_4(1+mA^*)^n(1+\alpha_0)^n(\|\varphi-\bar\varphi\|+|\mu-\bar\mu|)$$

$$+ \beta_0[(1+\alpha_0)^n(\|\varphi-\bar\varphi\|+|\mu-\bar\mu|)$$

$$+ |\mu-\bar\mu|] \leq K_5(1+mA^*)^n(1+\alpha_0)^n(\|\varphi-\bar\varphi\|+|\mu-\bar\mu|),$$

where $K_5 = 2(K_4 + \beta_0)$.

Two last estimates in the statement of Lemma 2.6 are proved analogously to two first ones with the replacement of the constant β_0 by the constant γ_0, where γ_0 is a coefficient, with which the function $c(\varphi,\mu)$ enters the set $C^1_{Lip}(\varphi,\mu)$, i.e., $K_6 = 2m(mC_* + \gamma_0)$ and $K_7 = 2(K_6 + \gamma_0)$, which completes the proof. $\qquad\square$

Lemma 2.7. *Under conditions of Lemma 2.5, the estimates*

$$\left\| \frac{\partial \overset{(s)}{\Omega_l^0}(p,\varphi,\mu)}{\partial \varphi^i} - \frac{\partial \overset{(s)}{\Omega_l^0}(p,\bar\varphi,\bar\mu)}{\partial \varphi^i} \right\|$$

$$\leq K_8 \lambda^l (1+mA^*)^l (1+\alpha_0)^l (\|\varphi-\bar\varphi\|+|\mu-\bar\mu|),$$

$$\left\| \frac{\partial \overset{(s)}{\Omega_l^0}(p,\varphi,\mu)}{\partial \mu} - \frac{\partial \overset{(s)}{\Omega_l^0}(p,\bar\varphi,\bar\mu)}{\partial \mu} \right\|$$

$$\leq K_9 \lambda^l (1+mA^*)^l (1+\alpha_0)^l (\|\varphi-\bar\varphi\|+|\mu-\bar\mu|),$$

where the positive constants K_8, K_9 are independent of $\{\varphi,\bar\varphi\} \subset \mathcal{T}_m$, $\{\mu,\bar\mu\} \subset \sigma, s \in Z^+, i = \overline{1,m}$, hold.

Proof.   The following inequalities hold:

$$\left\| \frac{\partial \overset{(s)}{\Omega}{}_l^0(p,\varphi,\mu)}{\partial \varphi^i} - \frac{\partial \overset{(s)}{\Omega}{}_l^0(p,\bar\varphi,\bar\mu)}{\partial \varphi^i} \right\|$$

$$\leq \sum_{k=1}^l \left\| \overset{(s)}{\Omega}{}_k^0(p,\varphi,\mu)\frac{\partial \overset{(s)}{P}(\varphi_{k+p-1}(\varphi,\mu),\mu)}{\partial \varphi^i} \right.$$

$$\times \overset{(s)}{\Omega}{}_l^{k-1}(p,\varphi,\mu) - \overset{(s)}{\Omega}{}_k^0(p,\bar\varphi,\bar\mu)\frac{\partial \overset{(s)}{P}(\varphi_{k+p-1}(\bar\varphi,\bar\mu),\bar\mu)}{\partial \varphi^i} \overset{(s)}{\Omega}{}_l^{k-1}(p,\bar\varphi,\bar\mu) \Big\|$$

$$\leq \sum_{k=1}^l \left\{ \left\| \overset{(s)}{\Omega}{}_k^0(p,\varphi,\mu) \right\| \Big[\sum_{\nu=1}^m \left\| \frac{\partial \overset{(s)}{P}(\varphi_{k+p-1}(\varphi,\mu),\mu)}{\partial \varphi_{k+p-1}^\nu(\varphi,\mu)} \right\| \left\| \frac{\partial \varphi_{k+p-1}(\varphi,\mu)}{\partial \varphi^i} \right\| \right.$$

$$\times \left\| \overset{(s)}{\Omega}{}_l^{k-1}(p,\varphi,\mu) - \overset{(s)}{\Omega}{}_l^{k-1}(p,\bar\varphi,\bar\mu) \right\| + \left\| \frac{\partial \overset{(s)}{P}(\varphi_{k+p-1}(\varphi,\mu),\mu)}{\partial \varphi^i} \right.$$

$$\left. - \frac{\partial \overset{(s)}{P}(\varphi_{k+p-1}(\bar\varphi,\bar\mu),\bar\mu)}{\partial \varphi^i} \right\| \left\| \overset{(s)}{\Omega}{}_l^{k-1}(p,\bar\varphi,\bar\mu) \right\| \Big]$$

$$+ \left\| \overset{(s)}{\Omega}{}_k^0(p,\varphi,\mu) - \overset{(s)}{\Omega}{}_k^0(p,\bar\varphi,\bar\mu) \right\|$$

$$\times \sum_{\nu=1}^m \left\| \frac{\partial \overset{(s)}{P}(\varphi_{k+p-1}(\bar\varphi,\bar\mu),\bar\mu)}{\partial \varphi_{k+p-1}^\nu(\bar\varphi,\bar\mu)} \right\| \left\| \frac{\partial \varphi_{k+p-1}(\bar\varphi,\bar\mu)}{\partial \varphi^i} \right\| \left\| \overset{(s)}{\Omega}{}_l^{k-1}(p,\bar\varphi,\bar\mu) \right\| \right\}$$

$$\leq \sum_{k=1}^l \left\{ M\lambda^k [mP^*(1+mA^*)^{k+p-1}K_3\lambda^{l-k+1}(1+\alpha_0)^{l-k+1} \right.$$

$$\times ((1+\alpha_0)^{k-1}(\|\varphi-\bar\varphi\| + |\mu-\bar\mu|) + |\mu-\bar\mu|)$$

$$+ K_4(1+\alpha_0)^{k+p-1}(1+mA^*)^{k+p-1}(\|\varphi-\bar\varphi\| + |\mu-\bar\mu|)M\lambda^{l-k+1}]$$

$$+ K_3\lambda^k(1+\alpha_0)^k(\|\varphi-\bar\varphi\| + |\mu-\bar\mu|)mP^*(1+mA^*)^{k+p-1}M\lambda^{l-k+1} \Big\}$$

$$\leq \sum_{k=1}^l \left\{ 2mMP^*K_3\lambda^{l+1}(1+mA^*)^{k+p-1}(1+\alpha_0)^l \right.$$

$$+ K_4 M^2 \lambda^{l+1}(1+\alpha_0)^{k+p-1}(1+mA^*)^{k+p-1}$$

$$+ K_3 mMP^*\lambda^{l+1}(1+\alpha_0)^k(1+mA^*)^{k+p-1} \Big\}(\|\varphi-\bar\varphi\| + |\mu-\bar\mu|)$$

$$\leq \left\{ \lambda^l(1+\alpha_0)^l(1+mA^*)^{p-1}2mMP^*K_3\sum_{k=1}^l(1+mA^*)^k \right.$$

$$+ K_4 M^2 \lambda^l (1+\alpha_0)^{p-1}(1+mA^*)^{p-1} \sum_{k=1}^{l}(1+mA^*)^k(1+\alpha_0)^k$$

$$+ K_3 m M P^* \lambda^l (1+mA^*)^{p-1}$$

$$\times \sum_{k=1}^{l}(1+mA^*)^k(1+\alpha_0)^k\}(\|\varphi - \bar\varphi\| + |\mu - \bar\mu|)$$

$$\leq K_8 \lambda^l (1+\alpha_0)^l (1+mA^*)^l (\|\varphi - \bar\varphi\| + |\mu - \bar\mu|).$$

Here,

$$K_8 = \max\{K_1^0, K_2^0, K_3^0\}, \quad K_1^0 = \frac{2}{A^*} M P^* K_3 (1+mA^*)^p,$$

$$K_2^0 = \frac{M^2 K_4 (1+mA^*)^p (1+\alpha_0)^p}{(1+\alpha_0)(1+mA^*)-1},$$

$$K_3^0 = \frac{mM K_3 P^* (1+mA^*)^p (1+\alpha_0)^p}{(1+\alpha_0)(1+mA^*)-1}.$$

Analogously, we have

$$\left\| \frac{\partial \overset{(s)}{\Omega}{}_l^0(p,\varphi,\mu)}{\partial \mu} - \frac{\partial \overset{(s)}{\Omega}{}_l^0(p,\bar\varphi,\bar\mu)}{\partial \mu} \right\|$$

$$\leq \sum_{k=1}^{l} \left\| \overset{(s)}{\Omega}{}_k^0(p,\varphi,\mu) \frac{\partial \overset{(s)}{P}(\varphi_{k+p-1}(\varphi,\mu),\mu)}{\partial \mu} \times \overset{(s)}{\Omega}{}_l^{k-1}(p,\varphi,\mu) \right.$$

$$\left. - \overset{(s)}{\Omega}{}_k^0(p,\bar\varphi,\bar\mu) \frac{\partial \overset{(s)}{P}(\varphi_{k+p-1}(\bar\varphi,\bar\mu),\bar\mu)}{\partial \mu} \overset{(s)}{\Omega}{}_l^{k-1}(p,\bar\varphi,\bar\mu) \right\|$$

$$\leq \sum_{k=1}^{l}\{M\lambda^k[(\sum_{\nu=1}^{m} P^*(1+mA^*)^{k+p-1} + P^*)2K_3\lambda^{l-k+1}$$

$$\times (1+\alpha_0)^l(\|\varphi - \bar\varphi\| + |\mu - \bar\mu|) + M\lambda^{l-k+1} K_5 (1+\alpha_0)^{k+p-1}$$

$$\times (1+mA^*)^{k+p-1}(\|\varphi - \bar\varphi\| + |\mu - \bar\mu|)] + K_3 \lambda^k (1+\alpha_0)^k$$

$$\times (\|\varphi - \bar\varphi\| + |\mu - \bar\mu|)M\lambda^{l-k+1}(\sum_{\nu=1}^{m} P^*(1+mA^*)^{k+p-1} + P^*)\}$$

$$\leq K_9 \lambda^l (1+\alpha_0)^l (1+mA^*)^l (\|\varphi - \bar\varphi\| + |\mu - \bar\mu|),$$

where

$$K_9 = \max\{2K_1^0, K_2^1, 2K_3^0\}, \quad K_2^1 = \frac{M^2 K_5 (1+\alpha_0)^p (1+mA^*)^p}{(1+\alpha_0)(1+mA^*)-1}.$$

Lemma 2.7 is proved. $\qquad\qquad\square$

Proof of Theorem 2.5

Proof.   First, we show that the function $u(p, g, \mu, \varphi)$ satisfies the Lipschitz condition with respect to the variables φ, μ. Indeed, since

$$\| \overset{(s)}{u}(p, g, \mu, \varphi) - \overset{(s)}{u}(p, g, \bar{\mu}, \bar{\varphi})\| \leq \sum_{l=1}^{\infty} \{\| \overset{(s)}{\Omega_l^0}(p, \varphi, \mu)\|$$

$$\times \| \overset{(s)}{c}(\varphi_{l+g}(\varphi, \mu), \mu) - \overset{(s)}{c}(\varphi_{l+g}(\bar{\varphi}, \bar{\mu}), \bar{\mu})\|$$

$$+ \| \overset{(s)}{\Omega_l^0}(p, \varphi, \mu) - \overset{(s)}{\Omega_l^0}(p, \bar{\varphi}, \bar{\mu})\|\| \overset{(s)}{c}(\varphi_{l+g}(\bar{\varphi}, \bar{\mu}), \bar{\mu})\|\}$$

$$\leq \sum_{l=1}^{\infty} \{M\lambda^l \gamma_0[(1 + \alpha_0)^{l+g}(\|\varphi - \bar{\varphi}\| + |\mu - \bar{\mu}|) + |\mu - \bar{\mu}|]$$

$$+ C^0 K_3 \lambda^l (1 + \alpha_0)^l \times (\|\varphi - \bar{\varphi}\| + |\mu - \bar{\mu}|)\}$$

$$\leq K_{10}(\|\varphi - \bar{\varphi}\| + |\mu - \bar{\mu}|),$$

where

$$K_{10} = \frac{\lambda(1 + \alpha_0)}{1 - \lambda(1 + \alpha_0)}\{2M\gamma_0(1 + \alpha_0)^g + C^0 K_3\}$$

is a constant independent of $\{\varphi, \bar{\varphi}\} \subset \mathcal{T}_m, \{\mu, \bar{\mu}\} \subset \sigma, s \in Z^+$, we have, for all $j \in Z^+$,

$$|u_j(p, g, \mu, \varphi) - u_j(p, g, \bar{\mu}, \bar{\varphi})|$$

$$= \lim_{s \to \infty} | \overset{(s)}{u}_j(p, g, \mu, \varphi) - \overset{(s)}{u}_j(p, g, \bar{\mu}, \bar{\varphi})|$$

$$\leq K_{10}(\|\varphi - \bar{\varphi}\| + |\mu - \bar{\mu}|).$$

Then $\|u(p, g, \mu, \varphi) - u(p, g, \bar{\mu}, \bar{\varphi})\| \leq K_{10}(\|\varphi - \bar{\varphi}\| + |\mu - \bar{\mu}|)$.

Consider the sequence of functions $\{\frac{\partial \overset{(s)}{u}(p,g,\mu,\varphi)}{\partial \varphi^i}\}_{s=1}^{\infty}$ and the sequence of their k-th coordinates $\{\frac{\partial \overset{(s)}{u}_k(p,g,\mu,\varphi)}{\partial \varphi^i}\}_{s=1}^{\infty}$, where $i = \overline{1, m}$. If the first sequence is uniformly bounded and equicontinuous in φ, μ in the norm, then the second sequence of scalar functions $\forall k \in Z^+$ possesses the same properties. This yields the existence of the derivative $\frac{\partial u_k(p,g,\mu,\varphi)}{\partial \varphi^i}$ $\forall k \in Z^+$ and, in turn, the existence of the derivative $\frac{\partial u(p,g,\mu,\varphi)}{\partial \varphi^i}, i = \overline{1, m}$.

An analogous situation takes place also for the derivative $\frac{\partial u_k(p,g,\mu,\varphi)}{\partial \mu}$.

We now show that the sequences $\{\frac{\partial\,\overset{(s)}{u}(p,g,\mu,\varphi)}{\partial\varphi^i}\}_{s=1}^{\infty}$ and $\{\frac{\partial\,\overset{(s)}{u}(p,g,\mu,\varphi)}{\partial\mu}\}_{s=1}^{\infty}$ are uniformly bounded. Indeed, for all $i=\overline{1,m}$, $s\in Z^+$, the inequalities

$$\left\|\frac{\partial\,\overset{(s)}{u}(p,g,\mu,\varphi)}{\partial\varphi^i}\right\| \le \sum_{l=1}^{\infty}\{\|\frac{\partial\,\overset{(s)}{\Omega_l^0}(p,\varphi,\mu)}{\partial\varphi^i}\|\|\,\overset{(s)}{c}(\varphi_{l+g}(\varphi,\mu),\mu)\|$$

$$+\|\overset{(s)}{\Omega_l^0}(p,\varphi,\mu)\|\|\frac{\partial\,\overset{(s)}{c}(\varphi_{l+g}(\varphi,\mu),\mu)}{\partial\varphi^i}\|\} \le \sum_{l=1}^{\infty}\{K_1 C^0\lambda^l(1+mA^*)^l$$

$$+M\lambda^l\sum_{\nu=1}^{m}C_*(1+mA^*)^{l+g}\}$$

$$\le (K_1 C^0 + MC_*m(1+mA^*)^g)\sum_{l=1}^{\infty}\lambda^l(1+mA^*)^l$$

hold, and the series $\sum_{l=1}^{\infty}\lambda^l(1+mA^*)^l$ converges, since $\lambda(1+mA^*)<1$. Analogously $\forall s\in Z^+$, we obtain the estimates

$$\left\|\frac{\partial\,\overset{(s)}{u}(p,g,\mu,\varphi)}{\partial\mu}\right\| \le \sum_{l=1}^{\infty}\{K_2 C^0\lambda^l(1+mA^*)^l + M\lambda^l[\sum_{\nu=1}^{m}C_*(1+mA^*)^{l+g}$$

$$+C_*]\} \le (K_2 C^0 + MC_*m(1+mA^*)^g + MC_*)\sum_{l=1}^{\infty}\lambda^l(1+mA^*)^l.$$

Let us establish the equicontinuity of these two sequences. It is easy to verify the validity of the inequality

$$\left\|\frac{\partial\,\overset{(s)}{u}(p,g,\mu,\varphi)}{\partial\varphi^i} - \frac{\partial\,\overset{(s)}{u}(p,g,\bar\mu,\bar\varphi)}{\partial\varphi^i}\right\| \le \sum_{l=1}^{\infty}(G_1+G_2+G_3+G_4), \qquad (2.49)$$

where

$$G_1 = \|\frac{\partial\,\overset{(s)}{\Omega_l^0}(p,\varphi,\mu)}{\partial\varphi^i}\|\|\,\overset{(s)}{c}(\varphi_{l+g}(\varphi,\mu),\mu) - \overset{(s)}{c}(\varphi_{l+g}(\bar\varphi,\bar\mu),\bar\mu)\|,$$

$$G_2 = \|\frac{\partial\,\overset{(s)}{\Omega_l^0}(p,\varphi,\mu)}{\partial\varphi^i} - \frac{\partial\,\overset{(s)}{\Omega_l^0}(p,\bar\varphi,\bar\mu)}{\partial\varphi^i}\|\|\,\overset{(s)}{c}(\varphi_{l+g}(\bar\varphi,\bar\mu),\bar\mu)\|,$$

$$G_3 = \|\overset{(s)}{\Omega_l^0}(p,\varphi,\mu)\|\|\frac{\partial\,\overset{(s)}{c}(\varphi_{l+g}(\varphi,\mu),\mu)}{\partial\varphi^i} - \frac{\partial\,\overset{(s)}{c}(\varphi_{l+g}(\bar\varphi,\bar\mu),\bar\mu)}{\partial\varphi^i}\|,$$

$$G_4 = \| \overset{(s)}{\Omega}{}^0_l(p,\varphi,\mu) - \overset{(s)}{\Omega}{}^0_l(p,\bar\varphi,\bar\mu)\| \| \frac{\partial \overset{(s)}{c}(\varphi_{l+g}(\bar\varphi,\bar\mu),\bar\mu)}{\partial\varphi^i}\|.$$

Estimating each of the expressions $G_i (i = \overline{1,4})$, we have

$$G_i \le K_{i+10}\lambda^l(1 + mA^*)^l(1 + \alpha_0)^l(\|\varphi - \bar\varphi\| + |\mu - \bar\mu|) \quad (i = \overline{1,4}),$$

where $K_{11} = 2K_1\gamma_0(1 + \alpha_0)^g$, $K_{12} = K_8C^0$, $K_{13} = MK_6(1 + \alpha_0)^g(1 + mA^*)^g$, $K_{14} = mK_3C_*(1 + mA^*)^g$.

Returning to (2.49), we obtain the estimate

$$\|\frac{\partial \overset{(s)}{u}(p,g,\mu,\varphi)}{\partial\varphi^i} - \frac{\partial \overset{(s)}{u}(p,g,\bar\mu,\bar\varphi)}{\partial\varphi^i}\| \le \sum_{j=11}^{14} K_j \sum_{l=1}^{\infty} \lambda^l(1 + mA^*)^l(1 + \alpha_0)^l$$

$$\times (\|\varphi - \bar\varphi\| + |\mu - \bar\mu|) \le K_{15}(\|\varphi - \bar\varphi\| + |\mu - \bar\mu|), \quad (2.50)$$

where

$$K_{15} = \frac{\lambda(1 + mA^*)(1 + \alpha_0)(K_{11} + K_{12} + K_{13} + K_{14})}{1 - \lambda(1 + mA^*)(1 + \alpha_0)}$$

is a constant independent of $\{\varphi,\bar\varphi\} \subset \mathcal{T}_m$, $\{\mu,\bar\mu\} \subset \sigma, s \in Z^+, i = \overline{1,m}$.

Together with (2.50), the inequality

$$\|\frac{\partial \overset{(s)}{u}(p,g,\mu,\varphi)}{\partial\mu} - \frac{\partial \overset{(s)}{u}(p,g,\bar\mu,\bar\varphi)}{\partial\mu}\| \le \sum_{l=1}^{\infty}(G_1^0 + G_2^0 + G_3^0 + G_4^0), \quad (2.51)$$

where G_i^0 is obtained from $G_i(i = \overline{1,4})$ by substituting $\partial\varphi^i$ for $\partial\mu$, holds.

With regard for the estimates

$$G_i^0 \le K_{i+10}^0\lambda^l(1 + mA^*)^l(1 + \alpha_0)^l(\|\varphi - \bar\varphi\| + |\mu - \bar\mu|) \quad (i = \overline{1,4}),$$

where $K_{11}^0 = 2K_2\gamma_0(1 + \alpha_0)^g$, $K_{12}^0 = K_9C^0$, $K_{13}^0 = MK_7(1 + \alpha_0)^g(1 + mA^*)^g$, $K_{14}^0 = 2K_{14}$, and relation (2.51), we obtain the inequality

$$\|\frac{\partial \overset{(s)}{u}(p,g,\mu,\varphi)}{\partial\mu} - \frac{\partial \overset{(s)}{u}(p,g,\bar\mu,\bar\varphi)}{\partial\mu}\| \le K_{16}(\|\varphi - \bar\varphi\| + |\mu - \bar\mu|), \quad (2.52)$$

where

$$K_{16} = \frac{\lambda(1 + mA^*)(1 + \alpha_0)(K_{11}^0 + K_{12}^0 + K_{13}^0 + K_{14}^0)}{1 - \lambda(1 + mA^*)(1 + \alpha_0)}$$

is a constant independent of $\{\varphi,\bar\varphi\} \subset \mathcal{T}_m$, $\{\mu,\bar\mu\} \subset \sigma, s \in Z^+$.

Thus, the sequences

$$\{\frac{\partial \overset{(s)}{u}(p,g,\mu,\varphi)}{\partial\varphi^i}\}_{s=1}^{\infty} \quad \text{and} \quad \{\frac{\partial \overset{(s)}{u}(p,g,\mu,\varphi)}{\partial\mu}\}_{s=1}^{\infty}$$

are equicontinuous in φ, μ $\forall i = \overline{1, m}$.

Let us fix $k \in Z^+$, $i \in \overline{1, m}$. By the Arzela–Ascoli theorem, the sequences

$$\left\{\frac{\partial \overset{(s)}{u}_k(p, g, \mu, \varphi)}{\partial \varphi^i}\right\}_{s=1}^{\infty} \quad \text{and} \quad \left\{\frac{\partial \overset{(s)}{u}_k(p, g, \mu, \varphi)}{\partial \mu}\right\}_{s=1}^{\infty}$$

contain the subsequences

$$\left\{\frac{\partial \overset{(r)}{u}_k(p, g, \mu, \varphi)}{\partial \varphi^i}\right\}_{r=1}^{\infty} \quad \text{and} \quad \left\{\frac{\partial \overset{(\nu)}{u}_k(p, g, \mu, \varphi)}{\partial \mu}\right\}_{\nu=1}^{\infty}$$

convergent to the functions

$$\frac{\partial u_k(p, g, \mu, \varphi)}{\partial \varphi^i} \quad \text{and} \quad \frac{\partial u_k(p, g, \mu, \varphi)}{\partial \mu},$$

respectively, as $r \to \infty$ and $\nu \to \infty$. Inequalities (2.50) and (2.52) yield

$$\left\|\frac{\partial u_k(p, g, \mu, \varphi)}{\partial \varphi^i} - \frac{\partial u_k(p, g, \bar{\mu}, \bar{\varphi})}{\partial \varphi^i}\right\| \leq K_{15}(\|\varphi - \bar{\varphi}\| + |\mu - \bar{\mu}|),$$

$$\left\|\frac{\partial u_k(p, g, \mu, \varphi)}{\partial \mu} - \frac{\partial u_k(p, g, \bar{\mu}, \bar{\varphi})}{\partial \mu}\right\| \leq K_{16}(\|\varphi - \bar{\varphi}\| + |\mu - \bar{\mu}|).$$

Since the constants K_{15}, K_{16} are independent of $k \in Z^+$, the derivatives

$$\frac{\partial u(p, g, \mu, \varphi)}{\partial \varphi^i} \quad \text{and} \quad \frac{\partial u(p, g, \mu, \varphi)}{\partial \mu}$$

satisfy the Cauchy–Lipschitz conditions with respect to φ, μ for all $i = \overline{1, m}$.

Thus, the function $u(p, g, \mu, \varphi) \in C^1_{Lip}(\varphi, \mu)$ with the coefficient

$$\delta = \max\{K_{10}, K_{15}, K_{16}\},$$

which completes the proof of Theorem 2.5. $\qquad\qquad\qquad\qquad \square$

Corollary 2.5. *Let $a(\varphi, \mu)$ be a constant vector, $\{P(\varphi, \mu), c(\varphi, \mu)\} \subset C^1_{Lip}(\varphi, \mu)$, let the conditions of Lemma 2.3 be satisfied, and let the first and third inequalities in (2.48) hold.*

Then, $\forall\{p, g\} \subset Z$, the function $u(p, g, \mu, \varphi) \in C^1_{Lip}(\varphi, \mu)$.

In the proof of this proposition, it should be taken into account that the mapping $\Phi(\varphi)$ is invertible under its conditions, the inverse mapping is differentiable with respect to $\varphi^i(i = \overline{1, m})$, and, for all $n \in Z$ and $i = \overline{1, m}$, the following equalities hold:

$$\frac{\partial \varphi_n(\varphi, \mu)}{\partial \varphi^i} = \frac{\partial \varphi}{\partial \varphi^i}; \quad \left\|\frac{\partial \varphi_n(\varphi, \mu)}{\partial \varphi^i}\right\| = 1; \quad \left\|\frac{\partial \varphi_n(\varphi, \mu)}{\partial \mu}\right\| = 0;$$

$$\left\|\frac{\partial \varphi_n(\varphi,\mu)}{\partial \varphi^i} - \frac{\partial \varphi_n(\bar\varphi,\bar\mu)}{\partial \varphi^i}\right\| = 0; \quad \left\|\frac{\partial \varphi_n(\varphi,\mu)}{\partial \mu} - \frac{\partial \varphi_n(\bar\varphi,\bar\mu)}{\partial \mu}\right\| = 0;$$

$$\|\varphi_n(\varphi,\mu) - \varphi_n(\bar\varphi,\bar\mu)\| = \|\varphi - \bar\varphi\|,$$

where $\{\varphi,\bar\varphi\} \subset \mathcal{T}_m$, $\{\mu,\bar\mu\} \subset \sigma$.

2^0. On the differentiability of an invariant torus up to the order $\rho \geq 2$. Consider the system of equations (2.4). The corresponding truncated system of equations reads

$$\varphi_{n+1} = \varphi_n + a(\varphi_n), \quad \overset{(s)}{x}_{n+1} = \overset{(s)}{P}(\varphi_{n+p})\overset{(s)}{x}_n + \overset{(s)}{c}(\varphi_{n+g+1}), \qquad (2.53)$$

and $\Omega_l^n(p,\varphi)$ and $\overset{(s)}{\Omega}{}_l^n(p,\varphi)$ are the matriciants of the homogeneous equations $x_{n+1} = P(\varphi_{n+p}(\varphi))x_n$ and $\overset{(s)}{x}_{n+1} = \overset{(s)}{P}(\varphi_{n+p}(\varphi))\overset{(s)}{x}_n$, respectively.

We agree that $D_\varphi^s(\Xi(\varphi^1,\varphi^2,\ldots,\varphi^m))$ stands for an arbitrary derivative of the s-th order of the function $\Xi(\varphi)$ with respect to $\varphi^1,\varphi^2,\ldots,\varphi^m$. In this case, we consider, as above, that the matrix and the vector-function are differentiated, respectively, elementwise and coordinatewise.

We say that $\Xi(\varphi) \in C_{Lip}^\rho(\varphi)$ with the coefficient α, if, for all $0 \leq s \leq \rho$, the inequality

$$\|D_\varphi^s \Xi(\varphi) - D_\varphi^s \Xi(\bar\varphi)\| \leq \alpha\|\varphi - \bar\varphi\|$$

holds. Here, α is a positive constant independent of s, and $\{\varphi,\bar\varphi\} \subset \mathcal{T}_m$.

We now formulate the main result of Subsection 2.3 as the following proposition.

Theorem 2.6. *Let, for all natural s and $\varphi \in \mathcal{T}_m$, the following conditions be satisfied:*

1) $\{a(\varphi), P(\varphi), c(\varphi)\} \subset C_{Lip}^\rho(\varphi)$, and $a(\varphi)$ enters with the coefficient α_0;

2) $\|D_\varphi^l(P(\varphi))\| \leq P^*$, $\|D_\varphi^l(a(\varphi))\| \leq A^*$, $\|D_\varphi^l(c(\varphi))\| \leq C^*$, where $1 \leq l \leq \rho$, and the positive constants P^*, A^*, C^* are independent of l and φ;

3) the matrix $\overset{(s)}{P}(\varphi)$ is nondegenerate;

4) for all integers $l > n$, the inequality $\|\overset{(s)}{\Omega}{}_l^n(0,\varphi)\| \leq M\lambda^{l-n}$, where the constants $M > 0$ and $0 < \lambda < 1$ are independent of s and φ, holds;

5) the sequence $\{\overset{(s)}{P}{}^{-1}(\varphi)\}_{s=1}^\infty$ is proper, and $\|\overset{(s)}{P}{}^{-1}(\varphi)\| \leq P_1$, where the positive constant P_1 is independent of s and φ.

In this case, if $\lambda(1+\alpha_0)(1+mA^*)^\rho < 1$, then the function $u(p, g, \varphi)$ generating the invariant torus of the system of equations (2.4) belongs to $C^\rho_{Lip}(\varphi)$ for any integers $g \geq -1, p \geq 0$.

Before the proof of Theorem 2.6, we formulate several auxiliary propositions.

Lemma 2.8. *Let $a(\varphi) \in C^\rho(\varphi)$, and let $\|D^s_\varphi(a(\varphi))\| \leq A^*$, where A^* is a positive constant independent of $\varphi \in \mathcal{T}_m$, and $1 \leq s \leq \rho$. Then, for all natural n and $1 \leq s \leq \rho$, the inequality*

$$\|D^s_\varphi(\varphi_n(\varphi))\| \leq K_s(1 + mA^*)^{sn}, \tag{2.54}$$

where the constants K_s are independent of n and $\varphi \in \mathcal{T}_m$, holds.

Proof. For $s = 1$, inequality (2.54) is proved in i. 1^0 of this section. In this case, $K_1 = 1$. We assume that this inequality holds for $1 < s \leq k$ $(k < \rho)$ and will prove its validity for $s = k + 1$.

It is easy to verify that $D^k_\varphi(\varphi_n(\varphi)) = D^k_\varphi(\varphi_{n-1}(\varphi)) + L$, where L is the sum of a finite number of terms of two types:

$$D^p_{\varphi_{n-1}(\varphi)}(a(\varphi_{n-1}(\varphi)))D^{\lambda_1}_\varphi(\varphi^{k_1}_{n-1}(\varphi)) \cdots D^{\lambda_r}_\varphi(\varphi^{k_r}_{n-1}(\varphi)) \tag{2.55}$$

and

$$D^1_{\varphi_{n-1}(\varphi)}(a(\varphi_{n-1}(\varphi)))D^k_\varphi(\varphi^\sigma_{n-1}(\varphi)),$$

where

$$p \in \{1, 2, \ldots, k\}, \lambda_i \in \{1, 2, \ldots, k-1\}, \{k_i, \sigma\} \subset \{1, 2, \ldots, m\},$$

$i = 1, 2, \ldots, r, \sum_{j=1}^r \lambda_j = k$.

Since

$$D^{k+1}_\varphi(\varphi_n(\varphi)) = D^1_\varphi(D^k_\varphi(\varphi_n(\varphi))),$$

we have

$$D^{k+1}_\varphi(\varphi_n(\varphi)) = D^{k+1}_\varphi(\varphi_{n-1}(\varphi)) + L_1,$$

where L_1 is the sum of a finite number of terms of the form (2.55) and of the form

$$D^1_{\varphi_{n-1}(\varphi)}(a(\varphi_{n-1}(\varphi)))D^{k+1}_\varphi(\varphi^\sigma_{n-1}(\varphi)),$$

where

$$p \in \{1, 2, \ldots, k+1\}, \lambda_i \in \{1, 2, \ldots, k\}, \{k_i, \sigma\} \subset \{1, 2, \ldots, m\},$$

$i = 1, 2, \ldots, r, \sum_{j=1}^r \lambda_j = k + 1$.

In this case,

$$\|D_\varphi^{k+1}(\varphi_n(\varphi))\| \leq \Delta_1(1+mA^*)^{(\lambda_1+\cdots+\lambda_r)(n-1)} + \Delta_2\|D_\varphi^{k+1}(\varphi_{n-1}(\varphi))\|$$
$$= \Delta_1(1+mA^*)^{(k+1)(n-1)} + \Delta_2\|D_\varphi^{k+1}(\varphi_{n-1}(\varphi))\|,$$

where the constants Δ_1 and Δ_2 are independent of n and φ.

By continuing the process, we obtain the relations

$$\|D_\varphi^{k+1}(\varphi_n(\varphi))\| \leq \sum_{i=1}^{2} Z_i(1+mA^*)^{(k+1)(n-i)} + Z_3\|D_\varphi^{k+1}(\varphi_{n-2}(\varphi))\| \leq \cdots$$

$$\leq Z_0 \sum_{i=1}^{n-1}(1+mA^*)^{(k+1)i} \leq Z_0 \frac{(1+mA^*)^{k+1}((1+mA^*)^{(k+1)(n-1)}-1)}{mA^*}$$

$$\leq K_{k+1}(1+mA^*)^{(k+1)n},$$

where the constant K_{k+1} is independent of n and φ. Lemma 2.8 is proved.
$\square$

Lemma 2.9. *Let the conditions of Lemma 2.8 be satisfied, and let* $\{P(\varphi), c(\varphi)\} \subset C^\rho(\varphi)$. *Moreover, let* $\|D_\varphi^s(P(\varphi))\| \leq P^*$, $\|D_\varphi^s(c(\varphi))\| \leq C^*$, *where* P^* *and* C^* *are positive constants independent of* $\varphi \in \mathcal{T}_m$ *and* $1 \leq s \leq \rho$. *Then, for all natural* n *and* $1 \leq s \leq \rho$ *for any integers* $l+g \geq 0, p \geq 0$, *the inequalities*

$$\|D_\varphi^s(\overset{(n)}{c}(\varphi_{l+g}(\varphi)))\| \leq C_s^*(1+mA^*)^{s(l+g)}, \tag{2.56}$$

$$\|D_\varphi^s(\overset{(n)}{\Omega_l^0}(p,\varphi))\| \leq Z_s\lambda^l(1+mA^*)^{sl} \tag{2.57}$$

hold. Here, the positive constants C_s^* *and* Z_s *are independent of* n *and* $\varphi \in \mathcal{T}_m$.

Proof. It is easy to verify that $D_\varphi^s(\overset{(n)}{c}(\varphi_{l+g}(\varphi)))$ is the sum of a finite number of terms of the form

$$\Phi_\eta(\varphi) = D_{\varphi_{l+g}(\varphi)}^\eta(\overset{(n)}{c}(\varphi_{l+g}(\varphi)))D_\varphi^{\lambda_1}(\varphi_{l+g}^{l_1}(\varphi)) \cdots D_\varphi^{\lambda_m}(\varphi_{l+g}^{l_m}(\varphi)), \tag{2.58}$$

where $\eta \in \{1, 2, \ldots, s\}$, $\sum_{i=1}^{m} \lambda_i = s$, $\lambda_i \in \{0, 1, 2, \ldots, s\}$, $i \in \{1, 2, \ldots, m\}$, and the expression $D^0(\varphi_{l+g}^{l_i}(\varphi))$ in (2.58) is equal to 1.

Using estimate (2.54), setting $K_{\lambda_i} = 1$ for $\lambda_i = 0$, and denoting $C^* \prod_{i=1}^{m} K_{\lambda_i}$ by Y_η, we obtain that the norm of relation (2.55) does not exceed $Y_\eta(1+mA^*)^{s(l+g)}$, where Y_η are independent of n and $\varphi \in \mathcal{T}_m$. This yields the validity of estimate (2.56).

To prove inequality (2.57), we apply the method of mathematical induction. For $s = 1$, it was proved in i. 1^0 of this subsection. We assume that this inequality holds for all natural $s < d < \rho$ and will prove its validity for $s = d$.

Using the relation

$$\frac{\partial \overset{(n)}{\Omega}{}^0_l(p,\varphi)}{\partial \varphi^i} = -\sum_{k=1}^{l} \overset{(n)}{\Omega}{}^0_k(p,\varphi) \frac{\partial \overset{(n)}{P}(\varphi_{k+p-1}(\varphi))}{\partial \varphi^i} \overset{(n)}{\Omega}{}^{k-1}_l(p,\varphi),$$

which was proved in i. 1^0 of this subsection for $i \in \{1, 2, \ldots, m\}$, we write down the equality

$$D^d_\varphi(\overset{(n)}{\Omega}{}^0_l(p,\varphi)) = -\sum_{k=1}^{l} \Gamma, \tag{2.59}$$

where Γ is the sum of a finite number of terms of the form

$$\Gamma^*(p_0, p_1, p_2, \varphi) = D^{p_0}_\varphi(\overset{(n)}{\Omega}{}^0_k(p,\varphi)) D^{p_1}_\varphi(\overset{(n)}{P}(\varphi_{k+p-1}(\varphi))) D^{p_2}_\varphi(\overset{(n)}{\Omega}{}^{k-1}_l(p,\varphi)),$$

where $\{p_0, p_2\} \subset \{0, 1, \ldots, d-1\}$, $p_1 \in \{1, 2, \ldots, d\}$, $p_0 + p_1 + p_2 = d$.

First, we estimate the expression $D^{p_1}_\varphi(\overset{(n)}{P}(\varphi_{k+p-1}(\varphi)))$, by representing it in the form of a sum of a finite number of terms:

$$D^\eta_{\varphi_{k+p-1}(\varphi)}(\overset{(n)}{P}(\varphi_{k+p-1}(\varphi))) D^{\lambda_1}_\varphi(\varphi^{l_1}_{k+p-1}(\varphi)) \cdots D^{\lambda_m}_\varphi(\varphi^{l_m}_{k+p-1}(\varphi)), \tag{2.60}$$

where η takes values from 1 to p_1 inclusively, $\sum_{i=1}^{m} \lambda_i = p_1$, $\lambda_i \in \{0, 1, \ldots, p_1\}$, $l_i \in \{1, 2, \ldots, m\}$, $i = \overline{1, m}$, and $D^0(\varphi^{l_i}_{k+p-1}(\varphi)) = 1$ by agreement.

Using Lemma 2.8 and denoting $P^* \prod_{i=1}^{m} K_{\lambda_i}$ by $\tilde{X}(\eta)$, we conclude that the norm of expression (2.60) does not exceed

$$\tilde{X}(\eta) \prod_{i=1}^{m} (1 + mA^*)^{\lambda_i(k+p-1)} = \tilde{X}(\eta)(1 + mA^*)^{p_1(k+p-1)},$$

where $\tilde{X}(\eta)$ is independent of n and $\varphi \in \mathcal{T}_m$.

This yields

$$\|D^{p_1}_\varphi(\overset{(n)}{P}(\varphi_{k+p-1}(\varphi)))\| \leq X(p_1)(1 + mA^*)^{p_1(k+p-1)}, \tag{2.61}$$

where $X(p_1)$ is independent of n and $\varphi \in \mathcal{T}_m$ and is equal to the product of $\tilde{X}(\eta)$ by the number of terms of the form (2.60) in the representation $D^{p_1}_\varphi(\overset{(n)}{P}(\varphi_{k+p-1}(\varphi)))$.

Finally, we estimate the expression

$$D_\varphi^{p_2}(\overset{(n)}{\Omega}{}_l^{k-1}(p,\varphi)) = D_\varphi^{p_2}(\overset{(n)}{\Omega}{}_{l-k+1}^0(p,\varphi_{k-1}(\varphi))),$$

by representing it in the form of a sum of a finite number of terms:

$$D_{\varphi_{k-1}(\varphi)}^\eta(\overset{(n)}{\Omega}{}_{l-k+1}^0(p,\varphi_{k-1}(\varphi)))D_\varphi^{\lambda_1}(\varphi_{k-1}^{l_1}(\varphi))\cdots D_\varphi^{\lambda_m}(\varphi_{k-1}^{l_m}(\varphi)). \quad (2.62)$$

Here, η takes values from 1 to p_2 inclusively, $\sum_{i=1}^m \lambda_i = p_2$, $\lambda_i \in \{0,1,\ldots,p_2\}$, $l_i \in \{1,2,\ldots,m\}$, and $D_\varphi^0(\varphi_{k-1}^{l_i}(\varphi)) = 1$.

The norm of expression (2.62) is bounded by the product

$$\|D_{\varphi_{k-1}(\varphi)}^\eta(\overset{(n)}{\Omega}{}_{l-k+1}^0(p,\varphi_{k-1}(\varphi)))\|\prod_{i=1}^m K_{\lambda_i}(1+mA^*)^{\lambda_i(k-1)}$$

$$\leq Z_\eta\lambda^{l-k+1}(1+mA^*)^{\eta(l-k+1)}(1+mA^*)^{p_2(k-1)}\prod_{i=1}^m K_{\lambda_i}.$$

In this case,

$$\|D_\varphi^{p_2}(\overset{(n)}{\Omega}{}_l^{k-1}(p,\varphi))\| \leq \tilde{Z}(p_2)\lambda^{l-k+1}(1+mA^*)^{p_2 l}, \quad (2.63)$$

where $\tilde{Z}(p_2)$ is independent of n and $\varphi \in \mathcal{T}_m$.

Then

$$\|\Gamma^*(p_0,p_1,p_2,\varphi)\| \leq Z_{p_0}\lambda^k(1+mA^*)^{p_0 k}X(p_1)(1+mA^*)^{p_1(k+p-1)}$$

$$\times \tilde{Z}(p_2)\lambda^{l-k+1}(1+mA^*)^{l(d-p_1-p_0)}$$

$$\leq B(p_0,p_1,p_2)\lambda^l(1+mA^*)^{(p_0+p_1)k+ld-l(p_0+p_1)},$$

where $B(p_0,p_1,p_2)$ is also independent of n and $\varphi \in \mathcal{T}_m$.

Hence,

$$\sum_{k=1}^l \|\Gamma^*(p_0,p_1,p_2,\varphi)\| \leq B(p_0,p_1,p_2)\lambda^l(1+mA^*)^{ld-l(p_0+p_1)}$$

$$\times \sum_{k=1}^l(1+mA^*)^{(p_0+p_1)k} \leq B(p_0,p_1,p_2)\lambda^l(1+mA^*)^{ld-l(p_0+p_1)}$$

$$\times \frac{(1+mA^*)^{p_1+p_0+l(p_0+p_1)}}{(1+mA^*)^{p_1+p_0}-1} \leq \tilde{B}\lambda^l(1+mA^*)^{ld},$$

where $\tilde{B}$ is independent of n and $\varphi \in \mathcal{T}_m$.

Taking equality (2.59) into account, we complete the proof of Lemma 2.9. $\qquad\square$

Lemma 2.10. *Let the conditions of Lemma 2.8 be satisfied, and let $a(\varphi) \in C^\rho_{Lip}(\varphi)$ with the coefficient α_0. Then, for all natural n and $1 \le s \le \rho$, the inequality*

$$\|D^s_\varphi(\varphi_n(\varphi) - \varphi_n(\bar\varphi))\| \le \bar K_s (1 + \alpha_0)^n (1 + mA^*)^{sn}\|\varphi - \bar\varphi\| \qquad (2.64)$$

where the constants $\bar K_s$ are independent of n and $\{\varphi, \bar\varphi\} \subset \mathcal{T}_m$, holds.

Proof. For $s = 1$, inequality (2.64) was proved in i. 1^0 of this section. Assume that it holds for $1 < s \le k(k < \rho)$ and will substitute its validity for $s = k + 1$.

It follows from the proof of Lemma 2.8 that

$$\|D^{k+1}_\varphi(\varphi_n(\varphi) - \varphi_n(\bar\varphi))\| \le \|D^{k+1}_\varphi(\varphi_{n-1}(\varphi) - \varphi_{n-1}(\bar\varphi))\|+$$

$$+\|R_p(\varphi, \bar\varphi)\| + \|\bar R_p(\varphi, \bar\varphi)\|, \qquad (2.65)$$

where $R_p(\varphi, \bar\varphi)$ is the sum of a finite number of terms of the form

$$D^p = D^p_{\varphi_{n-1}(\varphi)}(a(\varphi_{n-1}(\varphi)))D^{\lambda_1}_\varphi(\varphi^{k_1}_{n-1}(\varphi)) \cdots D^{\lambda_r}_\varphi(\varphi^{k_r}_{n-1}(\varphi))$$

$$- D^p_{\varphi_{n-1}(\varphi)}(a(\varphi_{n-1}(\bar\varphi)))D^{\lambda_1}_\varphi(\varphi^{k_1}_{n-1}(\bar\varphi)) \cdots D^{\lambda_r}_\varphi(\varphi^{k_r}_{n-1}(\bar\varphi)),$$

and $\bar R_p(\varphi, \bar\varphi)$ is the sum of m terms of the form

$$D^1 = D^1_{\varphi_{n-1}(\varphi)}(a(\varphi_{n-1}(\varphi)))D^{k+1}_\varphi(\varphi^\sigma_n(\varphi))$$

$$- D^1_{\varphi_{n-1}(\varphi)}(a(\varphi_{n-1}(\bar\varphi)))D^{k+1}_\varphi(\varphi^\sigma_n(\bar\varphi)),$$

and $p \in \{1, 2, \ldots, k+1\}$, $\lambda_i \in \{1, 2, \ldots, k\}$, $k_i \in \{1, 2, \ldots, m\}$, $\sigma \in \{1, 2, \ldots, m\}$, $i = 1, 2, \ldots, r$, $\sum_{j=1}^r \lambda_j = k + 1$.

We write down the inequality

$$\|D^p\| \le \alpha_0\|\varphi_{n-1}(\varphi) - \varphi_{n-1}(\bar\varphi)\| \prod_{i=1}^r K_{\lambda_i}(1 + mA^*)^{\lambda_i(n-1)}$$

$$+ A^* \sum_{i=1}^r \{\|D^{\lambda_i}_\varphi(\varphi^{k_i}_{n-1}(\varphi) - \varphi^{k_i}_{n-1}(\bar\varphi))\| \prod_{j=1}^{i-1} K_{\lambda_j}(1 + mA^*)^{\lambda_j(n-1)}$$

$$\times \prod_{j=i+1}^r K_{\lambda_j}(1 + mA^*)^{\lambda_j(n-1)}\},$$

where the symbols $\prod_{j=1}^0(\cdot)$, $\prod_{j=r+1}^r(\cdot)$ are equal to 1.

We introduce the notation

$$\overset{(0)}{A} = \max\{K_{\lambda_2} \cdots K_{\lambda_r}, K_{\lambda_1}K_{\lambda_3}K_{\lambda_4} \cdots K_{\lambda_r}, \ldots, K_{\lambda_1}K_{\lambda_2} \cdots K_{\lambda_{r-1}}\};$$

$$\bar{D}^p = \alpha_0 \prod_{i=1}^{r} K_{\lambda_i} + A^* \overset{(0)}{A} \sum_{i=1}^{r} \bar{K}_{\lambda_i}, \quad D_1^1 = m\alpha_0 K_{k+1},$$

$$\bar{D}_1^p = \bar{D}^p \xi, \quad \tilde{D} = \bar{D}_1^p + D_1^1,$$

where ξ is the number of terms in the sum $R_p(\varphi, \bar{\varphi})$.

We now write down the chain of inequalities

$$\|D^p\| \leq \alpha_0 (1+\alpha_0)^{n-1}(1+mA^*)^{(k+1)(n-1)} \prod_{i=1}^{r} K_{\lambda_i} \|\varphi - \bar{\varphi}\|$$

$$+ A^*(1+mA^*)^{(k+1)(n-1)} \overset{(0)}{A} \sum_{i=1}^{r} \|D_\varphi^{\lambda_i}(\varphi_{n-1}^{k_i}(\varphi) - \varphi_{n-1}^{k_i}(\bar{\varphi}))\|$$

$$\leq \alpha_0 (1+\alpha_0)^{n-1}(1+mA^*)^{(k+1)(n-1)} \prod_{i=1}^{r} K_{\lambda_i} \|\varphi - \bar{\varphi}\|$$

$$+ A^*(1+mA^*)^{(k+1)(n-1)} \overset{(0)}{A} \sum_{i=1}^{r} \bar{K}_{\lambda_i}(1+\alpha_0)^{n-1}(1+mA^*)^{\lambda_i(n-1)} \|\varphi - \bar{\varphi}\|$$

$$\leq (1+\alpha_0)^{n-1}(1+mA^*)^{(k+1)(n-1)} \|\varphi - \bar{\varphi}\| \{\alpha_0 \prod_{i=1}^{r} K_{\lambda_i} + A^* \overset{(0)}{A} \sum_{i=1}^{r} \bar{K}_{\lambda_i}\}$$

$$\leq \bar{D}^p (1+\alpha_0)^{n-1}(1+mA^*)^{(k+1)(n-1)} \|\varphi - \bar{\varphi}\|.$$

Analogously, we have

$$\|D^1\| \leq \alpha_0 K_{k+1}(1+\alpha_0)^{n-1}(1+mA^*)^{(k+1)(n-1)} \|\varphi - \bar{\varphi}\|$$
$$+ A^* \|D_\varphi^{k+1}(\varphi_{n-1}(\varphi) - \varphi_{n-1}(\bar{\varphi}))\|.$$

Then relation (2.65) yields

$$\|D_\varphi^{k+1}(\varphi_n(\varphi) - \varphi_n(\bar{\varphi}))\| \leq \|D_\varphi^{k+1}(\varphi_{n-1}(\varphi) - \varphi_{n-1}(\bar{\varphi}))\|$$
$$+ \bar{D}_1^p (1+\alpha_0)^{n-1}(1+mA^*)^{(k+1)(n-1)} \|\varphi - \bar{\varphi}\| + D_1^1 (1+\alpha_0)^{n-1}$$
$$\times (1+mA^*)^{(k+1)(n-1)} \|\varphi - \bar{\varphi}\| + mA^* \|D_\varphi^{k+1}(\varphi_{n-1}(\varphi) - \varphi_{n-1}(\bar{\varphi}))\|$$
$$= (1+mA^*) \|D_\varphi^{k+1}(\varphi_{n-1}(\varphi) - \varphi_{n-1}(\bar{\varphi}))\|$$
$$+ \tilde{D}(1+\alpha_0)^{n-1}(1+mA^*)^{(k+1)(n-1)} \|\varphi - \bar{\varphi}\|,$$

where $\tilde{D}$ is independent of n and $\{\varphi, \bar{\varphi}\} \subset \mathcal{T}_m$.

The obtained recurrence relation leads to the inequalities

$$\|D_\varphi^{k+1}(\varphi_n(\varphi) - \varphi_n(\bar\varphi))\| \leq (1 + mA^*)^n \|D_\varphi^{k+1}(\varphi_0(\varphi) - \varphi_0(\bar\varphi))\|$$

$$+ \tilde{D} \sum_{i=0}^{n-1} (1 + \alpha_0)^i (1 + mA^*)^{ik+n-1} \|\varphi - \bar\varphi\|$$

$$= \tilde{D} \sum_{i=0}^{n-1} (1 + \alpha_0)^i (1 + mA^*)^{ik+n-1} \|\varphi - \bar\varphi\|$$

$$= \frac{\tilde{D}(1 + mA^*)^{n-1}((1 + \alpha_0)^n (1 + mA^*)^{kn} - 1)}{(1 + \alpha_0)(1 + mA^*)^k - 1}$$

$$\leq \bar{K}_{k+1}(1 + \alpha_0)^n (1 + mA^*)^{n(k+1)},$$

where the constant

$$\bar{K}_{k+1} = \frac{\tilde{D}}{((1 + \alpha_0)(1 + mA^*) - 1)(1 + mA^*)}$$

is independent of n and $\{\varphi, \bar\varphi\} \subset \mathcal{T}_m$.

The last result proves estimate (2.64) for $s = k + 1$ and also Lemma 2.10. $\qquad\square$

Lemma 2.11. *Let the conditions of Lemma 2.10 be satisfied, and* $\{P(\varphi), c(\varphi)\} \subset C_{Lip}^\rho(\varphi)$. *Then, for all natural n and $1 \leq s \leq \rho$ for any integers $l + g \geq 0, p \geq 0$, the inequalities*

$$\|D_\varphi^s(\overset{(n)}{c}(\varphi_{l+g}(\varphi)) - \overset{(n)}{c}(\varphi_{l+g}(\bar\varphi)))\| \leq C_s(1 + \alpha_0)^{l+g}(1 + mA^*)^{s(l+g)} \|\varphi - \bar\varphi\|,$$
$$(2.66)$$

$$\|D_\varphi^s(\overset{(n)}{\Omega}{}_l^0(p, \varphi) - \overset{(n)}{\Omega}{}_l^0(p, \bar\varphi))\| \leq \Omega_s \lambda^l (1 + \alpha_0)^l (1 + mA^*)^{sl} \|\varphi - \bar\varphi\| \quad (2.67)$$

hold. Here, the constants C_s and Ω_s are independent of n, and $\{\varphi, \bar\varphi\} \subset \mathcal{T}_m$.

Proof. Relation (2.58) implies that $\|D_\varphi^s(\overset{(n)}{c}(\varphi_{l+g}(\varphi)) - \overset{(n)}{c}(\varphi_{l+g}(\bar\varphi)))\|$ does not exceed the sum of a finite number of terms of the form $\|\Phi_\eta(\varphi) - \Phi_\eta(\bar\varphi)\|$.

Let $c(\varphi)$ and $P(\varphi)$ belong to $C_{Lip}^\rho(\varphi)$ with coefficients β and γ, respectively.

Taking Lemmas 2.8 and 2.10 into account and introducing the notation

$$\bar{C}_s = \beta \prod_{i=1}^m K_{\lambda_i} + C^* \{ \bar{K}_{\lambda_1} K_{\lambda_2} K_{\lambda_3} \cdots K_{\lambda_m} + K_{\lambda_1} \bar{K}_{\lambda_2} K_{\lambda_3} \cdots K_{\lambda_m} + \ldots$$

$$+ K_{\lambda_1} K_{\lambda_2} K_{\lambda_3} \cdots K_{\lambda_{m-1}} \bar{K}_{\lambda_m} \},$$

we write down the inequalities

$$\|\Phi_\eta(\varphi) - \Phi_\eta(\bar\varphi)\| \le \beta\|\varphi_{l+g}(\varphi) - \varphi_{l+g}(\bar\varphi)\| \prod_{i=1}^{m} K_{\lambda_i}(1 + mA^*)^{\lambda_i(l+g)}$$

$$+ C^*\|D_\varphi^{\lambda_1}(\varphi_{l+g}^{l_1}(\varphi)) \cdots D_\varphi^{\lambda_m}(\varphi_{l+g}^{l_m}(\varphi)) - D_\varphi^{\lambda_1}(\varphi_{l+g}^{l_1}(\bar\varphi)) \cdots D_\varphi^{\lambda_m}(\varphi_{l+g}^{l_m}(\bar\varphi))\|$$

$$\le \beta(1+\alpha_0)^{l+g}(1 + mA^*)^{s(l+g)} \prod_{i=1}^{m} K_{\lambda_i}\|\varphi - \bar\varphi\| + C^*\{\bar K_{\lambda_1} K_{\lambda_2} K_{\lambda_3} \cdots K_{\lambda_m}$$

$$+ K_{\lambda_1}\bar K_{\lambda_2} K_{\lambda_3} \cdots K_{\lambda_m} + \cdots + K_{\lambda_1} K_{\lambda_2} K_{\lambda_3} \cdots K_{\lambda_{m-1}}\bar K_{\lambda_m}\}(1+\alpha_0)^{l+g}$$

$$\times (1 + mA^*)^{s(l+g)}\|\varphi - \bar\varphi\| \le \bar C_s(1+\alpha_0)^{l+g}(1 + mA^*)^{s(l+g)}\|\varphi - \bar\varphi\|.$$

For $\lambda_i = 0$, we set $K_{\lambda_i} = \bar K_{\lambda_i} = 1$, and the constants $\bar C_s$ are independent of n and $\{\varphi, \bar\varphi\} \subset \mathcal{T}_m$, which proves inequality (2.66).

To prove estimate (2.67), we use the method of mathematical induction. For $s = 1$, the estimate was proved in i. 1^0 of this section. Assume that estimate (2.67) holds for all natural $s < d < \rho$ and will prove its validity for $s = d$.

Using equality (2.59), we conclude that $\|D_\varphi^d(\overset{(n)}{\Omega}{}_l^0(p,\varphi) - \overset{(n)}{\Omega}{}_l^0(p,\bar\varphi))\|$ does not exceed the expression $\sum_{k=1}^{l} \bar\Gamma$, where $\bar\Gamma$ is the sum of a finite number of terms of the form $\bar\Gamma^* = \|\Gamma^*(p_0, p_1, p_2, \varphi) - \Gamma^*(p_0, p_1, p_2, \bar\varphi)\|$. But

$$\bar\Gamma^* \le \|D_\varphi^{p_0}(\overset{(n)}{\Omega}{}_k^0(p,\varphi) - \overset{(n)}{\Omega}{}_k^0(p,\bar\varphi))\|\|D_\varphi^{p_1}(\overset{(n)}{P}(\varphi_{k+p-1}(\varphi)))\|$$

$$\times \|D_\varphi^{p_2}(\overset{(n)}{\Omega}{}_l^{k-1}(p,\varphi))\| + \|D_\varphi^{p_0}(\overset{(n)}{\Omega}{}_k^0(p,\bar\varphi))\|\|D_\varphi^{p_1}(\overset{(n)}{P}(\varphi_{k+p-1}(\varphi))$$

$$- \overset{(n)}{P}(\varphi_{k+p-1}(\bar\varphi))\|\|D_\varphi^{p_2}(\overset{(n)}{\Omega}{}_l^{k-1}(p,\varphi))\|$$

$$+ \|D_\varphi^{p_0}(\overset{(n)}{\Omega}{}_k^0(p,\bar\varphi))\|\|D_\varphi^{p_1}(\overset{(n)}{P}(\varphi_{k+p-1}(\bar\varphi)))\|$$

$$\times \|D_\varphi^{p_2}(\overset{(n)}{\Omega}{}_l^{k-1}(p,\varphi) - \overset{(n)}{\Omega}{}_l^{k-1}(p,\bar\varphi))\|.$$

In view of (2.57), (2.61), and (2.63), we have

$$\bar\Gamma^* \le \|D_\varphi^{p_0}(\overset{(n)}{\Omega}{}_k^0(p,\varphi) - \overset{(n)}{\Omega}{}_k^0(p,\bar\varphi))\|X(p_1)(1 + mA^*)^{p_1(k+p-1)}$$

$$\times \tilde Z(p_2)\lambda^{l-k+1}(1 + mA^*)^{p_2 l}$$

$$+ Z_{p_0}\lambda^k(1 + mA^*)^{p_0 k}\tilde Z(p_2)\lambda^{l-k+1}(1 + mA^*)^{p_2 l}\|D_\varphi^{p_1}(\overset{(n)}{P}(\varphi_{k+p-1}(\varphi))$$

$$- \overset{(n)}{P}(\varphi_{k+p-1}(\bar\varphi))\| + Z_{p_0}\lambda^k(1 + mA^*)^{p_0 k}X(p_1)(1 + mA^*)^{p_1(k+p-1)}$$

$$\times \|D_\varphi^{p_2}(\overset{(n)}{\Omega}{}_l^{k-1}(p,\varphi) - \overset{(n)}{\Omega}{}_l^{k-1}(p,\bar\varphi))\|.$$

With regard for representation (2.60), we may assert that the expression

$$\|D_\varphi^{p_1}(\overset{(n)}{P}(\varphi_{k+p-1}(\varphi)) - \overset{(n)}{P}(\varphi_{k+p-1}(\bar\varphi)))\|$$

does not exceed the sum of a finite number of terms of the form:

$$\gamma(1+\alpha_0)^{k+p-1}\prod_{i=1}^{m}K_{\lambda_i}(1+mA^*)^{\lambda_i(k+p-1)}\|\varphi-\bar\varphi\|$$

$$+ mP^*\{\bar K_{\lambda_1}K_{\lambda_2}K_{\lambda_3}\cdots K_{\lambda_m} + K_{\lambda_1}\bar K_{\lambda_2}K_{\lambda_3}\cdots K_{\lambda_m} + \cdots + K_{\lambda_1}$$

$$\times K_{\lambda_2}\cdots K_{\lambda_{m-1}}\bar K_{\lambda_m}\}(1+\alpha_0)^{k+p-1}(1+mA^*)^{p_1(k+p-1)}\|\varphi-\bar\varphi\|.$$

By designating their number by m_1, we obtain the estimate

$$\|D_\varphi^{p_1}(\overset{(n)}{P}(\varphi_{k+p-1}(\varphi)) - \overset{(n)}{P}(\varphi_{k+p-1}(\bar\varphi)))\|$$

$$\leq \gamma_0(1+\alpha_0)^{k+p-1}(1+mA^*)^{p_1(k+p-1)}\|\varphi-\bar\varphi\|,$$

where

$$\gamma_0 = \{\gamma\prod_{i=1}^{m}K_{\lambda_i} + mP^*(\bar K_{\lambda_1}K_{\lambda_2}K_{\lambda_3}\cdots K_{\lambda_m}$$

$$+ K_{\lambda_1}\bar K_{\lambda_2}K_{\lambda_3}\cdots K_{\lambda_m} + \cdots + K_{\lambda_1}K_{\lambda_2}\cdots K_{\lambda_{m-1}}\bar K_{\lambda_m})\}m_1$$

is a constant independent of n and $\{\varphi,\bar\varphi\}\subset\mathcal{T}_m$.

Then, in view of the equality

$$\|D_\varphi^{p_2}(\overset{(n)}{\Omega}{}_l^{k-1}(p,\varphi) - \overset{(n)}{\Omega}{}_l^{k-1}(p,\bar\varphi))\|$$

$$= \|D_\varphi^{p_2}(\overset{(n)}{\Omega}{}_{l-k+1}^{0}(p,\varphi_{k-1}(\varphi)) - \overset{(n)}{\Omega}{}_{l-k+1}^{0}(p,\varphi_{k-1}(\bar\varphi)))\|,$$

we have

$$\bar\Gamma^* \leq X(p_1)\tilde Z(p_2)\Omega_{p_0}\lambda^k(1+\alpha_0)^k(1+mA^*)^{p_0 k}$$

$$\times (1+mA^*)^{p_1(k+p-1)}\lambda^{l-k+1}(1+mA^*)^{p_2 l}\|\varphi-\bar\varphi\|$$

$$+ Z(p_0)\tilde Z(p_2)\gamma_0\lambda^k(1+mA^*)^{p_0 k}\lambda^{l-k+1}(1+mA^*)^{p_2 l}$$

$$\times (1+\alpha_0)^{k+p-1}(1+mA^*)^{p_1(k+p-1)}\|\varphi-\bar\varphi\|$$

$$+ Z_{p_0}X(p_1)\Omega_{p_2}\lambda^k(1+mA^*)^{p_0 k}(1+mA^*)^{p_1(k+p-1)}$$

$$\times \lambda^{l-k+1}(1+\alpha_0)^{l-k+1}(1+mA^*)^{p_2(l-k+1)}\|\varphi-\bar\varphi\|$$

$$\leq \lambda X(p_1)\tilde Z(p_2)\Omega_{p_0}(1+mA^*)^{pd}\lambda^l(1+\alpha_0)^k(1+mA^*)^{ld}\|\varphi-\bar\varphi\|$$

$$+ \lambda(1+\alpha_0)^p Z_{p_0}\tilde Z(p_2)\gamma_0(1+mA^*)^{pd}\lambda^l(1+\alpha_0)^k(1+mA^*)^{ld}\|\varphi-\bar\varphi\|$$

$$+\lambda(1+mA^*)^{pd}Z_{p_0}X(p_1)\Omega_{p_2}(1+mA^*)^{p_2 l}\lambda^l(1+\alpha_0)^l(1+mA^*)^{(p_0+p_1)k}\|\varphi-\bar\varphi\|.$$

Introducing the notation

$$\lambda X(p_1)\tilde{Z}(p_2)\Omega_{p_0}(1+mA^*)^{pd} = \bar{\Gamma}_1^*,$$

$$\lambda(1+\alpha_0)^p Z_{p_0}\tilde{Z}(p_2)\gamma_0(1+mA^*)^{pd} = \bar{\Gamma}_2^*,$$

$$\lambda(1+mA^*)^{pd} Z_{p_0}X(p_1)\Omega_{p_2} = \bar{\Gamma}_3^*,$$

we obtain the estimate

$$\bar{\Gamma}^* \le \lambda^l\{(1+\alpha_0)^k(1+mA^*)^{dl}(\bar{\Gamma}_1^* + \bar{\Gamma}_2^*)$$
$$+ \bar{\Gamma}_3^*(1+\alpha_0)^l(1+mA^*)^{p_2 l}(1+mA^*)^{(p_0+p_1)k}\}\|\varphi - \bar{\varphi}\|.$$

Whence we have

$$\sum_{k=1}^{l} \bar{\Gamma}^* \le \lambda^l\{(1+mA^*)^{dl}\frac{1}{\alpha_0}(\bar{\Gamma}_1^* + \bar{\Gamma}_2^*)(1+\alpha_0)^{l+1}$$
$$+ \bar{\Gamma}_3^*(1+\alpha_0)^l(1+mA^*)^{p_2 l}\frac{1}{(1+mA^*)^{p_0+p_1} - 1}$$
$$\times (1+mA^*)^{p_0+p_1}(1+mA^*)^{(p_0+p_1)l}\}\|\varphi - \bar{\varphi}\|.$$

Designating, in turn, the expression

$$\frac{1+\alpha_0}{\alpha_0}(\bar{\Gamma}_1^* + \bar{\Gamma}_2^*) + \bar{\Gamma}_3^*\frac{(1+mA^*)^d}{mA^*}$$

by $\bar{\Gamma}_4^*$, we obtain the inequality

$$\sum_{k=1}^{l} \bar{\Gamma}^* \le \bar{\Gamma}_4^*\lambda^l(1+\alpha_0)^l(1+mA^*)^{dl}\|\varphi - \bar{\varphi}\|,$$

where $\bar{\Gamma}_4^*$ is a constant independent of n, l and $\{\varphi, \bar{\varphi}\} \subset \mathcal{T}_m$. This completes the proof of Lemma 2.11. $\qquad\square$

Proof of Theorem 2.6.

Proof.   As was shown above, the functions

$$u(p, g, \varphi) = -\sum_{l=1}^{\infty} \Omega_l^0(p, \varphi)c(\varphi_{l+g}(\varphi))$$

and

$$\overset{(s)}{u}(p, g, \varphi) = -\sum_{l=1}^{\infty} \overset{(s)}{\Omega_l^0}(p, \varphi)\overset{(s)}{c}(\varphi_{l+g}(\varphi))$$

generate, under conditions of the theorem, the invariant tori of the systems of equations (2.4) and (2.53), respectively, and

$$\lim_{s \to \infty} \overset{(s)}{u}(p, g, \varphi) = u(p, g, \varphi)$$

in the coordinatewise meaning.

We now show that the sequence

$$D_\varphi^s(\overset{(1)}{u}(p, g, \varphi)), D_\varphi^s(\overset{(2)}{u}(p, g, \varphi)), \ldots, D_\varphi^s(\overset{(n)}{u}(p, g, \varphi)), \ldots \qquad (2.68)$$

is bounded uniformly in n and is equicontinuous in φ for every natural $s \le \rho$.

It is obvious that

$$D_\varphi^s(\overset{(n)}{u}(p, g, \varphi)) = D_\varphi^s(-\sum_{l=1}^{\infty} \overset{(n)}{\Omega_l^0}(p, \varphi) \overset{(n)}{c}(\varphi_{l+g}(\varphi))).$$

If the series

$$-\sum_{l=1}^{\infty} D_\varphi^k(\overset{(n)}{\Omega_l^0}(p, \varphi)) D_\varphi^{s-k}(\overset{(n)}{c}(\varphi_{l+g}(\varphi))) \quad (k = 0, 1, 2, \ldots, s) \qquad (2.69)$$

converge in the norm uniformly in $\varphi \in \mathcal{T}_m$ for each indicated k, then the equality

$$D_\varphi^s(\overset{(n)}{u}(p, g, \varphi)) = -\sum_{l=1}^{\infty} D_\varphi^s(\overset{(n)}{\Omega_l^0}(p, \varphi) \overset{(n)}{c}(\varphi_{l+g}(\varphi))) = \varepsilon(s, \varphi)$$

holds. Here, $\varepsilon(s, \varphi)$ is the sum of a finite number of terms of the form (2.69).

Using Lemma 2.9, we write down the estimates

$$\sum_{l=1}^{\infty} \|D_\varphi^k(\overset{(n)}{\Omega_l^0}(p, \varphi))\| \|D_\varphi^{s-k}(\overset{(n)}{c}(\varphi_{l+g}(\varphi)))\|$$

$$\le \sum_{l=1}^{\infty} Z_k \lambda^l (1 + mA^*)^{kl} C_{s-k}^* (1 + mA^*)^{(s-k)(l+g)}$$

$$\le Z_k C_{s-k}^* (1 + mA^*)^{(s-k)g} \sum_{l=1}^{\infty} \lambda^l (1 + mA^*)^{sl}$$

$$\le Z_k C_{s-k}^* (1 + mA^*)^{sg} \frac{\lambda(1 + mA^*)^s}{1 - \lambda(1 + mA^*)^s}.$$

Hence, series (2.69) converge in the norm for each $(k = 0, 1, 2, \ldots, s)$ uniformly in $\varphi \in \mathcal{T}_m$, and sequence (2.68) is uniformly bounded.

The expression

$$\|D_\varphi^s(\overset{(n)}{u}(p,g,\varphi) - \overset{(n)}{u}(p,g,\bar\varphi))\| = \|\varepsilon(s,\varphi) - \varepsilon(s,\bar\varphi)\|$$

does not exceed the sum of a finite number of terms of the form

$$I(s) = \sum_{l=1}^{\infty} \|D_\varphi^k(\overset{(n)}{\Omega^0_l}(p,\varphi))D_\varphi^{s-k}(\overset{(n)}{c}(\varphi_{l+g}(\varphi)))$$

$$- D_\varphi^k(\overset{(n)}{\Omega^0_l}(p,\bar\varphi))D_\varphi^{s-k}(\overset{(n)}{c}(\varphi_{l+g}(\bar\varphi)))\|, \quad k = 0,1,2,\ldots,s.$$

Using Lemmas 2.9, 2.11 and denoting the expression

$$\Omega_k C^*_{s-k}(1+mA^*)^{sg} + Z_k(1+\alpha_0)^g(1+mA^*)^{sg}C_{s-k}$$

by $\Omega(s,k)$, we obtain the relations

$$I(s) \le \sum_{l=1}^{\infty}\{\|D_\varphi^k(\overset{(n)}{\Omega^0_l}(p,\varphi) - \overset{(n)}{\Omega^0_l}(p,\bar\varphi))\|\|D_\varphi^{s-k}(\overset{(n)}{c}(\varphi_{l+g}(\varphi)))\|$$

$$+ \|D_\varphi^k(\overset{(n)}{\Omega^0_l}(p,\bar\varphi))\|\|D_\varphi^{s-k}(\overset{(n)}{c}(\varphi_{l+g}(\varphi)) - \overset{(n)}{c}(\varphi_{l+g}(\bar\varphi)))\|\}$$

$$\le \sum_{l=1}^{\infty}\{\Omega_k\lambda^l(1+\alpha_0)^l(1+mA^*)^{kl}C^*_{s-k}(1+mA^*)^{(s-k)(l+g)}\|\varphi - \bar\varphi\|$$

$$+ Z_k\lambda^l(1+mA^*)^{kl}C_{s-k}(1+\alpha_0)^{l+g}(1+mA^*)^{(s-k)(l+g)}\|\varphi - \bar\varphi\|$$

$$\le \|\varphi - \bar\varphi\|\Omega(s,k)\sum_{l=1}^{\infty}\lambda^l(1+\alpha_0)^l(1+mA^*)^{sl}$$

$$\le \Omega(s,k)\frac{\lambda(1+\alpha_0)(1+mA^*)^s}{1-\lambda(1+\alpha_0)(1+mA^*)^s}\|\varphi - \bar\varphi\|.$$

This guarantees the equicontinuity of sequence (2.68) in $\varphi \in \mathcal{T}_m$ for every $s \le \rho$.

In view of the coordinatewise convergence of the sequence $\{\overset{(n)}{u}(p,g,\varphi)\}$ to $u(p,g,\varphi)$ as $n \to \infty$, we apply the Arzela–Ascoli theorem the required number of times, which completes the proof of the theorem. $\square$

Corollary 2.6. *If $a(\varphi) = \omega$ in the system of equations (2.4), where ω is a constant vector, then the assertion of Theorem 2.6 is valid for all integers p and g.*

Performing the analogous reasoning, it is possible to obtain the sufficient conditions of the differentiability of the invariant torus of the system of equations (2.1) with respect to the parameter μ up to the order $\rho > 1$.

For example, let us consider this system in the case where the function $a(\varphi,\mu) = a(\varphi)$, i.e., it is independent of μ.

Corollary 2.7. *Consider the system of equations*

$$\varphi_{n+1} = \varphi_n + a(\varphi_n), \quad x_{n+1} = P(\varphi_{n+p},\mu)x_n + c(\varphi_{n+g+1},\mu), \qquad (2.70)$$

which is a partial case of system (2.1).

For all $s \in Z^+$, $\mu \in \sigma$, $\varphi \in \mathcal{T}_m$, let the following conditions be satisfied:

1) $\{P(\varphi,\mu), c(\varphi,\mu)\} \subset C^\rho_{Lip}(\mu);$

2) $\|D^l_\mu(P(\varphi,\mu))\| \leq P^*$, $\|D^l_\mu(c(\varphi,\mu))\| \leq C^*$,

where $0 \leq l \leq \rho$, and the positive constants P^*, C^* are independent of l, μ, and φ;

3) the matrix $\overset{(s)}{P}(\varphi,\mu)$ is nondegenerate;

4) for all integers $l > n$,

$$\|\overset{(s)}{\Omega^n_l}(0,\varphi,\mu)\| \leq M\lambda^{l-n},$$

where the constants $M > 0$ and $0 < \lambda < 1$ are independent of s, μ, and φ;

5) the sequence $\{\overset{(s)}{P^{-1}}(\varphi,\mu)\}^\infty_{s=1}$ is proper, and $\|\overset{(s)}{P^{-1}}(\varphi,\mu)\| \leq P_1$, where the positive constant P_1 is independent of s, μ, and φ.

Then, $\forall\{p,g\} \subset Z$, the function $u(p,g,\mu,\varphi)$, which generates the invariant torus of the system of equations (2.70), belongs to $C^\rho_{Lip}(\mu)$.

Proof. It is sufficient to show that the sequence

$$D^s_\mu(\overset{(1)}{u}(p,g,\mu,\varphi)), D^s_\mu(\overset{(2)}{u}(p,g,\mu,\varphi)), \ldots, D^s_\mu(\overset{(n)}{u}(p,g,\mu,\varphi)), \ldots \qquad (2.71)$$

is bounded uniformly in n and is equicontinuous in μ for every natural $s \leq \rho$. Here,

$$\overset{(s)}{u}(p,g,\mu,\varphi) = -\sum_{l=1}^\infty \overset{(s)}{\Omega^0_l}(p,\varphi,\mu) \overset{(s)}{c}(\varphi_{l+g}(\varphi),\mu).$$

First, we prove that, for all natural n and $1 \leq s \leq \rho$, the inequalities

$$\|D^s_\mu(\overset{(n)}{\Omega^0_l}(p,\varphi,\mu))\| \leq Z_{s*}\lambda^l(l+s)^s, \qquad (2.72)$$

$$\|D^s_\mu(\overset{(n)}{\Omega^0_l}(p,\varphi,\mu) - \overset{(n)}{\Omega^0_l}(p,\varphi,\bar\mu))\| \leq \Omega_{s*}\lambda^l(l+s)^s\|\mu - \bar\mu\| \qquad (2.73)$$

hold. Here, the positive constants Z_{s*} and Ω_{s*} are independent of n, $\varphi \in \mathcal{T}_m$, and $\{\mu,\bar\mu\} \subset \sigma$.

For $s = 1$, inequality (2.72) holds, since

$$\|D_\mu^1(\overset{(n)}{\Omega}\,^0_l(p, \varphi, \mu))\| \le M^2 P^* \lambda^l l.$$

We assume that it holds for all natural $s < d \le \rho$ and will prove its validity for $s = d$. Itis easy to see that

$$D_\mu^d(\overset{(n)}{\Omega}\,^0_l(p, \varphi, \mu)) = -\sum_{k=1}^{l} \Gamma_\mu,$$

where Γ_μ is the sum of a finite number of terms of the form

$$\Gamma_\mu^* = D_\mu^{p_0}(\overset{(n)}{\Omega}\,^0_k(p, \varphi, \mu))D_\mu^{p_1}(\overset{(n)}{P}(\varphi_{k+p-1}(\varphi), \mu))D_\mu^{p_2}(\overset{(n)}{\Omega}\,^{k-1}_l(p, \varphi, \mu)),$$

$\{p_0, p_2\} \subset \{0, 1, \ldots, d-1\}$, $p_1 \in \{1, 2, \ldots, d\}$, $p_0 + p_1 + p_2 = d$.

Since $p_0 + p_2 \le d - 1$ and $1 \le k \le l$, it is easy to obtain the estimates

$$\|\Gamma_\mu^*\| \le Z_{p_0*} \lambda^k (k + p_0)^{p_0} P^* \lambda^{l-k+1} (l - k + 1 + p_2)^{p_2} Z_{p_2*}$$

$$\le Z_{p_0*} Z_{p_2*} P^* \lambda^l (1 + d)^{p_0 + p_2},$$

which yield the inequality $\sum_{k=1}^{l} \|\Gamma_\mu^*\| \le Z_{p_0*} Z_{p_2*} P^* \lambda^l (1 + d)^d$. This inequality ensures the validity of estimate (2.72).

For $s = 1$, inequality (2.73) can be prove in the same way as the analogous estimate in Lemma 2.7. Indeed, the relations

$$\|D_\mu^1(\overset{(n)}{\Omega}\,^0_l(p, \varphi, \mu) - \overset{(n)}{\Omega}\,^0_l(p, \varphi, \bar\mu))\| \le \Omega_{1*} \lambda^l l \|\mu - \bar\mu\| < \Omega_{1*} \lambda^l (l + 1) \|\mu - \bar\mu\|$$

hold. Then we again use the method of complete mathematical induction. We assume that inequality (2.73) holds for all natural $s < d \le \rho$ and will prove its validity for $s = d$. It is easy to see that

$$\bar\Gamma_\mu^* = \|\Gamma_\mu^*(\mu) - \Gamma_\mu^*(\bar\mu)\| \le \Omega_{d*}^0 \lambda^l (1 + d)^{d-1} \|\mu - \bar\mu\|,$$

where the positive constant Ω_{d*}^0 is independent of $n, \mu, \bar\mu$, and φ. Then the inequality $\sum_{k=1}^{l} \bar\Gamma_\mu^* \le \Omega_{d*} \lambda^l (l + d)^d \|\mu - \bar\mu\|$ ensuring the validity of estimate (2.73) holds.

It remains to carry out the consideration analogous to the proof of Theorem 2.6.

We note that

$$D_\mu^s(\overset{(n)}{u}(p, g, \mu, \varphi)) = D_\mu^s(-\sum_{l=1}^{\infty} \overset{(n)}{\Omega}\,^0_l(p, \varphi, \mu) \overset{(n)}{c}(\varphi_{l+g}(\varphi), \mu)).$$

Therefore, if the series

$$-\sum_{l=1}^{\infty} D_\mu^k(\overset{(n)}{\Omega}\,^0_l(p, \varphi, \mu))D_\mu^{s-k}(\overset{(n)}{c}(\varphi_{l+g}(\varphi), \mu)) \quad (k = 0, 1, 2, \ldots, s) \quad (2.74)$$

converge in the norm uniformly in $\mu \in \sigma$ for each indicated k, then the equality

$$D_\mu^s(\overset{(n)}{u}(p,g,\mu,\varphi)) = -\sum_{l=1}^{\infty} D_\mu^s(\overset{(n)}{\Omega_l^0}(p,\varphi,\mu)\,\overset{(n)}{c}(\varphi_{l+g}(\varphi),\mu)) = \varepsilon_0(s,\mu),$$

where $\varepsilon_0(s,\mu)$ is the sum of a finite number of terms of the form (2.74), holds.

The estimate

$$\sum_{l=1}^{\infty} \|D_\mu^k(\overset{(n)}{\Omega_l^0}(p,\varphi,\mu))\|\,\|D_\mu^{s-k}(\overset{(n)}{c}(\varphi_{l+g}(\varphi),\mu))\| \leq \sum_{l=1}^{\infty} Z_{k*}\lambda^l(l+k)^k C^*$$

indicates that series (2.74) converge in the norm uniformly in $\mu \in \sigma$ for each $(k = 0,1,2,\ldots,s)$, and sequence (2.71) is uniformly bounded, since the series $\sum_{l=1}^{\infty} \lambda^l(l+k)^k$ is convergent.

The expression

$$\|D_\mu^s(\overset{(n)}{u}(p,g,\mu,\varphi) - \overset{(n)}{u}(p,g,\bar{\mu},\varphi))\| = \|\varepsilon_0(s,\mu) - \varepsilon_0(s,\bar{\mu})\|$$

does not exceed the sum of a finite number of terms of the form

$$I_0(s) = \sum_{l=1}^{\infty} \|D_\mu^k(\overset{(n)}{\Omega_l^0}(p,\varphi,\mu))D_\mu^{s-k}(\overset{(n)}{c}(\varphi_{l+g}(\varphi),\mu))$$

$$- D_\mu^k(\overset{(n)}{\Omega_l^0}(p,\varphi,\bar{\mu}))D_\mu^{s-k}(\overset{(n)}{c}(\varphi_{l+g}(\varphi),\bar{\mu}))\|, \quad k = 0,1,2,\ldots,s.$$

The estimate

$$I_0(s) \leq (\Omega_{k*}C^* + Z_{k*}\delta)\|\mu - \bar{\mu}\| \sum_{l=1}^{\infty} \lambda^l(l+k)^k,$$

where δ is a coefficient, with which $c(\varphi,\mu)$ enters $C_{Lip}^\rho(\mu)$, guarantees the equicontinuity of sequence (2.71) in $\mu \in \sigma$ for every $s \leq \rho$ and the differentiability of the invariant torus of system (2.70) with respect to the parameter μ up to the order $\rho \geq 2$. $\qquad\square$

2.4   Case of linear and quasilinear systems defined on the infinite-dimensional tori

In three previous sections, we have considered the questions of the existence and the smoothness of the invariant tori of difference systems defined on finite-dimensional tori. Theorems 2.1 and 2.2 concerning the continuity of invariant tori are easily transferred to the case of a countable number of angular variables. But Theorems 2.3 and 2.4 of the differentiability of invariant tori cannot be transferred to this case, since the methods of their proof cannot be used.

Consider the system of equations

$$\varphi_{n+1} = \varphi_n + a(\varphi_n, \mu), \quad x_{n+1} = P(\varphi_{n+p}, \mu)x_n + c(\varphi_{n+g+1}, \mu), \qquad (2.75)$$

where $\varphi = (\varphi^1, \varphi^2, \varphi^3, \dots) \in \mathfrak{M}$, $x = (x^1, x^2, x^3, \dots) \in \mathfrak{M}$; the functions

$$a(\varphi, \mu) = \{a_1(\varphi, \mu), a_2(\varphi, \mu), a_3(\varphi, \mu), \dots\},$$

$$c(\varphi, \mu) = \{c_1(\varphi, \mu), c_2(\varphi, \mu), c_3(\varphi, \mu), \dots\}$$

and the infinite matrix $P(\varphi, \mu) = [p_{ij}(\varphi, \mu)]_{i,j=1}^{\infty}$ are real and 2π-periodic in $\varphi^i (i = 1, 2, 3, \dots)$; $n \in Z$; p and g are integer-valued parameters, which characterize a deviation of the discrete argument; $\mu \in S \subset \mathfrak{M}$ is a parameter, and S is an open ball in $\mathfrak{M}$.

By interpreting φ^i as angular coordinates, we consider that the system of equations (2.75) is defined on the infinite-dimensional torus $\mathcal{T}_\infty$.

In what follows in this subsection, we consider that, for every $\mu \in S$, the mapping $\Phi(\varphi, \mu) = \varphi + a(\varphi, \mu) : \mathfrak{M} \to \mathfrak{M}$ is invertible,

$$\|a(\varphi, \mu)\| \le A^0, \|c(\varphi, \mu)\| \le C^0, \|P(\varphi, \mu)\| = \sup_i \sum_{j=1}^{\infty} |p_{ij}(\varphi, \mu)| \le P^0,$$

and A^0, P^0, C^0 are positive constants, which are independent of $\mu \in S$, $\varphi \in \mathcal{T}_\infty$.

Remark 2.2. As for the mapping $\Phi(\varphi, \mu) : \mathfrak{M} \times S \to \mathfrak{M}$, it is not invertible, even if the mentioned condition is satisfied, and the mapping $\Phi(\varphi, \mu) : S \to \Phi(\varphi, S)$ is invertible for every $\varphi \in \mathfrak{M}$.

The invariant torus $\mathcal{T}(p, g, \mu)$ of the system of equations (2.75) is called Fréchet-differentiable with respect to (φ, μ), if the function $u(p, g, \mu, \varphi)$ generating it has such a property at every point $(\varphi, \mu) \in \Lambda = \mathfrak{M} \times S$. Here, $\mathfrak{M} \times S$ is a Cartesian product.

By $C_\Lambda^1(\varphi, \mu)$, we denote the set of mappings, which are defined and continuously Fréchet-differentiable with respect to (φ, μ) on Λ, and agree that, in what follows, the derivatives are exclusively the Fréchet derivatives.

Like in the first chapter, we denote, by $\boldsymbol{\Gamma}$, the set of all infinite matrices bounded in the matrix norm $\| \cdot \|$ and recall that $\boldsymbol{\Gamma}$ forms a vector normed space over the field of real numbers. It is understandable that the set Λ is open in the set $\mathfrak{M} \times \mathfrak{M}$. Moreover, the norm of an element $h = (\varphi, \mu) \in \Lambda$

is given by the expression $\|h\| = \max\{\|\varphi\|, \|\mu\|\}$, where $\|\varphi\|$ and $\|\mu\|$ are norms in the space $\mathfrak{M}$.

First, we prove several auxiliary propositions.

Lemma 2.12. *Let every of the matrices $P_i(h)(i = \overline{1, n})$ maps $\Lambda \to \Gamma$ and belongs to $C^1_\Lambda(h)$. Then the product $\prod_{i=1}^n P_i(h) \in C^1_\Lambda(h)$.*

Proof. First, consider the case $n = 2$. The operation of multiplication of matrices define a bilinear mapping $B^2 : \Gamma \times \Gamma \to \Gamma$. By Theorem 52 in [136], the mapping B^2 is continuous, because the inequality $\|A_1 A_2\| \leq \|A_1\|\|A_2\|$ holds for any $\{A_1, A_2\} \subset \Gamma$. Then, by Theorem 12 in [136], the mapping $P_1(h)P_2(h) : \Lambda \to \Gamma$ is continuously Fréchet-differentiable. Moreover, the equality

$$\frac{d(P_1(h)P_2(h))}{dh}(\xi) = \frac{dP_1(h)}{dh}(\xi)P_2(h) + P_1(h)\frac{dP_2(h)}{dh}(\xi), \qquad (2.76)$$

where $h \in \Lambda, \xi \in \mathfrak{M} \times \mathfrak{M}$, holds.

It is easy to extend formula (2.76) to the case of the multilinear mapping $B^n : \Gamma^n \to \Gamma$, which is defined by a product of n matrices from Γ, and to obtain the assertion of Lemma 2.12. In this case, the equality

$$\frac{d(P_1(h)P_2(h) \cdots P_n(h))}{dh}(\xi) = \frac{dP_1(h)}{dh}(\xi)P_2(h) \cdots P_n(h)$$

$$+P_1(h)\frac{dP_2(h)}{dh}(\xi)P_3(h) \cdots P_n(h) + \cdots + P_1(h) \cdots P_{n-1}(h)\frac{dP_n(h)}{dh}(\xi)$$
$$(2.77)$$

holds. $\qquad\square$

Lemma 2.13. *Let $\{a(\varphi, \mu), \Phi^{-1}(\varphi, \mu)\} \subset C^1_\Lambda(\varphi, \mu)$, and let the inequalities*

$$\left\|\frac{da(\varphi, \mu)}{d(\varphi, \mu)}\right\| \leq A^*; \quad \left\|\frac{d\Phi^{-1}(\varphi, \mu)}{d(\varphi, \mu)}\right\| \leq \Phi_*,$$

where A^* and Φ_* are positive constants, be valid $\forall(\varphi, \mu) \in \Lambda$. Then, for all $n \in Z$, the function $\varphi_n(\varphi, \mu) \in C^1_\Lambda(\varphi, \mu)$, and

$$\left\|\frac{d\varphi_n(\varphi, \mu)}{d(\varphi, \mu)}\right\| \leq \begin{cases} (1 + A^*)^n & for \quad n \geq 0; \\ \Phi_*^{-n} & for \quad n < 0, \end{cases}$$

if $\Phi_* > 1$.

Proof. It is obvious that, for $n \in \{0, 1\}$, the assertion of Lemma 2.13 is valid. Indeed, we have

$$\frac{d\varphi_0(\varphi, \mu)}{d(\varphi, \mu)}(\varphi^*, \mu^*) = E\varphi^* + 0\mu^*,$$

where $(\varphi, \mu) \in \Lambda, (\varphi^*, \mu^*) \in \mathfrak{M} \times \mathfrak{M}$, E is the identity operator, 0 is the zero operator; and

$$\frac{d\varphi_1(\varphi, \mu)}{d(\varphi, \mu)} = \frac{d\varphi_0(\varphi, \mu)}{d(\varphi, \mu)} + \frac{da(\varphi, \mu)}{d(\varphi, \mu)}.$$

In addition,

$$\left\|\frac{d\varphi_0(\varphi, \mu)}{d(\varphi, \mu)}\right\| = \sup_{\|(\varphi^*, \mu^*)\|=1} \|E\varphi^* + 0\mu^*\| = 1;$$

$$\left\|\frac{d\varphi_1(\varphi, \mu)}{d(\varphi, \mu)}\right\| \le 1 + A^*.$$

For $n = 2$, we have $\varphi_2(\varphi, \mu) = \varphi_1(\varphi, \mu) + a(\varphi_1(\varphi, \mu), \mu)$. We represent the function $a(\varphi_1(\varphi, \mu), \mu)$ as a mapping $a(h) : \Lambda \to \mathfrak{M}$, where the mapping $h : \Lambda \to \Lambda$ is composed of two components: $\varphi_1(\varphi, \mu) : \Lambda \to \mathfrak{M}$ and $\mu(\varphi, \mu) : \Lambda \to S$ by the law $\mu(\varphi, \mu) = \mu$. Since these components are continuously Fréchet-differentiable on Λ, the mappings h and $a(\varphi_1(\varphi, \mu), \mu)$ have the same property, i.e.,

$$\frac{da(\varphi_1(\varphi, \mu), \mu)}{d(\varphi, \mu)} = \frac{da(h)}{dh}\left(\frac{dh}{d(\varphi, \mu)}\right), \quad \frac{da(h)}{dh} = \frac{da(\varphi, \mu)}{d(\varphi, \mu)}\Big|_{h(\varphi, \mu)},$$

and

$$\frac{da(\varphi_1(\varphi, \mu), \mu)}{d(\varphi, \mu)}(\varphi^*, \mu^*) = \frac{da(h)}{dh}\left(\frac{d\varphi_1(\varphi, \mu)}{d(\varphi, \mu)}(\varphi^*, \mu^*); \frac{d\mu(\varphi, \mu)}{d(\varphi, \mu)}(\varphi^*, \mu^*)\right),$$

where

$$\frac{d\mu(\varphi, \mu)}{d(\varphi, \mu)}(\varphi^*, \mu^*) = \frac{d\mu}{d(\varphi, \mu)}(\varphi^*, \mu^*) = 0\varphi^* + E\mu^* = \mu^*.$$

Then

$$\left\|\frac{da(\varphi_1(\varphi, \mu), \mu)}{d(\varphi, \mu)}\right\|$$

$$\le \left\|\frac{da(h)}{dh}\right\| \sup_{\max\{\|\varphi^*\|, \|\mu^*\|\}=1} \max\left\{\left\|\frac{d\varphi_1(\varphi, \mu)}{d(\varphi, \mu)}(\varphi^*, \mu^*)\right\|; \|\mu^*\|\right\}$$

$$\le A^*(1 + A^*),$$

whence we have

$$\left\|\frac{d\varphi_2(\varphi, \mu)}{d(\varphi, \mu)}\right\| \le 1 + A^* + A^*(1 + A^*) = (1 + A^*)^2.$$

For $n \ge 2$, the assertion of Lemma 2.13 can be easily substantiated by the method of complete mathematical induction.

For $n < 0$, the analogous reasoning yields the relations

$$\varphi_{-1}(\varphi,\mu) = \Phi^{-1}(\varphi,\mu), \quad \left\|\frac{d\varphi_{-1}(\varphi,\mu)}{d(\varphi,\mu)}\right\| \leq \Phi_*;$$

$$\left\|\frac{d\varphi_{-2}(\varphi,\mu)}{d(\varphi,\mu)}\right\| = \left\|\frac{d\Phi^{-1}(\varphi_{-1}(\varphi,\mu),\mu)}{d(\varphi,\mu)}\right\| \leq \Phi_*\Phi_* = \Phi_*^2,$$

and so on. The lemma is proved. $\qquad\qquad\square$

Let $\Omega_l^n(p,\varphi,\mu)$ be the matriciant of the equation

$$x_{n+1} = P(\varphi_{n+p}(\varphi,\mu),\mu)x_n.$$

Lemma 2.14. *Assume that, $\forall\mu \in S$, $\varphi \in \mathcal{T}_\infty$, there exists the matrix $P^{-1}(\varphi,\mu)$ inverse to the matrix $P(\varphi,\mu)$, and $\|P^{-1}(\varphi,\mu)\| \leq P_1 = const > 0$. Let, in addition, $\{P^{-1}(\varphi,\mu), c(\varphi,\mu)\} \subset C_\Lambda^1(\varphi,\mu)$, let the conditions of Lemma 2.13 be satisfied, and let*

$$\left\|\frac{dP^{-1}(\varphi,\mu)}{d(\varphi,\mu)}\right\| \leq P_*, \quad \left\|\frac{dc(\varphi,\mu)}{d(\varphi,\mu)}\right\| \leq C_*,$$

where P_* and C_* are positive constants, which are independent of $(\varphi,\mu) \in \Lambda$.

Then, for all $l > 0$, $\{p,g\} \subset Z$, the inclusion

$$\{\Omega_l^0(p,\varphi,\mu), c(\varphi_{l+g}(\varphi,\mu),\mu)\} \subset C_\Lambda^1(\varphi,\mu)$$

holds, and the inequalities

$$\left\|\frac{dc(\varphi_{l+g}(\varphi,\mu),\mu)}{d(\varphi,\mu)}\right\| \leq \begin{cases} C_*(1+A^*)^{l+g} & for \quad l+g \geq 0; \\ C_*\Phi_*^{-(l+g)} & for \quad l+g < 0, \end{cases} \qquad (2.78)$$

$$\left\|\frac{d\Omega_l^0(p,\varphi,\mu)}{d(\varphi,\mu)}\right\|$$
$$< \begin{cases} \xi_1(p)P_1^{l-1}(1+A^*)^l & for \quad p \geq 0; \\ \xi_2(p)P_1^{l-1} & for \quad p < 0, 0 < l < 1-p; \\ P_1^{l-1}(\xi_2(p) + \xi_1(p)(1+A^*)^l) & for \quad p < 0, l \geq 1-p, \end{cases} \qquad (2.79)$$

where $\xi_1(p)$, $\xi_2(p)$ are independent of l, φ, μ, hold.

Proof.   Analogously to the proof of Lemma 2.13, it is easy to verify that, for all $n \in Z$, the matrix $P^{-1}(\varphi_n(\varphi, \mu), \mu) \in C_\Lambda^1(\varphi, \mu)$, and

$$\left\| \frac{dP^{-1}(\varphi_n(\varphi, \mu), \mu)}{d(\varphi, \mu)} \right\|$$

$$\leq P_* \sup_{\max\{\|\varphi^*\|, \|\mu^*\|\}=1} \max\left\{ \left\| \frac{d\varphi_n(\varphi, \mu)}{d(\varphi, \mu)}(\varphi^*, \mu^*) \right\|; \|\mu^*\| \right\}$$

$$\leq \begin{cases} P_*(1 + A^*)^n & \text{for} \quad n \geq 0; \\ P_* \Phi_*^{-n} & \text{for} \quad n < 0, \Phi_* > 1. \end{cases}$$

The same reasoning leads to inequalities (2.78).

For the matrix $\Omega_l^0(p, \varphi, \mu)$ for any $l > 0$, the representation

$$\Omega_l^0(p, \varphi, \mu) = \prod_{i=p}^{l+p-1} P^{-1}(\varphi_i(\varphi, \mu), \mu) \tag{2.80}$$

is valid.

Equality (2.80) allows us to write down the following inequality:

$$\Omega \overset{(df)}{=} \left\| \frac{d\Omega_l^0(p, \varphi, \mu)}{d(\varphi, \mu)} \right\| \leq P_1^{l-1} \sum_{i=p}^{p+l-1} \left\| \frac{dP^{-1}(\varphi_i(\varphi, \mu), \mu)}{d(\varphi, \mu)} \right\|.$$

In the case where $p \geq 0$, we have the estimate

$$\Omega \leq P_1^{l-1} \sum_{i=p}^{p+l-1} P_*(1 + A^*)^i$$

$$< P_1^{l-1} P_* \frac{1}{A^*}(1 + A^*)^{p+l} = \xi_1(p) P_1^{l-1}(1 + A^*)^l,$$

where $\xi_1(p) = P_* \frac{1}{A^*}(1 + A^*)^p$.

In the case where $p < 0, 0 < l < 1 - p$, we obtain

$$\Omega \leq P_1^{l-1} \sum_{i=p}^{p+l-1} P_* \Phi_*^{-i} = P_1^{l-1} P_* \frac{\Phi_*^{-(p+l-1)}(\Phi_*^l - 1)}{\Phi_* - 1} < \xi_2(p) P_1^{l-1},$$

where

$$\xi_2(p) = \frac{P_* \Phi_*^{-p+1}}{\Phi_* - 1}, \Phi_* > 1.$$

Finally, let $p < 0, l \geq 1 - p$. In this case,

$$\Omega \leq P_1^{l-1} P_* \left\{ \sum_{i=1}^{-p} \Phi_*^i + \sum_{k=0}^{p+l-1} (1 + A^*)^k \right\}$$

$$= P_1^{l-1} P_* \left\{ \frac{\Phi_*(\Phi_*^{-p} - 1)}{\Phi_* - 1} + \frac{(1 + A^*)^{l+p} - 1}{A^*} \right\}$$

$$< P_1^{l-1}(\xi_2(p) + \xi_1(p)(1 + A^*)^l),$$

which completes the proof of estimates (2.79) and, hence, Lemma 2.14. $\square$

Using Lemmas 2.12 – 2.14, we will prove the following proposition.

Theorem 2.7. *Let, $\forall \mu \in S$ and $\forall \varphi \in \mathcal{T}_\infty$, there exist the matrix $P^{-1}(\varphi, \mu)$, $\|P^{-1}(\varphi, \mu)\| \leq P_1 = const > 0$, and let the following conditions hold:*

1) $\{a(\varphi, \mu), c(\varphi, \mu), P^{-1}(\varphi, \mu), \Phi^{-1}(\varphi, \mu)\} \subset C^1_\Lambda(\varphi, \mu)$, and

$$\left\|\frac{da(\varphi, \mu)}{d(\varphi, \mu)}\right\| \leq A^*; \quad \left\|\frac{d\Phi^{-1}(\varphi, \mu)}{d(\varphi, \mu)}\right\| \leq \Phi_*,$$

$$\left\|\frac{dP^{-1}(\varphi, \mu)}{d(\varphi, \mu)}\right\| \leq P_*, \quad \left\|\frac{dc(\varphi, \mu)}{d(\varphi, \mu)}\right\| \leq C_*,$$

where A^*, Φ_*, P_*, and C_* are positive constants, which are independent of $(\varphi, \mu) \in \Lambda$;

2) $P_1 < \frac{1}{1+A^*}$.

Then, for all $\{p, g\} \subset Z$, the system of equations (2.75) has the invariant torus Fréchet-differentiable with respect to (φ, μ) on the set Λ.

Proof. Condition 2 of Theorem 2.7 and equality (2.80) yield the estimate $\|\Omega^0_l(p, \varphi, \mu)\| \leq P_1^l$, where $l > 0, P_1 < 1$. In this case, the Green–Samoilenko function of the problem of the invariant torus of the system of equations (2.75) takes the form

$$G_0(l, p, \mu, \varphi) = \begin{cases} 0, & \text{if } l \leq 0; \\ -\Omega^0_l(p, \varphi, \mu), & \text{if } l > 0, \end{cases}$$

and the function generating the invariant torus $\mathcal{T}(p, g, \mu)$ of this system is defined by the equality

$$u(p, g, \mu, \varphi) = -\sum_{l=1}^{\infty} \Omega^0_l(p, \varphi, \mu) c(\varphi_{l+g}(\varphi, \mu), \mu). \tag{2.81}$$

The series on the right-hand side of equality (2.81) converges uniformly in $\{p, g\} \subset Z$ and $(\varphi, \mu) \in \Lambda$ in the norm of the space $\mathfrak{M}$.

The Fréchet derivative

$$u_l \overset{(df)}{=} \frac{d\{\Omega^0_l(p, \varphi, \mu) c(\varphi_{l+g}(\varphi, \mu), \mu)\}}{d(\varphi, \mu)}$$

exists at every point of the set Λ open in the affine normed space $\mathfrak{M} \times \mathfrak{M}$ and realizes the mapping of every point $(h_1, h_2) \in \mathfrak{M} \times \mathfrak{M}$ into $\mathfrak{M}$ by the law

$$u_l(h_1, h_2) = \frac{d\Omega^0_l(p, \varphi, \mu)}{d(\varphi, \mu)}(h_1, h_2) c(\varphi_{l+g}(\varphi, \mu), \mu)$$

$$+ \Omega^0_l(p, \varphi, \mu) \frac{dc(\varphi_{l+g}(\varphi, \mu), \mu)}{d(\varphi, \mu)}(h_1, h_2).$$

In view of equality (2.81) and Theorem 111 in [136], in order to prove the theorem, it is sufficient to prove that the series $\sum_{l=1}^{\infty} \|u_l\|$ converges uniformly in $(\varphi, \mu) \in \Lambda$. The following inequalities hold:

$$\|u_l\| \leq \left\| \frac{d\Omega_l^0(p, \varphi, \mu)}{d(\varphi, \mu)} \right\| \|c(\varphi_{l+g}(\varphi, \mu), \mu)\| + \|\Omega_l^0(p, \varphi, \mu)\|$$

$$\times \left\| \frac{dc(\varphi_{l+g}(\varphi, \mu), \mu)}{d(\varphi, \mu)} \right\| \leq \left\| \frac{d\Omega_l^0(p, \varphi, \mu)}{d(\varphi, \mu)} \right\| \|C^0 + P_1^l \left\| \frac{dc(\varphi_{l+g}(\varphi, \mu), \mu)}{d(\varphi, \mu)} \right\|.$$

It is obvious that $\xi_1(p)$ and $\xi_2(p)$ have meaning for any $p \in Z$. Then, denoting the expression $2(\xi_1(p) + \xi_2(p))$ by $\eta(p)$, we write down the estimate

$$\left\| \frac{d\Omega_l^0(p, \varphi, \mu)}{d(\varphi, \mu)} \right\| < P_1^{l-1}(1 + A^*)^l \eta(p)$$

for all $p \in Z$. In this case, for all $p \in Z$ and $g \geq -1$, the following inequalities hold:

$$\sum_{l=1}^{\infty} \|u_l\| < \sum_{l=1}^{\infty} \{C^0 P_1^{l-1}(1 + A^*)^l \eta(p) + P_1^l C_*(1 + A^*)^{l+g}\}$$

$$\leq C^0 P_1^{-1} \eta(p) \sum_{l=1}^{\infty}(P_1(1 + A^*))^l + C_*(1 + A^*)^g \sum_{l=1}^{\infty}(P_1(1 + A^*))^l.$$

Under condition 2 of Theorem 2.7, $P_1(1 + A^*) < 1$. Hence,

$$\sum_{l=1}^{\infty} \|u_l\| < \frac{C^0 \eta(p)(1 + A^*)}{1 - P_1(1 + A^*)} + \frac{P_1 C_*(1 + A^*)^{g+1}}{1 - P_1(1 + A^*)},$$

i.e., the last series converges uniformly in $(\varphi, \mu) \in \Lambda$.

But if $g < -1$, then

$$\sum_{l=1}^{\infty} \|u_l\| < \sum_{l=1}^{-g-1} \{C^0 P_1^{l-1}(1 + A^*)^l \eta(p) + P_1^l C_* \Phi_*^{-l-g}\}$$

$$+ \sum_{l=-g}^{\infty} \{C^0 P_1^{l-1}(1 + A^*)^l \eta(p) + P_1^l C_*(1 + A^*)^{l+g}\}.$$

We obtain an analogous result, since the right-hand side of the last inequality includes a finite sum and the convergent series. The theorem is proved. $\qquad\square$

We note that, for nonnegative p and for $g \geq -1$, we should not require the differentiability of the mapping $\Phi^{-1}(\varphi, \mu)$ on Λ. The following proposition is valid.

Corollary 2.8. *Let condition 2 of Theorem 2.7 be satisfied,*

$$\{a(\varphi,\mu), c(\varphi,\mu), P^{-1}(\varphi,\mu)\} \subset C^1_\Lambda(\varphi,\mu),$$

and

$$\|\frac{da(\varphi,\mu)}{d(\varphi,\mu)}\| \le A^*; \quad \|\frac{dP^{-1}(\varphi,\mu)}{d(\varphi,\mu)}\| \le P_*, \quad \|\frac{dc(\varphi,\mu)}{d(\varphi,\mu)}\| \le C_*,$$

where $A^*, P_*,$ and C_* are positive constants, which are independent of $(\varphi,\mu) \in \Lambda$.

Then, for all $\{p \ge 0, g \ge -1\} \subset Z$, the system of equations (2.75) has the invariant torus Fréchet-differentiable with respect to (φ,μ) on the set Λ.

We now assume that the matrix $P(\varphi,\mu)$ is not invertible on Λ or condition 2 of Theorem 2.7 is not satisfied. In this case, we formulate the following proposition.

Corollary 2.9. *Let condition 1 of Theorem 2.7, where the matrix $P^{-1}(\varphi,\mu)$ is replaced by the matrix $P(\varphi,\mu)$, be satisfied, and*

$$P^0 < \frac{1}{\Phi_*} \quad (\Phi_* > 1). \tag{2.82}$$

Then, for all $\{p, g\} \subset Z$, the system of equations (2.75) has the invariant torus Fréchet-differentiable with respect to (φ,μ) on the set Λ.

Proof. For all $l < 0$, we set

$$\Omega^0_l(p,\varphi,\mu) = \prod_{i=p-1}^{l+p} P(\varphi_i(\varphi,\mu),\mu).$$

Whence, with regard for (2.82), we obtain the inequalities

$$\|\Omega^0_l(p,\varphi,\mu)\| \le \prod_{i=p-1}^{l+p} \|P(\varphi_i(\varphi,\mu),\mu)\| \le (P^0)^{-l}.$$

Then the GSF of the system of equations (2.75) is represented as

$$G_0(l,p,\mu,\varphi) = \begin{cases} \Omega^0_l(p,\varphi,\mu), & \text{if } l \le 0; \\ 0, & \text{if } l > 0, \end{cases}$$

and, respectively,

$$u(p,g,\mu,\varphi) = \sum_{l=-\infty}^{0} \Omega^0_l(p,\varphi,\mu) c(\varphi_{l+g}(\varphi,\mu),\mu).$$

Analogously to the proof of Theorem 2.7, we verify the validity of the inequalities

$$\|\frac{d\Omega_l^0(p,\varphi,\mu)}{d(\varphi,\mu)}\| \leq (P^0)^{-l-1}P_* \sum_{i=p-1}^{p+l} \|\frac{d\varphi_i(\varphi,\mu)}{d(\varphi,\mu)}\|$$

$$< \begin{cases} \xi_2(p)(P^0)^{-l-1}\Phi_*^{-l} & \text{for} \quad p \leq 0, l < 0; \\ \xi_1(p)(P^0)^{-l-1} & \text{for} \quad p > 0, -p \leq l < 0; \\ (P^0)^{-l-1}(\xi_1(p) + \xi_2(p)\Phi_*^{-l}) & \text{for} \quad p > 0, l < -p \end{cases}$$

$$< (P^0)^{-l-1}\Phi_*^{-l}\eta(p), \quad l < 0, p \in Z.$$

Then, for all $p \in Z$ and $g < 1$, the series

$$\sum_{l=-\infty}^{-1} \|u_l\| < \sum_{l=-\infty}^{-1} \{C^0(P^0)^{-l-1}\Phi_*^{-l}\eta(p) + (P^0)^{-l}C_*\Phi_*^{-(l+g)}\}$$

converges uniformly in $(\varphi,\mu) \in \Lambda$, since condition (2.82) yields $P_0\Phi_* < 1$. But if $g \geq 1$, then the right-hand side of the inequality

$$\sum_{l=-\infty}^{-1} \|u_l\| < \sum_{l=-\infty}^{-g-1} \{C^0(P^0)^{-l-1}\Phi_*^{-l}\eta(p) + (P^0)^{-l}C_*\Phi_*^{-(l+g)}\}$$

$$+ \sum_{l=-g}^{-1} \{C^0(P^0)^{-l-1}\Phi_*^{-l}\eta(p) + (P^0)^{-l}C_*(1 + A^*)^{l+g}\}$$

includes the convergent series and the finite sum. Hence, for all $\{p, g\} \subset Z$, the series $\sum_{l=-\infty}^{0} \|u_l\|$ converges uniformly in $(\varphi,\mu) \in \Lambda$, which completes the proof of Corollary 2.9. $\qquad\qquad\square$

We note that, under conditions of Theorem 2.7 and Corollary 2.9, the differentiability of the invariant torus of the system of equations (2.75) is defined by the limitations imposed only on the functions $a(\varphi,\mu), P(\varphi,\mu), c(\varphi,\mu)$. In the general case, we fail to make it.

Indeed, let the conditions of Theorem 2.7 and Corollary 2.9 be not satisfied, but let the GSF of the system of equations (2.75) exist. This means that there exists the infinite matrix $C(\varphi,\mu)$, which is 2π-periodic in $\varphi^i(i = 1, 2, 3, \ldots)$, bounded in the norm, and such that the function

$$G_0(l,p,\mu,\varphi) = \begin{cases} \Omega_l^0(p,\varphi,\mu)C(\varphi_{l+p}(\varphi,\mu),\mu), & \text{if} \quad l \leq 0; \\ \Omega_l^0(p,\varphi,\mu)[C(\varphi_{l+p}(\varphi,\mu),\mu) - E], & \text{if} \quad l > 0 \end{cases} \qquad (2.83)$$

satisfies the inequality $\|G_0(l,p,\mu,\varphi)\| \leq M\lambda^{|l|}$ for all $\{p,l\} \subset Z, \mu \in S,$ $\varphi \in \mathcal{T}_\infty$, where M and $\lambda < 1$ are positive constants independent of $p, l, \varphi, \mu,$ and E is the infinite identity matrix.

Corollary 2.10. *Let the following conditions hold:*

1) for $p = 0$, there exists the GSF of the system of equations (2.75);

2) $\{a(\varphi, \mu), c(\varphi, \mu), \Phi^{-1}(\varphi, \mu), G_0(l, p, \mu, \varphi)\} \subset C_\Lambda^1(\varphi, \mu)$, and

$$\left\| \frac{da(\varphi, \mu)}{d(\varphi, \mu)} \right\| \le A^*; \quad \left\| \frac{d\Phi^{-1}(\varphi, \mu)}{d(\varphi, \mu)} \right\| \le \Phi_*, \quad \left\| \frac{dc(\varphi, \mu)}{d(\varphi, \mu)} \right\| \le C_*,$$

where A^*, Φ_*, and C_* are positive constants, which are independent of $(\varphi, \mu) \in \Lambda$;

3) $\lambda < \min\{\frac{1}{\Phi_*}; \frac{1}{1+A^*}\}$;

4) the series $\sum_{l=-\infty}^{\infty} \left\| \frac{dG_0(l,p,\mu,\varphi)}{d(\varphi,\mu)} \right\|$ converges uniformly in $(\varphi, \mu) \in \Lambda$.

Then, for all $\{p, g\} \subset Z$, the system of equations (2.75) has the invariant torus Fréchet-differentiable with respect to (φ, μ) on the set Λ.

Proof. Since, under conditions of Corollary 2.10,

$$u(p, g, \mu, \varphi) = \sum_{l=-\infty}^{\infty} G_0(l, p, \varphi, \mu) c(\varphi_{l+g}(\varphi, \mu), \mu),$$

it is sufficient for the inclusion $u(p, g, \mu, \varphi) \in C_\Lambda^1$ to exist that the series

$$I_1 = \sum_{l=-\infty}^{\infty} \left\| \frac{dG_0(l, p, \varphi, \mu)}{d(\varphi, \mu)} \right\| \|c(\varphi_{l+g}(\varphi, \mu), \mu)\|,$$

$$I_2 = \sum_{l=-\infty}^{\infty} \|G_0(l, p, \varphi, \mu)\| \left\| \frac{dc(\varphi_{l+g}(\varphi, \mu), \mu)}{d(\varphi, \mu)} \right\|$$

converge uniformly in $(\varphi, \mu) \in \Lambda$. Obviously, the series I_1 satisfies this requirement. For the series I_2, we write down the following inequalities, by setting $\Phi_* > 1$:

$$I_2 \le \sum_{l=-\infty}^{\infty} M\lambda^{|l|} \left\| \frac{dc(\varphi_{l+g}(\varphi, \mu), \mu)}{d(\varphi, \mu)} \right\|$$

$$\le MC_*\Phi_*^{-g} \sum_{l=-\infty}^{-g-1} \lambda^{|l|}\Phi_*^{-l} + MC_*(1 + A^*)^g \sum_{l=-g}^{\infty} \lambda^{|l|}(1 + A^*)^l.$$

It is easy to verify that, for all $g \in Z$, the series $\sum_{l=-\infty}^{-g-1} \lambda^{|l|}\Phi_*^{-l}$ and $\sum_{l=-g}^{\infty} \lambda^{|l|}(1 + A^*)^l$ converge, since $\lambda\Phi_* < 1$ and $\lambda(1 + A^*) < 1$. The corollary is proved. $\qquad\square$

Consider now the quasilinear system of equations

$$\varphi_{n+1} = \varphi_n + a(\varphi_n, \mu),$$

$$x_{n+1} = P(\varphi_{n+p}, \mu)x_n + c(\varphi_{n+g+1}, \mu, x_{n+1}), \quad n \in Z, \qquad (2.84)$$

where the function $c(\varphi, \mu, x)$ is defined on the set $\mathfrak{M} \times S \times D, \quad D = \{x \in \mathfrak{M} | \|x\| \le d = const > 0\}$, and $\|c(\varphi, \mu, x)\| \le C^0$ on this set.

We say that the function $u(p, g, \mu, \varphi)$ generates the invariant torus $\mathcal{T}(p, g, \mu)$ of the system of equations (2.84), if, for any $\{p, g\} \subset Z, \mu \in S$, it is 2π-periodic in $\varphi^i (i = 1, 2, 3, \dots)$, bounded in the norm $\| \cdot \|$, and satisfies the equality

$$u(p, g, \mu, \varphi_{n+1}(\varphi, \mu)) = P(\varphi_{n+p}(\varphi, \mu), \mu)u(p, g, \mu, \varphi_n(\varphi, \mu))$$
$$+ c(\varphi_{n+g+1}(\varphi, \mu), \mu, u(p, g, \mu, \varphi_{n+1}(\varphi, \mu))) \quad (2.85)$$

for any $\varphi \in \mathcal{T}_\infty$.

Lemma 2.15. *Let the GSF defined by equality (2.83) exist, and let*

$$\|c(\varphi, \mu, x) - c(\varphi, \mu, x_1)\| \le K\|x - x_1\|,$$

where K is a positive constant independent of $\{\varphi, \mu\} \in \mathfrak{M} \times S, \{x, x_1\} \subset D$. If the inequalities

$$C^0 \le \frac{d(1 - \lambda)}{1 + \lambda}; \quad MK\frac{1 + \lambda}{1 - \lambda} < 1,$$

hold, then the system of equations (2.84) has an invariant torus.

Proof. Consider the sequence of functions

$$u_{i+1}(p, g, \mu, \varphi)$$

$$= \sum_{l=-\infty}^{\infty} G_0(l, p, \mu, \varphi)c(\varphi_{l+g}(\varphi, \mu), \mu, u_i(p, g, \mu, \varphi_l(\varphi, \mu))), \quad (2.86)$$

where $i = 0, 1.2, 3, \dots$, $u_0(p, g, \mu, \varphi)$ generates the invariant torus $\mathcal{T}_0(p, g, \mu)$ of the system of equations

$$\varphi_{n+1} = \varphi_n + a(\varphi_n, \mu), \quad x_{n+1} = P(\varphi_{n+p}, \mu)x_n + c(\varphi_{n+g+1}, \mu, 0),$$

and

$$\|u_0(p, g, \mu, \varphi)\| \le \sum_{l=-\infty}^{\infty} \|G_0(l, p, \mu, \varphi)c(\varphi_{l+g}(\varphi, \mu), \mu, 0)\| \le d.$$

We now show that the function $u_{i+1}(p, g, \mu, \varphi)$ generates, for all $i \in \{1, 2, 3, \ldots\}$, the invariant torus $\mathcal{T}_{i+1}(p, g, \mu)$ of the system of equations

$$\varphi_{n+1} = \varphi_n + a(\varphi_n, \mu),$$

$$x_{n+1} = P(\varphi_{n+p}, \mu)x_n + c(\varphi_{n+g+1}, \mu, u_i(p, g, \mu, \varphi_{n+1})). \qquad (2.87)$$

It is obvious that, for all natural i, $\|u_i(p, g, \mu, \varphi)\| \le d$, and the function $u_i(p, g, \mu, \varphi)$ is 2π-periodic in $\varphi^j (j = 1, 2, 3, \ldots)$.

The function

$$x_n = u_{i+1}(p, g, \mu, \varphi_n(\varphi, \mu))$$

$$= \sum_{l=-\infty}^{\infty} G_n(l, p, \mu, \varphi)c(\varphi_{l+g}(\varphi, \mu), \mu, u_i(p, g, \mu, \varphi_l(\varphi, \mu)))$$

defines a family of bounded solutions of the system of equations (2.87).

Indeed, the following equalities hold:

$$V_1 = \Omega_l^{n+1}(p, \varphi, \mu)C(\varphi_{p+l}(\varphi, \mu), \mu)c(\varphi_{l+g}(\varphi, \mu), \mu, u_i(p, g, \mu, \varphi_l(\varphi, \mu)))$$
$$= P(\varphi_{n+p}(\varphi, \mu), \mu)\Omega_l^n(p, \varphi, \mu)C(\varphi_{p+l}(\varphi, \mu), \mu)$$
$$\times c(\varphi_{l+g}(\varphi, \mu), \mu, u_i(p, g, \mu, \varphi_l(\varphi, \mu))),$$

$$V_2 = \Omega_l^{n+1}(p, \varphi, \mu)[C(\varphi_{p+l}(\varphi, \mu), \mu) - E]$$
$$\times c(\varphi_{l+g}(\varphi, \mu), \mu, u_i(p, g, \mu, \varphi_l(\varphi, \mu)))$$
$$= P(\varphi_{n+p}(\varphi, \mu), \mu)\Omega_l^n(p, \varphi, \mu)[C(\varphi_{p+l}(\varphi, \mu), \mu) - E]$$
$$\times c(\varphi_{l+g}(\varphi, \mu), \mu, u_i(p, g, \mu, \varphi_l(\varphi, \mu))).$$

Then

$$x_{n+1} = \sum_{l=-\infty}^{n+1} V_1 + \sum_{l=n+2}^{\infty} V_2 = \sum_{l=-\infty}^{n} V_1 + \sum_{l=n+1}^{\infty} V_2$$
$$+ C(\varphi_{n+p+1}(\varphi, \mu), \mu)c(\varphi_{n+g+1}(\varphi, \mu), \mu, u_i(p, g, \mu, \varphi_{n+1}(\varphi, \mu)))$$
$$- [C(\varphi_{n+p+1}(\varphi, \mu), \mu) - E]c(\varphi_{n+g+1}(\varphi, \mu), \mu, u_i(p, g, \mu, \varphi_{n+1}(\varphi, \mu)))$$
$$= P(\varphi_{n+p}(\varphi, \mu), \mu)x_n + c(\varphi_{n+g+1}(\varphi, \mu), \mu, u_i(p, g, \mu, \varphi_{n+1}(\varphi, \mu))).$$

The inductive reasoning leads to the chain of inequalities:

$$\|u_{i+1}(p, g, \mu, \varphi) - u_i(p, g, \mu, \varphi)\| \leq \sum_{l=-\infty}^{\infty} M\lambda^{|l|} K\|u_i(p, g, \mu, \varphi_l(\varphi, \mu))$$

$$- u_{i-1}(p, g, \mu, \varphi_l(\varphi, \mu))\| \leq \sum_{l=-\infty}^{\infty} MK\lambda^{|l|} \sum_{k=-\infty}^{\infty} MK\lambda^{|k-l|}$$

$$\times \|u_{i-1}(p, g, \mu, \varphi_k(\varphi, \mu)) - u_{i-2}(p, g, \mu, \varphi_k(\varphi, \mu))\|$$

$$\leq \sum_{l=-\infty}^{\infty} MK\lambda^{|l|} \sum_{k=-\infty}^{\infty} MK\lambda^{|k-l|} \sum_{s=-\infty}^{\infty} MK\lambda^{|s-k|} \cdots \sum_{r=-\infty}^{\infty} MK\lambda^{|r-\rho|}$$

$$\times \|u_1(p, g, \mu, \varphi_r(\varphi, \mu)) - u_0(p, g, \mu, \varphi_r(\varphi, \mu))\| \leq 2d(MK)^i$$

$$\times \sum_{l=-\infty}^{\infty} \lambda^{|l|} \sum_{k=-\infty}^{\infty} \lambda^{|k-l|} \cdots \sum_{r=-\infty}^{\infty} \lambda^{|r-\rho|} = 2d(MK\frac{1+\lambda}{1-\lambda})^i. \quad (2.88)$$

In this case, the estimate

$$\|u_{s+m}(p, g, \mu, \varphi) - u_s(p, g, \mu, \varphi)\|$$

$$\leq \sum_{l=1}^{m} \|u_{s+i}(p, g, \mu, \varphi) - u_{s+i-1}(p, g, \mu, \varphi)\|$$

$$\leq 2d \sum_{i=1}^{\infty} (MK\frac{1+\lambda}{1-\lambda})^{s+i-1} = 2d\frac{(MK\frac{1+\lambda}{1-\lambda})^s}{1 - MK\frac{1+\lambda}{1-\lambda}} \to 0$$

proves the fundamentality of the sequence $\{u_i(p, g, \mu, \varphi)\}_{i=0}^{\infty} \subset \mathfrak{M}$ as $s \to \infty$. The sequence converges to some function $u(p, g, \mu, \varphi)$, since the space $\mathfrak{M}$ is complete.

Passing to the limit as $i \to \infty$ in the equality

$$u_{i+1}(p, g, \mu, \varphi_{n+1}(\varphi, \mu)) = P(\varphi_{n+p}(\varphi, \mu), \mu)u_{i+1}(p, g, \mu, \varphi_n(\varphi, \mu))$$

$$+ c(\varphi_{n+g+1}(\varphi, \mu), \mu, u_i(p, g, \mu, \varphi_{n+1}(\varphi, \mu))),$$

we obtain equality (2.85), which completes the proof of Lemma 2.15, since the invariant torus of the system of equations (2.84) is generated by the function $u(p, g, \mu, \varphi)$. $\qquad \square$

By D_ρ and D_0, we denote the sets $\{x \in \mathfrak{M} | \|x\| < d + \rho\}$ and $\mathfrak{M} \times S \times D_\rho$, respectively. Here, ρ is an arbitrarily small positive constant. It is obvious that D_0 is a set open in $\mathfrak{M} \times \mathfrak{M} \times \mathfrak{M}$, if we set $\|(\varphi, \mu, x) \in D_0\| = \max\{\|\varphi\|, \|\mu\|, \|x\|\}$, where the symbol $\|\cdot\|$ on the right-hand side of the last equality stands for the norm in $\mathfrak{M}$.

The following proposition gives the sufficient conditions of differentiability of the invariant torus of the system of equations (2.84).

Theorem 2.8. *Let the following conditions hold:*

1) for $p = 0$, there exists the GSF defined by equality (2.83);

2) $\{a(\varphi, \mu), \Phi^{-1}(\varphi, \mu), G_0(l, p, \mu, \varphi)\} \subset C_\Lambda^1(\varphi, \mu)$,
$c(\varphi, \mu, x) \in C_{D_0}^1(\varphi, \mu, x)$, *and*

$$\left\| \frac{da(\varphi, \mu)}{d(\varphi, \mu)} \right\| \leq A^*, \quad \left\| \frac{d\Phi^{-1}(\varphi, \mu)}{d(\varphi, \mu)} \right\| \leq \Phi_*, \quad \left\| \frac{dc(\varphi, \mu, x)}{d(\varphi, \mu, x)} \right\| \leq C_*,$$

$$\left\| \frac{dG_0(l, p, \mu, \varphi)}{d(\varphi, \mu)} \right\| \leq M_1(p)\lambda_1^{|l|}, \tag{2.89}$$

$$\left\| \frac{dc(\varphi, \mu, x)}{d(\varphi, \mu, x)} - \frac{dc(\varphi, \mu, \bar{x})}{d(\varphi, \mu, x)} \right\| \leq L_0 \|x - \bar{x}\|,$$

where $A^*, \Phi_* > 1, C_*, M_1(p)$, and L_0 are positive constants independent of $(\varphi, \mu) \in \Lambda, \{x, \bar{x}\} \subset D_\rho$; $0 < \lambda_1 = \text{const} < 1$ and is independent of l, p, μ, φ;

3) $C_0 \leq \frac{d(1-\lambda)}{1+\lambda}$; $\max\{MC_* \frac{1+\delta\lambda}{1-\delta\lambda}; \delta\lambda\} < 1$, where $\delta = \max\{\Phi_*, 1+A^*\}$.

Then the system of equations (2.84) has the invariant torus $\mathcal{T}(p, g, \mu)$, whose generating function $u(p, g, \mu, \varphi) \in C_\Lambda^1(\varphi, \mu)$ for all $\{p, g\} \subset Z$.

Proof. It is easy to see that, under conditions of the formulated theorem, Corollary 2.10 with $c(\varphi, \mu, 0) = c(\varphi, \mu)$ and Lemma 2.15 with $K = C_*$ are valid.

First, we show that, for all $n = 0, 1, 2, \ldots, u_n(p, g, \mu, \varphi) \in C_\Lambda^1(\varphi, \mu)$, and

$$\left\| \frac{du_n(p, g, \mu, \varphi_l(\varphi, \mu))}{d(\varphi, \mu)} \right\| \leq Z_n \delta^{|l|},$$

where Z_n is a positive constant independent of $(\varphi, \mu) \in \Lambda, l \in Z$. To this end, we use the method of complete mathematical induction. We take into account that, by Lemma 2.13,

$$\left\| \frac{d\varphi_l(\varphi, \mu)}{d(\varphi, \mu)} \right\| \leq \delta^{|l|}$$

for all $l \in Z$. For $n = 0$,

$$\left\| \frac{du_0(p, g, \mu, \varphi)}{d(\varphi, \mu)} \right\| \leq Z_0,$$

which follows from the proof of Corollary 2.10. Then

$$\left\|\frac{du_0(p,g,\mu,\varphi_l(\varphi,\mu))}{d(\varphi,\mu)}\right\| \leq \left\|\frac{du_0(p,g,\mu,\varphi)}{d(\varphi,\mu)}\right\| \sup_{\max\{\|\varphi^*\|,\|\mu^*\|\}=1}$$

$$\max\{\|\frac{d\varphi_l(\varphi,\mu)}{d(\varphi,\mu)}(\varphi^*,\mu^*)\|,\|\mu^*\|\} \leq Z_0\delta^{|l|}, \quad (\varphi^*,\mu^*) \in \mathfrak{M} \times \mathfrak{M}.$$

We now assume that the formulated proposition is valid for $n \leq i$ and will prove that it is satisfied for $n = i + 1$.

The function $c(\varphi_{l+g}(\varphi,\mu),\mu,u_i(p,g,\mu,\varphi_l(\varphi,\mu)))$ is a mapping $c(h)$: $\Phi \to \mathfrak{M}$, where the mapping h is composed of three components: $\varphi_{l+g}(\varphi,\mu) : \Lambda \to \mathfrak{M}$; $\mu(\varphi,\mu) = \mu : \Lambda \to S$, and $u_i(p,g,\mu,\varphi_l(\varphi,\mu))$: $\Lambda \to D_\rho$. Each component is continuously Fréchet-differentiable on the set Λ.

Then $c(\varphi_{l+g}(\varphi,\mu),\mu,u_i(p,g,\mu,\varphi_l(\varphi,\mu))) \in C^1_\Lambda(\varphi,\mu)$, and

$$\left\|\frac{dc(\varphi_{l+g}(\varphi,\mu),\mu,u_i(p,g,\mu,\varphi_l(\varphi,\mu)))}{d(\varphi,\mu)}\right\|$$

$$\leq \left\|\frac{dc(h)}{dh}\right\| \max\{\|\frac{d\varphi_{l+g}(\varphi,\mu)}{d(\varphi,\mu)}\|, 1, \|\frac{du_i(p,g,\mu,\varphi_l(\varphi,\mu))}{d(\varphi,\mu)}\|\}$$

$$\leq C_* \max\{\delta^{|l+g|}, Z_i\delta^{|l|}\} \leq \delta^{|l|}r_i,$$

where $r_i = C_* \max\{\delta^{|g|}, Z_i\}$.

With regard for the inequalities

$$\sum_{l=-\infty}^{\infty} \left\|\frac{dG_0(l,p,\varphi,\mu)}{d(\varphi,\mu)}\right\|\|c(\varphi_{l+g}(\varphi,\mu),\mu,u_i(p,g,\mu,\varphi_l(\varphi,\mu)))\|$$

$$\leq \sum_{l=-\infty}^{\infty} M_1(p)\lambda_1^{|l|}C^0,$$

$$\sum_{l=-\infty}^{\infty} \|G_0(l,p,\varphi,\mu)\|\|\frac{dc(\varphi_{l+g}(\varphi,\mu),\mu,u_i(p,g,\mu,\varphi_l(\varphi,\mu)))}{d(\varphi,\mu)}\|$$

$$\leq \sum_{l=-\infty}^{\infty} M\lambda^{|l|}\delta^{|l|}r_i = Mr_i \sum_{l=-\infty}^{\infty} (\lambda\delta)^{|l|}, \quad (2.90)$$

equality (2.86), and the estimate $\lambda\delta < 1$, we conclude that $u_{i+1}(p,g,\mu,\varphi) \in C^1_\Lambda(\varphi,\mu)$, and

$$\left\|\frac{du_{i+1}(p,g,\mu,\varphi_l(\varphi,\mu))}{d(\varphi,\mu)}\right\| \leq Z_{i+1}\delta^{|l|},$$

where
$$Z_{i+1} = M_1(p)C^0\frac{1+\lambda_1}{1-\lambda_1} + Mr_i\frac{1+\lambda\delta}{1-\lambda\delta},$$
which proves the above-posed proposition.

It remains to prove that the sequence of functions
$$\{\frac{du_i(p,g,\mu,\varphi)}{d(\varphi,\mu)}\}_{i=0}^{\infty}$$
defined on Λ with values in the space of linear operators $\mathbf{L}(\mathfrak{M}\times\mathfrak{M},\mathfrak{M})$ converges uniformly in $(\varphi,\mu)\in\Lambda$. Since the indicated space is complete, it is sufficient to show that the mentioned sequence is fundamental uniformly in $(\varphi,\mu)\in\Lambda$.

We introduce the notation
$$\bar{u}_{i+1} = \frac{du_{i+1}(p,g,\mu,\varphi)}{d(\varphi,\mu)} - \frac{du_i(p,g,\mu,\varphi)}{d(\varphi,\mu)}.$$
In view of equality (2.86) and inequalities (2.90), we obtain the relation

$$\|\bar{u}_{i+1}\| \le \sum_{l=-\infty}^{\infty} M_1(p)\lambda_1^{|l|}\|c(\varphi_{l+g}(\varphi,\mu),\mu,u_i(p,g,\mu,\varphi_l(\varphi,\mu)))$$

$$- c(\varphi_{l+g}(\varphi,\mu),\mu,u_{i-1}(p,g,\mu,\varphi_l(\varphi,\mu)))\| + \sum_{l=-\infty}^{\infty} M\lambda^{|l|}$$

$$\times \|\frac{dc(\varphi_{l+g}(\varphi,\mu),\mu,u_i(p,g,\mu,\varphi_l(\varphi,\mu)))}{d(\varphi,\mu)}$$

$$- \frac{dc(\varphi_{l+g}(\varphi,\mu),\mu,u_{i-1}(p,g,\mu,\varphi_l(\varphi,\mu)))}{d(\varphi,\mu)}\|$$

$$\le \sum_{l=-\infty}^{\infty} M_1(p)\lambda_1^{|l|}C_*\|u_i(p,g,\mu,\varphi_l(\varphi,\mu)) - u_{i-1}(p,g,\mu,\varphi_l(\varphi,\mu))\|$$

$$+ MC_* \sum_{l=-\infty}^{\infty} \|\bar{u}_{il}\|\lambda^{|l|} + \sum_{l=-\infty}^{\infty} ML_0\lambda^{|l|}\delta^{|l|}\max\{\delta^{|g|},Z_i\}\|$$

$$\times u_i(p,g,\mu,\varphi_l(\varphi,\mu)) - u_{i-1}(p,g,\mu,\varphi_l(\varphi,\mu))\|,$$

where
$$\bar{u}_{il} = \frac{du_i(p,g,\mu,\varphi_l(\varphi,\mu))}{d(\varphi,\mu)} - \frac{du_{i-1}(p,g,\mu,\varphi_l(\varphi,\mu))}{d(\varphi,\mu)}.$$

For any points $(\varphi^*,\mu^*)\in\mathfrak{M}\times\mathfrak{M}$,

$$\|\bar{u}_{il}\| = \sup_{\max\{\|\varphi^*\|,\|\mu^*\|\}=1} \|\bar{u}_i|_{(\varphi_l(\varphi,\mu),\mu)}(\frac{d\varphi_l(\varphi,\mu)}{d(\varphi,\mu)}(\varphi^*,\mu^*),\mu^*)\|$$

$$\le \sup_{\max\{\|\varphi^*\|,\|\mu^*\|\}=1} \|\bar{u}_i\|_0 \max\{\|\frac{d\varphi_l(\varphi,\mu)}{d(\varphi,\mu)}(\varphi^*,\mu^*)\|,\|\mu^*\|\} \le \|\bar{u}_i\|_0\delta^{|l|},$$

$$\|\bar{u}_i\|_0 = \sup_{(\varphi,\mu)} \|\bar{u}_i\|.$$

We introduce the notation:

$$\gamma_1 = 2dC_*M_1(p)\frac{1+\lambda_1}{1-\lambda_1} + ML_0\max\{\delta^{|g|}, Z^*\}2d\frac{1+\delta\lambda}{1-\delta\lambda};$$

$$\gamma_2 = MC_*\frac{1+\lambda}{1-\lambda}; \quad \gamma_3 = MC_*\frac{1+\delta\lambda}{1-\delta\lambda} \quad (\gamma_3 > \gamma_2, Z^* = \sup_{i\in Z_0^+}\{Z_i\}).$$

Taking (2.88) and (2.89) into account, we write down the inductive inequality

$$\|\bar{u}_{i+1}\|_0 \leq \gamma_1\gamma_2^{i-1} + \gamma_3\|\bar{u}_i\|_0, \quad i = 1, 2, 3, \ldots,$$

which allows us to obtain the relations

$$\|\bar{u}_{i+1}\| \leq \gamma_1\{\gamma_2^{i-1}\gamma_3^0 + \gamma_2^{i-2}\gamma_3 + \cdots + \gamma_2^0\gamma_3^{i-1}\} + \gamma_3^i(Z_0 + Z_1)$$
$$< \gamma_1\frac{\gamma_2^{i-1}(\frac{\gamma_3}{\gamma_2})^i\gamma_2}{\gamma_3 - \gamma_2} + \gamma_3^i(Z_0 + Z_1) = F\sigma^i.$$

Here, we set $\sigma = \gamma_3$, $\quad F = \gamma_1/(\gamma_3 - \gamma_2) + Z_0 + Z_1$.

Since $\sigma = const < 1$, and F is independent of $(\varphi, \mu) \in \Lambda$, the estimate

$$\left\|\frac{du_{s+m}(p,g,\mu,\varphi)}{d(\varphi,\mu)} - \frac{du_s(p,g,\mu,\varphi)}{d(\varphi,\mu)}\right\| \leq \sum_{i=1}^{\infty} F\sigma^{s+i-1} = F\frac{\sigma^s}{1-\sigma} \to 0$$

as $s \to \infty$ proves that that the sequence

$$\left\{\frac{du_i(p,g,\mu,\varphi)}{d(\varphi,\mu)}\right\}_{i=0}^{\infty}$$

is fundamental uniformly in $(\varphi, \mu) \in \Lambda$. Theorem 2.8 is proved. $\square$

We now write down the system of equations

$$\varphi_{n+1} = \varphi_n + a(\varphi_n, \mu),$$

$$x_{n+1} = P(\varphi_{n+p}, \mu)x_n + c(\varphi_{n+g}, \mu, x_n), \quad n \in Z, \qquad (2.91)$$

which satisfies the same requirements as system (2.84).

As above, we set the GSF by equality (2.83). It is easy to verify that the function

$$u_0(p,g,\mu,\varphi) = \sum_{l=-\infty}^{\infty} G_0(l,p,\mu,\varphi)c(\varphi_{l+g-1}(\varphi,\mu),\mu,0)$$

generates the invariant torus of the system of equations

$$\varphi_{n+1} = \varphi_n + a(\varphi_n, \mu), \quad x_{n+1} = P(\varphi_{n+p}, \mu)x_n + c(\varphi_{n+g}, \mu, 0),$$

and the function

$$u_{i+1}(p, g, \mu, \varphi)$$
$$= \sum_{l=-\infty}^{\infty} G_0(l, p, \mu, \varphi)c(\varphi_{l+g-1}(\varphi, \mu), \mu, u_i(p, g, \mu, \varphi_{l-1}(\varphi, \mu)))$$

is the invariant torus of the system of equations

$$\varphi_{n+1} = \varphi_n + a(\varphi_n, \mu),$$

$$x_{n+1} = P(\varphi_{n+p}, \mu)x_n + c(\varphi_{n+g}, \mu, u_i(p, g, \mu, \varphi_n))$$

for all $i \in \{1, 2, 3, \dots\}$.

It is easy to verify also that the sequence $\{u_i(p, g, \mu, \varphi)\}$ as $i \to \infty$ converges to the function, which generates the invariant torus of the system of equations (2.91), and this torus in the conditions of Theorem 2.8 is differentiable on Λ in the Fréchet meaning with respect to (φ, μ) for all $\{p, g\} \subset Z$.

Remark 2.3. Under the conditions of Theorem 2.7 and Corollary 2.9, inequality (2.89) holds automatically. For example, the proof of Theorem 2.7 allows us to set $\lambda_1 = P_1(1 + A^*)$, $M_1(p) = \eta(p)/P_1$. For the existence of the GSF defined by equality (2.83) for all $p \in Z$, it is sufficient that it exist for $p = 0$. The constant M is independent of $p \in Z$.

2.5   On the existence of the invariant tori of nonlinear systems

Consider the system of equations

$$\varphi_{n+1} = \varphi_n + a(\varphi_n, \mu),$$

$$x_{n+1} = P(\varphi_{n+p}, \mu, x_{n+k})x_n + c(\varphi_{n+g+1}, \mu), \quad n \in Z, \tag{2.92}$$

which satisfies the requirements imposed on system (2.75), $k \in Z$ is a parameter, and

$$\|P(\varphi, \mu, x)\| = \sup_i \sum_{j=1}^{\infty} |p_{ij}(\varphi, \mu, x)| \leq P^0$$

for all $x \in D$.

The set of points $x \in \mathfrak{M}$ is called the invariant torus $\mathcal{T}(p, g, k, \mu)$ of the system of equations (2.92),

$$x = u(p, g, k, \mu, \varphi) = (u_1(p, g, k, \mu, \varphi), u_2(p, g, k, \mu, \varphi), \dots), \quad \varphi \in \mathcal{T}_\infty,$$

if the function $u(p, g, k, \mu, \varphi)$ is defined for any $\{p, g, k\} \subset Z, \mu \in S$, and $\varphi \in \mathfrak{M}$, 2π is periodic in $\varphi^i (i = 1, 2, 3, \dots)$, bounded in the norm $\|\cdot\|$, and satisfies the equality

$$u(p, g, k, \mu, \varphi_{n+1}(\varphi, \mu)) = P(\varphi_{n+p}(\varphi, \mu), \mu, u(p, g, k, \mu, \varphi_{n+k}(\varphi, \mu)))$$
$$\times u(p, g, k, \mu, \varphi_n(\varphi, \mu)) + c(\varphi_{n+g+1}(\varphi, \mu), \mu), \quad n \in Z, \quad (2.93)$$

for any $\varphi \in \mathcal{T}_\infty$.

Theorem 2.9. *Let the matrix $P(\varphi, \mu, x)$ be invertible on the set $D^0 = \Lambda \times D$, $\|P^{-1}(\varphi, \mu, x)\| \leq P_1$, and let the following conditions be satisfied:*

1) the matrix $P(\varphi, \mu, x)$ has the Lipschitz property in the variable x with the coefficient independent of φ, μ, x;

2) for all $\{x, \bar{x}\} \subset D, (\varphi, \mu) \in \Lambda$, the inequality

$$\|P^{-1}(\varphi, \mu, x) - P^{-1}(\varphi, \mu, \bar{x})\| \leq P_*\|x - \bar{x}\|,$$

where P_* is a positive constant independent of $\varphi, \mu, x, \bar{x}$, holds;

3) the estimates

$$P_1 < 1, \quad \eta = C^0 P_* \sum_{l=1}^{\infty} P_1^{l-1} l < 1, \quad \frac{C^0 P_1}{1 - P_1} \leq d$$

are valid. Then, for all $\{p, g, k\} \subset Z, \mu \in S$, the system of equations (2.92) has the invariant torus $\mathcal{T}(p, g, k, \mu)$.

Proof. Let us write down the formal sequence of functions

$$u^{(s)}(p, g, k, \mu, \varphi) \quad (s \in Z_0^+ = \{0, 1, 2, \dots\}, u^{(0)}(p, g, k, \mu, \varphi) = 0 \in \mathfrak{M}),$$

each of which defines, for $s \in Z^+$, the invariant torus of the system of equations

$$\varphi_{n+1} = \varphi_n + a(\varphi_n, \mu),$$

$$x_{n+1} = P(\varphi_{n+p}, \mu, u^{(s-1)}(p, g, k, \mu, \varphi_{n+k}))x_n + c(\varphi_{n+g+1}, \mu). \quad (2.94)$$

Such a sequence does exist. By the inductive reasoning, we can establish that the matriciant $\Omega_l^n(p, g, k, \mu, \varphi)_s$ of the homogeneous equation

$$x_{n+1} = P(\varphi_{n+p}(\varphi, \mu), \mu, u^{(s-1)}(p, g, k, \mu, \varphi_{n+k}(\varphi, \mu)))x_n \quad (2.95)$$

is defined for $l > n$ by the equality

$$\Omega_l^n(p, g, k, \mu, \varphi)_s = \prod_{i=n}^{l-1} P^{-1}(\varphi_{p+i}(\varphi, \mu), \mu, u^{(s-1)}(p, g, k, \mu, \varphi_{i+k}(\varphi, \mu))).$$

Since $P_1 < 1$, the GSF of Eq. (2.95) takes the form

$$G_0(l, p, g, k, \mu, \varphi)_s = \begin{cases} 0, & \text{if } l \le 0; \\ -\Omega_l^0(p, g, k, \mu, \varphi)_s, & \text{if } l > 0. \end{cases}$$

Then the function

$$u^{(s)}(p, g, k, \mu, \varphi)$$
$$= -\sum_{l=1}^{\infty} \prod_{i=0}^{l-1} P^{-1}(\varphi_{p+i}(\varphi, \mu), \mu, u^{(s-1)}(p, g, k, \mu, \varphi_{i+k}(\varphi, \mu)))c(\varphi_{l+g}(\varphi, \mu), \mu)$$

defines the invariant torus $\mathcal{T}^{(s)}(p, g, k, \mu)$ of the system of equations (2.94). In this case,

$$\|u^{(s)}(p, g, k, \mu, \varphi)\| \le \frac{C^0 P_1}{1 - P_1}.$$

This means that the equality

$$u^{(s)}(p, g, k, \mu, \varphi_{n+1}(\varphi, \mu))$$
$$= P(\varphi_{n+p}(\varphi, \mu), \mu, u^{(s-1)}(p, g, k, \mu, \varphi_{n+k}(\varphi, \mu)))$$
$$\times u^{(s)}(p, g, k, \mu, \varphi_n(\varphi, \mu)) + c(\varphi_{n+g+1}(\varphi, \mu), \mu), \quad n \in Z, \quad (2.96)$$

is valid.

We now show that the sequence $\{u^{(s)}(p, g, k, \mu, \varphi)\}_{s=1}^{\infty}$ converges in $\mathfrak{M}$ as $s \to \infty$ uniformly in p, g, k, μ, φ.

The inequalities

$$\|u^{(s)}(p, g, k, \mu, \varphi) - u^{(s-1)}(p, g, k, \mu, \varphi)\| \le C^0 P_* \sum_{l=1}^{\infty} P_1^{l-1}$$

$$\times \sum_{i=0}^{l-1} \|u^{(s-1)}(p, g, k, \mu, \varphi_{k+i}(\varphi, \mu)) - u^{(s-2)}(p, g, k, \mu, \varphi_{k+i}(\varphi, \mu))\|$$

$$\le C^0 P_* \sum_{l=1}^{\infty} P_1^{l-1} l \sup_{p, g, k, \mu, \varphi} \|u^{(s-1)}(p, g, k, \mu, \varphi) - u^{(s-2)}(p, g, k, \mu, \varphi)\| \le \dots$$

$$\le \eta^{s-2} \sup_{p, g, k, \mu, \varphi} \|u^{(2)}(p, g, k, \mu, \varphi) - u^{(1)}(p, g, \mu, \varphi)\| \le \eta^{s-1} \frac{C^0 P_1}{1 - P_1}$$

prove the fundamentality of the sequence $\{u^{(s)}(p, g, k, \mu, \varphi)\}_{s=1}^{\infty}$, since $\eta^{s-1} \to 0$ as $s \to \infty$. Its convergence to some function $u(p, g, k, \mu, \varphi)$ 2π-periodic in $\varphi^i (i \in Z^+)$ follows from the completeness of the space $\mathfrak{M}$.

Taking condition 1 of Theorem 2.9 into account, we pass to the limit in Eq. (2.96) as $s \to \infty$. We obtain equality (2.93), which completes the proof of the theorem. $\qquad\qquad\square$

We now assume that the condition $P_1 < 1$ is not satisfied, but $P^0 < 1$. In this case, the following proposition holds.

Theorem 2.10. *Let the following conditions be satisfied on the set D^0:*

1) for all $\{x, \bar{x}\} \subset D$, $\|P(\varphi, \mu, x) - P(\varphi, \mu, \bar{x})\| \le P^* \|x - \bar{x}\|$, where P^* is a positive constant, which is independent of $\varphi, \mu, x, \bar{x}$;

2) the inequalities

$$P^0 < 1, \quad \eta^* = C^0 P^* \sum_{l=1}^{\infty} P^{0^{l-1}} l < 1, \quad \frac{C^0}{1 - P^0} \le d$$

hold.

Then, for all $\{p, g, \} \subset Z, \mu \in S$, the system of equations (2.92) has the invariant torus $T(p, g, k, \mu)$.

Proof.   Proof is performed analogously to the previous one. But, in this case, the matriciant of Eq. (2.95) for $l < n$ is defined by the equality

$$\Omega_l^n(p, g, k, \mu, \varphi)_s^* = \prod_{i=n-1}^{l} P(\varphi_{p+i}(\varphi, \mu), \mu, u^{(s-1)}(p, g, k, \mu, \varphi_{i+k}(\varphi, \mu))),$$

and its GSF is given by the relation

$$G_0^*(l, p, g, k, \mu, \varphi)_s = \begin{cases} \Omega_l^0(p, g, k, \mu, \varphi)_s^*, & \text{if } l \le 0; \\ 0, & \text{if } l > 0. \end{cases}$$

Then the function

$$u_*^{(s)}(p, g, k, \mu, \varphi) = c(\varphi_g(\varphi, \mu), \mu)$$

$$+ \sum_{l=-\infty}^{-1} \prod_{i=-1}^{l} P(\varphi_{p+i}(\varphi, \mu), \mu, u_*^{(s-1)}(p, g, k, \mu, \varphi_{i+k}(\varphi, \mu))) c(\varphi_{l+g}(\varphi, \mu), \mu)$$

defines the invariant torus $T_*^{(s)}(p, g, k, \mu)$ of the system of equations (2.94). In this case,

$$\|u_*^{(s)}(p, g, k, \mu, \varphi)\| \le \frac{C^0}{1 - P^0}.$$

This means that equality (2.96), where $u^{(s)}$ is replaced by $u_*^{(s)}$, is valid. Not writing the last equality, we denote it by the symbol (*).

The chain of inequalities

$$\|u_*^{(s)}(p,g,k,\mu,\varphi) - u_*^{(s-1)}(p,g,k,\mu,\varphi)\| \le C^0 P^* \sum_{l=-\infty}^{-1} (P^0)^{-l-1}$$

$$\times \sum_{i=-1}^{l} \|u_*^{(s-1)}(p,g,k,\mu,\varphi_{k+i}(\varphi,\mu)) - u_*^{(s-2)}(p,g,k,\mu,\varphi_{k+i}(\varphi,\mu))\|$$

$$\le C^0 P^* \sum_{l=-\infty}^{-1} (P^0)^{-l-1}(-l) \sup_{p,g,k,\mu,\varphi} \|u_*^{(s-1)}(p,g,k,\mu,\varphi)$$

$$- u_*^{(s-2)}(p,g,k,\mu,\varphi)\| \le \cdots \le \eta^{*s-1}\frac{C^0}{1-P^0}$$

implies that the sequence $\{u_*^{(s)}(p,g,k,\mu,\varphi)\}_{s=1}^{\infty}$ converges uniformly in p,g,k,μ,φ as $s \to \infty$ to some function $u(p,g,k,\mu,\varphi)$. Passing to the limit as $s \to \infty$ in equality (*), we obtain equality (2.93), i.e., the function $u(p,g,k,\mu,\varphi)$ generates the invariant torus of the system of equations (2.92). Theorem 2.10 is proved. $\qquad\square$

We now assume that none of the conditions $P_1 < 1, P^0 < 1$ is satisfied, but the GSF of the system of equations

$$x_{n+1} = P_1(\varphi_{n+p}(\varphi,\mu),\mu)x_n, \quad n \in Z, \tag{2.97}$$

where $P_1(\varphi_{n+p}(\varphi,\mu),\mu) = P(\varphi_{n+p}(\varphi,\mu),\mu,0)$, exists. Like in the previous section, we define it by equality (2.83). Hence, for all $\{p,l\} \subset Z$, $\mu \in S$, $\varphi \in \mathcal{T}_\infty$, the inequality $\|G_0(l,p,\mu,\varphi)\| \le M\lambda^{|l|}$, where M and $\lambda < 1$ are positive constants independent of p,l,φ,μ, holds. As was mentioned above, the existence of the GSF for $p = 0$ guarantees its existence for any $p \in Z$. Of course, $\Omega_l^n(p,\mu,\varphi)$ in formula (2.83) is the matricant of Eq. (2.97).

We now use the ideas of a method proposed in [105] and consider the system of equations

$$\varphi_{n+1} = \varphi_n + a(\varphi_n,\mu),$$

$$x_{n+1} = P_1(\varphi_{n+p},\mu)x_n + P^1(\varphi_{n+p},\mu)x_n + c(\varphi_{n+g+1},\mu), \quad n \in Z, \tag{2.98}$$

where $P^1(\varphi,\mu)$ is an infinite matrix, which is bounded in the norm and 2π-periodic in $\varphi^i (i = 1,2,3,\dots)$. Moreover, $\|P^1(\varphi,\mu)\| \le P_2 = const > 0$.

Lemma 2.16. *Let the GSF of Eq. (2.97) exist for $p = 0$, and let the inequality $P_2 < \frac{1-\lambda}{M(1+\lambda)}$ hold. Then, for any $\{p,g\} \subset Z, \mu \in S$, the invariant torus of the system of equations (2.98) exists.*

Proof.   We write down the equation

$$x_{n+1} = P_1(\varphi_{n+p}(\varphi, \mu), \mu)x_n$$
$$+ P^1(\varphi_{n+p}(\varphi, \mu), \mu)u(p, g, \mu, \varphi_n(\varphi, \mu)) + c(\varphi_{n+g+1}(\varphi, \mu), \mu), \quad n \in Z,$$

where $u(p, g, \mu, \varphi)$ is a function, which is bounded and 2π-periodic in $\varphi^i (i \in Z^+)$.

The invariant torus of this system exists and is generated by the function

$$\bar{u}(p, g, \mu, \varphi) = \sum_{l=-\infty}^{\infty} G_0(l, p, \mu, \varphi)P^1(\varphi_{l+p-1}(\varphi, \mu), \mu)u(p, g, \mu, \varphi_{l-1}(\varphi, \mu))$$
$$+ \sum_{l=-\infty}^{\infty} G_0(l, p, \mu, \varphi)c(\varphi_{l+g}(\varphi, \mu), \mu).$$

In this case, the following equality holds:

$$\bar{u}(p, g, \mu, \varphi_{n+1}(\varphi, \mu)) = P_1(\varphi_{n+p}(\varphi, \mu), \mu)\bar{u}(p, g, \mu, \varphi_n(\varphi, \mu))$$
$$+ P^1(\varphi_{n+p}(\varphi, \mu), \mu)u(p, g, \mu, \varphi_n(\varphi, \mu)) + c(\varphi_{n+g+1}(\varphi, \mu), \mu), \quad n \in Z.$$

If $\bar{u}(p, g, \mu, \varphi) = u(p, g, \mu, \varphi)$, then $u(p, g, \mu, \varphi)$ defines the invariant torus of the system of equations (2.98).

Consider the space $\mathfrak{M}^{\mathcal{T}_\infty}$ of bounded mappings $f(\varphi)$ of the torus $\mathcal{T}_\infty$ in the Banach space $\mathfrak{M}$. The space $\mathfrak{M}^{\mathcal{T}_\infty}$ is, in turn, a Banach one with the norm $\|f(\varphi)\|_0 = \sup_{\varphi \in \mathcal{T}_\infty} \|f(\varphi)\|$.

We set the operator G_{P_1}, which acts on $f(\varphi) \in \mathfrak{M}^{\mathcal{T}_\infty}$ in the following way:

$$G_{P_1}f(\varphi) = \sum_{l=-\infty}^{\infty} G_0(l, p, \mu, \varphi)P^1(\varphi_{l+p-1}(\varphi, \mu), \mu)f(\varphi_{l-1}(\varphi)).$$

It is obvious that $G_{P_1}f(\varphi) \in \mathfrak{M}^{\mathcal{T}_\infty}$ for all $p \in Z, \mu \in S$, since

$$\|G_{P_1}f(\varphi)\|_0 \leq P_2 M\|f(\varphi)\|_0 \frac{1+\lambda}{1-\lambda},$$

i.e., $G_{P_1}f(\varphi) : \mathfrak{M}^{\mathcal{T}_\infty} \to \mathfrak{M}^{\mathcal{T}_\infty}$.

Hence, for the determination of the function $u(p, g, \mu, \varphi)$, we obtain the equation

$$(I - G_{P_1})u(p, g, \mu, \varphi) = G_E c(\varphi_{g+1}(\varphi, \mu), \mu), \tag{2.99}$$

where I is the identity operator on $\mathfrak{M}^{\mathcal{T}_\infty}$, and E is the infinite identity matrix.

It is easy to see that G_{P_1} is a linear bounded operator, whose norm

$$\|G_{P_1}\|_0 = \sup_{\|f(\varphi)\|_0=1} \|G_{P_1}f(\varphi)\|_0 \le P_2 M \frac{1+\lambda}{1-\lambda} < 1.$$

In this case, the operator $(I - G_{P_1})$ is invertible, and Eq. (2.99) has, in the space $\mathfrak{M}^{T\infty}$, the single solution

$$u(p, g, \mu, \varphi) = \sum_{k=0}^{\infty} G_{P_1}^k G_E c(\varphi_{g+1}(\varphi, \mu), \mu),$$

which generates the invariant torus of the system of equations (2.98). In this case, uniformly in $\{p, g\} \subset Z, \mu \in S$, the estimate

$$\|u(p, g, \mu, \varphi)\|_0 \le \frac{C^0 M(1 + \lambda)}{1 - \lambda - P_2 M(1 + \lambda)}$$

holds. $\qquad\square$

We write down the system of equations (2.92) as one equation

$$x_{n+1} = P_1(\varphi_{n+p}(\varphi, \mu), \mu)x_n$$
$$+ P_*^1(\varphi_{n+p}(\varphi, \mu), \mu, x_{n+k})x_n + c(\varphi_{n+g+1}(\varphi, \mu), \mu), \quad n \in Z, \quad (2.100)$$

where

$$P_*^1(\varphi_{n+p}(\varphi, \mu), \mu, x_{n+k})$$
$$= P(\varphi_{n+p}(\varphi, \mu), \mu, x_{n+k}) - P_1(\varphi_{n+p}(\varphi, \mu), \mu).$$

Theorem 2.11. *Let the following conditions be satisfied on the set D^0:*
1) for $p = 0$, the GSF of Eq. (2.97) exists;
3) for all $\{x, \bar{x}\} \subset D$, $\quad \|P(\varphi, \mu, x) - P(\varphi, \mu, \bar{x})\| \le P^*\|x - \bar{x}\|$, where P^* is a positive constant, which is independent of $\varphi, \mu, x, \bar{x}$;
3) the inequalities

$$d_0 = 2P^0 M \frac{1+\lambda}{1-\lambda} < 1, \quad \frac{d_1}{1 - d_0} \le d, \quad \xi = \frac{P^* d_1^2 \eta_0}{C^0} < 1,$$

where $d_1 = C^0 M \frac{1+\lambda}{1-\lambda}$, and η_0 is the sum of the convergent series $\sum_{r=1}^{\infty} r d_0^{r-1}$, hold.
Then, for any $\{p, g, k\} \subset Z, \mu \in S$, the invariant torus of the system of equations (2.92) (or of the equation (2.100)) exists.

Proof. By the method of complete mathematical induction, we now show that there exists a sequence of functions $u^{(s)}(p, g, k, \mu, \varphi)$ $(s \in Z_0^+, u^{(0)} =$

$0 \in \mathfrak{M}$), each of which defines, for all $s \in Z^+$, the invariant torus of the system of equations

$$\varphi_{n+1} = \varphi_n + a(\varphi_n, \mu),$$

$$x_{n+1} = P_1(\varphi_{n+p}, \mu)x_n + P_*^1(\varphi_{n+p}, \mu, u^{(s-1)}(p, g, k, \mu, \varphi_{n+k}))x_n$$
$$+ c(\varphi_{n+g+1}, \mu), \quad n \in Z. \quad (2.101)$$

It is obvious that the function $u^{(1)}(p, g, \mu, \varphi)$ generates the invariant torus of the system of equations (2.101), where P_*^1 is the zero matrix.

In this case,

$$u^{(1)}(p, g, \mu, \varphi) = \sum_{l=-\infty}^{\infty} G_0(l, p, \mu, \varphi)c(\varphi_{l+g}(\varphi, \mu), \mu)$$

and

$$\|u^{(1)}(p, g, \mu, \varphi)\|_0 \le \frac{C^0 M(1+\lambda)}{1-\lambda} < d.$$

Assume that, for $s = \overline{2, m}$, the functions $u^{(s)}(p, g, k, \mu, \varphi)$ define the invariant tori of the relevant systems, and their values do not leave the domain D. Using Lemma 2.16, we now prove the validity of the last assertion for $s = m + 1$.

Indeed, the equation

$$x_{n+1} = P_1(\varphi_{n+p}(\varphi, \mu), \mu)x_n$$
$$+ P_*^1(\varphi_{n+p}(\varphi, \mu), \mu, u^{(m)}(p, g, k, \mu, \varphi_{n+k}(\varphi, \mu)))x_n$$
$$+ c(\varphi_{n+g+1}(\varphi, \mu), \mu), \quad n \in Z,$$

has the invariant torus, which is generated by the function

$$u^{(m+1)}(p, g, k, \mu, \varphi) = \sum_{r=0}^{\infty} G_{P_1^*(m)}^r G_E c(\varphi_{l+g}(\varphi, \mu), \mu),$$

since

$$\|G_{P_1^*(m)}\| \le 2MP^0 \frac{1+\lambda}{1-\lambda} < 1.$$

By $P_1^*(m)$, we denote the expression

$$P_*^1(\varphi_{l+p-1}, \mu, u^{(m)}(p, g, k, \mu, \varphi_{l+k-1})).$$

In this case,

$$\|u^{(m+1)}(p, g, k, \mu, \varphi)\|_0 \le \frac{C^0 M(1+\lambda)}{1-\lambda-2MP^0(1+\lambda)} \le d.$$

Hence, for all $\{n, p, g, k\} \subset Z, \mu \in S, \varphi \in \mathcal{T}_\infty, s \in Z_0^+$, the equality

$$u^{(s+1)}(p, g, k, \mu, \varphi_{n+1}(\varphi, \mu))$$
$$= P(\varphi_{n+p}(\varphi, \mu), \mu, u^{(s)}(p, g, k, \mu, \varphi_{n+k}(\varphi, \mu)))$$
$$\times u^{(s+1)}(p, g, k, \mu, \varphi_n(\varphi, \mu)) + c(\varphi_{n+g+1}(\varphi, \mu), \mu), \quad n \in Z, \quad (2.102)$$

holds.

We now show that the sequence $\{u^{(s)}(p, g, k, \mu, \varphi)\}_{s=1}^\infty$ converges in the space $\mathfrak{M}^{\mathcal{T}_\infty}$ as $s \to \infty$ uniformly in p, g, k, μ, φ to some function $u(p, g, k, \mu, \varphi)$.

By ω_{s+1}, we denote the difference

$$u^{(s+1)}(p, g, k, \mu, \varphi) - u^{(s)}(p, g, k, \mu, \varphi),$$

and will estimate it in the norm.

The following inequalities hold:

$$\|\omega_{s+1}\|_0 \leq \sum_{r=0}^\infty \|G_{P_1^*(s)}^r - G_{P_1^*(s-1)}^r\|_0 \|G_E c(\varphi_{1+g}(\varphi, \mu), \mu)\|_0$$

$$\leq d_1 \sum_{r=1}^\infty \|G_{P_1^*(s)}^r - G_{P_1^*(s-1)}^r\|_0$$

$$\leq d_1 \sum_{r=1}^\infty \sum_{i=0}^{r-1} \|G_{P_1^*(s)}^{r-1-i} (G_{P_1^*(s)} - G_{P_1^*(s-1)}) G_{P_1^*(s-1)}^i\|_0$$

$$\leq d_1 \sum_{r=1}^\infty \sum_{i=0}^{r-1} d_0^{r-1} \|G_{P_1^*(s)} - G_{P_1^*(s-1)}\|_0$$

$$\leq d_1 \eta_0 \sum_{l=-\infty}^\infty \|G_0(l, p, \mu, \varphi)\|_0 \|P_*^1(s) - P_*^1(s - 1)\|_0$$

$$\leq d_1 \eta_0 \frac{M(1 + \lambda)}{1 - \lambda} P^* \|\omega_{s+1}\|_0.$$

They yield the inductive estimate

$$\|\omega_{s+1}\|_0 \leq \xi \|\omega_s\|_0, \quad s \in Z^+.$$

Hence, $\|\omega_{s+1}\|_0 \leq \xi^{s-1} \|\omega_2\|_0$. Since $\xi < 1$, the sequence of functions $\{u^{(s)}(p, g, k, \mu, \varphi)\}_{s=1}^\infty$ is fundamental in the space $\mathfrak{M}^{\mathcal{T}_\infty}$ and, therefore, converges as $s \to \infty$ uniformly in p, g, k, μ, φ. It is clear that the limiting function $u(p, g, k, \mu, \varphi)$ is 2π-periodic in $\varphi^i (i \in Z^+)$.

Passing to the limit in equality (2.102) as $s \to \infty$, we obtain equality (2.93), which proves Theorem 2.11. $\square$

We now write down the system of equations

$$\varphi_{n+1} = \varphi_n + a(\varphi_n, \mu),$$

$$x_{n+1} = P(\varphi_{n+p}, \mu, x_{n+k})x_n + c(\varphi_{n+g+1}, \mu, x_{n+1}) \quad (n \in Z), \qquad (2.103)$$

such that the estimates

$$\|c(\varphi, \mu, x)\| \leq C^0,$$

$$\|c(\varphi, \mu, x) - c(\varphi, \mu, \bar{x})\| \leq C_L \|x - \bar{x}\| \quad \forall \{x, \bar{x}\} \subset D,$$

where C_L is a positive constant, hold on the set D^0.

Corollary 2.11. *The conditions of each of Theorems 2.9 – 2.11 are sufficient for the existence of the invariant torus of the system of equations (2.103), if the estimates $\eta < 1$, $\eta^* < 1$, and $\xi < 1$ entering them are replaced by the estimates*

$$\eta + \frac{C_L P_1}{1 - P_1} < 1, \quad \eta^* + \frac{C_L}{1 - P^0} < 1, \quad \xi + \frac{MC_L(1 + \lambda)}{(1 - \lambda)(1 - d_0)} < 1,$$

respectively.

Proof. Let us start from the conditions of Theorem 2.10. It is easy to verify that, $\forall s \in Z^+$, the functions

$$u^{(s)}(\varphi) = c(\varphi_g, \mu, u^{(s-1)}(\varphi))$$

$$+ \sum_{l=-\infty}^{-1} \prod_{i=-1}^{l} P(\varphi_{p+i}, \mu, u^{(s-1)}(\varphi_{i+k}))c(\varphi_{l+g}, \mu, u^{(s-1)}(\varphi_l)), \quad u^{(0)}(\varphi) = 0$$

do not exceed d in the norm and define the invariant tori of the systems of equations

$$\varphi_{n+1} = \varphi_n + a(\varphi_n, \mu),$$

$$x_{n+1} = P(\varphi_{n+p}, \mu, u^{(s-1)}(\varphi_{n+k}))x_n + c(\varphi_{n+g+1}, \mu, u^{(s-1)}(\varphi_{n+1})),$$

$n \in Z$, i.e.,

$$u^{(s)}(\varphi_{n+1}) = P(\varphi_{n+p}, \mu, u^{(s-1)}(\varphi_{n+k}))u^{(s)}(\varphi_n)$$

$$+ c(\varphi_{n+g+1}, \mu, u^{(s-1)}(\varphi_{n+1}))$$

identically in $\varphi \in T_\infty, \mu \in S, \{p, g.k\} \subset Z$. Here, by $u^{(s)}(\varphi)$, we denote $u^{(s)}(p, g, k, \mu, \varphi)$.

For all $s \in Z^+\backslash\{1\}$, the following relations hold:

$$\|u^{(s)}(\varphi) - u^{(s-1)}(\varphi)\| \le C_L\|u^{(s-1)}(\varphi) - u^{(s-2)}(\varphi)\|$$

$$+ \sum_{l=-\infty}^{-1} \| \prod_{i=-1}^{l} P(\varphi_{p+i}, \mu, u^{(s-1)}(\varphi_{k+i}))c(\varphi_{l+g}, \mu, u^{(s-1)}(\varphi_l))$$

$$- \prod_{i=-1}^{l} P(\varphi_{p+i}, \mu, u^{(s-2)}(\varphi_{k+i}))c(\varphi_{l+g}, \mu, u^{(s-2)}(\varphi_l))\|$$

$$\le C_L\|u^{(s-1)}(\varphi) - u^{(s-2)}(\varphi)\|_0$$

$$+ \sum_{l=-\infty}^{-1} (P^0)^{-l}C_L\|u^{(s-1)}(\varphi_l) - u^{(s-2)}(\varphi_l)\|$$

$$+ \sum_{l=-\infty}^{-1} C^0\| \prod_{i=-1}^{l} P(\varphi_{p+i}, \mu, u^{(s-1)}(\varphi_{k+i}))$$

$$- \prod_{i=-1}^{l} P(\varphi_{p+i}, \mu, u^{(s-1)}(\varphi_{k+i}))\|$$

$$\le (C_L + \frac{C_L P^0}{1 - P^0} + C^0 P^* \sum_{l=-\infty}^{-1} (P^0)^{-l-1}(-l))\|u^{(s-1)}(\varphi) - u^{(s-2)}(\varphi)\|_0$$

$$= (\frac{C_L}{1 - P^0} + \eta^*)\|u^{(s-1)}(\varphi) - u^{(s-2)}(\varphi)\|_0.$$

The recurrence formula

$$\|u^{(s)}(\varphi) - u^{(s-1)}(\varphi)\|_0 \le (\frac{C_L}{1 - P^0} + \eta^*)\|u^{(s-1)}(\varphi) - u^{(s-2)}(\varphi)\|_0$$

guarantees that the sequence $\{u^{(s)}(\varphi)\}_{s=1}^\infty$ converges uniformly in p, g, k, μ, φ under the condition

$$\frac{C_L}{1 - P^0} + \eta^* < 1.$$

Just it is the estimate, by which the condition $\eta^* < 1$ of Theorem 2.10 should be replaced, in order that the system of equations (2.103) have the invariant torus defined by the function $u(\varphi) = \lim_{s\to\infty} u^{(s)}(\varphi)$, where the limiting transition is realized in the norm of the space $\mathfrak{M}$.

By an analogous reasoning, it is easy to verify that, under the conditions of Theorem 2.9, where the inequality $\eta < 1$ is replaced by the estimate $\frac{C_L P_1}{1 - P_1} + \eta < 1$, the system of equations (2.103) has the invariant torus, whose generating function is the limit of the sequence

$$\{-\sum_{l=1}^{\infty}\prod_{i=0}^{l-1} P^{-1}(\varphi_{p+i}, \mu, u^{(s-1)}(\varphi_{i+k}))c(\varphi_{l+g}, \mu, u^{(s-1)}(\varphi_l))\}_{s=1}^\infty,$$

$$u^{(0)}(\varphi) = 0$$

as $s \to \infty$.

Under the conditions of Theorem 2.11, there exists the invariant torus of a system of equations of the form (2.103), where $c(\varphi_{n+g+1}, \mu, x_{n+1})$ is replaced by $c(\varphi_{n+g+1}, \mu, 0)$. This torus is defined by the function

$$u_0(\varphi) = u_0(p, g, k, \mu, \varphi) = \lim_{m \to \infty} \sum_{r=0}^{\infty} G^r_{P_1^*(m)} G_E c(\varphi_{g+1}, \mu, 0).$$

By an inductive reasoning, we can verify that, for all $s \in Z_0^+$, there exists the invariant torus of the system of equations of the form (2.103), where $c(\varphi_{n+g+1}, \mu, x_{n+1})$ is replaced by $c(\varphi_{n+g+1}, \mu, u_s(\varphi_{n+1}))$.

This torus is generated by the function $u_{s+1}(\varphi) = \lim_{m \to \infty} u_s^{(m+1)}(\varphi)$, where

$$u_s^{(m+1)}(\varphi) = \sum_{r=0}^{\infty} G^r_{P_1^*(m,s)} G_E c(\varphi_{g+1}, \mu, u_s(\varphi_1)),$$

$$P_1^*(m, s) = P_*^1(\varphi_{l+p-1}, \mu, u_s^{(m)}(\varphi_{l+k-1})).$$

The function $u_s^{(m)}$ can be constructed analogously to the function $u^{(m)}$ (see the proof of Theorem 2.11), so that $\|u_s(\varphi)\| \le d \quad \forall s \in Z_0^+$.

This means that

$$u_{s+1}(\varphi_{n+1}) = P(\varphi_{n+p}, \mu, u_{s+1}(\varphi_{n+k}))u_{s+1}(\varphi_n) + c(\varphi_{n+g+1}, \mu, u_s(\varphi_{n+1}))$$

for all $\varphi \in \mathcal{T}_\infty, \mu \in S, \{p, g.k\} \subset Z$.

It remains to prove the convergence of the sequence $\{u_s(\varphi)\}$ as $s \to \infty$ in the norm of the space $\mathfrak{M}$ and to pass to the limit as $s \to \infty$ in the last identity. For this purpose, it is sufficient to show that the indicated sequence is fundamental.

The following estimates hold:

$$\|G_{P_1^*(m,s)} - G_{P_1^*(m,s-1)}\| \le \sum_{l=-\infty}^{\infty} M\lambda^{|l|}\|P_*^1(\varphi_{l+p-1}, \mu, u_s^{(m)}(\varphi_{l+k-1}))$$

$$- P_*^1(\varphi_{l+p-1}, \mu, u_{s-1}^{(m)}(\varphi_{l+k-1}))\| \le MP^* \frac{1+\lambda}{1-\lambda}\|u_s^{(m)}(\varphi) - u_{s-1}^{(m)}(\varphi)\|_0;$$

$$\|G^r_{P_1^*(m,s)} - G^r_{P_1^*(m,s-1)}\| \le \sum_{r=0}^{\infty} r(2MP^0 \frac{1+\lambda}{1-\lambda})^{r-1}$$

$$\times \|G_{P_1^*(m,s)} - G_{P_1^*(m,s-1)}\| \le \sum_{r=0}^{\infty} r d_0^{r-1} MP^* \frac{1+\lambda}{1-\lambda}\|u_s^{(m)}(\varphi) - u_{s-1}^{(m)}(\varphi)\|_0$$

$$= \eta^0 MP^* \frac{1+\lambda}{1-\lambda}\|u_s^{(m)}(\varphi) - u_{s-1}^{(m)}(\varphi)\|_0.$$

Then

$$\|u_s^{(m+1)}(\varphi) - u_{s-1}^{(m+1)}(\varphi)\|$$

$$\leq \sum_{r=0}^{\infty} [\|G^r_{P_1^*(m,s)} - G^r_{P_1^*(m,s-1)}\| \|G_E c(\varphi_{g+1}, \mu, u_s(\varphi_1))\|$$

$$+ \|G^r_{P_1^*(m,s-1)}\| \|G_E c(\varphi_{g+1}, \mu, u_s(\varphi_1)) - G_E c(\varphi_{g+1}, \mu, u_{s-1}(\varphi_1))\|]$$

$$\leq MC^0 \frac{1+\lambda}{1-\lambda} \eta^0 MP^* \frac{1+\lambda}{1-\lambda} \|u_s^{(m)}(\varphi) - u_{s-1}^{(m)}(\varphi)\|_0$$

$$+ MC_L \frac{1+\lambda}{1-\lambda} \|u_s(\varphi) - u_{s-1}(\varphi)\|_0 \sum_{r=0}^{\infty} (2MP^0 \frac{1+\lambda}{1-\lambda})^r$$

$$= \frac{M^2 C^0 P^* \eta^0 (1+\lambda)^2}{(1-\lambda)^2} \|u_s^{(m)}(\varphi) - u_{s-1}^{(m)}(\varphi)\|_0$$

$$+ \frac{MC_L(1+\lambda)}{(1-\lambda)(1-d_0)} \|u_s(\varphi) - u_{s-1}(\varphi)\|_0.$$

Passing to the limit as $m \to \infty$, we obtain

$$\|u_{s+1}(\varphi) - u_s(\varphi)\|_0 \leq (\frac{M^2 C^0 P^* \eta^0 (1+\lambda)^2}{(1-\lambda)^2}$$

$$+ \frac{MC_L(1+\lambda)}{(1-\lambda)(1-d_0)}) \|u_s(\varphi) - u_{s-1}(\varphi)\|_0.$$

The last inequality ensures the fundamentality of the sequence $\{u_s(\varphi)\}_{s=0}^{\infty}$, since

$$\frac{M^2 C^0 P^* \eta^0 (1+\lambda)^2}{(1-\lambda)^2} = \xi.$$

Hence, the function $u(\varphi) = \lim_{s\to\infty} u_s(\varphi)$ generates the invariant torus of the system of equations (2.103). $\qquad\qquad\square$

2.6 Differentiability of the invariant tori of nonlinear systems in the Fréchet meaning

In this section, we continue the study of the system of equations (2.92). As above, the function generating its invariant torus is denoted by $u(p, g, k, \mu, \varphi)$. First, we prove two important auxiliary propositions.

Lemma 2.17. *Let the following conditions hold on the set* $D_0 = \Lambda \times D_\rho$:

   1) $\{a(\varphi,\mu), c(\varphi,\mu), \Phi^{-1}(\varphi,\mu)\} \subset C_\Lambda^1(\varphi,\mu)$

and $P^{-1}(\varphi,\mu,x) \in C^1_{D_0}(\varphi,\mu,x)$; moreover,

$$\left\|\frac{da(\varphi,\mu)}{d(\varphi,\mu)}\right\| \le A^*, \quad \left\|\frac{d\Phi^{-1}(\varphi,\mu)}{d(\varphi,\mu)}\right\| \le \Phi_*,$$

$$\left\|\frac{dP^{-1}(\varphi,\mu,x)}{d(\varphi,\mu,x)}\right\| \le P_*, \quad \left\|\frac{dc(\varphi,\mu)}{d(\varphi,\mu)}\right\| \le C_*,$$

where A^*, Φ_*, P_*, and C_* are positive constants;

2) the inequalities

$$P_1 < \frac{1}{1+A^*}, \quad \eta = C^0 P_* \sum_{l=1}^{\infty} P_1^{l-1} l < 1, \quad \frac{C^0 P_1}{1-P_1} \le d$$

hold.

Then, for all $\{p,g,k\} \subset Z, s \in Z^+$, the functions $u^{(s)} = u^{(s)}(p,g,k,\mu,\varphi)$ constructed in Theorem 2.9 belong to $C^1_\Lambda(\varphi,\mu)$, and

$$\left\|\frac{du^{(s)}}{d(\varphi,\mu)}\right\| < \kappa_s(p,g,k) \quad \forall(\varphi,\mu) \in \Lambda. \tag{2.104}$$

Proof.   Proof will be executed by the method of complete mathematical induction. For $s = 1$, the function $u^{(1)}$ is independent of $k \in Z$, and the proof of Lemma repeats that of Theorem 2.7. Indeed, by introducing the notation

$$\frac{C^0\eta(p)(1+A^*)}{1-P_1(1+A^*)} = W_0; \quad \frac{P_1 C_*(1+A^*)^{g+1}}{1-P_1(1+A^*)} = W_1 \quad (g \ge -1);$$

$$C_* \sum_{l=1}^{-g-1} \Phi_*^{-(l+g)} P_1^l + \frac{C_* P_1^{-g}}{1-P_1(1+A^*)} = W_2 \quad (g < -1);$$

$$\kappa_1(p,g) = \begin{cases} W_0 + W_1, & \text{if } g \ge -1; \\ W_0 + W_2, & \text{if } g < -1, \end{cases}$$

we obtain the estimate

$$\left\|\frac{du^{(1)}}{d(\varphi,\mu)}\right\| < \kappa_1(p,g) \quad \forall\{p,g\} \subset Z, (\varphi,\mu) \in \Lambda.$$

We now assume that the formulated proposition is proper for all $s = \overline{2,q}$ and will demonstra-te its validity for $s = q+1$.

We introduce the following formal notation:

$$\kappa_s(p,g,k) = \kappa_s, \quad \Omega_l^0(p,g,k,\mu,\varphi)_s = \Omega_{l,s}^0, \quad \left\|\frac{d\Omega_{l,s}^0}{d(\varphi,\mu)}\right\| = \Omega_s,$$

$$u^{(s)}(p, g, k, \mu, \varphi_{k+i}(\varphi, \mu)) = u^{(s)}(k + i),$$

$$\frac{d\Omega^0_{l,s}c(\varphi_{l+g}(\varphi, \mu), \mu)}{d(\varphi, \mu)} = u^{(s)}_l,$$

$$\left\| \frac{dP^{-1}\varphi_{p+i}(\varphi, \mu), \mu, u^{(s)}(k + i)}{d(\varphi, \mu)} \right\| = \Delta P^{-1}_s(i).$$

In this case,

$$u^{(q+1)} = -\sum_{l=1}^{\infty} \prod_{i=0}^{l-1} P^{-1}(\varphi_{p+i}(\varphi, \mu), \mu, u^{(q)}(i + k))c(\varphi_{l+g}(\varphi, \mu), \mu).$$

It is easy to verify that the following inequalities hold:

$$\Delta P^{-1}_q(i) \le P_* \max\left\{ \left\| \frac{d\varphi_{p+i}(\varphi, \mu)}{d(\varphi, \mu)} \right\|, 1, \left\| \frac{du^{(q)}(k + i)}{d(\varphi, \mu)} \right\| \right\}$$

$$< P_* \max\left\{ \left\| \frac{d\varphi_{p+i}(\varphi, \mu)}{d(\varphi, \mu)} \right\|, \kappa_q \left\| \frac{d\varphi_{k+i}(\varphi, \mu)}{d(\varphi, \mu)} \right\| \right\}$$

$$\le \begin{cases} P_* \max\{(1 + A^*)^{p+i}, \kappa_q(1 + A^*)^{k+i}\}, & \text{if } p + i \ge 0, k + i \ge 0; \\[2mm] P_* \max\{\Phi_*^{-(p+i)}, \kappa_q\Phi_*^{-(k+i)}\}, & \text{if } p + i < 0, k + i < 0; \\[2mm] P_* \max\{(1 + A^*)^{p+i}, \kappa_q\Phi_*^{-(k+i)}\}, & \text{if } p + i \ge 0, k + i < 0; \\[2mm] P_* \max\{\Phi_*^{-(p+i)}, \kappa_q(1 + A^*)^{k+i}\}, & \text{if } p + i < 0, k + i \ge 0; \end{cases}$$

$$\tag{2.105}$$

We recall that $\Phi_* > 1$. Without loss of generality, we consider that $\kappa_q \ge 1$ for all $\{p, g, k\} \subset Z$.

We reduce the number of cases, which can occur, to four.

$1^0. \quad \underline{p \ge 0, k \ge 0}.$

Since $i \ge 0$, we have $p + i \ge 0, k + i \ge 0$. In this case, relation (2.105) yields

$$\Delta P^{-1}_q(i) < P_* \kappa_q(1 + A^*)^{m+i}, \quad m = \max\{p, k\}.$$

The last inequality ensures the validity of the estimate

$$\Omega_{q+1} < P_1^{l-1} P_* \kappa_q \sum_{i=0}^{l-1}(1 + A^*)^{m+i}$$

$$= P_1^{l-1} P_* \kappa_q \sum_{i=m}^{m+l-1}(1 + A^*)^i < \kappa_q\xi_1(m)P_1^{l-1}(1 + A^*)^l,$$

where $\xi_1(m) = \frac{P_*}{A^*}(1 + A^*)^m$.

Then the series $\sum_{l=1}^{\infty} \|u_l^{(q+1)}\|$ converges uniformly in $(\varphi, \mu) \in \Lambda$, and its sum does not exceed $\kappa_{q+1}^1(p, g, k)$, where

$$
\kappa_{q+1}^1 = \begin{cases} \dfrac{C^0 \kappa_q \xi_1(m)(1 + A^*)}{1 - P_1(1 + A^*)} + W_1, & \text{if} \quad g \geq -1; \\[4mm] \dfrac{C^0 \kappa_q \xi_1(m)(1 + A^*)}{1 - P_1(1 + A^*)} + W_2, & \text{if} \quad g < -1. \end{cases}
$$

Hence, $u^{(q+1)} \in C_\Lambda^1(\varphi, \mu)$, and

$$
\|\frac{du^{(q+1)}}{d(\varphi, \mu)}\| < \kappa_{q+1}^1 \quad \forall g \in Z.
$$

$2^0.\ \underline{p \geq 0, k < 0.}$

In this case, the inequalities $p + i \geq 0, k + i < 0$ hold for $l < 1 - k, i = \overline{0, l - 1}$. Then relation (2.105) yields the estimates

$$
\Delta P_q^{-1}(i) < P_* \max\{(1 + A^*)^{p+i}, \kappa_q \Phi_*^{-(k+i)}\};
$$

$$
\Omega_{q+1} \leq P_1^{l-1} \sum_{i=0}^{l-1} \Delta P_q^{-1}(i) < P_1^{l-1} P_* \sum_{i=0}^{l-1} \{(1 + A^*)^{p+i} + \kappa_q \Phi_*^{-k}\}
$$

$$
< \xi_1(p) P_1^{l-1}(1 + A^*)^l + P_* \kappa_q \Phi_*^{-k} P_1^{l-1} l. \quad (2.106)
$$

But if $l \geq 1 - k$, then $\forall i = \overline{0, l - 1}\ p + i \geq 0, \forall i = \overline{0, -k - 1}\ k + i < 0$, $\forall i = \overline{-k, l - 1}\ k + i \geq 0$. Again, using (2.105), we write down the inequalities

$$
\Omega_{q+1} < P_1^{l-1} \sum_{i=0}^{-k-1} P_* \max\{(1 + A^*)^{p+i}, \kappa_q \Phi_*^{-(k+i)}\}
$$

$$
+ P_1^{l-1} \sum_{i=-k}^{l-1} P_* \max\{(1 + A^*)^{p+i}, \kappa_q(1 + A^*)^{k+i}\}
$$

$$
< P_1^{l-1} P_* \{Z_1 + \sum_{i=-k}^{l-1} \kappa_q(1 + A^*)^{p+i}\}
$$

$$
< P_1^{l-1} P_* \{Z_1 + \frac{\kappa_q(1 + A^*)^{p+l}}{A^*}\}, \quad (2.107)
$$

where

$$
Z_1 = \sum_{i=0}^{-k-1} \{(1 + A^*)^{p+i} + \kappa_q \Phi_*^{-k}\}-
$$

is a finite sum independent of φ, μ, l.

Relations (2.105) and (2.107) yield the inequalities

$$\sum_{l=1}^{\infty} \|u_l^{(q+1)}\| = \sum_{l=1}^{-k} \|u_l^{(q+1)}\| + \sum_{l=-k+1}^{\infty} \|u_l^{(q+1)}\|$$

$$< \sum_{l=1}^{-k} \{C^0[\xi_1(p)P_1^{l-1}(1+A^*)^l + P_*\kappa_q\Phi_*^{-k}P_1^{l-1}l] + P_1^l C_* \|\frac{d\varphi_{l+g}(\varphi,\mu)}{d(\varphi,\mu)}\|\}$$

$$+ \sum_{l=-k+1}^{\infty} \{P_1^{l-1}C^0 P_*[Z_1 + \frac{\kappa_q}{A^*}(1+A^*)^{p+l}] + P_1^l C_* \|\frac{d\varphi_{l+g}(\varphi,\mu)}{d(\varphi,\mu)}\|\}$$

$$= \sum_{l=1}^{-k} C^0[\xi_1(p)P_1^{l-1}(1+A^*)^l + P_*\kappa_q\Phi_*^{-k}P_1^{l-1}l]$$

$$+ \sum_{l=-k+1}^{\infty} P_1^{l-1}C^0 P_*[Z_1 + \frac{\kappa_q(1+A^*)^{p+l}}{A^*}]$$

$$+ \sum_{l=1}^{\infty} P_1^l C_* \|\frac{d\varphi_{l+g}(\varphi,\mu)}{d(\varphi,\mu)}\| < \kappa_{q+1}^2(p,g,k),$$

where

$$\kappa_{q+1}^2 = Z_2 + \frac{C^0 P_* Z_1}{1 - P_1} + \frac{C^0 P_* \kappa_q (1+A^*)^{p+1}}{A^*(1 - P_1(1+A^*))} + \begin{cases} W_1, & \text{if} \quad g \geq -1; \\ W_2, & \text{if} \quad g < -1, \end{cases}$$

$$Z_2 = \sum_{l=1}^{-k} C^0\{\xi_1(p)P_1^{l-1}(1+A^*)^l + P_*\kappa_q\Phi_*^{-k}P_1^{l-1}l\}$$

is a finite sum independent of φ, μ.

$3^0.$ $\underline{p < 0, k \geq 0.}$

This case is analogous to the previous one. For $l < 1 - p, i = \overline{0, l-1}$, the inequalities $k + i \geq 0, p + i < 0$ hold. Then

$$\Delta P_q^{-1}(i) < P_*\{\Phi_*^{-p} + \kappa_q(1+A^*)^{k+i}\},$$

$$\Omega_{q+1} < P_*\Phi_*^{-p}P_1^{l-1}l + \xi_1(k)\kappa_q P_1^{l-1}(1+A^*)^l. \tag{2.108}$$

But if $l \geq 1 - p$, then $k + i \geq 0$ for all $i = \overline{0, l-1}$, $p + i < 0$ for $i = \overline{0, -p-1}$, $p + i \geq 0$ and for $i = \overline{-p, l-1}$. In this case, we obtain the estimate

$$\Omega_{q+1} < P_1^{l-1}P_*\{Z_3 + \frac{\kappa_q}{A^*}(1+A^*)^{k+l}\}, \tag{2.109}$$

where

$$Z_3 = \sum_{i=0}^{-p-1} \{\Phi_*^{-p} + \kappa_q(1 + A^*)^{k+i}\}$$

is a finite sum independent of φ, μ, l.

Relations (2.108) and (2.109) yield an inequality of the form (2.104),

$$\sum_{l=1}^{\infty} \|u_l^{q+1}\| < \kappa_{q+1}^3(p, g, k),$$

where

$$\kappa_{q+1}^3 = Z_4 + \frac{C^0 P_* Z_3}{1 - P_1} + \frac{C^0 P_* \kappa_q(1 + A^*)^{k+1}}{A^*(1 - P_1(1 + A^*))} + \begin{cases} W_1, & \text{if } g \geq -1; \\ W_2, & \text{if } g < -1, \end{cases}$$

$$Z_4 = \sum_{l=1}^{-p} C^0 \{P_1^{l-1} P_* \Phi_*^{-p} l + \xi_1(k) \kappa_q P_1^{l-1}(1 + A^*)^l\}$$

is a finite sum independent of φ, μ.

$4^0.$ $\underline{p < 0, k < 0.}$

It can happen that, in this case: a) $p < k$, b) $p > k$, c) $p = k$. These three subcases are of the same type. Therefore, we consider only the first one.

For $l < 1 - k$ and $i = \overline{0, l-1}$, $k + i < 0, p + i < 0$. Then relation (2.105) yields $\Delta P_q^{-1}(i) < P_* \kappa_q \Phi_*^n$, where $n = \max\{|p|, |k|\} = -p$. Then, in turn,

$$\Omega_{q+1} < P_1^{l-1} P_* \kappa_q \Phi_*^n \sum_{i=0}^{l-1} 1 = P_1^{l-1} P_* \kappa_q \Phi_*^n l. \tag{2.110}$$

If $1 - k \leq l < 1 - p$, then $p + i < 0$ for all $i = \overline{0, l-1}$; $k + i < 0$ for $i = \overline{0, -k-1}$; and $k + i \geq 0$ for $i = \overline{-k, l-1}$. In this case,

$$\Omega_{q+1} < P_1^{l-1} P_* \sum_{i=0}^{-k-1} \max\{\Phi_*^{-(p+i)}, \kappa_q \Phi_*^{-(k+i)}\}$$

$$+ P_1^{l-1} P_* \sum_{i=-k}^{l-1} \max\{\Phi_*^{-(p+i)}, \kappa_q(1 + A^*)^{k+i}\}$$

$$< P_* \kappa_q \Phi_*^n(-k) P_1^{l-1} + P_1^{l-1} P_* \Phi_*^{-p}(l + k) + \frac{P_*}{A^*} \kappa_q(1 + A^*)^{k+l} P_1^{l-1}. \tag{2.111}$$

Let, eventually, $l \geq 1 - p$. Then we obtain

$$k + i < 0,\ p + i < 0 \text{ for } i = \overline{0, -k-1};$$

$$k + i \geq 0,\ p + i < 0 \text{ for } i = \overline{-k, -p-1};$$

$$k + i > 0,\ p + i \geq 0 \text{ for } i = \overline{-p, l-1}.$$

The last relations and (2.105) yield the inequalities

$$\Omega_{q+1} < P_1^{l-1} P_* \Big\{ \sum_{i=0}^{-k-1} \max\{\Phi_*^{-(p+i)}, \kappa_q \Phi_*^{-(k+i)}\}$$

$$+ \sum_{i=-k}^{-p-1} \max\{\Phi_*^{-(p+i)}, \kappa_q (1 + A^*)^{k+i}\}$$

$$+ \sum_{i=-p}^{l-1} \max\{(1 + A^*)^{p+i}, \kappa_q (1 + A^*)^{k+i}\}\Big\}$$

$$< P_1^{l-1} P_* \Big\{ \sum_{i=0}^{-k-1} \kappa_q \Phi_*^n + \sum_{i=-k}^{-p-1} (\Phi_*^{-p} + \kappa_q (1 + A^*)^{k+i})$$

$$+ \sum_{i=-p}^{l-1} \kappa_q (1 + A^*)^{k+i} \Big\} < P_1^{l-1} P_* \Big\{ Z_5 + Z_6 + \frac{\kappa_q (1 + A^*)^{k+l}}{A^*} \Big\}, \quad (2.112)$$

where

$$Z_5 = \sum_{i=0}^{-k-1} \kappa_q \Phi_*^n, \quad Z_6 = \sum_{i=-k}^{-p-1} (\Phi_*^{-p} + \kappa_q (1 + A^*)^{k+i})$$

are finite sums, which are independent of φ, μ, l.

Relations $(2.110) - (2.112)$ yield an inequality of the form (2.104),

$$\sum_{l=1}^{\infty} \|u_l^{(q+1)}\| < \kappa_{q+1}^4 (p, g, k),$$

in which

$$\kappa_{q+1}^4 = Z_7 + Z_8 + C^0 P_* (Z_5 + Z_6) \frac{P_1^{-p}}{1 - P_1}$$

$$+ \frac{C^0 P_* P_1^{-p} \kappa_q (1 + A^*)^{k-p+1}}{A^* (1 - P_1 (1 + A^*))} + \begin{cases} W_1, & \text{if } g \geq -1; \\ W_2, & \text{if } g < -1, \end{cases}$$

where

$$Z_7 = \sum_{l=1}^{-k} C^0 P_* \kappa_q \Phi_*^n P_1^{l-1} l,$$

$$Z_8 = \sum_{l=-k+1}^{-p} C^0 \{ P_* \kappa_q \Phi_*^n(-k) P_1^{l-1}$$

$$+ P_1^{l-1} P_* \Phi_*^{-p}(l+k) + P_1^{l-1} P_* \kappa_q \frac{(1+A^*)^{k+l}}{A^*} \}$$

are finite sums, which are independent of φ, μ. The lemma is proved. $\square$

Lemma 2.18. *Let the conditions of Lemma 2.17 be satisfied, and let,* $\forall (\varphi, \mu) \in \Lambda$, $\{x, \bar{x}\} \subset D_\rho$,

$$\| \frac{dP^{-1}(\varphi, \mu, x)}{d(\varphi, \mu, x)} - \frac{dP^{-1}(\varphi, \mu, \bar{x})}{d(\varphi, \mu, \bar{x})} \| \le L \|x - \bar{x}\|,$$

where $L = const > 0$. *Then, for all* $s \in Z^+$, *the inequalities*

$$I_1 = \sum_{l=1}^{\infty} \| \Omega_{l,s+1}^0 - \Omega_{l,s}^0 \| \| \frac{dc(\varphi_{l+g}(\varphi, \mu), \mu)}{d(\varphi, \mu)} \| \le B \eta^{s-1}, \tag{2.113}$$

$$\| \Delta \Omega_s \| = \| \frac{d\Omega_{l,s+1}^0}{d(\varphi, \mu)} - \frac{d\Omega_{l,s}^0}{d(\varphi, \mu)} \| < I_2 + I_3 + I_4, \tag{2.114}$$

where

$$I_2 = P_1^{l-1} L \eta^{s-1} \frac{C^0 P_1}{1 - P_1} \sum_{i=0}^{l-1} \max \{ \| \frac{d\varphi_{i+p}(\varphi, \mu)}{d(\varphi, \mu)} \|, \kappa_s \| \frac{d\varphi_{i+k}(\varphi, \mu)}{d(\varphi, \mu)} \| \},$$

$$I_3 = P_1^{l-1} P_* \sum_{i=0}^{l-1} \| \frac{d\varphi_{i+k}(\varphi, \mu)}{d(\varphi, \mu)} \| \| \frac{du^{(s)}}{d(\varphi, \mu)} |_{\varphi=\varphi_{k+i}} - \frac{du^{(s-1)}}{d(\varphi, \mu)} |_{\varphi=\varphi_{k+i}} \|,$$

$$I_4 = P_* \eta^{s-1} \frac{C^0 P_1}{1 - P_1} P_1^{l-2} \{ \sum_{i=1}^{l-1} i \Delta P_s^{-1}(i) + \sum_{i=0}^{l-2} (l-1-i) \Delta P_{s-1}^{-1}(i) \},$$

hold, and the series in (2.113) converges uniformly in $(\varphi, \mu) \in \Lambda$, *the positive constant* B *is independent of* $(\varphi, \mu) \in \Lambda$, *and the symbols* $\sum_{i=1}^{0}$ *and* $\sum_{i=0}^{-1}$ *stands for zeros.*

Proof.   We denote $P^{-1}(\varphi_{p+i}(\varphi, \mu), \mu, u^{(s)}(k+i))$ by $P_{s,i}^{-1}$. The convexity of the set D_0 allows us to write down the inequalities

$$\| P_{s,i}^{-1} - P_{s-1,i}^{-1} \| \le P_* \|u^{(s)}(k+i) - u^{(s-1)}(k+i)\|$$

$$\le P_* \sup_{(p,g,k,\mu,\varphi)} \|u^{(s)} - u^{(s-1)}\| \le P_* \eta^{(s-1)} \frac{C^0 P_1}{1 - P_1}, \quad (s \in Z^+, i = \overline{0, l-1}),$$

$$\tag{2.115}$$

which ensure the validity of the relations

$$\|\Omega^0_{l,s+1} - \Omega^0_{l,s}\| = \|\prod_{i=0}^{l-1} P_{s,i}^{-1} - \prod_{i=0}^{l-1} P_{s-1,i}^{-1}\|$$

$$= \|(P_{s,0}^{-1} - P_{s-1,0}^{-1})P_{s,1}^{-1} \cdots P_{s,l-1}^{-1}$$

$$+ P_{s-1,0}^{-1}(P_{s,1}^{-1} - P_{s-1,1}^{-1})P_{s,2}^{-1}P_{s,3}^{-1} \cdots P_{s,l-1}^{-1}$$

$$+ P_{s-1,0}^{-1}P_{s-1,1}^{-1}(P_{s,2}^{-1} - P_{s-1,2}^{-1})P_{s,3}^{-1} \cdots P_{s,l-1}^{-1} + \cdots$$

$$+ P_{s-1,0}^{-1} \cdots P_{s-1,l-2}^{-1}(P_{s,l-1}^{-1} - P_{s-1,l-1}^{-1})\|$$

$$\leq P_1^{l-1} \sum_{i=0}^{l-1} \|P_{s,i}^{-1} - P_{s-1,i}^{-1}\| \leq P_* P_1^{l-1} \sum_{i=0}^{l-1} \sup_{(p,g,k,\mu,\varphi)} \|u^{(s)} - u^{(s-1)}\|$$

$$\leq P_* P_1^{l-1} l \frac{C^0 P_1}{1 - P_1} \eta^{s-1} \quad (s \in Z^+).$$

Then we have

$$I_1 \leq P_* \sum_{l=1}^{\infty} P_1^{l-1} l \frac{C^0 P_1}{1 - P_1} \eta^{s-1} \|\frac{dc(\varphi_{l+g}(\varphi,\mu))}{d(\varphi,\mu)}\| \leq B\eta^{s-1},$$

where

$$B = \frac{P_* C^0 C_* (1 + A^*)^g}{1 - P_1} \sum_{l=1}^{\infty} (P_1(1 + A^*))^l l$$

for $g \geq -1$ and

$$B = \frac{P_* C_* C^0 P_1}{1 - P_1} \{ \sum_{l=1}^{-g-1} P_1^{l-1} l \Phi^{-(l+g)} + \sum_{l=-g}^{\infty} P_1^{l-1} l (1 + A^*)^{l+g} \}$$

for $g < -1$. In both cases, B is a positive constant, which is independent of $(\varphi,\mu) \in \Lambda$, since $P_1(1 + A^*) < 1$, and the series $\sum_{l=1}^{\infty}(P_1(1 + A^*))^l l$ converges. Estimate (2.113) is proved.

We now estimate the difference $\Delta\Omega_s$ in the norm. The inequality

$$\|\Delta\Omega_s\| \leq \|\frac{dP_{s,0}^{-1}}{d(\varphi,\mu)} - \frac{dP_{s-1,0}^{-1}}{d(\varphi,\mu)}\| \|P_{s,1}^{-1}\| \cdots \|P_{s,l-1}^{-1}\|$$

$$+ \|P_{s,0}^{-1} - P_{s-1,0}^{-1}\| \|\frac{dP_{s,1}^{-1}}{d(\varphi,\mu)}\| \|P_{s,2}^{-1}\| \cdots \|P_{s,l-1}^{-1}\| + \cdots$$

$$+ \|P_{s,0}^{-1} - P_{s-1,0}^{-1}\| \|P_{s,1}^{-1}\| \cdots \|P_{s,l-2}^{-1}\| \|\frac{dP_{s,l-1}^{-1}}{d(\varphi,\mu)}\|$$

$$+ \left\| \frac{dP_{s-1,0}^{-1}}{d(\varphi,\mu)} \right\| \|P_{s,1}^{-1} - P_{s-1,1}^{-1}\| \|P_{s,2}^{-1}\| \cdots \|P_{s,l-1}^{-1}\|$$

$$+ \|P_{s-1,0}^{-1}\| \left\| \frac{dP_{s,1}^{-1}}{d(\varphi,\mu)} - \frac{dP_{s-1,1}^{-1}}{d(\varphi,\mu)} \right\| \|P_{s,2}^{-1}\| \cdots \|P_{s,l-1}^{-1}\| + \cdots$$

$$+ \|P_{s-1,0}^{-1}\| \|P_{s,1}^{-1} - P_{s-1,1}^{-1}\| \|P_{s,2}^{-1}\| \cdots \|P_{s,l-2}^{-1}\| \left\| \frac{dP_{s,l-1}^{-1}}{d(\varphi,\mu)} \right\|$$

$$+ \left\| \frac{dP_{s-1,0}^{-1}}{d(\varphi,\mu)} \right\| \|P_{s-1,1}^{-1}\| \cdots \|P_{s-1,l-2}^{-1}\| \|P_{s,l-1}^{-1} - P_{s-1,l-1}^{-1}\|$$

$$+ \|P_{s-1,0}^{-1}\| \left\| \frac{dP_{s-1,1}^{-1}}{d(\varphi,\mu)} \right\| \|P_{s-1,2}^{-1}\| \cdots \|P_{s,l-1}^{-1} - P_{s-1,l-1}^{-1}\| + \cdots$$

$$+ \|P_{s-1,0}^{-1}\| \cdots \|P_{s-1,l-2}^{-1}\| \left\| \frac{dP_{s,l-1}^{-1}}{d(\varphi,\mu)} - \frac{dP_{s-1,l-1}^{-1}}{d(\varphi,\mu)} \right\| \qquad (2.116)$$

holds. Obviously, the right-hand side of this inequality includes l^2 terms.

Consider the mapping $h : \Lambda \to D_0$, which is composed of three components: $\xi_1 : \Lambda \to \mathfrak{M}$, $\xi_2 : \Lambda \to S$, and $\xi_3 : \Lambda \to D_\rho$ acting by the laws $\xi_1(\varphi,\mu) = \varphi_{i+p}(\varphi,\mu)$, $\xi_2(\varphi,\mu) = \mu$, and $\xi_3(\varphi,\mu) = u^{(s)}(k+i)$.

Let the mapping h_1 act analogously to h, only $\xi_3(\varphi,\mu) = u^{(s-1)}(k+i)$.

Since the components $\xi_i(i = \overline{1,3})$ are continuously differentiable on Λ, we have

$$\frac{dP_{s,i}^{-1}}{d(\varphi,\mu)} = \frac{dP^{-1}(\varphi,\mu,x)}{d(\varphi,\mu,x)} \Big|_{h(\varphi,\mu)} \cdot \frac{dh(\varphi,\mu)}{d(\varphi,\mu)},$$

$$\frac{dP_{s-1,i}^{-1}}{d(\varphi,\mu)} = \frac{dP^{-1}(\varphi,\mu,x)}{d(\varphi,\mu,x)} \Big|_{h_1(\varphi,\mu)} \cdot \frac{dh_1(\varphi,\mu)}{d(\varphi,\mu)},$$

where the sign "$\cdot$" means a superposition of mappings. Then the inequalities

$$\left\| \frac{dP_{s,i}^{-1}}{d(\varphi,\mu)} - \frac{dP_{s-1,i}^{-1}}{d(\varphi,\mu)} \right\|$$

$$\leq \left\| \frac{dP^{-1}(\varphi,\mu,x)}{d(\varphi,\mu,x)} \Big|_{h(\varphi,\mu)} - \frac{dP^{-1}(\varphi,\mu,x)}{d(\varphi,\mu,x)} \Big|_{h_1(\varphi,\mu)} \right\|$$

$$\times \left\| \frac{dh(\varphi,\mu)}{d(\varphi,\mu)} \right\| + \left\| \frac{dP^{-1}(\varphi,\mu,x)}{d(\varphi,\mu,x)} \Big|_{h_1(\varphi,\mu)} \right\| \left\| \frac{dh(\varphi,\mu)}{d(\varphi,\mu)} - \frac{dh_1(\varphi,\mu)}{d(\varphi,\mu)} \right\|$$

$$\leq L\|u^{(s)}(k+i)-u^{(s-1)}(k+i)\|\|\|\frac{dh(\varphi,\mu)}{d(\varphi,\mu)}\|$$

$$+P_*\|\frac{du^{(s)}(k+i)}{d(\varphi,\mu)}-\frac{du^{(s-1)}(k+i)}{d(\varphi,\mu)}\|$$

$$<L\eta^{s-1}\frac{C^0P_1}{1-P_1}\max\{\|\frac{d\varphi_{i+p}(\varphi,\mu)}{d(\varphi,\mu)}\|,\kappa_s\|\frac{d\varphi_{i+k}(\varphi,\mu)}{d(\varphi,\mu)}\|\}$$

$$+P_*\|\frac{d\varphi_{i+k}(\varphi,\mu)}{d(\varphi,\mu)}\|\|\frac{du^{(s)}}{d(\varphi,\mu)}|_{\varphi=\varphi_{k+i}}-\frac{du^{(s-1)}}{d(\varphi,\mu)}|_{\varphi=\varphi_{k+i}}\| \qquad (2.117)$$

hold.

In view of the relation

$$\|\frac{dP_{s,i}^{-1}}{d(\varphi,\mu)}\|=\Delta P_s^{-1}(i) \quad (i=\overline{0,l-1})$$

and $(2.115)-(2.117)$, we obtain the estimate

$$\|\Delta\Omega_s\|\leq P_1^{l-1}\sum_{i=0}^{l-1}\|\frac{dP_{s,i}^{-1}}{d(\varphi,\mu)}-\frac{dP_{s-1,i}^{-1}}{d(\varphi,\mu)}\|$$

$$+P_*\eta^{s-1}\frac{C^0P_1}{1-P_1}P_1^{l-2}\{\sum_{i=1}^{l-1}i\|\frac{dP_{s,i}^{-1}}{d(\varphi,\mu)}\|+\sum_{i=0}^{l-2}(l-1-i)\|\frac{dP_{s-1,i}^{-1}}{d(\varphi,\mu)}\|\},$$

which yields inequality (2.114). Lemma 2.18 is proved. $\qquad\square$

We note that, $\forall s\in Z^+$, the mapping $\frac{du^{(s)}}{d(\varphi,\mu)}$ belongs to the Banach space $\mathbf{L}(\mathfrak{M}\times\mathfrak{M},\mathfrak{M})$ of linear operators. Therefore, in order that the sequence $\{\frac{du^{(s)}}{d(\varphi,\mu)}\}_{s=1}^{\infty}$ converge uniformly in $(\varphi,\mu)\in\Lambda$, it is sufficient that it be fundamental uniformly in $(\varphi,\mu)\in\Lambda$.

The uniform convergence of this sequence yields the existence and the continuity in $(\varphi,\mu)\in\Lambda$ of the Fréchet derivative of the function $u(p,g,k,\mu,\varphi)$.

The verification of the fundamentality of the indicated sequence should be performed separa-tely in each of the cases considered in Lemma 2.17. Due to the awkwardness of calculations, we consider only case 1^0. The study of the other cases can be made, by following the below-presented scheme with the use of Lemmas 2.17 and 2.18.

Theorem 2.12. *Let the matrix $P(\varphi,\mu,x)$ have the Lipschitz property in x on the set D_0, and let the following conditions be satisfied:*

1) $\{a(\varphi,\mu),c(\varphi,\mu),\Phi^{-1}(\varphi,\mu)\}\subset C_\Lambda^1(\varphi,\mu)$ and

$P^{-1}(\varphi, \mu, x) \in C^1_{D_0}(\varphi, \mu, x)$; *moreover,*

$$\left\|\frac{da(\varphi, \mu)}{d(\varphi, \mu)}\right\| \leq A^*, \quad \left\|\frac{d\Phi^{-1}(\varphi, \mu)}{d(\varphi, \mu)}\right\| \leq \Phi_*,$$

$$\left\|\frac{dP^{-1}(\varphi, \mu, x)}{d(\varphi, \mu, x)}\right\| \leq P_*, \quad \left\|\frac{dc(\varphi, \mu)}{d(\varphi, \mu)}\right\| \leq C_*,$$

$$\left\|\frac{dP^{-1}(\varphi, \mu, x)}{d(\varphi, \mu, x)} - \frac{dP^{-1}(\varphi, \mu, \bar{x})}{d(\varphi, \mu, \bar{x})}\right\| \leq L\|x - \bar{x}\|,$$

where $\{x, \bar{x}\} \subset D_\rho$, A^*, Φ_*, P_*, *and* C_*, L *are positive constants;*
2) *the inequalities*

$$P_1 < \frac{1}{1 + A^*}, \quad \eta = C^0 P_* \sum_{l=1}^{\infty} P_1^{l-1} l < 1, \quad \frac{C^0 P_1}{1 - P_1} \leq d$$

hold.

Then, $\forall g \in Z$ *and* $\{p, k\} \subset Z_0^+$ *such that*

$$\max\{p, k\} < \log_{(1+A^*)} \frac{A^*(1 - P_1(1 + A^*))}{C^0 P_*} - 1, \tag{2.118}$$

the function $u(p, g, k, \mu, \varphi) \in C^1_\Lambda(\varphi, \mu)$.

Proof.   First, we show that, under the conditions of Theorem 2.12, the set $\kappa_s(p, g, k)$ $(s \in Z^+)$ is bounded for fixed p, g, k, i.e.,

$$\left\|\frac{du^{(s)}}{d(\varphi, \mu)}\right\| < \kappa(p, g, k) \quad \forall s \in Z^+. \tag{2.119}$$

Assume that $\kappa_1 = \kappa_1(p, g) \geq 1$ (in the opposite case, we set $\kappa_1 = 1$). First, let $g \geq -1$.

We denote the fraction $C^0 \xi_1(m)(1 + A^*)/(1 - P_1(1 + A^*))$ by A and write down the recurrence formula

$$\kappa_s = \kappa_{s-1} A + W_1 \quad (s = 2, 3, 4, \dots).$$

If all $\kappa_s(s = 2, 3, 4, \dots)$ are not less than 1, then it is easy to obtain the estimate

$$\left\|\frac{du^{(s)}}{d(\varphi, \mu)}\right\| < W_1 \sum_{i=0}^{s-2} A^i + A^{s-1}\kappa_1 \quad (s = 2, 3, 4, \dots),$$

by carrying out the inductive reasoning.

We note that relation (2.118) yields $A < 1$. Therefore, by setting $\kappa(p, g, k) = W_1 \frac{1}{1-A} + \kappa_1$, we obtain estimate (2.119).

But if $\kappa_s \geq 1(s = 2, 3, 4, \ldots, n - 1)$, and $\kappa_n < 1$, we replace κ_n by κ_1. Then we obtain

$$\left\| \frac{du^{(n+1)}}{d(\varphi, \mu)} \right\| < \kappa_1 A + W_1 = \kappa_2,$$

and the process becomes cyclic.

For $g < -1$, the situation is analogous, only W_1 must be replaced by W_2.

Hence, by setting

$$\kappa(p, g, k) = \begin{cases} W_1 \dfrac{1}{1 - A} + \kappa_1, & \text{if} \quad g \geq -1; \\[3mm] W_2 \dfrac{1}{1 - A} + \kappa_1, & \text{if} \quad g < -1, \end{cases}$$

we obtain estimate (2.119) $\forall g \in Z$.

Using Lemma 2.18, we can write down the inequalities

$$\sum_{l=1}^{\infty} \|\Delta\Omega_s\| \|c(\varphi_{l+g}(\varphi, \mu), \mu)\|$$

$$< L\eta^{s-1} \frac{C^{0^2}(1 + A^*)^m \kappa}{(1 - P_1)A^*} \sum_{l=1}^{\infty} P_1^l (1 + A^*)^l$$

$$+ P_*^2 \kappa \eta^{s-1} \frac{C^{0^2}(1 + A^*)^m}{(1 - P_1)A^*} \sum_{l=1}^{\infty} P_1^{l-1}(1 + A^*)^l (l - 1)$$

$$+ \frac{P_* C^0}{A^* P_1}(1 + A^*)^k \left\| \frac{du^{(s)}}{d(\varphi, \mu)} - \frac{du^{(s-1)}}{d(\varphi, \mu)} \right\|_0 \sum_{l=1}^{\infty} P_1^l (1 + A^*)^l$$

$$< B_1 \eta^{s-1} + B_2 \left\| \frac{du^{(s)}}{d(\varphi, \mu)} - \frac{du^{(s-1)}}{d(\varphi, \mu)} \right\|_0,$$

where

$$B_1 = \frac{\kappa C^{0^2}(1 + A^*)^{m+1}}{A^*(1 - P_1)} \left(\frac{L}{1 - P_1(1 + A^*)} \right.$$

$$\left. + P_*^2 \sum_{l=1}^{\infty} (P_1(1 + A^*))^{l-1}(l - 1) \right)$$

and

$$B_2 = \frac{P_* C^0 (1 + A^*)^{k+1}}{A^*(1 - P_1(1 + A^*))}$$

are positive constants, which are independent of $(\varphi, \mu) \in \Lambda$, $\| \cdot \|_0 = \sup_{(\varphi, \mu) \in \Lambda} \| \cdot \|$.

Thus, we obtain the recurrence formula

$$\|\frac{du^{(s+1)}}{d(\varphi,\mu)} - \frac{du^{(s)}}{d(\varphi,\mu)}\|_0 < B\eta^{s-1} + B_1\eta^{s-1} + B_2\|\frac{du^{(s)}}{d(\varphi,\mu)} - \frac{du^{(s-1)}}{d(\varphi,\mu)}\|_0.$$

We denote $B + B_1$ by B_3. Then

$$\Gamma_s = \|\frac{du^{(s+1)}}{d(\varphi,\mu)} - \frac{du^{(s)}}{d(\varphi,\mu)}\|_0 < B_3\eta^{s-1} + B_2\|\frac{du^{(s)}}{d(\varphi,\mu)} - \frac{du^{(s-1)}}{d(\varphi,\mu)}\|_0$$

$$< B_3(\eta^{s-1} + B_2\eta^{s-2} + B_2^2\eta^{s-3} + \cdots + B_2^{s-1}) + B_2^{s-1}\|\frac{du^{(1)}}{d(\varphi,\mu)} - \frac{du^{(0)}}{d(\varphi,\mu)}\|_0$$

$$= B_3\frac{\eta^{s-1}((\eta^{-1}B_2)^s - 1)}{\eta^{-1}B_2 - 1} + B_2^{s-1}\|\frac{du^{(1)}}{d(\varphi,\mu)}\|_0.$$

In view of the inequality $\eta < 1$ and the fact that inequality (2.118) ensures the estimate $B_2 < 1$, we obtain the relations

$$\Gamma_s < \begin{cases} \eta^{s-1}(\dfrac{B_3}{1 - \eta^{-1}B_2} + \kappa(p,g,k)), & \text{if} \quad B_2 < \eta; \\[2ex] B_2^{s-1}(\dfrac{B_3}{\eta(\eta^{-1}B_2 - 1)} + \kappa(p,g,k)), & \text{if} \quad \eta < B_2; \\[2ex] \eta^{s-1}(B_3 s + \kappa(p,g,k)), & \text{if} \quad \eta = B_2, \end{cases}$$

which guarantee that the sequence $\{\frac{du^{(s)}}{d(\varphi,\mu)}\}_{s=1}^\infty$ is fundamental uniformly in $(\varphi,\mu) \in \Lambda$. The theorem is proved. $\square$

Remark 2.4. Inequality (2.118) cannot be satisfied for any $\{p,k\} \subset Z^+$. In order that it hold for $p = k = 0$, it is sufficient that the estimate $C^0 P_*(1 + A^*)/A^*(1 - P_1(1+A^*)) < 1$ hold. It is clear that this should be attained due to the smallness of the constant C^0. It is also clear that condition (2.118) is changed in cases $2^0 - 4^0$.

We now assume that the matrix $P(\varphi,\mu,x)$ is not invertible or the condition $P_1 < 1/(1+A^*)$ does not hold. Then Theorem 2.12 is invalid. But it can happen that, in this case, $P^0 < 1/\Phi_*$. Let us consider this case. To this end, we present an analog of the above-constructed theory. Since we meet no basic differences in this case, we restrict ourselves by the consideration of one of the possible cases, for example, $p \leq 0, k \leq 0, g \in Z$.

Corollary 2.12. *Let the following conditions hold on the set D_0:*

1) $\{a(\varphi,\mu), c(\varphi,\mu), \Phi^{-1}(\varphi,\mu)\} \subset C_\Lambda^1(\varphi,\mu)$ *and* $P(\varphi,\mu,x) \in C_{D_0}^1(\varphi,\mu,x)$; *moreover,*

$$\|\frac{da(\varphi,\mu)}{d(\varphi,\mu)}\| \leq A^*, \quad \|\frac{d\Phi^{-1}(\varphi,\mu)}{d(\varphi,\mu)}\| \leq \Phi_*,$$

$$\left\|\frac{dP(\varphi,\mu,x)}{d(\varphi,\mu,x)}\right\| \le P^*, \quad \left\|\frac{dc(\varphi,\mu)}{d(\varphi,\mu)}\right\| \le C_*,$$

$$\left\|\frac{dP(\varphi,\mu,x)}{d(\varphi,\mu,x)} - \frac{dP(\varphi,\mu,\bar{x})}{d(\varphi,\mu,\bar{x})}\right\| \le L^*\|x - \bar{x}\|,$$

where $\{x,\bar{x}\} \subset D_\rho,\ A^*,\Phi_* > 1, P^*, C_*,$ and L^* are positive constants;

2) the inequalities

$$P^0 < \frac{1}{\Phi_*}, \quad \eta^* = C^0 P^* \sum_{l=1}^{\infty}(P^0)^{l-1}l < 1, \quad \frac{C^0}{1 - P^0} \le d$$

hold.

Then, $\forall g \subset Z$ and $\{p,k\} \subset Z_0^- = Z \setminus Z^+$ such that

$$\max\{-p, -k\} < \log_{\Phi_*}\frac{(\Phi_* - 1)(1 - P^0\Phi_*)}{C^0 P^*} - 2, \tag{2.120}$$

the function $u(p,g,k,\mu,\varphi) \in C_\Lambda^1(\varphi,\mu)$.

Proof. Proof is analogous to that of Theorem 2.12. Therefore, we give it only schematically.

Taking the formula for $u_*^{(s)} = u_*^{(s)}(p,g,k,\mu,\varphi)$ from Theorem 2.10 into account and applying the inductive reasoning, we verify that $u_*^{(s)} \in C_\Lambda^1(\varphi,\mu)$ for all $s \in Z^+$, and

$$\left\|\frac{du_*^{(s+1)}}{d(\varphi,\mu)}\right\| < \kappa_{s+1}^*(p,g,k) = \begin{cases} \dfrac{C^0 \kappa_s^* \xi_2(m^*)\Phi_*}{1 - P^0\Phi_*} + W_1^*, & \text{if}\quad g < 1; \\[3mm] \dfrac{C^0 \kappa_s^* \xi_2(m^*)\Phi_*}{1 - P^0\Phi_*} + W_2^*, & \text{if}\quad g \ge 1, \end{cases}$$

where we set

$$\kappa_s^* \ge 1(s \in Z^+), \quad m^* = \min\{p,k\}, \quad \xi_2(m^*) = \frac{P^*\Phi_*^{-m^*+1}}{\Phi_* - 1}, \quad \Phi_* > 1,$$

$$W_1^* = C_*\Phi_*^{-g} + \frac{C_* P^0 \Phi_*^{-g+1}}{1 - P^0\Phi_*},$$

$$W_2^* = C_*(1 + A^*)^g + \sum_{l=-g}^{-1}(P^0)^{-l}C_*(1 + A^*)^{l+g} + \frac{C_*(P^0)^{g+1}\Phi_*}{1 - P^0\Phi_*}.$$

It is easy to verify that, under condition (2.120), $\frac{C^0 \xi_2(m^*)\Phi_*}{1 - P^0\Phi_*} < 1$. Hence, the sequence $\{\frac{du_*^{(s+1)}}{d(\varphi,\mu)}\}_{s=0}^{\infty}$ is uniformly bounded, i.e.,

$$\left\|\frac{du_*^{(s)}}{d(\varphi,\mu)}\right\| < \kappa^*(p,g,k), \quad s \in Z^+,$$

where $\kappa^*(p, g, k)$ is independent of $(\varphi, \mu) \in \Lambda$.

This allows us to obtain the recurrence formula

$$\left\|\frac{du_*^{(s+1)}}{d(\varphi, \mu)} - \frac{du_*^{(s)}}{d(\varphi, \mu)}\right\|_0 < B_3^* \eta^{*^{s-1}} + B_2^* \left\|\frac{du_*^{(s)}}{d(\varphi, \mu)} - \frac{du_*^{(s-1)}}{d(\varphi, \mu)}\right\|_0,$$

where B_3^* and B_2^* are independent of $(\varphi, \mu) \in \Lambda$, moreover,

$$B_2^* = \frac{C^0 P^* \Phi_*^{-k+2}}{(\Phi_* - 1)(1 - P^0 \Phi_*)} \leq \frac{C^0 \Phi_* \xi_2(m^*)}{(1 - P^0 \Phi_*)} < 1.$$

This guarantees that the sequence $\left\{\frac{du_*^{(s)}}{d(\varphi,\mu)}\right\}_{s=1}^{\infty}$ is fundamental uniformly in $(\varphi, \mu) \in \Lambda$, which completes the proof of Corollary 2.12. $\square$

Using the above-obtained results as examples, we can study the differentiability of the invariant tori of the system of equations (2.103) for various values of deviations $\{p, g, k\} \subset Z$.

Example 2.3. Assume that $c(\varphi, \mu, x) \in C_{D_0}^1(\varphi, \mu, x)$ and, $\forall (\varphi, \mu) \in \Lambda$, $\{x, \bar{x}\} \subset D_\rho$, the inequalities

$$\left\|\frac{dc(\varphi, \mu, x)}{d(\varphi, \mu, x)}\right\| \leq C_*,$$

$$\left\|\frac{dc(\varphi, \mu, x)}{d(\varphi, \mu, x)} - \frac{dc(\varphi, \mu, \bar{x})}{d(\varphi, \mu, \bar{x})}\right\| \leq L_* \|x - \bar{x}\|,$$

where C_* and L_* are positive constants, hold.

How must the conditions of Theorem 2.12 be changed in order that the invariant torus of the system of equations (2.103) for nonnegative deviations p, g, k be Fréchet-differentiable on the set Λ ?

By Corollary 2.11, if the inequality $\eta < 1$ in the conditions of Theorem 2.12 is replaced by $\psi = C_* P_1/(1 - P_1) + \eta < 1$, then the sequence

$$\{u^{(s+1)} = -\sum_{l=1}^{\infty} \Omega_{l,s+1}^0 c(\varphi_{l+g}, \mu, u^{(s)}(l))\}_{s=0}^{\infty}$$

converges in the norm of the space $\mathfrak{M}$ to a function, which generates the invariant torus of the system of equations (2.103).

By $\bar{A}(p, g, k)$, we denote the expression

$$\frac{P_* C^0 (1 + A^*)^{m+1}}{A^*(1 - P_1(1 + A^*))} + \frac{P_1 C_* (1 + A^*)^{g+1}}{1 - P_1(1 + A^*)} \quad (m = \max\{p, k\})$$

and assume that the deviations $\{p, g, k\} \in Z_0^+$ satisfy the inequality

$$\bar{A}(p, g, k) < 1. \tag{2.121}$$

We can directly verify that

$$\|\frac{du^{(1)}}{d(\varphi,\mu)}\| < \bar{A}(p,g,k) < 1.$$

Using the inequalities

$$\|\frac{du^{(s+1)}}{d(\varphi,\mu)}\| \le \sum_{l=1}^{\infty}\{\Omega_{s+1}C^0 + P_1^l C_* \max\{(1+A^*)^{l+g},(1+A^*)^l\|\frac{du^{(s)}}{d(\varphi,\mu)}\|\}\}$$

$$\le \sum_{l=1}^{\infty}\{P_1^{l-1}P_*C^0 \sum_{i=0}^{l-1}\max\{(1+A^*)^{p+i},(1+A^*)^{k+i}\|\frac{du^{(s)}}{d(\varphi,\mu)}\|\}$$

$$+ P_1^l C_* \max\{(1+A^*)^{l+g},(1+A^*)^l\|\frac{du^{(s)}}{d(\varphi,\mu)}\|\},$$

it it easy to prove by the method of complete mathematical induction that

$$\|\frac{du^{(s)}}{d(\varphi,\mu)}\| < 1$$

for all $s \in Z^+$.

Analogously to the previous consideration, we obtain the estimates

$$\|u^{(s+1)} - u^{(s)}\|_0 \le \psi^s \frac{C^0 P_1}{1 - P_1};$$

$$\|\Omega^0_{l,s+1} - \Omega^0_{l,s}\| \le P_* P_1^{l-1} l\psi^{s-1}\frac{C^0 P_1}{1 - P_1};$$

$$\|\frac{dc(\varphi_{l+g},\mu,u^{(s)}(l))}{d(\varphi,\mu)} - \frac{dc(\varphi_{l+g},\mu,u^{(s-1)}(l))}{d(\varphi,\mu)}\|_0$$

$$< C_*\|\frac{du^{(s)}}{d(\varphi,\mu)} - \frac{du^{(s-1)}}{d(\varphi,\mu)}\|_0 + L_*\|u^{(s)} - u^{(s-1)}\|_0(1+A^*)^{l+g};$$

$$\|\Delta\Omega_s\| < \bar{I}_2 + I_3 + \bar{I}_4,$$

where $\bar{I}_2$ and $\bar{I}_4$ are obtained from I_2 and I_4, respectively, if we replace η by ψ and κ_s by 1 in the expressions for them (see (2.114)).

Then we have

$$\|\frac{du^{(s+1)}}{d(\varphi,\mu)} - \frac{du^{(s)}}{d(\varphi,\mu)}\|_0 \le \sum_{l=1}^{\infty}\{\|\frac{d\Omega^0_{l,s+1}}{d(\varphi,\mu)}\|\|c(\varphi_{l+g},\mu,u^{(s)}(l))$$

$$- c(\varphi_{l+g},\mu,u^{(s-1)}(l))\| + \|\Omega^0_{l,s+1}\|\|\frac{dc(\varphi_{l+g},\mu,u^{(s)}(l))}{d(\varphi,\mu)}$$

$$-\frac{dc(\varphi_{l+g},\mu,u^{(s-1)}(l))}{d(\varphi,\mu)}\| + \|\Delta\Omega_s\|\|c(\varphi_{l+g},\mu,u^{(s-1)}(l))\|$$

$$+ \|\Omega^0_{l,s+1} - \Omega^0_{l,s}\|\|\frac{dc(\varphi_{l+g},\mu,u^{(s-1)}(l))}{d(\varphi,\mu)}\|\}$$

$$< \sum_{l=1}^{\infty}\{\xi_1(m)P_1^{l-1}(1+A^*)^l C_*\psi^{s-1}\frac{C^0 P_1}{1-P_1} + P_1^l[C_*\|\frac{du^{(s)}}{d(\varphi,\mu)} - \frac{du^{(s-1)}}{d(\varphi,\mu)}\|_0$$

$$+L_*(1+A^*)^{l+g}\psi^{s-1}\frac{C^0 P_1}{1-P_1}]+(\bar{I}_2+\bar{I}_4)C^0+P_*P_1^l l\psi^{s-1}C^0\frac{C_*}{1-P_1}(1+A^*)^{l+g}$$

$$+\frac{C^0}{A^*}P_1^{l-1}P_*(1+A^*)^{k+l}\|\frac{du^{(s)}}{d(\varphi,\mu)} - \frac{du^{(s-1)}}{d(\varphi,\mu)}\|_0\}$$

$$< E_3\psi^{s-1} + E_2\|\frac{du^{(s)}}{d(\varphi,\mu)} - \frac{du^{(s-1)}}{d(\varphi,\mu)}\|_0,$$

where the coefficients E_2 and E_3 are independent of $(\varphi,\mu) \in \Lambda$, and

$$E_2 = \frac{P_*C^0(1+A^*)^{k+1}}{A^*(1-P_1(1+A^*))} + \frac{P_1C_*}{1-P_1}.$$

It is easy to verify that $E_2 < \bar{A}$. Hence, $E_2 < 1$, which ensures that the sequence $\{\frac{du^{(s)}}{d(\varphi,\mu)}\}_{s=1}^{\infty}$ is fundamental uniformly in $(\varphi,\mu) \in \Lambda$. This is sufficient for the Fréchet-differentiability of the invariant torus of the system of equations (2.103).

Thus, under the conditions of Theorem 2.12, it is sufficient to take into account that $g \geq 0$, to replace the inequality $\eta < 1$ by $\psi < 1$, and to replace estimate (2.118) by (2.121). The membership condition $\Phi^{-1} \in C^1_\Lambda(\varphi,\mu)$ can be omitted.

2.7   Conditions of existence of the Green–Samoilenko function for a linear system defined on the set $\mathfrak{M} \times \mathcal{T}_\infty$. Reduction of the problem of construction of the invariant torus of this system to an analogous problem in the space $R^s \times \mathcal{T}_m$

Consider the system of equations

$$\varphi_{n+1} = \varphi_n + \omega, \quad x_{n+1} = P(\varphi_{n+p})x_n + c(\varphi_n + \omega), \tag{2.122}$$

where $\varphi = (\varphi^1,\varphi^2,\varphi^3,\dots)$, $\omega = (\omega^1,\omega^2,\omega^3,\dots)$, $x = (x^1,x^2,x^3,\dots)$ belong to the space $\mathfrak{M}$, the infinite matrix $P(\varphi)$ and the function $c(\varphi) = (c_1(\varphi),c_2(\varphi),c_3(\varphi),\dots)$ are real and 2π-periodic in $\varphi^i, i \in Z^+, n \in Z$, and

p is an integer parameter. As above, we consider that the initial values φ of the solutions $\varphi_n = \varphi_n(\varphi)$ *and* $\varphi_0(\varphi) = \varphi$ of the first equation of system (2.122) belong to the infinite-dimensional torus $\mathcal{T}_\infty$.

Setting $\varphi_n = \varphi_n(\varphi)$ in the second equation of system (2.122), we obtain the equation

$$x_{n+1} = P(\varphi_{n+p}(\varphi))x_n + c(\varphi_{n+1}(\varphi)). \tag{2.123}$$

By $x_n(p) = x_n(p, \varphi, x_k)$, we denote a solution of Eq. (2.123) such that $x_k(p, \varphi, x_k) = x_k \in \mathfrak{M}, k \in Z$. We also consider that, for all $\varphi \in \mathcal{T}_\infty$,

$$\|P(\varphi)\| \leq P^0 = const < \infty, \quad \|c(\varphi)\| \leq C = const < \infty,$$

the matrix $P(\varphi)$ is invertible, and $\|P^{-1}(\varphi)\| \leq P_1 = const < \infty$. Then, for any $x_k \in \mathfrak{M}, \varphi \in \mathcal{T}_\infty, p \in Z$, such a solution exists, is unique, and belongs to $\mathfrak{M}$ for all $n \in Z$.

We present the matriciant of the homogeneous equation

$$x_{n+1} = P(\varphi_{n+p}(\varphi))x_n \tag{2.124}$$

in the form

$$\Omega_l^n(\varphi, p) = \begin{cases} \displaystyle\prod_{i=n+p-1}^{l+p} P(\varphi_i(\varphi)) & \text{for} \quad n > l; \\[2ex] E & \text{for} \quad n = l; \\[2ex] \displaystyle\prod_{i=n+p}^{l+p-1} P^{-1}(\varphi_i(\varphi)) & \text{for} \quad n < l, \end{cases} \tag{2.125}$$

where E is the identity matrix.

For any $\varphi \in \mathcal{T}_\infty$, let the space $\mathfrak{M}$ can be presented as the direct sum of subspaces $E_1(\varphi) \ E_2(\varphi)$ so that the solution $x_n(p, \varphi, x_0)$ of Eq. (2.124), which takes the value $x_0 \in E_1(\varphi)$ for $n = 0$, satisfies the estimate

$$\|x_n(p, \varphi, x_0)\| \leq K\lambda^n \|x_0\|, \quad \{p, n\} \subset Z, n \geq 0.$$

Let also the solution, which takes the value $x_0 \in E_2(\varphi)$, satisfies the estimate

$$\|x_n(p, \varphi, x_0)\| \leq K\lambda^{-n} \|x_0\|, \quad \{p, n\} \subset Z, n \leq 0,$$

where K and λ are positive constants independent of φ, p, and $\lambda < 1$. In this case, if there exists the matrix $C_1(\varphi)$, which is 2π-periodic in $\varphi^i, i \in Z^+$, bounded in the norm for all $\varphi \in \mathcal{T}_\infty$, and projects the space $\mathfrak{M}$ onto $E_1(\varphi)$ with the help of the multiplication of a matrix by a vector, we call Eq. (2.124) Z-dichotomous on $\mathcal{T}_\infty$.

Taking into account that the operation of multiplication of a bounded infinite matrix by a vector from $\mathfrak{M}$ defines, on this set, a linear bounded operator, whose norm coincides with the matrix norm of this matrix, it is easy to verify that, for Eq. (2.124) Z-dichotomous on $\mathcal{T}_\infty$, there exists a GSF of the form

$$G_0(l, \varphi, p) = \begin{cases} \Omega_l^0(\varphi, p)C_1(\varphi_l(\varphi)) & \text{for} \quad l \le 0; \\ \Omega_l^0(\varphi, p)[C_1(\varphi_l(\varphi)) - E] & \text{for} \quad l > 0. \end{cases}$$

This GSF satisfies the inequality

$$\|G_0(l, \varphi, p)\| \le M\lambda^{|l|}, \quad l \in Z$$

uniformly in $p \in Z$ and $\varphi \in \mathcal{T}_\infty$. Here, M and λ are positive constants, and $\lambda < 1$.

In this case, the system of equations (2.122) has the unique invariant torus $\mathcal{T}(p)$, which is defined by the function

$$u(p, \varphi) = \sum_{l=-\infty}^{-\infty} G_0(l, \varphi, p)c(\varphi_l(\varphi)).$$

We now formulate the criterion of Z-dichotomy of Eq. (2.124) on $\mathcal{T}_\infty$.

Theorem 2.13. *In order that Eq. (2.124) be Z-dichotomous on $\mathcal{T}_\infty$, it is necessary and sufficient that the matrix $C_1(\varphi)$, which is bounded in the norm for all $\varphi \in \mathcal{T}_\infty$, 2π-periodic in $\varphi^i, i \in Z^+$, and satisfies the equalities*

$$C_1(\varphi_n(\varphi)) = \Omega_0^n(\varphi, p)C_1(\varphi)\Omega_n^0(\varphi, p), \tag{2.126}$$

$$C_1^2(\varphi) = C_1(\varphi), \tag{2.127}$$

and the positive constants $T \in Z, d < 1$ such that

$$\sup_{\varphi \in \mathcal{T}_\infty, p \in Z} \{\|\Omega_0^T(\varphi, p)C_1(\varphi)\|, \|\Omega_0^{-T}(\varphi, p)(C_1(\varphi) - E)\|\} \le d \tag{2.128}$$

exist.

Proof.   *Necessity.* Since Eq. (2.124) has the single GSF defined by the matrix projector $C_1(\varphi)$, the latter satisfies equalities (2.126) and (2.127). The proof of this proposition is analogous to that of Lemma 1 in [105].

The estimates

$$\|\Omega_0^n(\varphi, p)C_1(\varphi)\| \le M\lambda^n, \quad n \in Z_0^+,$$

$$\|\Omega_0^n(\varphi, p)(C_1(\varphi) - E)\| \le M\lambda^{-n}, \quad n \in Z^-,$$

yield the existence of constants $T \in Z$ and $d < 1$, with which inequality (2.128) holds.

Sufficiency. Let $n \in Z^+$. Then there exist $t, k \in Z^+, t \in [0, T]$ such that $n = kT + t$. We note that, for any $l, n, p, k \in Z$, the equality

$$\Omega_{l+k}^{n+k}(\varphi, p) = \Omega_l^n(\varphi_k(\varphi), p),$$

holds. Taking conditions (2.126) and (2.127) and estimate (2.128) into account, we obtain the inequalities

$$\|\Omega_0^n(\varphi, p)C_1(\varphi)\| \le \|\Omega_0^t(\varphi_{kT}(\varphi), p)d^{-t/T}\|d^{n/T} \le K_0\lambda^n, \tag{2.129}$$

where

$$\lambda = d^{1/T} < 1, \quad K_0 = \sup_{t\in[0,T], p\in Z, \varphi\in\mathcal{T}_\infty} \|\Omega_0^t(\varphi, p)d^{-t/T}\|.$$

We now assume that $n \in Z^-$ and give it in the form $n = -kT + t$. Considering that the matrix $C_2(\varphi) = E - C_1(\varphi)$ satisfies also conditions (2.126) and (2.127), we obtain the estimate

$$\|\Omega_0^n(\varphi, p)C_2(\varphi)\| \le \|\Omega_0^t(\varphi_{-kT}(\varphi), p)d^{t/T}\|d^{-n/T} \le K_0\lambda^{-n}. \tag{2.130}$$

Estimates (2.129) and (2.130) prove the Z-dichotomy of Eq. (2.124) on $\mathcal{T}_\infty$.

$\square$

Consider the system of equations

$$\overset{(m)}{\varphi}_{n+1} = \overset{(m)}{\varphi}_n + \overset{(m)}{\omega}, \quad \overset{(s)}{x}_{n+1} = P(\overset{(m)}{\varphi}_{n+p})\overset{(s)}{x}_n + \overset{(s)}{c}(\overset{(m)}{\varphi}_{n+1}) \tag{2.131}$$

obtained from system (2.122) by the truncation with respect to φ and x up to the m-th and s-th orders, respectively, i.e.,

$$\overset{(m)}{\varphi} = (\varphi^1, \varphi^2, \ldots, \varphi^m), \quad \overset{(m)}{\omega} = (\omega^1, \omega^2, \ldots, \omega^m), \quad \overset{(s)}{x} = (x^1, x^2, \ldots, x^s),$$

$$\overset{(s)}{c}(\overset{(m)}{\varphi}) = (c_1(\overset{(m)}{\varphi}, 0, 0, \ldots), c_2(\overset{(m)}{\varphi}, 0, 0, \ldots), \ldots, c_s(\overset{(m)}{\varphi}, 0, 0, \ldots)),$$

$$\overset{(s)}{P}(\overset{(m)}{\varphi}) = [p_{ij}(\overset{(m)}{\varphi}, 0, 0, \ldots)]_{i,j=1}^s.$$

If $m = \infty$, then system (2.122) is truncated only with respect to x. For $s = \infty$, it is truncated only with respect to φ. For $m = \infty$ and $s = \infty$, systems (2.122) and (2.131) coincide with each other ($\varphi = \overset{(\infty)}{\varphi}, x = \overset{(\infty)}{x}$).

We denote the matriciant of the equation

$$\overset{(s)}{x}_{n+1} = \overset{(s)}{P}(\overset{(m)}{\varphi}_{n+p}(\overset{(m)}{\varphi}))\overset{(s)}{x}_n, \tag{2.132}$$

which corresponds to Eq. (2.124), by $\Omega_l^n(\overset{(m)}{\varphi}, p)_s$ and the invariant torus of system (2.131) by $\mathcal{T}_{sm}(\varphi): \quad \overset{(s)}{x} = u_s(p, \overset{(m)}{\varphi}), \; \overset{(m)}{\varphi} \in \mathcal{T}_m$.

Below, we present the conditions, under which

$$u(p, \varphi) = \lim_{s \to \infty} \lim_{m \to \infty} u_s(p, \overset{(m)}{\varphi}), \tag{2.133}$$

where $u(p, \varphi)$ is a function defining the invariant torus $\mathcal{T}(p)$ of the system of equations (2.122).

Theorem 2.14. *Let $\{P(\varphi), c(\varphi)\} \subset L_{Lip}(\varphi)$, and let, for all integers $l \leq 0$, the estimate*

$$\|\Omega_l^0(\overset{(m)}{\varphi}, 0)_s\| \leq K\lambda^{-l}, \tag{2.134}$$

where K and $\lambda < \min\{1; \frac{1}{P^0}\}$ are positive constants independent of $\varphi \in \mathcal{T}_\infty$, $\{s, m\} \subset Z^+$, be satisfied. Then relation (2.133), where the internal and external limiting transitions are understood in the meaning of the norm and in the coordinatewise meaning, respectively, holds.

Proof.   In (2.131), we set $m = \infty$. If the estimate

$$\|\Omega_l^0(\varphi, p)_s\| \leq K\lambda^{-l}, \quad l \leq 0, \tag{2.135}$$

where K and $\lambda < 1$ are positive constants independent of $s \in Z^+, \varphi \in \mathcal{T}_\infty, p \in Z$, holds, then, analogously to the proof of Lemma 2.3, we obtain the relation

$$\lim_{s \to \infty} u_s(p, \varphi) = u(p, \varphi), \tag{2.136}$$

where the convergence is understood in the coordinatewise meaning.

Since the matrix $P(\varphi) \in L_{Lip}(\varphi)$ with the coefficient $\varepsilon(m)$, the estimate

$$\|\Omega_l^0(\varphi, p)_s - \Omega_l^0(\overset{(m)}{\varphi}, p)_s\|$$
$$\leq -l(P^0)^{-(l+1)}(\|\varphi_{l+p}(\varphi)\| + \|\omega\|)\varepsilon(m) \tag{2.137}$$

holds uniformly in $s \in Z^+$ for all $l < 0, l \in Z$.

We will prove the last proposition by the method of complete mathematical induction. In view of (2.125), we present the left-hand side of inequality (2.137) in the form

$$I(l, p, s) = \| \prod_{i=p-1}^{l+p} \overset{(s)}{P}(\varphi_i(\varphi)) - \prod_{i=p-1}^{l+p} \overset{(s)}{P}(\overset{(m)}{\varphi}_i(\overset{(m)}{\varphi}))\|, \quad l < 0, \tag{2.138}$$

and note that

$$\varphi_i(\varphi) = (\varphi_i^1(\varphi), \varphi_i^2(\varphi), \dots), \ \overset{(m)}{\varphi}_i(\overset{(m)}{\varphi}) = (\overset{(m)}{\varphi}{}_i^1(\overset{(m)}{\varphi}), \dots, \overset{(m)}{\varphi}{}_i^m(\overset{(m)}{\varphi}), 0, 0, \dots),$$

where $\overset{(m)}{\varphi}{}_i^g(\overset{(m)}{\varphi}) = \varphi_i^g(\varphi)$ for all $g \in \{1, 2, \dots, m\}$. In addition,

$$\| \overset{(s)}{P}(\varphi_i(\varphi)) - \overset{(s)}{P}(\overset{(m)}{\varphi}_i(\overset{(m)}{\varphi}))\|$$
$$\leq \varepsilon(m) \sup\{|\varphi_i^{m+1}(\varphi)|, |\varphi_i^{m+2}(\varphi)|, \dots\} \leq \varepsilon(m)\|\varphi_i(\varphi)\| \quad (2.139)$$

uniformly in $s \in Z^+$.

Let $l = -1$. Relations (2.138) and (2.139) yield directly the inequality

$$I(-1, p, s) \leq \varepsilon(m)\|\varphi_{p-1}(\varphi)\|,$$

which ensures estimate (2.137) for $l = -1$.

We assume that inequality (2.137) holds for $l = -k, k \in Z^+$ and prove its validity for $l = -(k+1)$. We set

$$\prod_{i=p-1}^{-k+p} \overset{(s)}{P}(\varphi_i(\varphi)) = Z_k(\varphi), \qquad \prod_{i=p-1}^{-k+p} \overset{(s)}{P}(\overset{(m)}{\varphi}_i(\overset{(m)}{\varphi})) = Z_k(\overset{(m)}{\varphi}).$$

Then

$$I(-(k+1), p, s) = \|Z_k(\varphi)\, \overset{(s)}{P}(\varphi_{-(k+1)+p}(\varphi))$$
$$- Z_k(\overset{(m)}{\varphi})\, \overset{(s)}{P}(\overset{(m)}{\varphi}_{-(k+1)+p}(\overset{(m)}{\varphi}))\|$$
$$\leq k(P^0)^{-(-k+1)}(\|\varphi_{-k+p}(\varphi)\| + \|\omega\|)\varepsilon(m)P^0$$
$$+ (P^0)^k \varepsilon(m)\|\varphi_{-(k+1)+p}(\varphi)\|$$
$$\leq (k+1)(P^0)^k(\|\varphi_{-(k+1)+p}(\varphi)\| + \|\omega\|)\varepsilon(m),$$

which completes the proof of the above-formulated proposition.

Estimates (2.134) and (2.137) yield estimate (2.135), which leads to the inequality $\|\Omega_l^0(\varphi, p)\| \leq K\lambda^{-l}, l \leq 0$. Then, for any $\{m, s\} \subset Z^+$, Eqs. (2.124) and (2.132) are Z-dichotomous on the tori $\mathcal{T}_\infty$ and $\mathcal{T}_m$, respectively.

In this case, for any $\{m, s\} \subset Z^+$, each of the systems (2.122) and (2.131) has the unique invariant torus ($\mathcal{T}(p)$ and $\mathcal{T}_{sm}(p)$), and these tori are defined by the functions

$$u(p, \varphi) = \sum_{l=-\infty}^{0} \Omega_l^0(\varphi, p)c(\varphi_l(\varphi))$$

and

$$u_s(p, \overset{(m)}{\varphi}) = \sum_{l=-\infty}^{0} \Omega_l^0(\overset{(m)}{\varphi}, p)_s\, \overset{(s)}{c}(\overset{(m)}{\varphi}_l(\overset{(m)}{\varphi})),$$

respectively. To complete the proof of the theorem, it is sufficient to show that

$$\sigma_m(s,p,\varphi) = \sum_{l=-\infty}^{0} \|\Omega_l^0(\varphi,p)_s \overset{(s)}{c}(\varphi_l(\varphi)) - \Omega_l^0(\overset{(m)}{\varphi},p)_s \overset{(s)}{c}(\overset{(m)}{\varphi}_l(\overset{(m)}{\varphi}))\| \to 0$$

as $m \to \infty$. Assume that $c(\varphi) \in L_{Lip}(\varphi)$ with the coefficient $\delta(m)$, and let $\varphi_n(\varphi), n \in Z$ be a trajectory on the torus, i.e., we consider that $\|\varphi_n(\varphi)\| \le 2\pi, n \in Z$.

For all integers $l \le 0$, the estimate

$$\|\Omega_l^0(\varphi,p)_s - \Omega_l^0(\overset{(m)}{\varphi},p)_s\| \le (2K\varepsilon(m)\zeta\lambda^{-l})^{1/2},$$

where $\zeta = (-l)(P^0)^{-(l+1)}(2\pi + \|\omega\|)$, holds. This allows us to write down the chain of inequalities

$$\sigma_m(s,p,\varphi) \le \sum_{l=-\infty}^{0} \|\Omega_l^0(\varphi,p)_s - \Omega_l^0(\overset{(m)}{\varphi},p)_s\|\| \overset{(s)}{c}(\varphi_l(\varphi))\|$$

$$+ \sum_{l=-\infty}^{0} \|\Omega_l^0(\overset{(m)}{\varphi},p)_s\|\| \overset{(s)}{c}(\varphi_l(\varphi)) - \overset{(s)}{c}(\overset{(m)}{\varphi}_l(\overset{(m)}{\varphi}))\|$$

$$\le \sum_{l=-\infty}^{0} \{C(2K\varepsilon(m)\zeta\lambda^{-l})^{1/2} + K\lambda^{-l}\delta(m)2\pi\}$$

$$\le C(2K\varepsilon(m))^{1/2} \sum_{l=-\infty}^{0} (\zeta\lambda^{-l})^{1/2} + \frac{2\pi K}{1-\lambda}\delta(m).$$

For $\lambda P^0 < 1$, the series $\sum_{l=-\infty}^{0}(\zeta\lambda^{-l})^{1/2}$ converges to some positive number η. Then

$$\sigma_m(s,p,\varphi) \le \sqrt{2K}C\eta\sqrt{\varepsilon(m)} + \frac{2\pi K}{1-\lambda}\delta(m),$$

and the last relation is satisfied uniformly in all $\varphi \in \mathcal{T}_\infty, s \in Z^+$ and $p \in Z$. This result together with (2.136) complete the proof of the theorem. $\square$

2.8   On the existence of the smooth bounded semi-invariant manifold of a degenerate nonlinear system

First, we consider the linear system

$$\varphi_{n+1} = \varphi_n + a(\varphi_n), \quad x_{n+1} = P(\varphi_n)x_n + c(\varphi_n), \tag{2.140}$$

where $\varphi \in \mathbf{W}$, $\quad x = (x^1, x^2, x^3, \dots) \in \mathfrak{M}$, $\mathbf{W}$ is any Banach space, $P(\varphi)$ is an infinite matrix with real elements, and the functions $a(\varphi)$ and $c(\varphi) = (c_1(\varphi), c_2(\varphi), c_3(\varphi), \dots)$ are defined on $\mathbf{W}$ and take values from the spaces $\mathbf{W}$ and $\mathfrak{M}$, respectively.

In this section, let us agree to designate the norm of the matrix $P(\varphi)$ and the norms in the spaces $\mathbf{W}$ and $\mathfrak{M}$ by the same symbol $\| \cdot \|$ and to distinguish these norms by context. We also consider that $\|a(\varphi)\| \leq A^0$, $\|c(\varphi)\| \leq C^0$, $\|P(\varphi)\| \leq P^0$, where A^0, C^0, P^0 are positive constants independent of $\varphi \in \mathbf{W}$.

Definition 2.2. By a bounded invariant manifold of the system of equations (2.140), we call the set of points $x \in \mathfrak{M} : x = u(\varphi) = (u_1(\varphi), u_2(\varphi), \dots)$, if the function $u(\varphi)$ is defined $\forall \varphi \in \mathbf{W}$, bounded in the norm, and, $\forall \varphi \in \mathbf{W}$, satisfies the equality

$$u(\varphi_{n+1}(\varphi)) = P(\varphi_n(\varphi))u(\varphi_n(\varphi)) + c(\varphi_n(\varphi)), \quad n \in Z, \qquad (2.141)$$

where $\varphi_n(\varphi)$ is the solution of the first equation of system (2.140) such that $\varphi_0(\varphi) = \varphi \in \mathbf{W}$.

Definition 2.3. We call the set defined in the previous definition by a bounded semi-invariant manifold of the system of equations (2.140), if equality (2.141) holds $\forall n \in Z_0^+$.

For the construction of the invariant manifold of system (2.140) by the method of GSF, the necessary condition is the existence of the matriciant of the equation

$$x_{n+1} = P(\varphi_n(\varphi))x_n, \quad n \in Z. \qquad (2.142)$$

If the propositions

a) $\forall \varphi \in \mathbf{W}$, the matrix $P(\varphi)$ is invertible, and the matrix $P^{-1}(\varphi)$ inverse to it is bounded in the norm;

b) the mapping $\Phi(\varphi) = \varphi + a(\varphi) : \mathbf{W} \to \mathbf{W}$ is invertible,

hold, then, as was shown above, the matriciant $\Omega_l^n(\varphi)$ exists $\forall l \in Z, \varphi \in \mathbf{W}$. If at least one of these propositions does not hold, then the system of equations (2.140) is called degenerate. It is clear that, in this case, Eq. (2.142) can have no matriciant, since the solutions needed for its construction can turn out to be nonextensible "to the left."

Lemma 2.19. *The following propositions hold:*

1^0) *if condition b) holds, and condition a) does not hold, and, for all integer $n > l$,*

$$\left\| \prod_{i=n-1}^{l} P(\varphi_i(\varphi)) \right\| \leq K\lambda^{n-l}, \tag{2.143}$$

then the system of equations (2.140) has a bounded invariant manifold;

2^0) *if condition a) holds, and condition b) does not hold, and, for all integer n and l, which satisfy the inequality $0 \leq n < l$, the estimate*

$$\left\| \prod_{i=n}^{l-1} P^{-1}(\varphi_i(\varphi)) \right\| \leq K\lambda^{l-n} \tag{2.144}$$

holds, then the system of equations (2.140) has a bounded semi-invariant manifold. Here, K and $\lambda < 1$ are positive constants independent of $\varphi \in \mathbf{W}$.

Proof.　Proof of the lemma is not associated with difficulties. Therefore, we give only its scheme. In case 1^0, we choose the function

$$G_0(l, \varphi) = \begin{cases} \prod_{i=-1}^{l} P(\varphi_i(\varphi)) & \text{for} \quad l < 0; \\ E & \text{for} \quad l = 0, \\ 0 & \text{for} \quad l > 0, \end{cases}$$

where E is the infinite identity matrix, as the GSF of Eq. (2.142). It is obvious that if condition a) does not hold, but condition b) holds, then, $\forall n > l$, the matrix $\Omega_l^n(\varphi)$ can be unambiguously constructed, and $\Omega_l^n(\varphi) = \prod_{i=n-1}^{l} P(\varphi_i(\varphi))$　$\forall \varphi \in \mathbf{W}$. Then the bounded invariant manifold of the system of equations (2.140) exists and is defined by the function

$$u(\varphi) = c(\varphi_{-1}(\varphi)) + \sum_{l=-\infty}^{-1} \prod_{i=-1}^{l} P(\varphi_i(\varphi))c(\varphi_{l-1}(\varphi)),$$

since

$$x_n = c(\varphi_{n-1}(\varphi)) + \sum_{l=-\infty}^{n-1} \prod_{i=n-1}^{l} P(\varphi_i(\varphi))c(\varphi_{l-1}(\varphi)) = u(\varphi_n(\varphi))$$

is a solution of the equation

$$x_{n+1} = P(\varphi_n(\varphi))x_n + c(\varphi_n(\varphi)), \quad n \in Z, \varphi \in \mathbf{W}.$$

The boundedness of the function $u(\varphi)$ in the norm, which is uniform in $\varphi \in \mathbf{W}$, is obvious.

In case 2^0, we choose the function

$$
G_0(l, \varphi) = \begin{cases} 0 & \text{for} \quad l \le 0; \\ -\displaystyle\prod_{i=0}^{l-1} P^{-1}(\varphi_i(\varphi)) & \text{for} \quad l > 0 \end{cases}
$$

as the GSF of Eq. (2.142). If condition a) holds, but condition b) does not hold, then, $\forall \{n, l\} \subset Z$ such that $0 \le n < l$, the matrix $\Omega_l^n(\varphi)$ is uniquely defined by the equality $\Omega_l^n(\varphi) = \prod_{i=n}^{l-1} P^{-1}(\varphi_i(\varphi)) \; \forall \varphi \in \mathbf{W}$. Then there exists the bounded semi-invariant manifold of the system of equations (2.140), which is defined by the function

$$
u(\varphi) = -\sum_{l=1}^{\infty} \prod_{i=0}^{l-1} P^{-1}(\varphi_i(\varphi)) c(\varphi_{l-1}(\varphi)),
$$

since

$$
x_n = -\sum_{l=n+1}^{\infty} \prod_{i=n}^{l-1} P^{-1}(\varphi_i(\varphi)) c(\varphi_{l-1}(\varphi)) = u(\varphi_n(\varphi))
$$

is a solution of the equation

$$
x_{n+1} = P(\varphi_n(\varphi)) x_n + c(\varphi_n(\varphi)), \quad n \in Z_0^+, \varphi \in \mathbf{W}.
$$

It is clear that the validity of inequalities (2.143) and (2.144) is ensured, for example, by the estimates $\|P(\varphi)\| \le g$ and $\|P^{-1}(\varphi)\| \le g$ respectively, where $g = const < 1, \varphi \in \mathbf{W}$. $\qquad\square$

The subsequent reasoning will concern the construction of the bounded semi-invariant manifold of a nonlinear system of difference equations, which has a degeneration of the form 2^0. Consider the system of equations

$$
\varphi_{n+1} = \varphi_n + a(\varphi_n, x_n), \quad x_{n+1} = P(\varphi_n, x_n) x_n + c(\varphi_n), n \in Z, \quad (2.145)
$$

where φ, x are the same as above, the function $a(\varphi, x)$ and the matrix $P(\varphi, x)$ are defined on the set $\mathfrak{D}^0 = \mathbf{W} \times D = \mathbf{W} \times \{x \in \mathfrak{M} | \|x\| \le d = const > 0\}$, and $\|a(\varphi, x)\| \le A^0$, $\|P(\varphi, x)\| \le P^0$, and $\|c(\varphi)\| \le C^0$ on this set, where A^0, P^0, C^0 are positive constants. We also consider that, $\forall (\varphi, x) \in \mathfrak{D}^0$, the matrix $P(\varphi, x)$ is invertible, and the matrix $P^{-1}(\varphi, x)$ inverse to it is bounded uniformly in $(\varphi, x) \in \mathfrak{D}^0$ in the norm by a positive constant P_1.

Definition 2.4. By a bounded semi-invariant manifold of the system of equations (2.145), we call the set of points $x = u(\varphi)$, which satisfies the conditions of Definition 2.2, if Eq. (2.141) in it is replaced by

$$
u(\varphi_{n+1}(\varphi)) = P(\varphi_n(\varphi), u(\varphi_n(\varphi))) u(\varphi_n(\varphi)) + c(\varphi_n(\varphi)), \quad n \in Z_0^+,
$$
$$
(2.146)
$$

where $\varphi_n(\varphi)$ is a solution of the equation

$$\varphi_{n+1}(\varphi) = \varphi_n(\varphi) + a(\varphi_n(\varphi), u(\varphi_n(\varphi))), \quad n \in Z_0^+, \tag{2.147}$$

and $\varphi_0(\varphi) = \varphi \in \mathbf{W}$.

We assume that the inequalities

$$P_1 < 1, \quad C^0 P_1 \le d(1 - P_1) \tag{2.148}$$

hold and write down the system of equations

$$\varphi_{n+1} = \varphi_n + a(\varphi_n, 0), \quad x_{n+1} = P(\varphi_n, 0)x_n + c(\varphi_n), \quad n \in Z. \tag{2.149}$$

It is obvious that, for $P_1 < 1$, conditions 2^0 of Lemma 2.19 are satisfied. In this case, system (2.149) has the bounded semi-invariant manifold, which is defined by the function

$$u^{(0)}(\varphi) = - \sum_{l=1}^{\infty} \prod_{i=0}^{l-1} P^{-1}(\varphi_i^{(0)}(\varphi), 0)c(\varphi_{l-1}^{(0)}(\varphi)). \tag{2.150}$$

This means that the equalities

$$\varphi_0^{(0)}(\varphi) = \varphi \in \mathbf{W}, \quad \varphi_{n+1}^{(0)}(\varphi) = \varphi_n^{(0)}(\varphi) + a(\varphi_n^{(0)}(\varphi), 0),$$

$$u^{(0)}(\varphi_{n+1}^{(0)}(\varphi)) = P(\varphi_n^{(0)}(\varphi), 0)u^{(0)}(\varphi_n^{(0)}(\varphi)) + c(\varphi_n^{(0)}(\varphi)), \quad n \in Z_0^+,$$

hold. In this case, relations (2.148) and (2.150) yield the estimates $\|u^{(0)}(\varphi)\| \le \frac{C^0 P_1}{1 - P_1} \le d$, i.e., $u^{(0)}(\varphi) \in D \ \forall \varphi \in \mathbf{W}$.

By means of the inductive reasoning, it is easy to verify the possibility to construct the sequence of functions $\{u^{(s)}(\varphi)\}_{s=1}^{\infty}$, each of which defines the bounded semi-invariant manifold of the system of equations

$$\varphi_{n+1} = \varphi_n + a(\varphi_n, u^{(s-1)}(\varphi_n)),$$

$$x_{n+1} = P(\varphi_n, u^{(s-1)}(\varphi_n))x_n + c(\varphi_n), \quad n \in Z.$$

This means that, $\forall n \in Z_0^+, s \in Z^+, \varphi \in \mathbf{W}$, the equalities

$$\varphi_{n+1}^{(s)}(\varphi) = \varphi_n^{(s)}(\varphi) + a(\varphi_n^{(s)}(\varphi), u^{(s-1)}(\varphi_n^{(s)}(\varphi))), \quad \varphi_0^{(s)}(\varphi) = \varphi \in \mathbf{W},$$

$$u^{(s)}(\varphi_{n+1}^{(s)}(\varphi)) = P(\varphi_n^{(s)}(\varphi), u^{(s-1)}(\varphi_n^{(s)}(\varphi)))u^{(s)}(\varphi_n^{(s)}(\varphi)) + c(\varphi_n^{(s)}(\varphi)) \tag{2.151}$$

hold. Here,

$$u^{(s)}(\varphi) = - \sum_{l=1}^{\infty} \prod_{i=0}^{l-1} P^{-1}(\varphi_i^{(s)}(\varphi), u^{(s-1)}(\varphi_i^{(s)}(\varphi)))c(\varphi_{l-1}^{(s)}(\varphi)), \tag{2.152}$$

and $\forall s \in Z^+$

$$\|u^{(s)}(\varphi)\| \leq \frac{C^0 P_1}{1 - P_1} \leq d.$$

Lemma 2.20. *Let inequalities (2.148) hold, and let, $\forall \varphi \in \mathbf{W}, \{x, \bar{x}\} \subset D$,*

$$\|a(\varphi, x) - a(\varphi, \bar{x})\| \leq \alpha \|x - \bar{x}\|,$$

$$\|P^{-1}(\varphi, x) - P^{-1}(\varphi, \bar{x})\| \leq \beta \|x - \bar{x}\|,$$

where the positive constants α and β are independent of $\varphi, x, \bar{x}$. If, $\forall \{\varphi, \bar{\varphi}\} \subset \mathbf{W}$ and $s \in Z_0^+$,

$$\|u^{(s)}(\varphi) - u^{(s)}(\bar{\varphi})\| \leq L^{(s)} \|\varphi - \bar{\varphi}\|,$$

where the positive constants $L^{(s)}$ are independent of $\varphi, \bar{\varphi}$ and are bounded from above by a constant L, and the estimate

$$\gamma = \frac{P_1 L\alpha + \beta d + \beta C^0}{1 - P_1} < 1$$

holds, then the sequence $\{u^{(s)}(\varphi)\}_{s=0}^{\infty}$ converges in $\mathfrak{M}$ uniformly in $\varphi \in \mathbf{W}$.

Proof. We write down equalities (2.151) for $n = 0$ as

$$\varphi_1^{(s)}(\varphi) = \varphi + a(\varphi, u^{(s-1)}(\varphi)),$$

$$u^{(s)}(\varphi_1^{(s)}(\varphi)) = P(\varphi, u^{(s-1)}(\varphi))u^{(s)}(\varphi) + c(\varphi), \quad s \in Z^+.$$

With regard for the invertibility of the matrix $P(\varphi, x)$ on $\mathfrak{D}^0$, we obtain the equality

$$u^{(s)}(\varphi) = P^{-1}(\varphi, u^{(s-1)}(\varphi))(u^{(s)}(\varphi_1^{(s)}(\varphi)) - c(\varphi)), \quad s \in Z^+.$$

In this case, we have

$$u^{(s+1)}(\varphi) - u^{(s)}(\varphi = P^{-1}(\varphi, u^{(s)}(\varphi))(u^{(s+1)}(\varphi_1^{(s+1)}(\varphi)) - c(\varphi))$$
$$- P^{-1}(\varphi, u^{(s-1)}(\varphi))(u^{(s)}(\varphi_1^{(s)}(\varphi)) - c(\varphi))$$
$$= P^{-1}(\varphi, u^{(s)}(\varphi))(u^{(s+1)}(\varphi_1^{(s+1)}(\varphi)) - u^{(s)}(\varphi_1^{(s)}(\varphi)))$$
$$+ (P^{-1}(\varphi, u^{(s)}(\varphi)) - P^{-1}(\varphi, u^{(s-1)}(\varphi)))(u^{(s)}(\varphi_1^{(s)}(\varphi)) - c(\varphi)).$$

Let us denote the difference $u^{(s+1)}(\varphi) - u^{(s)}(\varphi)$ by $w^{(s+1)}$ and set $\|w^{(s+1)}\|_0 = \sup_{\varphi \in \mathbf{W}} \|w^{(s+1)}\|$. In view of the estimate

$$\|\varphi_1^{(s+1)}(\varphi) - \varphi_1^{(s)}(\varphi)\| \leq \alpha \|u^{(s)}(\varphi) - u^{(s-1)}(\varphi)\| = \alpha \|w^{(s)}\|,$$

we write down the chain of inequalities

$$\|w^{(s+1)}\| \le P_1 \|u^{(s+1)}(\varphi_1^{(s+1)}(\varphi)) - u^{(s)}(\varphi_1^{(s)}(\varphi))\| + \beta\|u^{(s)}(\varphi) - u^{(s-1)}(\varphi)\|$$
$$\times (d + C^0) \le P_1 \|u^{(s+1)}(\varphi_1^{(s+1)}(\varphi)) - u^{(s+1)}(\varphi_1^{(s)}(\varphi))\| + P_1 \|u^{(s+1)}(\varphi_1^{(s)}(\varphi))$$
$$- u^{(s)}(\varphi_1^{(s)}(\varphi))\| + \beta\|w^{(s)}\|(d + C^0) \le P_1 L^{(s+1)}\|\varphi_1^{(s+1)}(\varphi) - \varphi_1^{(s)}(\varphi)\|$$
$$+ P_1\|w^{(s+1)}\|_0 + \beta\|w^{(s)}\|(d + C^0)$$
$$\le P_1\|w^{(s+1)}\|_0 + (P_1 L^{(s+1)}\alpha + \beta(d + C^0))\|w^{(s)}\|,$$

which yield the recurrence estimate

$$\|w^{(s+1)}\|_0 \le \gamma\|w^{(s)}\|, \quad s \in Z^+.$$

The obvious inductive reasoning leads to the inequalities

$$\|w^{(s+1)}\|_0 \le \gamma^s\|w^{(1)}\| \le 2d\gamma^s,$$

which ensure the fundamentality of the sequence $\{u^{(s)}(\varphi)\}_{s=0}^{\infty}$ in the space $\mathfrak{M}$. The completeness of this space allows us to end the proof of Lemma 2.20. $\square$

We take the expression $\max\{\|\varphi\|, \|x\|\}$ as the norm of an element $(\varphi, x) \in \mathfrak{D}^0$ and denote it by $\|(\varphi, x)\|$. Here, $\|\varphi\|$ and $\|x\|$ are the norms of φ and x in the spaces $\mathbf{W}$ and $\mathfrak{M}$, respectively.

Lemma 2.21. *Assume that the functions $a(\varphi, x)$ and $c(\varphi)$ are continuous on $\mathfrak{D}^0$ and $\mathbf{W}$, respectively, the matrix function $P(\varphi, x)$ is continuous on $\mathfrak{D}^0$, and the conditions of Lemma 2.20 are satisfied. Then the system of equations (2.145) has a bounded semi-invariant manifold.*

Proof. We fix any $\varphi \in \mathbf{W}$ and pass to the limit in the first equality in (2.151) successfully for $n = 0, n = 1, n = 2, \ldots$ as $s \to \infty$. It is obvious that $\lim_{s\to\infty} \varphi_0^{(s)}(\varphi) = \varphi$. We denote $\lim_{s\to\infty} \varphi_1^{(s)}(\varphi) = \varphi + a(\varphi, u(\varphi))$ by $\varphi_1(\varphi)$. For $n = 1$,

$$\lim_{s\to\infty} \varphi_2^{(s)}(\varphi) = \lim_{s\to\infty} \varphi_1^{(s)}(\varphi) + \lim_{s\to\infty} a(\varphi_1^{(s)}(\varphi), u^{(s-1)}(\varphi_1^{(s)}(\varphi)))$$

$$= \varphi_1(\varphi) + a(\varphi_1(\varphi), u(\varphi_1(\varphi))).$$

This follows from the continuity of the function $a(\varphi, x)$ on $\mathfrak{D}^0$ and from the convergence of the sequence $\{(\varphi_1^{(s)}(\varphi), u^{(s-1)}(\varphi_1^{(s)}(\varphi)))\}_{s=1}^{\infty}$ in the norm of the space $\mathfrak{D}^0$, which is ensured by the estimate

$$\|u^{(s)}(\varphi) - u^{(s)}(\bar{\varphi})\| \le L\|\varphi - \bar{\varphi}\| \quad \forall s \in Z^+, \{\varphi, \bar{\varphi}\} \subset \mathbf{W}.$$

Indeed,

$$V_1 = \|(\varphi_1^{(s)}(\varphi), u^{(s-1)}(\varphi_1^{(s)}(\varphi))) - (\varphi_1(\varphi), u(\varphi_1(\varphi)))\|$$
$$= \|(\varphi_1^{(s)}(\varphi) - \varphi_1(\varphi), u^{(s-1)}(\varphi_1^{(s)}(\varphi)) - u(\varphi_1(\varphi)))\|$$
$$= \max\{\|\varphi_1^{(s)}(\varphi) - \varphi_1(\varphi)\|, \|u^{(s-1)}(\varphi_1^{(s)}(\varphi)) - u(\varphi_1(\varphi))\|\}$$
$$\leq \max\{\|\varphi_1^{(s)}(\varphi) - \varphi_1(\varphi)\|, L\|\varphi_1^{(s)}(\varphi) - \varphi_1(\varphi)\|$$
$$+ \|u^{(s-1)}(\varphi_1(\varphi)) - u(\varphi_1(\varphi))\|\}.$$

Since, for any arbitrarily small positive real number ε, there exist natural numbers $N_1, N_2,$ and N_3 such that

$$\|\varphi_1^{(s)}(\varphi) - \varphi_1(\varphi)\| \leq \varepsilon \quad \forall s \geq N_1, \quad \|\varphi_1^{(s)}(\varphi) - \varphi_1(\varphi)\| \leq \frac{\varepsilon}{2L} \quad \forall s \geq N_2,$$

$$\|u^{(s-1)}(\varphi_1(\varphi)) - u(\varphi_1(\varphi))\| \leq \frac{\varepsilon}{2} \quad \forall s \geq N_3,$$

the estimate $V_1 \leq \varepsilon$ holds $\forall s \geq \max\{N_1, N_2, N_3\}$.

We denote $\lim_{s\to\infty} \varphi_2^{(s)}(\varphi)$ by $\varphi_2(\varphi)$. Continuing this process, we obtain the function $\varphi_n(\varphi)$, which is a solution of Eq. (2.147) $\forall n \in Z_0^+$, $\varphi_0(\varphi) = \varphi \in \mathbf{W}$.

It remain to prove equality (2.146) $\forall n \in Z_0^+$. For this purpose, it is sufficient to pass to the limit as $s \to \infty$ in the last equality in (2.151) for each $n \in Z_0^+$. This process induces no difficulties. For example, we show that

$$P(\varphi_n^{(s)}(\varphi), u^{(s-1)}(\varphi_n^{(s)}(\varphi)))u^{(s)}(\varphi_n^{(s)}(\varphi)) \to P(\varphi_n(\varphi), u(\varphi_n(\varphi)))u(\varphi_n(\varphi))$$

as $s \to \infty$ in the norm of the space $\mathfrak{M}$.

We denote the norm of the difference of two last products by V_2. It is easy to verify the validity of the inequalities

$$V_2 = \|P(\varphi_n^{(s)}(\varphi), u^{(s-1)}(\varphi_n^{(s)}(\varphi)))\|\|u^{(s)}(\varphi_n^{(s)}(\varphi)) - u(\varphi_n(\varphi))\|$$
$$+ \|P(\varphi_n^{(s)}(\varphi), u^{(s-1)}(\varphi_n^{(s)}(\varphi))) - P(\varphi_n(\varphi), u(\varphi_n(\varphi)))\|\|u(\varphi_n(\varphi))\|$$
$$\leq P^0(I_1 + I_2) + dI_3,$$

where

$$I_1 = \|u^{(s)}(\varphi_n^{(s)}(\varphi)) - u^{(s)}(\varphi_n(\varphi))\|, \quad I_2 = \|u^{(s)}(\varphi_n(\varphi)) - u(\varphi_n(\varphi))\|,$$

$$I_3 = \|P(\varphi_n^{(s)}(\varphi), u^{(s-1)}(\varphi_n^{(s)}(\varphi))) - P(\varphi_n(\varphi), u(\varphi_n(\varphi)))\|.$$

As was shown above, $\|\varphi_n^{(s)}(\varphi) - \varphi_n(\varphi)\| \to 0$ as $s \to \infty$. Therefore, the estimate $I_1 \leq L\|\varphi_n^{(s)}(\varphi) - \varphi_n(\varphi)\|$ yields the equality $\lim_{s\to\infty} I_1 =$

0; $\lim_{s\to\infty} I_2 = 0$ by Lemma 2.20; and $\lim_{s\to\infty} I_3 = 0$, since the matrix $P(\varphi, x)$ is continuous on $\mathfrak{D}^0$, and the sequence

$$\{(\varphi_n^{(s)}(\varphi), u^{(s-1)}(\varphi_n^{(s)}(\varphi)))\}_{s=1}^{\infty} \subset \mathfrak{D}^0$$

converges as $s \to \infty$ to the point $(\varphi_n(\varphi), u(\varphi_n(\varphi))) \in \mathfrak{D}^0$. Lemma 2.21 is proved. We note only that all limiting transitions considered in the last proof are realized uniformly in $\varphi \in \mathbf{W}$, if the functions $a(\varphi, x), P(\varphi, x), c(\varphi)$ in the conditions of this Lemma are uniformly continuous on the appropriate sets. $\qquad\square$

By D_ρ and $\mathfrak{D}_0$, where ρ is an arbitrarily small positive number, we denote the sets $\{x \in \mathfrak{M} | \|x\| < d + \rho\}$ and $\mathbf{W} \times D_\rho$, respectively. Moreover, by $C_{\mathbf{W}}^1(\varphi)$ and $C_{\mathfrak{D}_0}^1(\varphi, x)$, we denote the sets, whose elements are the functions continuously Fréchet-differentiable with respect to φ on $\mathbf{W}$ and with respect to (φ, x) on $\mathfrak{D}_0$, respectively.

The symbol $P(\varphi, x)$ stands for not only a matrix, but also a mapping (matrix function) of $\mathfrak{D}_0$ onto the set $\mathbf{\Gamma}$ of infinite matrices, which are bounded in the norm $\| \cdot \|$ and have real elements. We will distinguish the mapping $P(\varphi, x)$ and its value (matrix $P(\varphi, x)$) by context.

We consider the semi-invariant manifold of the system of equations (2.145) to be smooth, if the function $u(\varphi)$ generating it is continuously Fréchet-differentiable on $\mathbf{W}$.

Theorem 2.15. *Let the matrix function $P(\varphi, x)$ is uniformly continuous on $\mathfrak{D}_0$, and let the following conditions be satisfied:*

1) $\{a(\varphi, x), P^{-1}(\varphi, x)\} \subset C_{\mathfrak{D}_0}^1(\varphi, x), \ c(\varphi) \in C_{\mathbf{W}}^1(\varphi), \ and, \ \forall \varphi \in$ $\mathbf{W}, (\varphi, x) \in \mathfrak{D}_0$, *the estimates*

$$\left\|\frac{da(\varphi, x)}{d(\varphi, x)}\right\| \le A^*, \quad \left\|\frac{dP^{-1}(\varphi, x)}{d(\varphi, x)}\right\| \le P_*, \quad \left\|\frac{dc(\varphi)}{d\varphi}\right\| \le C_*,$$

where A^*, P_*, C_* are positive constants independent of φ, x, hold;

2) the inequalities

$$P < \frac{1}{1 + A^*}, \quad \frac{C^0 P_1}{1 - P_1} \le d, \quad \eta = \frac{P_* C^0(1 + A^*) + C_* P_1 A^*}{A^*(1 - P_1(1 + A^*))} \le 1,$$

$$\gamma = \frac{P_1 \eta A^* + P_*(d + C^0)}{1 - P_1} < 1$$

are satisfied. Then the system of equations (2.145) has the bounded semi-invariant manifold, whose generating function $u(\varphi)$ satisfies the Lipschitz condition on $\mathbf{W}$:

$$\|u(\varphi) - u(\bar{\varphi})\| \le \eta \|\varphi - \bar{\varphi}\| \quad \forall \{\varphi, \bar{\varphi}\} \subset \mathbf{W}.$$

Proof. It is obvious that the functions $a(\varphi, x)$ and $\varphi(x)$ in the conditions of the formulated theorem are uniformly continuous on $\mathfrak{D}_0$ and $\mathbf{W}$, respectively. The convexity of the sets $\mathfrak{D}_0$ and $\mathbf{W}$ yields the existence of constants α and β, which are present in the condition of Lemma 2.20; moreover, $\alpha = A^*$ and $\beta = P_*$. Hence, the proof of the theorem is reduced to the substantiation of the inequality

$$\|u^{(s)}(\varphi) - u^{(s)}(\bar{\varphi})\| \leq \eta \|\varphi - \bar{\varphi}\| \tag{2.153}$$

$\forall s \in Z_0^+, \{\varphi, \bar{\varphi}\} \subset \mathbf{W}$. We separate the proof into two parts.

1. We show that inequality (2.153) holds for $s = 0$. Using the method of complete mathematical induction, it is easy to verify that, $\forall n \in Z_0^+, \varphi \in \mathbf{W}$, there exists a linear operator $\frac{d\varphi_n^{(0)}(\varphi)}{d\varphi}$, which belongs to the space $\mathbf{L}(\mathbf{W}, \mathbf{W})$, and

$$\left\|\frac{d\varphi_n^{(0)}(\varphi)}{d\varphi}\right\| \leq (1 + A^*)^n. \tag{2.154}$$

The substantiation of this proposition will be given in the second part of the proof of the theorem.

We note that the mapping $P^{-1}(\varphi, x)$ transfers the set $\mathfrak{D}_0$ open in the Cartesian product $\mathbf{W} \times \mathfrak{M}$ in a linear normalized space $\boldsymbol{\Gamma}$. Therefore, $\frac{dP^{-1}(\varphi, x)}{d(\varphi, x)}$ is a linear operator from the space $\mathbf{L}(\mathbf{W} \times \mathfrak{M}, \boldsymbol{\Gamma})$. It is obvious that, $\forall i \in Z_0^+$, the inequalities

$$\left\|\frac{dP^{-1}(\varphi_i^{(0)}(\varphi), 0)}{d\varphi}\right\| \leq P_* \left\|\frac{d\varphi_i^{(0)}(\varphi)}{d\varphi}\right\| \leq P_*(1 + A^*)^i,$$

$$\left\|\frac{dc(\varphi_i^{(0)}(\varphi))}{d\varphi}\right\| \leq C_*(1 + A^*)^i \tag{2.155}$$

hold. Since the product of n matrices from the space $\boldsymbol{\Gamma}$ defines a multilinear mapping $\boldsymbol{\Gamma}^{\mathbf{n}} \to \boldsymbol{\Gamma}$, there exists the derivative of the product

$$\prod_{i=0}^{l-1} P^{-1}(\varphi_i^{(0)}(\varphi), 0)$$

with respect to $\varphi \in \mathbf{W}$, and

$$\left\|\frac{d\prod_{i=0}^{l-1} P^{-1}(\varphi_i^{(0)}(\varphi), 0)}{d\varphi}\right\| \leq P_1^{l-1} \sum_{i=0}^{l-1} \left\|\frac{dP^{-1}(\varphi_i^{(0)}(\varphi), 0)}{d\varphi}\right\|$$

$$\leq P_1^{l-1} P_* \sum_{i=0}^{l-1}(1 + A^*)^i \leq \frac{P_*}{A^*} P_1^{l-1}(1 + A^*)^l, \quad l \in Z^+. \tag{2.156}$$

By $u_l^{(0)}$, we denote the expression under the sign of a sum in equality (2.150). Relations (2.155) and (2.156) allow us to write down the estimate

$$\left\|\frac{du_l^{(0)}}{d\varphi}\right\| < \frac{P_*C^0}{A^*}P_1^{l-1}(1+A^*)^l + P_1^l C_*(1+A^*)^{l-1}, \quad l \in Z^+, \varphi \in \mathbf{W}.$$

In order to complete the first stage, it is sufficient to show that the series

$$\sum_{l=1}^{\infty}\left\|\frac{du_l^{(0)}}{d\varphi}\right\|$$

converges uniformly in $\varphi \in \mathbf{W}$. The last assertion follows from the relations

$$\sum_{l=1}^{\infty}\left\|\frac{du_l^{(0)}}{d\varphi}\right\| < \frac{P_*C^0}{A^*P_1}\sum_{l=1}^{\infty}(P_1(1+A^*))^l + P_1 C_*\sum_{l=1}^{\infty}(P_1(1+A^*))^{l-1}$$

$$= \frac{P_*C^0(1+A^*)P_1}{A^*P_1(1-P_1(1+A^*))} + \frac{P_1 C_*}{1-P_1(1+A^*)} = \eta.$$

Hence, $u^{(0)}(\varphi) \in C_{\mathbf{W}}^1(\varphi)$ and $\forall \varphi \in \mathbf{W} \left\|\frac{du^{(0)}(\varphi)}{d\varphi}\right\| \leq \eta$. Due to the connvexity of $\mathbf{W}$, this yields the estimate

$$\|u^{(0)}(\varphi) - u^{(0)}(\bar\varphi)\| \leq \eta\|\varphi - \bar\varphi\|.$$

2. In what follows, we will prove that inequality (2.153) holds for all $s \in Z^+$. To this end, we use the method of complete mathematical induction. Let $s = 1$. First, we show that, $\forall i \in Z_0^+$,

$$\left\|\frac{d\varphi_i^{(1)}(\varphi)}{d\varphi}\right\| \leq (1+A^*)^i \tag{2.157}$$

for $\eta \leq 1$. It is obvious that $\left\|\frac{d\varphi_0^{(1)}(\varphi)}{d\varphi}\right\| = 1$. For $i = 1$, the relations

$$\left\|\frac{d\varphi_1^{(1)}(\varphi)}{d\varphi}\right\| = \left\|\frac{d\varphi}{d\varphi} + \frac{da(\varphi_0^{(1)}(\varphi), u^{(0)}(\varphi_0^{(1)}(\varphi)))}{d\varphi}\right\|$$

$$\leq 1 + A^* \max\left\{\left\|\frac{d\varphi}{d\varphi}\right\|, \left\|\frac{du^{(0)}(\varphi)}{d\varphi}\right\|\right\} \leq 1 + A^* \max\{1, \eta\}$$

hold, i.e., inequality (2.157) is satisfied. Assume that it is satisfied $\forall i \leq k(k \in Z^+)$. Then

$$\left\|\frac{d\varphi_{k+1}^{(1)}(\varphi)}{d\varphi}\right\| \leq (1+A^*)^k + A^* \max\{(1+A^*)^k, \eta(1+A^*)^k\}$$

$$\leq (1+A^*)^k + A^*(1+A^*)^k = (1+A^*)^{k+1},$$

since $\eta \leq 1$. Hence, by the principle of complete mathematical induction, estimate (2.157) holds $\forall i \in Z^+$. Inequality (2.154) can be proved similarly.

Analogously to (2.155) and (2.156), we obtain the estimates

$$\left\| \frac{d \prod_{i=0}^{l-1} P^{-1}(\varphi_i^{(1)}(\varphi), u^{(0)}(\varphi_i^{(1)}(\varphi)))}{d\varphi} \right\| < \frac{P_*}{A^*} P_1^{l-1}(1+A^*)^l,$$

$$\left\| \frac{dc(\varphi_{l-1}^{(1)}(\varphi))}{d\varphi} \right\| \leq C_*(1+A^*)^{l-1}$$

$\forall \varphi \in \mathbf{W}$. We now write down equality (2.152) for $s = 1$ as

$$u^{(1)}(\varphi) = -\sum_{l=1}^{\infty} \prod_{i=0}^{l-1} P^{-1}(\varphi_i^{(1)}(\varphi), u^{(0)}(\varphi_i^{(1)}(\varphi)))c(\varphi_{l-1}^{(1)}(\varphi))$$

and denote the expression under the sign of a sum by $u_l^{(1)}$. It is easy to verify that, $\forall l \in Z^+, \varphi \in \mathbf{W}$, the inequality

$$\left\| \frac{du_l^{(1)}}{d\varphi} \right\| < \frac{P_* C^0}{A^*} P_1^{l-1}(1+A^*)^l + P_1^l C_*(1+A^*)^{l-1}$$

holds, since $\eta \leq 1$. Then

$$\sum_{l=1}^{\infty} \left\| \frac{du_l^{(1)}(\varphi)}{d\varphi} \right\| \leq \eta.$$

Hence, $u^{(1)}(\varphi) \in C_{\mathbf{W}}^1(\varphi)$, and

$$\left\| \frac{du^{(1)}(\varphi)}{d\varphi} \right\| \leq \eta \quad \forall \varphi \in \mathbf{W}.$$

Whence, $\forall\{\varphi, \bar{\varphi}\} \subset \mathbf{W}$, we have

$$\|u^{(1)}(\varphi) - u^{(1)}(\bar{\varphi})\| \leq \eta \|\varphi - \bar{\varphi}\|.$$

Thus, inequality (2.153) holds for $s = 1$.

Assume that, $\forall \varphi \in \mathbf{W}$, the inequality

$$\left\| \frac{du^{(s)}(\varphi)}{d\varphi} \right\| \leq \eta \tag{2.158}$$

holds $\forall s \leq k \in Z^+$. Writing equality (2.152) for $s = k + 1$, we obtain estimate (2.158) for $s = k+1$ analogously to the previous transformations. By the principle of complete mathematical induction, this estimate holds for all $s \in Z^+$, which guarantees the validity of the estimate

$$\|u^{(s)}(\varphi) - u^{(s)}(\bar{\varphi})\| \leq \eta \|\varphi - \bar{\varphi}\| \quad \forall\{\varphi, \bar{\varphi}\} \subset \mathbf{W}, s \in Z_0^+.$$

The limiting transition in the last inequality as $s \to \infty$ proves the assertion of Theorem 2.15. $\qquad\qquad\square$

By $C^1_{Lip}(\varphi)$ and $C^1_{Lip}(\varphi, x)$, we denote the subsets from $C^1_{\mathbf{W}}(\varphi)$ and $C^1_{\mathfrak{D}_0}(\varphi, x)$, whose elements have Lipschitz derivatives with respect to φ and (φ, x), respectively. This means that the function $f(\varphi) \in C^1_{Lip}(\varphi)$ with the coefficient δ, if, $\forall \{\varphi, \bar{\varphi}\} \subset \mathbf{W}$,

$$\left\| \frac{df(\varphi)}{d\varphi} - \frac{df(\bar{\varphi})}{d\bar{\varphi}} \right\| \leq \delta \|\varphi - \bar{\varphi}\|,$$

and $f(\varphi, x) \in C^1_{Lip}(\varphi, x)$ with the coefficient δ, if, $\forall \{\varphi, \bar{\varphi}\} \subset \mathbf{W}$, $\{x, \bar{x}\} \subset D_\rho$,

$$\left\| \frac{df(\varphi, x)}{d(\varphi, x)} - \frac{df(\bar{\varphi}, \bar{x})}{d(\bar{\varphi}, \bar{x})} \right\| \leq \delta \|(\varphi, x) - (\bar{\varphi}, \bar{x})\|,$$

where δ is a positive constant independent of $\varphi, \bar{\varphi}, x, \bar{x}$.

Lemma 2.22. *Let all conditions of Theorem 2.15 be satisfied, and let $\{a(\varphi, x), P^{-1}(\varphi, x)\} \subset C^1_{Lip}(\varphi, x)$, $c(\varphi) \in C^1_{Lip}(\varphi)$ and $P_1(1 + A^*)^2 < 1$. Then, $\forall s \in Z^+_0$, $u^{(s)}(\varphi) \in C^1_{Lip}(\varphi)$.*

Proof. Using equality (2.152), we write down the inequality

$$\left\| \frac{du^{(s)}(\varphi)}{d\varphi} - \frac{du^{(s)}(\bar{\varphi})}{d\bar{\varphi}} \right\| \leq \sum_{i=1}^{4} I_i^{(s)}, \tag{2.159}$$

where

$$I_1^{(s)} = \sum_{l=1}^{\infty} \left\| \frac{d\Pi^{(s)}(\varphi)}{d\varphi} \right\| \| c(\varphi^{(s)}_{l-1}(\varphi)) - c(\varphi^{(s)}_{l-1}(\bar{\varphi})) \|,$$

$$I_2^{(s)} = \sum_{l=1}^{\infty} \| \Pi^{(s)}(\varphi) - \Pi^{(s)}(\bar{\varphi}) \| \left\| \frac{dc(\varphi^{(s)}_{l-1}(\bar{\varphi}))}{d\bar{\varphi}} \right\|,$$

$$I_3^{(s)} = \sum_{l=1}^{\infty} \| \Pi^{(s)}(\varphi) \| \left\| \frac{dc(\varphi^{(s)}_{l-1}(\varphi))}{d\varphi} - \frac{dc(\varphi^{(s)}_{l-1}(\bar{\varphi}))}{d\bar{\varphi}} \right\|,$$

$$I_4^{(s)} = \sum_{l=1}^{\infty} \left\| \frac{d\Pi^{(s)}(\varphi)}{d\varphi} - \frac{d\Pi^{(s)}(\bar{\varphi})}{d\bar{\varphi}} \right\| \| c(\varphi^{(s)}_{l-1}(\bar{\varphi})) \|,$$

$$\Pi^{(s)}(\varphi) = \prod_{i=0}^{l-1} P^{-1}(\varphi^{(s)}_i(\varphi), u^{(s-1)}(\varphi^{(s)}_i(\varphi))).$$

This inequality holds $\forall s \in Z_0^+, \{\varphi, \bar{\varphi}\} \subset \mathbf{W}$, if the symbol $u^{(-1)}(\varphi_i^{(0)}(\varphi))$ means $0 \in \mathfrak{M}$. By the reasoning inductive relative to $i \in Z_0^+$, it is easy to verify the validity of the inequalities

$$\left\| \frac{d\varphi_i^{(s)}(\varphi)}{d\varphi} \right\| \leq (1 + A^*)^i,$$

$$\left\| \varphi_i^{(s)}(\varphi) - \varphi_i^{(s)}(\bar{\varphi}) \right\| \leq (1 + A^*)^i \|\varphi - \bar{\varphi}\| \qquad (2.160)$$

for all $s \in Z_0^+$.

In this case, $\forall \{i, s\} \subset Z_0^+, l \in Z^+$, the following estimates hold:

$$\left\| \frac{dc(\varphi_{l-1}^{(s)}(\varphi))}{d\varphi} \right\| \leq C_* \left\| \frac{d\varphi_{l-1}^{(s)}(\varphi)}{d\varphi} \right\| \leq C_*(1 + A^*)^{l-1};$$

$$\|P^{-1}(\varphi_i^{(s)}(\varphi), u^{(s-1)}(\varphi_i^{(s)}(\varphi))) - P^{-1}(\varphi_i^{(s)}(\bar{\varphi}), u^{(s-1)}(\varphi_i^{(s)}(\bar{\varphi})))\|$$
$$\leq P_* \max\{\|\varphi_i^{(s)}(\varphi) - \varphi_i^{(s)}(\bar{\varphi})\|, \|u^{(s-1)}(\varphi_i^{(s)}(\varphi)) - u^{(s-1)}(\varphi_i^{(s)}(\bar{\varphi}))\|\}$$
$$= P_*(1 + A^*)^i \|\varphi - \bar{\varphi}\|;$$

$$\|\Pi^{(s)}(\varphi) - \Pi^{(s)}(\bar{\varphi})\| \leq P_1^{l-1} \sum_{i=0}^{l-1} \|P^{-1}(\varphi_i^{(s)}(\varphi), u^{(s-1)}(\varphi_i^{(s)}(\varphi)))$$

$$- P^{-1}(\varphi_i^{(s)}(\bar{\varphi}), u^{(s-1)}(\varphi_i^{(s)}(\bar{\varphi})))\| \leq P_1^{l-1} \sum_{i=0}^{l-1} P_*(1 + A^*)^i \|\varphi - \bar{\varphi}\|$$

$$= \frac{P_*}{A^*} P_1^{l-1}(1 + A^*)^l \|\varphi - \bar{\varphi}\|;$$

$$\left\| \frac{d\Pi^{(s)}(\varphi)}{d\varphi} \right\| \leq P_1^{l-1} \sum_{i=0}^{l-1} \left\| \frac{dP^{-1}(\varphi_i^{(s)}(\varphi), u^{(s-1)}(\varphi_i^{(s)}(\varphi)))}{d\varphi} \right\|$$

$$\leq P_* P_1^{l-1} \sum_{i=0}^{l-1} \left\| \frac{d\varphi_i^{(s)}(\varphi)}{d\varphi} \right\| \leq P_* P_1^{l-1} \sum_{i=0}^{l-1} (1 + A^*)^i < \frac{P_*}{A^*} P_1^{l-1}(1 + A^*)^l.$$

For $I_1^{(s)}$ and $I_2^{(s)}$, they give possibility to write down the inequality

$$\max\{I_1^{(s)}, I_2^{(s)}\} \leq \frac{P_* C_*}{A^* P_1 (1 + A^*)} \sum_{l=1}^{\infty} P_1^l (1 + A^*)^{2l} \|\varphi - \bar{\varphi}\|$$

$$= \frac{P_* C_* (1 + A^*)}{A^*(1 - P_1(1 + A^*)^2)} \|\varphi - \bar{\varphi}\|,$$

which is independent of $s \in Z_0^+$.

The estimation of $I_3^{(s)}$ and $I_4^{(s)}$ is somewhat more complicated. Relations (2.151) and (2.160) yield the estimates

$$\left\|\frac{d\varphi_i^{(s)}(\varphi)}{d\varphi} - \frac{d\varphi_i^{(s)}(\bar\varphi)}{d\bar\varphi}\right\| \leq \left\|\frac{d\varphi_{i-1}^{(s)}(\varphi)}{d\varphi} - \frac{d\varphi_{i-1}^{(s)}(\bar\varphi)}{d\bar\varphi}\right\|$$

$$+ A^* \max\left\{\left\|\frac{d\varphi_{i-1}^{(s)}(\varphi)}{d\varphi} - \frac{d\varphi_{i-1}^{(s)}(\bar\varphi)}{d\bar\varphi}\right\|, \left\|\frac{du^{(s-1)}(\varphi_{i-1}^{(s)}(\varphi))}{d\varphi} - \frac{du^{(s-1)}(\varphi_{i-1}^{(s)}(\bar\varphi))}{d\bar\varphi}\right\|\right\}$$

$$+ A_* \max\left\{\left\|\varphi_{i-1}^{(s)}(\varphi) - \varphi_{i-1}^{(s)}(\bar\varphi)\right\|, \left\|u^{(s-1)}(\varphi_{i-1}^{(s)}(\varphi)) - u^{(s-1)}(\varphi_{i-1}^{(s)}(\bar\varphi))\right\|\right\}$$

$$\times \max\left\{\left\|\frac{d\varphi_{i-1}^{(s)}(\bar\varphi)}{d\bar\varphi}\right\|, \left\|\frac{du^{(s-1)}(\varphi_{i-1}^{(s)}(\bar\varphi))}{d\bar\varphi}\right\|\right\}$$

$$\leq (1 + A^*)\left\|\frac{d\varphi_{i-1}^{(s)}(\varphi)}{d\varphi} - \frac{d\varphi_{i-1}^{(s)}(\bar\varphi)}{d\bar\varphi}\right\| + A_*(1 + A^*)^{2(i-1)}\|\varphi - \bar\varphi\|$$

$$+ A^*(1 + A^*)^{i-1}\left\|\frac{du^{(s-1)}(\varphi_{i-1}^{(s)}(\varphi))}{d\varphi_{i-1}^{(s)}(\varphi)} - \frac{du^{(s-1)}(\varphi_{i-1}^{(s)}(\bar\varphi))}{d\varphi_{i-1}^{(s)}(\bar\varphi)}\right\|; \quad (2.161)$$

$$\left\|\frac{dc(\varphi_{l-1}^{(s)}(\varphi))}{d\varphi} - \frac{dc(\varphi_{l-1}^{(s)}(\bar\varphi))}{d\bar\varphi}\right\| = \left\|\frac{dc(\varphi_{l-1}^{(s)}(\varphi))}{d\varphi_{l-1}^{(s)}(\varphi)}\frac{d\varphi_{l-1}^{(s)}(\varphi)}{d\varphi}\right.$$

$$\left. - \frac{dc(\varphi_{l-1}^{(s)}(\bar\varphi))}{d\varphi_{l-1}^{(s)}(\bar\varphi)}\frac{d\varphi_{l-1}^{(s)}(\bar\varphi)}{d\bar\varphi}\right\|$$

$$\leq C_*\left\|\frac{d\varphi_{l-1}^{(s)}(\varphi)}{d\varphi} - \frac{d\varphi_{l-1}^{(s)}(\bar\varphi)}{d\bar\varphi}\right\| + C^*\left\|\varphi_{l-1}^{(s)}(\varphi) - \varphi_{l-1}^{(s)}(\bar\varphi)\right\|\left\|\frac{d\varphi_{l-1}^{(s)}(\bar\varphi)}{d\bar\varphi}\right\|$$

$$\leq C_*\left\|\frac{d\varphi_{l-1}^{(s)}(\varphi)}{d\varphi} - \frac{d\varphi_{l-1}^{(s)}(\bar\varphi)}{d\bar\varphi}\right\| + C^*(1 + A^*)^{2(l-1)}\|\varphi - \bar\varphi\|, \quad (2.162)$$

where A_* and C^* are the coefficients, with which the functions $a(\varphi, x)$ and $c(\varphi)$ enter the spaces $C_{Lip}^1(\varphi, x)$ and $C_{Lip}^1(\varphi)$, respectively.

In turn, we write down the chain of inequalities

$$\left\|\frac{d\Pi^{(s)}(\varphi)}{d\varphi} - \frac{d\Pi^{(s)}(\bar\varphi)}{d\bar\varphi}\right\|$$

$$\leq \left\{P_1^{l-1}\left\|\frac{dP^{-1}(\varphi_0)}{d\varphi} - \frac{dP^{-1}(\bar\varphi_0)}{d\bar\varphi}\right\| + P_1^{l-2}\left\|\frac{dP^{-1}(\bar\varphi_0)}{d\bar\varphi}\right\|\right.$$

$$\times \|P^{-1}(\varphi_1) - P^{-1}(\bar\varphi_1)\| + P_1^{l-2}\left\|\frac{dP^{-1}(\bar\varphi_0)}{d\bar\varphi}\right\|\|P^{-1}(\varphi_2) - P^{-1}(\bar\varphi_2)\| + \cdots$$

$$+\, P_1^{l-2}\left\|\frac{dP^{-1}(\bar\varphi_0)}{d\bar\varphi}\right\|\left\|P^{-1}(\varphi_{l-1}) - P^{-1}(\bar\varphi_{l-1})\right\|\}$$

$$+\, \{\|P^{-1}(\varphi_0) - P^{-1}(\bar\varphi_0)\|\left\|\frac{dP^{-1}(\varphi_1)}{d\varphi}\right\|P_1^{l-2} + P_1^{l-1}\left\|\frac{dP^{-1}(\varphi_1)}{d\varphi}\right.$$

$$\left. -\, \frac{dP^{-1}(\bar\varphi_1)}{d\bar\varphi}\right\| + P_1^{l-2}\left\|\frac{dP^{-1}(\bar\varphi_1)}{d\bar\varphi}\right\|\left\|P^{-1}(\varphi_2) - P^{-1}(\bar\varphi_2)\right\| + \cdots$$

$$+\, P_1^{l-2}\left\|\frac{dP^{-1}(\bar\varphi_1)}{d\bar\varphi}\right\|\left\|P^{-1}(\varphi_{l-1}) - P^{-1}(\bar\varphi_{l-1})\right\|\} + \cdots$$

$$+\, \{\|P^{-1}(\varphi_0) - P^{-1}(\bar\varphi_0)\|P_1^{l-2}\left\|\frac{dP^{-1}(\varphi_{l-1})}{d\varphi}\right\| + P_1^{l-2}\|P^{-1}(\varphi_1)$$

$$-\, P^{-1}(\bar\varphi_1)\|\left\|\frac{dP^{-1}(\varphi_{l-1})}{d\varphi}\right\| + \cdots + P_1^{l-1}\left\|\frac{dP^{-1}(\varphi_{l-1})}{d\varphi} - \frac{dP^{-1}(\bar\varphi_{l-1})}{d\bar\varphi}\right\|\}$$

$$\leq P_1^{l-1}\left\|\frac{dP^{-1}(\varphi_0)}{d\varphi} - \frac{dP^{-1}(\bar\varphi_0)}{d\bar\varphi}\right\| + P_1^{l-2}P_*^2(1 + A^*)K\|\varphi - \bar\varphi\|$$

$$+\, P_1^{l-1}\left\|\frac{dP^{-1}(\varphi_1)}{d\varphi} - \frac{dP^{-1}(\bar\varphi_1)}{d\bar\varphi}\right\| + P_1^{l-2}P_*^2(1 + A^*)^2 K\|\varphi - \bar\varphi\| + \cdots$$

$$+\, P_1^{l-1}\left\|\frac{dP^{-1}(\varphi_{l-1})}{d\varphi} - \frac{dP^{-1}(\bar\varphi_{l-1})}{d\bar\varphi}\right\| + P_1^{l-2}P_*^2(1 + A^*)^l K\|\varphi - \bar\varphi\|$$

$$=\, P_1^{l-1}\sum_{k=0}^{l-1}\left\|\frac{dP^{-1}(\varphi_k)}{d\varphi} - \frac{dP^{-1}(\bar\varphi_k)}{d\bar\varphi}\right\| + P_1^{l-2}P_*^2(1 + A^*)K^2\|\varphi - \bar\varphi\|$$

$$<\, P_1^{l-1}\sum_{k=0}^{l-1}\left\|\frac{dP^{-1}(\varphi_k)}{d\varphi} - \frac{dP^{-1}(\bar\varphi_k)}{d\bar\varphi}\right\| + \frac{P_1^{l-2}P_*^2}{(A^*)^2}(1 + A^*)^{2l+1}K^2\|\varphi - \bar\varphi\|,$$

where the expressions

$$P^{-1}(\varphi_k^{(s)}(\varphi), u^{(s-1)}(\varphi_k^{(s)}(\varphi)))$$

and

$$P^{-1}(\varphi_k^{(s)}(\bar\varphi), u^{(s-1)}(\varphi_k^{(s)}(\bar\varphi)))$$

are deno-ted by $P^{-1}(\varphi_k)$ and $P^{-1}(\bar\varphi_k)$, respectively, and K stands for the sum $\sum_{k=0}^{l-1}(1 + A^*)^k$.

In view of the relation

$$\left\|\frac{dP^{-1}(\varphi_k)}{d\varphi} - \frac{dP^{-1}(\bar\varphi_k)}{d\bar\varphi}\right\|$$

$$\leq P_*\left\|\frac{d\varphi_k^{(s)}(\varphi_k)}{d\varphi} - \frac{d\varphi_k^{(s)}(\bar\varphi_k)}{d\bar\varphi}\right\| + P^*(1 + A^*)^{2k}\|\varphi - \bar\varphi\|$$

$$+\, P_*(1 + A^*)^k\left\|\frac{du^{(s-1)}(\varphi_k^{(s)}(\varphi))}{d\varphi_k^{(s)}(\varphi)} - \frac{du^{(s-1)}(\varphi_k^{(s)}(\bar\varphi))}{d\varphi_k^{(s)}(\bar\varphi)}\right\|,$$

where P^* is the coefficient, with which $P^{-1}(\varphi, x)$ enters $C^1_{Lip}(\varphi, x)$, we have the estimate

$$\left\|\frac{d\Pi^{(s)}(\varphi)}{d\varphi} - \frac{d\Pi^{(s)}(\bar\varphi)}{d\bar\varphi}\right\|$$

$$< P_1^{l-1}\sum_{k=0}^{l-1}\left\{P_*\left\|\frac{d\varphi_k^{(s)}(\varphi)}{d\varphi} - \frac{d\varphi_k^{(s)}(\bar\varphi)}{d\bar\varphi}\right\|\right.$$

$$+ P^*(1+A^*)^{2k}\|\varphi-\bar\varphi\| + P_*(1+A^*)^k\left\|\frac{du^{(s-1)}(\varphi_k^{(s)}(\varphi))}{d\varphi_k^{(s)}(\varphi)} - \frac{du^{(s-1)}(\varphi_k^{(s)}(\bar\varphi))}{d\varphi_k^{(s)}(\bar\varphi)}\right\|\right\}$$

$$+ \frac{P_1^{l-2}P_*^2}{(A^*)^2}(1+A^*)^{2l+1}\|\varphi-\bar\varphi\|. \quad (2.163)$$

Then we use the method of complete mathematical induction to prove the assertion of Lemma 2.22 for $s = 0$. Relation (2.161) yields the inductive inequality

$$\left\|\frac{d\varphi_i^{(0)}(\varphi)}{d\varphi} - \frac{d\varphi_i^{(0)}(\bar\varphi)}{d\bar\varphi}\right\|$$

$$\leq (1+A^*)\left\|\frac{d\varphi_{i-1}^{(0)}(\varphi)}{d\varphi} - \frac{d\varphi_{i-1}^{(0)}(\bar\varphi)}{d\bar\varphi}\right\| + A_*(1+A^*)^{2(i-1)}\|\varphi-\bar\varphi\|,$$

which yields the estimates

$$\left\|\frac{d\varphi_i^{(0)}(\varphi)}{d\varphi} - \frac{d\varphi_i^{(0)}(\bar\varphi)}{d\bar\varphi}\right\|$$

$$\leq (1+A^*)\left\{(1+A^*)\left\|\frac{d\varphi_{i-2}^{(0)}(\varphi)}{d\varphi} - \frac{d\varphi_{i-2}^{(0)}(\bar\varphi)}{d\bar\varphi}\right\|\right.$$

$$+ A_*(1+A^*)^{2(i-2)}\|\varphi-\bar\varphi\|\Big\} + A_*(1+A^*)^{2(i-1)}\|\varphi-\bar\varphi\| \leq \cdots$$

$$\leq (1+A^*)^i\left\|\frac{d\varphi_0^{(0)}(\varphi)}{d\varphi} - \frac{d\varphi_0^{(0)}(\bar\varphi)}{d\bar\varphi}\right\| + A_*\{(1+A^*)^{i-1} + (1+A^*)^i + \cdots$$

$$+ (1+A^*)^{2(i-1)}\}\|\varphi-\bar\varphi\| < \frac{A_*}{A^*}(1+A^*)^{2i-1}\|\varphi-\bar\varphi\|.$$

Then, with the use of (2.162) and (2.163), we obtain the inequalities

$$I_3^{(0)} < \sum_{l=1}^{\infty} P_1^l\left\{\frac{A_*}{A^*}(1+A^*)^{2l-3} + C^*(1+A^*)^{2l-2}\right\}\|\varphi-\bar\varphi\|$$

$$< \frac{P_1(A_* + A^*C^*)}{A^*(1 - P_1(1+A^*)^2)}\|\varphi-\bar\varphi\|;$$

$$I_4^{(0)} < \sum_{l=1}^{\infty} C^0 \{ P_1^{l-1} \sum_{k=0}^{l-1} [P_* \| \frac{d\varphi_k^{(0)}(\varphi)}{d\varphi} - \frac{d\varphi_k^{(0)}(\bar\varphi)}{d\bar\varphi} \| + P^*(1+A^*)^{2k} \|\varphi - \bar\varphi \|]$$

$$+ \frac{P_*^2 P_1^{l-2}}{(A^*)^2}(1+A^*)^{2l+1} \|\varphi - \bar\varphi\| \} \le C^0 \{ \frac{P_* A_*}{A^*(1+A^*)} \sum_{l=1}^{\infty} P_1^{l-1} \sum_{k=0}^{l-1} (1+A^*)^{2k}$$

$$+ P^* \sum_{l=1}^{\infty} P_1^{l-1} \sum_{k=0}^{l-1} (1+A^*)^{2k} + \frac{P_*^2(1+A^*)}{(A^*)^2 P_1^2} \sum_{l=1}^{\infty} P_1^l (1+A^*)^{2l} \} \|\varphi - \bar\varphi\|$$

$$< C^0 \{ (P^* + \frac{P_* A_*}{A^*(1+A^*)}) \frac{(1+A^*)^2}{((1+A^*)^2 - 1)(1 - P_1(1+A^*)^2)}$$

$$+ \frac{P_*^2(1+A^*)^3}{(A^*)^2 P_1(1 - P_1(1+A^*)^2)} \} \|\varphi - \bar\varphi\|.$$

Hence, with regard for (2.159) and the estimates for $I_1^{(0)}$ and $I_2^{(0)}$, we conclude that $u^{(0)}(\varphi) \in C_{Lip}^1(\varphi)$ with the coefficient denoted by κ_0. The formula for this coefficient is quite awkward, and we omit it here.

We now assume that the assertion of Lemma 2.22 holds for all $s \in \{1, 2, 3, \ldots, m\}$. By κ_s, we denote the coefficients, with which the functions $u^{(s)}(\varphi)$ enter the set $C_{Lip}^1(\varphi)$. Relation (2.161) yields the inductive inequality

$$\| \frac{d\varphi_i^{(m+1)}(\varphi)}{d\varphi} - \frac{d\varphi_i^{(m+1)}(\bar\varphi)}{d\bar\varphi} \| \le (1+A^*) \| \frac{d\varphi_{i-1}^{(m+1)}(\varphi)}{d\varphi} - \frac{d\varphi_{i-1}^{(m+1)}(\bar\varphi)}{d\bar\varphi} \|$$

$$+ (A_* + A^* \kappa_m)(1+A^*)^{2(i-1)} \|\varphi - \bar\varphi\|,$$

which leads to the estimate

$$\| \frac{d\varphi_i^{(m+1)}(\varphi)}{d\varphi} - \frac{d\varphi_i^{(m+1)}(\bar\varphi)}{d\bar\varphi} \| \le \frac{A_* + A^* \kappa_m}{A^*}(1+A^*)^{2i-1} \|\varphi - \bar\varphi\|.$$

Using the last inequality, it is easy to verify that $I_3^{(m+1)}$ and $I_4^{(m+1)}$ have the Lipschitz property by analogy with the previous calculations. Since $I_1^{(s)}$ and $I_2^{(s)}$ are independent of $s \in Z_0^+$, we conclude that $u^{(m+1)}(\varphi) \in C_{Lip}^1(\varphi)$. The proof of Lemma 2.22 is completed. $\qquad\square$

We now formulate the main result of this subsection.

Theorem 2.16. *Let the matrix function $P(\varphi, x)$ be uniformly continuous on $\mathfrak{D}_0$, and let the following conditions be satisfied:*

1) $\{a(\varphi, x), P^{-1}(\varphi, x)\} \subset C^1_{Lip}(\varphi, x)$, $c(\varphi) \in C^1_{Lip}(\varphi)$; *moreover,* $\forall \varphi \in$ **W**, $(\varphi, x) \in \mathfrak{D}_0$, *the estimates*

$$\left\| \frac{da(\varphi, x)}{d(\varphi, x)} \right\| \leq A^*, \quad \left\| \frac{dP^{-1}(\varphi, x)}{d(\varphi, x)} \right\| \leq P_*, \quad \left\| \frac{dc(\varphi)}{d(\varphi)} \right\| \leq C_*,$$

where A^*, P_*, C_* *are positive constants independent of* φ, x, *hold;*

2) *the inequalities*

$$P_1 < \frac{1}{(1 + A^*)^2}, \quad \frac{C^0 P_1}{1 - P_1} \leq d, \quad \eta = \frac{P_* C^0 (1 + A^*) + C_* P_1 A^*}{A^* (1 - P_1(1 + A^*))} \leq 1,$$

$$p = \frac{P_* d + P_* C^0 + P_1 A^*}{1 - P_1(1 + A^*)}$$

hold. Then the system of equations (2.145) has the bounded semi-invariant manifold, whose generating function $u(\varphi)$ *satisfies the Lipschitz condition with the coefficient* η *on* **W**, *and the functions* $u^{(s)}(\varphi)$ *defined by equality (2.152) belong, for all* $s \in Z_0^+$, *to the set* $C^1_{Lip}(\varphi)$. *In this case, if* $u^{(s)}(\varphi) \in C^1_{Lip}(\varphi)$ *with the coefficient* G, *which is independent of* $s \in Z_0^+$, *then this manifold is smooth, and* $u(\varphi) \in C^1_{Lip}(\varphi)$.

Proof.   It is easy to see that, under conditions of Theorem 2.16, the assertions of Theorem 2.15 and Lemma 2.22 are valid. Therefore, only the condition of smoothness of the semi-invariant manifold of the system of equations (2.145) requires the substantiation. For convenience, we denote the difference

$$\frac{du^{(s+1)}(\varphi)}{d\varphi} - \frac{du^{(s)}(\varphi)}{d\varphi}$$

by $\omega_1^{(s+1)}$ $\forall s \in Z_0^+$.

Using the representation of the function $u^{(s)}(\varphi)$ from Lemma 2.20, we verify that its derivative with respect to $\varphi \in$ **W** acts on an arbitrary element $\xi \in$ **W** in the following way:

$$\frac{du^{(s)}(\varphi)}{d\varphi}\xi = \frac{dP^{-1}(\varphi, u^{(s-1)}(\varphi))}{d\varphi}\xi \cdot (u^{(s)}(\varphi_1^{(s)}(\varphi)) - c(\varphi))$$

$$+ P^{-1}(\varphi, u^{(s-1)}(\varphi)) \cdot \frac{d(u^{(s)}(\varphi_1^{(s)}(\varphi)) - c(\varphi))}{d\varphi}\xi, \quad s \in Z^+.$$

This gives directly the equality

$$\omega_1^{(s+1)}\xi = \frac{dP^{-1}(\varphi, u^{(s)}(\varphi))}{d\varphi}\xi\{u^{(s+1)}(\varphi_1^{(s+1)}(\varphi)) - u^{(s)}(\varphi_1^{(s)}(\varphi))\}$$

$$+ \{\frac{dP^{-1}(\varphi, u^{(s)}(\varphi))}{d\varphi} - \frac{dP^{-1}(\varphi, u^{(s-1)}(\varphi))}{d\varphi}\}\xi(u^{(s)}(\varphi_1^{(s)}(\varphi)) - c(\varphi))$$

$$+ P^{-1}(\varphi, u^{(s)}(\varphi))\{\frac{du^{(s+1)}(\varphi_1^{(s+1)}(\varphi))}{d\varphi} - \frac{du^{(s)}(\varphi_1^{(s)}(\varphi))}{d\varphi}\}\xi$$

$$+ \{P^{-1}(\varphi, u^{(s)}(\varphi)) - P^{-1}(\varphi, u^{(s-1)}(\varphi))\}\frac{d(u^{(s)}(\varphi_1^{(s)}(\varphi)) - c(\varphi))}{d\varphi}\xi,$$

which yields, for $\eta \le 1$ and $\forall s \in Z^+$, the inequality

$$\|\omega_1^{(s+1)}\| \le P_*\|u^{(s+1)}(\varphi_1^{(s+1)}(\varphi)) - u^{(s)}(\varphi_1^{(s)}(\varphi))\|$$

$$+\{P_*\|\omega_1^{(s)}\|+2P^*d\gamma^{s-1}\}(d+C^0)+P_1\|\frac{du^{(s+1)}(\varphi_1^{(s+1)}(\varphi))}{d\varphi} - \frac{du^{(s)}(\varphi_1^{(s)}(\varphi))}{d\varphi}\|$$

$$+ 2dP_*\gamma^{s-1}(\|\frac{d\varphi_1^{(s)}(\varphi)}{d\varphi}\| + C_*). \qquad (2.164)$$

In view of the inequalities

$$\|\frac{d\varphi_1^{(s)}(\varphi)}{d\varphi}\| \le 1 + \|\frac{da(\varphi, u^{(s-1)}(\varphi))}{d\varphi}\| \le 1 + A^*;$$

$$\|u^{(s+1)}(\varphi_1^{(s+1)}(\varphi)) - u^{(s)}(\varphi_1^{(s)}(\varphi))\|$$

$$\le \|u^{(s+1)}(\varphi_1^{(s+1)}(\varphi)) - u^{(s+1)}(\varphi_1^{(s)}(\varphi))\|$$

$$+ \|u^{(s+1)}(\varphi_1^{(s)}(\varphi)) - u^{(s)}(\varphi_1^{(s)}(\varphi))\| \le \|\varphi_1^{(s+1)}(\varphi) - \varphi_1^{(s)}(\varphi)\| + \|\omega^{(s+1)}\|_0$$

$$\le A^*\|\omega^{(s)}\| + \|\omega^{(s+1)}\|_0 \le 2d(\gamma + A^*)\gamma^{s-1};$$

$$\|\frac{d\varphi_1^{(s+1)}(\varphi)}{d\varphi} - \frac{d\varphi_1^{(s)}(\varphi)}{d\varphi}\| = \|\frac{da(\varphi, u^{(s)}(\varphi))}{d\varphi} - \frac{da(\varphi, u^{(s-1)}(\varphi))}{d\varphi}\|$$

$$\le A^*\|\omega_1^{(s)}\| + 2A_*d\gamma^{s-1};$$

$$\|\frac{du^{(s+1)}(\varphi_1^{(s+1)}(\varphi))}{d\varphi} - \frac{du^{(s)}(\varphi_1^{(s)}(\varphi))}{d\varphi}\|$$

$$= \|\frac{du^{(s+1)}(\varphi_1^{(s+1)}(\varphi))}{d\varphi_1^{(s+1)}(\varphi)}\frac{d\varphi_1^{(s+1)}(\varphi)}{d\varphi} - \frac{du^{(s)}(\varphi_1^{(s)}(\varphi))}{d\varphi_1^{(s)}(\varphi)}\frac{d\varphi_1^{(s)}(\varphi)}{d\varphi}\|$$

$$\leq \left\| \frac{du^{(s+1)}(\varphi_1^{(s+1)}(\varphi))}{d\varphi_1^{(s+1)}(\varphi)} \right\| \left\| \frac{d\varphi_1^{(s+1)}(\varphi)}{d\varphi} - \frac{d\varphi_1^{(s)}(\varphi)}{d\varphi} \right\|$$

$$+ \left\| \frac{du^{(s+1)}(\varphi_1^{(s+1)}(\varphi))}{d\varphi_1^{(s+1)}(\varphi)} - \frac{du^{(s)}(\varphi_1^{(s)}(\varphi))}{d\varphi_1^{(s)}(\varphi)} \right\| \left\| \frac{d\varphi_1^{(s)}(\varphi)}{d\varphi} \right\| \leq A^* \|\omega_1^{(s)}\|$$

$$+ 2A_* d\gamma^{s-1} + \left\| \frac{du^{(s+1)}(\varphi_1^{(s+1)}(\varphi))}{d\varphi_1^{(s+1)}(\varphi)} - \frac{du^{(s)}(\varphi_1^{(s)}(\varphi))}{d\varphi_1^{(s)}(\varphi)} \right\| (1 + A^*);$$

$$\left\| \frac{du^{(s+1)}(\varphi_1^{(s+1)}(\varphi))}{d\varphi_1^{(s+1)}(\varphi)} - \frac{du^{(s)}(\varphi_1^{(s)}(\varphi))}{d\varphi_1^{(s)}(\varphi)} \right\|$$

$$\leq G\|\varphi_1^{(s+1)}(\varphi) - \varphi_1^{(s)}(\varphi)\| + \|\omega_1^{(s+1)}\|_0 \leq 2dG\gamma^{s-1} + \|\omega_1^{(s+1)}\|_0$$

and (2.164), we obtain, for all $s \in Z^+, \varphi \in \mathbf{W}$, the inductive estimate

$$\|\omega_1^{(s+1)}\|_0 \leq \xi \gamma^{s-1} + p\|\omega_1^{(s)}\|_0, \tag{2.165}$$

where $\xi = 2d\{P_*(\gamma + A^*) + P^*(d + C^0) + A_* P_1 + P_1(1 + A^*)G + P_*(1 + A^* + C_*)\}$ is a positive constant independent of s, φ. By the inductive reasoning, it is easy to verify that relation (2.165) yields

$$\|\omega_1^{(s+1)}\|_0 \leq \xi \sum_{i=0}^{s-1} p^i \gamma^{s-i-1} + p^s\|\omega_1^{(1)}\|_0 = \xi \frac{\gamma^{s-1}((\frac{p}{\gamma})^s - 1)}{\frac{p}{\gamma} - 1}$$

$$+ p^s\|\omega_1^{(1)}\|_0 < \xi \frac{p^s}{p - \gamma} + p^s\|\omega_1^{(1)}\|_0 \leq (\frac{\xi}{p - \gamma} + 2)p^s \to 0$$

as $s \to \infty$, since $\gamma < p < 1$.

In this case, the sequence $\{\frac{du^{(s)}(\varphi)}{d\varphi}\}_{s=1}^{\infty}$ is fundamental in the Banach space $\mathbf{L}(\mathbf{W}, \mathfrak{M})$, which guarantees its convergence in its norm, and this convergence is uniform in $\varphi \in \mathbf{W}$. As is known [136], the limit of the indicated sequence is the derivative $\frac{du(\varphi)}{d\varphi}$. The inclusion $u(\varphi) \in C^1_{Lip}(\varphi)$ is obvious. Theorem 2.16 is proved. $\qquad\square$

Remark 2.5. If the space $\mathbf{W}$ is the space $\mathfrak{M}$, and if the functions $a(\varphi, x), P(\varphi, x), c(\varphi)$ are 2π-periodic in $\varphi^i(i = 1, 2, 3, \dots)$, where $(\varphi^1, \varphi^2, \varphi^3, \dots) = \varphi \in \mathfrak{M}$, then the semi-invariant manifold of the system of equations (2.145) is transformed in an infinite-dimensional semi-invariant torus.

Chapter 3

Periodic solutions of difference equations. Extension of solutions

The first section of this chapter is devoted to the construction of periodic solutions of nondegenerate and degenerate linear and quasilinear difference equations with periodic coefficients in the spaces of bounded number sequences in the nonresonance and resonance cases. In two following sections, we will realize an approximate construction of periodic solutions of nonlinear difference equations of the first and second orders in abstract Banach spaces. Then we give the conditions of asymptotic periodicity of solutions of a linear difference equation in a complex Banach space obtained by Yu.V. Tomilov [166]. In the final section, we study the problem of continuation "to the left" of solutions of nonlinear difference equations of the first and higher orders in Banach spaces. This problem is closely connected with the question of the invertibility of certain nonlinear difference operators, which was studied, for example, by V.Yu. Slyusarchuk [139; 140; 142]. Here, we consider the degenerate case where the appropriate difference operators are not invertible. The main methods used in our studies are the numerical-analytic method of construction of periodic solutions, which was proposed by A.M. Samoilenko for the systems of differential equations [99] – [101] and the modified Newton–Kantorovich method of solution of operator equations [38; 43].

3.1 On the periodic solutions of linear and quasilinear equations with periodic coefficients in the space $\mathfrak{M}$

1^0. **Nondegenerate nonresonance case.** Consider the equation

$$x_{n+1} = A(n)x_n, \quad n \in Z, \tag{3.1}$$

193

which is an equation of the form (1.1) with an N-periodic matrix $A(n)$, i.e., $A(n + N) = A(n)$ for all $n \in Z$, where $N \in Z^+$. Equation (3.1) is considered nondegenerate , i.e., the matrix $A(n)$ is invertible $\forall n \in Z$, and $\{A(n), A^{-1}(n)\} \subset \mathbf{\Gamma}$. By Ω_0^n, we denote the matriciant of Eq. (3.1). We recall that the eigenvalues of the monodromy matrix Ω_0^N are called the multipliers of this equation. We also note that, in section 1.1, we agree to call, for convenience, the set $[a, b]_Z = [a, b] \bigcap Z$ with $[a, b] \subset R^1$, on which the solution of the input equation is considered, as a segment (interval) and to denote it by $[a, b]$. If this will not lead to the misunderstanding, we use the mentioned agreement in what follows.

Lemma 3.1. *Equation (3.1) has a nontrivial solution $\xi_n = \xi(n)$, which satisfies $\forall n \in Z$ the condition $\xi_{n+N} = \rho\xi(n), \rho \in R^1$, if and only if ρ is the multiplier of this equation.*

Proof.   *Sufficiency.* Let v be the eigenvector of the monodromy matrix corresponding to its multiplier ρ. The obvious equalities

$$\Omega_0^{n+N+1} = A(n + N)\Omega_0^{n+N} = A(n)\Omega_0^{n+N}$$

yield $\Omega_0^{n+N} = \Omega_0^n\Omega_0^N$.

We denote the solution of Eq. (3.1) with initial values $0, v$ by ξ_n. Since $\xi_n = \Omega_0^n v$, we have

$$\xi_{n+N} = \Omega_0^{n+N} v = \Omega_0^n\Omega_0^N v = \rho\Omega_0^n v = \xi_n.$$

Necessity. Let ξ_n be the solution of Eq. (3.1), which satisfies the condition indicated in the statement of Lemma 3.1. Then $\xi_N = \rho\xi_0$, which yields $\Omega_0^N\xi_0 = \rho\xi_0$, i.e., ρ is the multiplier of Eq. (3.1). $\square$

From Lemma 3.1, we directly obtain the following proposition.

Corollary 3.1. *Equation (3.1) has a nontrivial N-periodic solution with initial value $x_0 \in \mathfrak{M}$ if and only if x_0 satisfies the equation*

$$(E - \Omega_0^N)x_0 = 0. \tag{3.2}$$

If the matrix $E - \Omega_0^N$ is invertible, and if $(E - \Omega_0^N)^{-1} \in \mathbf{\Gamma}$, then Eq. (3.1) is called a nonresonance one. It has no N-periodic solutions different from the trivial one. Since the point $\rho = 1$ is regular for the operator of multiplication of the matrix Ω_0^N by a vector from $\mathfrak{M}$, Eq. (3.2) has only the zero solution in the space $\mathfrak{M}$. In the opposite case, Eq. (3.1) is called a resonance one.

First, we consider the linear inhomogeneous equation

$$x_{n+1} = A(n)x_n + g_n, \quad n \in Z, \tag{3.3}$$

where, for all $n \in Z$ $\{A(n), A^{-1}(n)\} \subset \Gamma, \{x_n, g_n\} \subset \mathfrak{M}$; the matrix $A(n)$ and the function $g_n = g(n)$ are N-periodic in n.

The existence of N-periodic solutions of Eq. (3.3) is the content of the following theorem.

Theorem 3.1. *If Eq. (3.1) is not a resonance one, then Eq. (3.3) has the unique N-periodic solution defined on the set Z.*

Proof. Equation (3.3) cannot have more than one N-periodic solution, since the correspon-ding homogeneous equation (3.1) would have, in the opposite case, a nonzero N-periodic solution, which is impossible.

It is easy to verify that the solution $x_n = x(n, x_0), x_0 \in \mathfrak{M}$, of Eq. (3.3) can be presented in the form

$$x_n = \begin{cases} \Omega_0^n x_0 + \displaystyle\sum_{\nu=0}^{n-1} \Omega_0^n (\Omega_0^{\nu+1})^{-1} g_\nu & \text{for} \quad n \in Z^+; \\[2ex] x_0 & \text{for} \quad n = 0; \\[2ex] \Omega_0^n x_0 - \displaystyle\sum_{\nu=n}^{-1} \Omega_0^n (\Omega_0^{\nu+1})^{-1} g_\nu & \text{for} \quad n \in Z^-, \end{cases} \tag{3.4}$$

where Ω_0^n is the matriciant of Eq. (3.1).

Formula (3.4) yields the necessary and sufficient condition of N-periodicity of the solution of Eq. (3.3) with initial values $n = 0, x_0 \in \mathfrak{M}$ on the interval $[0, +\infty)$:

$$x_0 = \Omega_0^N x_0 + \sum_{\nu=0}^{N-1} \Omega_0^N (\Omega_0^{\nu+1})^{-1} g_\nu.$$

From this condition with regard for the invertibility of the matrix $E - \Omega_0^N$, we obtain the single required initial value

$$x_0 = (E - \Omega_0^N)^{-1} \sum_{\nu=0}^{N-1} \Omega_0^N (\Omega_0^{\nu+1})^{-1} g_\nu. \tag{3.5}$$

Then the solution of Eq. (3.3)

$$x_n = \Omega_0^n (E - \Omega_0^N)^{-1} \sum_{\nu=0}^{N-1} \Omega_0^N (\Omega_0^{\nu+1})^{-1} g_\nu + \sum_{\nu=0}^{n-1} \Omega_0^n (\Omega_0^{\nu+1})^{-1} g_\nu, \quad n \in Z^+, \tag{3.6}$$

is its unique N-periodic solution on the interval $[0, +\infty)$.

For the initial value $\bar{x}_0 \in \mathfrak{M}$ of the solution of Eq. (3.3) N-periodic on the interval $(-\infty, 0]$, we obtain the expression

$$\bar{x}_0 = (E - \Omega_0^N)^{-1}\Omega_0^N \sum_{\nu=-N}^{-1} \Omega_0^{-N}(\Omega_0^{\nu+1})^{-1}g_\nu. \qquad (3.7)$$

Hence, the solution of Eq. (3.3)

$$x_n = \Omega_0^n(E - \Omega_0^N)^{-1} \sum_{\nu=-N}^{-1} (\Omega_0^{\nu+1})^{-1}g_\nu - \sum_{\nu=n}^{-1} \Omega_0^n(\Omega_0^{\nu+1})^{-1}g_\nu, \quad n \in Z^-,$$

$$(3.8)$$

is its unique N-periodic solution on the interval $(-\infty, 0]$.

It is easy to verify that the initial values x_0 and $\bar{x}_0$ defined by equalities (3.5) and (3.7), respectively, coincide. Therefore, solutions (3.6) and (3.8) coincide at the point $n = 0$, which completes the proof of the theorem. $\square$

Let us consider the quasilinear equation

$$x_{n+1} = A(n)x_n + g(n) + \varepsilon F_n(x_n), \quad n \in Z_0^+, \qquad (3.9)$$

where $F_n(x)$ is an N-periodic vector-function, which takes values from the space $\mathfrak{M}$ for all $x \in \mathfrak{M}$, and ε is a real parameter. For $\varepsilon = 0$, Eq. (3.9) is transformed in Eq. (3.3) generating it.

We consider that the matrix $E - \Omega_0^N$ is invertible. Then, by Theorem 3.1, the generating equation on the interval $[0, +\infty)$ has the unique N-periodic solution defined by relations (3.5) and (3.6). We denote this solution by $\varphi_n = \varphi(n)$ and introduce the notation

$$K_1 = \max_{\nu=0,N-1}\{\|(\Omega_0^{\nu+1})^{-1}\|\},$$

$$K_2 = \|\Omega_0^N\|\|(E - \Omega_0^N)^{-1}\|,$$

$$\chi_n = \|A(n)\| + |\varepsilon|K, \quad n \in Z_0^+.$$

Theorem 3.2. *If the function $F_n(x)$ satisfies the inequality*

$$\|F_n(x) - F_n(\bar{x})\| \leq K\|x - \bar{x}\|,$$

$$\{x, \bar{x}\} \subset \mathfrak{M}; \quad K = const > 0; \quad n \in Z_0^+,$$

then, for any ε such that

$$|\varepsilon| < \frac{1}{KK_1K_2(1 + \sum_{\nu=0}^{N-2} \prod_{i=0}^{\nu} \chi_i)}, \qquad (3.10)$$

Eq. (3.9) has the unique N-periodic solution defined on the interval $[0, +\infty)$.

Proof. Any N-periodic solution $x_n = x(n, x_0), n \in Z_0^+$, of Eq. (3.9) must satisfy the system of equations

$$x_0 = (E - \Omega_0^N)^{-1} \sum_{\nu=0}^{N-1} \Omega_0^N (\Omega_0^{\nu+1})^{-1} (g(\nu) + \varepsilon F_\nu(x_\nu)),$$

$$x_n = \Omega_0^n (E - \Omega_0^N)^{-1} \sum_{\nu=0}^{N-1} \Omega_0^N (\Omega_0^{\nu+1})^{-1} (g(\nu) + \varepsilon F_\nu(x_\nu))$$

$$+ \sum_{\nu=0}^{n-1} \Omega_0^n (\Omega_0^{\nu+1})^{-1} (g(\nu) + \varepsilon F_\nu(x_\nu)), \quad n \in Z^+. \quad (3.11)$$

It is easy also to verify that the equalities

$$x_N = (\Omega_0^N (E - \Omega_0^N)^{-1} + E)(E - \Omega_0^N)x_0 = x_0,$$

$$x_{n+1} = A_n x_n + g(n) + \varepsilon F_n(x_n)$$

hold. In other words, any solution $x_n = x(n, x_0), n \in Z_0^+$, of the system of equations (3.11) is N-periodic and is a solution of Eq. (3.9).

Thus, the set of solutions of the system of equations (3.11) coincides with the set of N-periodic solutions of Eq. (3.9). Obviously, the number of these solutions coincides with the number of points $x_0 \in \mathfrak{M}$ that satisfy the first equation in (3.11).

Consider the operator

$$Lx_0 = (E - \Omega_0^N)^{-1} \sum_{\nu=0}^{N-1} \Omega_0^N (\Omega_0^{\nu+1})^{-1} (g(\nu) + \varepsilon F_\nu(x_\nu)),$$

where x_ν is expressed in terms of x_0 according to (3.9).

It is easy to verify that the operator L transfers $\mathfrak{M}$ in $\mathfrak{M}$, and the inequality

$$\rho(Lx_0, Lz_0) = \|Lx_0 - Lz_0\|$$

$$\leq |\varepsilon| \|\Omega_0^N\| \|(E - \Omega_0^N)^{-1}\| \sum_{\nu=0}^{N-1} \|(\Omega_0^{\nu+1})^{-1}\| \|F_\nu(x_\nu) - F_\nu(z_\nu)\|,$$

where x_0, z_0 are any points from $\mathfrak{M}$, holds. Then

$$\rho(Lx_0, Lz_0) \leq |\varepsilon| K K_1 K_2 \sum_{\nu=0}^{N-1} \|x_\nu - z_\nu\|$$

and

$$\|x_n - z_n\| \leq \|x_0 - z_0\| \prod_{i=0}^{n-1} \chi_i, \quad n \in Z^+.$$

Two last inequalities yield the estimate

$$\rho(Lx_0, Lz_0) \le |\varepsilon| KK_1K_2 (1 + \sum_{\nu=0}^{N-2} \prod_{i=0}^{\nu} \chi_i) \|x_0 - z_0\|,$$

which testifies that, for any ε satisfying inequality (3.10), the operator L is an operator of contraction. By virtue of the completeness of the space $\mathfrak{M}$, it has the single fixed point $x^* \in \mathfrak{M}$. This means that the solution $x_n = x(n, x^*)$, $x_0 = x^*$ of Eq. (3.9) is its unique N-periodic solution. $\square$

Remark 3.1. The solution $x_n = x(n, x^*)$ can be presented in the form of a sum $x(n, x^*) = \varphi_n + y_n$, where y_n is the unique N-periodic solution of the equation

$$y_{n+1} = A(n)y_n + \varepsilon F_n(\varphi_n + y_n), \quad n \in Z_0^+,$$

with the initial value y_0 satisfying the equation

$$y_0 = (E - \Omega_0^N)^{-1} \sum_{\nu=0}^{N-1} \Omega_0^N (\Omega_0^{\nu+1})^{-1} \varepsilon F_\nu(\varphi_\nu + y_\nu).$$

The initial values x^* and y_0 can be sought, by using method of successive approximations. As the initial point, $\varphi_0 \in \mathfrak{M}$ can be taken, for example.

The assertion of the last theorem can be easily extended to the case where $n \in Z$ in Eq. (3.9).

2^0. **Nondegenerate resonance case**. In the resonance case where the matrix $E - \Omega_0^N$ is not invertible, Theorem 3.1 does not hold. Below, we formulate the sufficient conditions of existence of N-periodic solutions of Eq. (3.3) in the resonance case. These conditions are obtained by the method of truncation of this equation. Equation (3.3) truncated to the order $s \in Z^+$ takes the form

$$\overset{(s)}{x}_{n+1} = \overset{(s)}{A}(n) \overset{(s)}{x}_n + \overset{(s)}{g}_n, \quad n \in Z, \tag{3.12}$$

where $\overset{(s)}{x} = (x^1, x^2, \dots, x^s)$, $\overset{(s)}{g}_n = (g_n^1, g_n^2, \dots, g_n^s)$, $\overset{(s)}{A}(n) = [a_{ij}(n)]_{i,j=1}^s$.

Theorem 3.3. *If, for a point $x_0 \in \mathfrak{M}$, there exists a sequence of natural numbers $m_1 < m_2 < \cdots < m_\nu < \dots$ $(m_\nu \to \infty$ as $\nu \to \infty)$ such that, $\forall s = m_\nu, \nu = 1, 2, \dots$, Eq. (3.12) has the N-periodic solution $\overset{(s)}{x} = \overset{(s)}{x}(n, \overset{(s)}{x}_0)$, then Eq. (3.3) has at least one N-periodic solution $x = x(n, x_0)$.*

Proof. It is sufficient to show that, as $\nu \to \infty$, $\overset{(m_\nu)}{x}{}_n \to x_n$ in the coordinatewise meaning. We prove it for the first coordinates of the vectors $\overset{(m_\nu)}{x}{}_n$ and x_n.

For $n = 1$,

$$\lim_{s\to\infty} \overset{(s)}{x}{}_1^1 = \lim_{s\to\infty} \sum_{j=1}^{s} a_{1j}(0)x_0^j + g^1(0) = x_1^1,$$

and $\| \overset{(s)}{x}{}_1 \| \le \|x_0\| \|A(0)\| + \|g(0)\| = P_0 = const$ uniformly in $s = m_\nu, \nu = 1, 2, \ldots$.

Setting $\overset{(s)}{x}{}_1^j = 0$ for $j > s$ and taking into account that the series

$$\sum_{j=1}^{\infty} a_{1j}(1) \overset{(s)}{x}{}_1^j$$

is majorized by the convergent number series

$$P_0 \sum_{j=1}^{\infty} |a_{1j}(1)|,$$

we obtain the equality

$$\lim_{s\to\infty} \overset{(s)}{x}{}_2^1 = \lim_{s\to\infty} \sum_{j=1}^{\infty} a_{1j}(1) \overset{(s)}{x}{}_1^j + g^1(1) = x_2^1,$$

and $\| \overset{(s)}{x}{}_2 \| \le \|A(1)\| P_0 + \|g(1)\| = P_1 = const$ for all $s = m_\nu, \nu = 1, 2, \ldots$.

By the method of complete mathematical induction, we can verify that, for all $n \in Z^+$, the equality

$$\lim_{\nu\to\infty} \overset{(m_\nu)}{x}{}_n = x_n \tag{3.13}$$

holds. In this case,

$$\| \overset{(s)}{x}{}_n \| \le \|g(n-1)\| + \|x_0\| \prod_{\alpha=0}^{n-1} \|A(\alpha)\|$$

$$+ \sum_{i=1}^{n-1} \|g(i-1)\| \prod_{\alpha=i}^{n-1} \|A(\alpha)\|, \quad n \in Z^+, n \neq 1.$$

By means of the analogous reasoning, it is easy to verify that relation (3.13) in the coordinatewise meaning holds also for all $n \in Z^-$. In this case, for all natural s, the estimate

$$\| \overset{(s)}{x}_n \| \leq \| x_0 - g(-1) \| \prod_{\alpha=n}^{-1} \| A^{-1}(\alpha) \|$$

$$+ \sum_{i=-2}^{n} \| g(i) \| \prod_{\alpha=n}^{i} \| A^{-1}(\alpha) \|, \quad n \in Z^-, n \neq -1,$$

holds.

This theorem reduces the problem of the search for an N-periodic solution of Eq. (3.3) to a finite-dimensional case. $\qquad\square$

We now pass to seeking the necessary conditions of existence of N-periodic solutions of Eq. (3.3) in the resonance case.

We denote the set of matrices A from the space $\mathbf{\Gamma}$, for which $\sup_j \sum_{i=1}^{\infty} |a_{ij}| < \infty$, by $\mathbf{B}$ and the set of vectors $x \in \mathfrak{M}$, for which $\sum_{i=1}^{\infty} |x^i| < \infty$, by $\mathbf{m}$.

Consider the equation

$$y_{n+1} = (A^T(n))^{-1} y_n, \quad n \in Z, \tag{3.14}$$

which is conjugate relative to (3.1). Here, the symbol T means the operation of transposition of a matrix.

Lemma 3.2. *Let $\{ A(n), A^{-1}(n) \} \subset \mathbf{B}$ $\quad \forall n \in Z$. Then*

$$(Y_0^n)^T \Omega_0^n = E, \quad n \in Z, \tag{3.15}$$

where Ω_0^n and Y_0^n are the matriciants of Eqs. (3.1) and (3.14), respectively.

Proof. If $y_n = y(n)$ is a solution of Eq. (3.14), then $y_n^T = y^T(n)$ (vector-row) is a solution of the equation

$$y_{n+1}^T A(n) = y_n^T, \quad n \in Z, \tag{3.16}$$

and conversely. Then $(Y_0^n)^T$ is is the matriciant of Eq. (3.16) in the meaning that its any solution can be presented in the form $c^T (Y_0^n)^T$, where c is a constant vector-column from $\mathfrak{M}$. The last means that $(Y_0^{n+1})^T = (Y_0^n)^T A^{-1}(n)$, which yields the representation

$$(Y_0^n)^T = \begin{cases} A^{-1}(0) A^{-1}(1) \cdots A^{-1}(n-1) & \text{for} \quad n > 0; \\ E & \text{for} \quad n = 0; \\ A(-1) A(-2) \cdots A(n) & \text{for} \quad n < 0, \end{cases}$$

which leads to (3.15). $\qquad\square$

Lemma 3.3. *For any solutions $x_n = x(n, x_0)$ and $y_n = y(n, y_0)$ of Eqs. (3.1) and (3.14), respectively, under conditions of Lemma 3.2, the equality $y_n^T x_n = y_0^T x_0$, $n \in Z$, holds if at least one of the initial values x_0, y_0 belongs to the set $\mathbf{m}$.*

The proof follows from the equality $y_n^T x_n = y_0^T (Y_0^n)^T \Omega_0^n x_0$, since Y_0^n and Ω_0^n belong to the set $\mathbf{B}$.

It is easy to see that the relation $x_0 \in \mathbf{m}$ yields the relation $x_n = x(n, x_0) \in \mathbf{m}$ for all $n \in Z$.

Theorem 3.4. *Let Eq. (3.3) satisfy the conditions of Lemma 3.2, let the monodromy matrix of the appropriate homogeneous equation (3.1) be symmetric, and let the unity be its multiplier. Then the necessary condition of existence of an N-periodic solution of Eq. (3.3) is the equality*

$$\sum_{\nu=1}^{N} y_\nu^T g_{\nu-1} = 0, \tag{3.17}$$

where $y_n = y(n)$ is any N-periodic solution of Eq. (3.14), which belongs to the space $\mathbf{m}$.

Proof.   Equation (3.2) with regard for Eq. (3.14) takes the form

$$(E - Y_0^N) y_0 = 0. \tag{3.18}$$

If $y_0 \in \mathfrak{M}$ is its solution, then y_0 is a solution of the equation $y_0^T (Y_0^N)^T = y_0^T$. In view of (3.15), we obtain the equation

$$y_0^T ((\Omega_0^N)^{-1} - E) \Omega_0^N = 0.$$

If the matrix Ω_0^N is symmetric, then the last equation is equivalent to Eq. (3.2). Hence, Eq. (3.18) has the same nontrivial solutions from $\mathfrak{M}$ as Eq. (3.2). Thus, the unity is the multiplier of Eqs. (3.1) and (3.14) with identical multiplicities, and each N-periodic solution of Eq. (3.1) corresponds to an N-periodic solution of Eq. (3.14) with the same initial values $0, x_0 = y_0$.

Let $x_n = x(n)$ be any N-periodic solution of Eq. (3.3), and let $y_n = y(n, y_0)$ be any N-periodic solution of Eq. (3.14) such that $y_0 \in \mathbf{m}$. Then $y_n \in \mathbf{m}$ for all $n \in Z$, and the equality

$$y_{n+1}^T x_{n+1} = y_{n+1}^T (A(n) x_n + g_n) = y_n^T x_n + y_{n+1}^T g_n$$

holds. Whence we have

$$y_{n+1}^T x_{n+1} - y_n^T x_n = y_{n+1}^T g_n. \tag{3.19}$$

Summing equalities (3.19) from $n = 0$ to $n = N - 1$, we obtain the equality

$$y_N^T x_N - y_0^T x_0 = \sum_{\nu=1}^{N} y_\nu^T g_{\nu-1},$$

which yields (3.17). The theorem is proved. $\qquad\square$

In order to study the problems of existence and approximate construction of periodic solutions of difference equations in the resonance case, it is convenient to use the numerical-analytic method developed by A.M. Samoilenko. We write the equation

$$x_{n+1} = x_n + \varepsilon f_n(x_n), \quad n \in Z_0^+, \tag{3.20}$$

where $x \in \mathfrak{M}$, $f_n(x)$ is a function N-periodic in n, which takes values from the set $\mathfrak{M}$ for all x from the domain of its definition, and ε is a real parameter.

We consider that the function $f_n(x)$ is defined on the set D^*,

$$D^* = Z_0^+ \times D = Z_0^+ \times \{x \in \mathfrak{M} | \|x\| \le R = const\},$$

is bounded on this set in the norm by a constant $M > 0$, and satisfies the Lipschitz condition in x uniformly relative to $n \in Z_0^+$

$$\|f_n(x) - f_n(\bar{x})\| \le \xi \|x - \bar{x}\|, \quad \xi = const > 0; \quad \{x, \bar{x}\} \subset D.$$

The monodromy matrix generating Eq. (3.20) of the linear homogeneous equation $x_{n+1} = x_n$, $n \in Z_0^+$, is the identity matrix. Therefore, the matrix $E - \Omega_0^N$ has no inverse matrix, which corresponds to the resonance case.

We denote the expression $\frac{1}{N} \sum_{\nu=0}^{N-1} f_\nu(x_\nu)$ by $\overline{f_\nu(x_\nu)}$ and set a sequence of functions $x_n^{(s)} = x^{(s)}(n, x_0)$, which are N-periodic in n and are defined on the set $\{0, 1, 2, \ldots, N - 1\}$ by the recurrence relation

$$x_0^{(s)} = x_0, \quad x_n^{(0)} = x_0, \quad s \in Z_0^+,$$

$$x_n^{(s)} = x_0 + \varepsilon \sum_{i=0}^{n-1} (f_i(x_i^{(s-1)}) - \overline{f_\nu(x_\nu^{(s-1)})}), \quad s \in Z^+. \tag{3.21}$$

We omit the proofs of the following theorem and its three corollaries, since their substantiation consists in the direct extension of the mentioned method to the case under consideration.

Theorem 3.5. *For any value of the parameter ε such that*

$$0 < |\varepsilon| < \min\{\frac{2R}{MN}; \frac{2}{\xi N}\}$$ (3.22)

and any initial values

$$x_0 \in D_f = \{x \in D | \|x\| < R - \frac{|\varepsilon|MN}{2}\} \subset D,$$

there exists the single control $u \in \mathfrak{M}$ such that the equation

$$x_{n+1} = x_n + \varepsilon f_n(x_n) - u, \quad n \in Z_0^+,$$ (3.23)

has an N-periodic solution $x_n = \tilde{x}(n, x_0), \tilde{x}(0, x_0) = x_0$. Moreover, $u = \overline{\varepsilon f_\nu(\tilde{x}_\nu)}$, and the estimates

$$\|\tilde{x}(n, x_0) - x^{(s)}(n, x_0)\| \leq \sigma(s), \quad \|u - u^{(s)}\| \leq \sigma(s)|\varepsilon|\xi,$$

where $x^{(s)}(n, x_0)$ is determined by the recurrence relation (3.21), and

$$u^{(s)} = \overline{\varepsilon f_\nu(x_\nu^{(s)})}, \quad \sigma(s) = \frac{|\varepsilon|^{s+1}\xi^s(\frac{N}{2})^{s+1}M}{1 - |\varepsilon|\xi\frac{N}{2}} \to 0,$$ (3.24)

hold as $s \to \infty$.

Corollary 3.2. *If Eq. (3.20), in which the parameter ε satisfies condition (3.22), has an N-periodic solution $x_n = x(n), x(0) = x_0 \in D_f$, then this solution is the limit of the sequence, which is uniformly convergent in the norm, of N-periodic functions $\{x_n^{(s)}\}, s \in Z^+$, defined by the recurrence relation (3.21).*

By $\Delta(x_0)$ and $\Delta_s(x_0)$, we denote the expressions $\overline{\varepsilon f_\nu(\tilde{x}_\nu(x_0))}$ and $\overline{\varepsilon f_\nu(x_\nu^{(s)}(x_0))}$, respectively.

It is obvious that Eq. (3.20) has an N-periodic solution $x_n = x(n)$, $x(0) = x_0 \in D_f$ if and only if $\Delta(x_0) = 0$. In this case, $x_n = \tilde{x}(n, x_0)$, where $\tilde{x}(n, x_0)$ is the equation of Eq. (3.23).

Corollary 3.3. *In order that the domain $D_1 \subset D_f$ include the point x^*, at which $\Delta(x^*) = 0$, it is necessary that, for all natural s and any $x_1 \in D_1$, the inequality*

$$\|\Delta_s(x_1)\| \leq |\varepsilon|\xi[\sigma(s) + \frac{2\sup_{x \in D_1}\|x - x_1\|}{2 - |\varepsilon|\xi N}]$$

be satisfied.

A mapping $L : X \subset \mathfrak{M} \to LX \subset \mathfrak{M}$ of a closed set X is called topological, if it is a homeomorphism on $IntX$ and transfers its boundary into the boundary of the image LX.

Corollary 3.4. *Let, for Eq. (3.20), the estimate (3.22) hold, and let the following conditions be satisfied:*

1) for some natural n, there exists a point $x^0 \in D_f$ such that $\Delta_n(x^0) = 0 \in \mathfrak{M}$;

2) there exists a closed set $D_1 \in D_f$ including the point x^0 such that Δ_n maps topologically D_1 onto $\Delta_n D_1$;

3) on the boundary Γ_{D_1} of the set D_1, the inequality

$$\inf_{\Gamma_{D_1}} \|\Delta_n(x)\| \geq |\varepsilon| \xi \sigma(n),$$

where $\sigma(n)$ is given by relation (3.24), is satisfied.

Then Eq. (3.20) has an N-periodic solution $x_n = x(n)$, for which $x_0 = x(0) \in D_1$.

Consider now the equation

$$\overset{(p)}{x}_{n+1} = \overset{(p)}{x}_n + \varepsilon \overset{(p)}{f}_n(\overset{(p)}{x}_n), \quad n \in Z_0^+, \tag{3.25}$$

which is truncated to the p-th order and corresponds to Eq. (3.20), i.e., $\overset{(p)}{x} = (x^1, x^2, \ldots, x^p, 0, 0, \ldots)$. For any natural p, there exists the single control $\overset{(p)}{u}$ such that the solution $\overset{(p)}{x}_n = \overset{(p)}{x}(n, \overset{(p)}{x}_0)$, $\overset{(p)}{x}(0, \overset{(p)}{x}_0) = \overset{(p)}{x}_0$ of the equation

$$\overset{(p)}{x}_{n+1} = \overset{(p)}{x}_n + \varepsilon \overset{(p)}{f}_n(\overset{(p)}{x}_n) - \overset{(p)}{u}, \quad n \in Z_0^+, \tag{3.26}$$

with initial values $0, \overset{(p)}{x}_0$ is N-periodic.

We recall that the function $f_n(x)$ belongs to the space $L_{Lip}(x)$ in a domain D^*, if it satisfies the sharpened Cauchy–Lipschitz condition in x in this domain. In other words, for all $n \in Z_0^+$, the inequality $\|f_n(x) - f_n(\bar{x})\| \leq \delta(s)\|x - \bar{x}\|$ holds, where x and $\bar{x}$ are any points of the domain D, whose first s corresponding coordinates coincide, and $\delta(s) \to 0$ as $s \to \infty$.

We now state the theorem reducing the periodic problem of control for Eq. (3.20) to an analogous problem for Eq. (3.25) in a finite-dimensional space.

Theorem 3.6. *Let condition (3.22) be satisfied, let $x_0 \in D_f$, and let $f_n(x) \in L_{Lip}(x)$ in the domain D^*. Then, coordinatewise,*

$$\lim_{p \to \infty} \overset{(p)}{x}_n = \tilde{x}_n(x_0), \quad \lim_{p \to \infty} \overset{(p)}{u} = u, \tag{3.27}$$

where $\tilde{x}_n(x_0)$ is the N-periodic solution of Eq. (3.23).

Proof. As is known, it is possible to choose a coordinatewise convergent subsequence from every bounded sequence in $\mathfrak{M}$ with the help of the process of diagonalization. For all $p \in Z_0^+$, the following estimates hold:

$$\| \overset{(p)}{u} \| = \lim_{s\to\infty} \| \overset{(p)}{u}{}^{(s)} \| = \lim_{s\to\infty} \frac{|\varepsilon|}{N} \| \sum_{\nu=0}^{N-1} \overset{(p)}{f}{}_\nu(\overset{(p)}{x}{}_\nu^{(s)}) \| \leq |\varepsilon| M,$$

$$\| \overset{(p)}{x}{}_n \| = \| \overset{(p)}{x}{}_0 + \varepsilon \sum_{i=0}^{N-1} (\overset{(p)}{f}{}_i(\overset{(p)}{x}{}_i) - \overline{\overset{(p)}{f}{}_\nu(\overset{(p)}{x}{}_\nu)}) \| \leq \|x_0\| + \frac{MN}{2}|\varepsilon| \leq R.$$

The functions $\overset{(p)}{x}{}_n$ are N-periodic in n. Therefore, we can choose an increasing sequence of indices $\{p_k\}_{k=0}^{+\infty}$ such that the sequences $\{\overset{(p_k)}{u}\}_{k=0}^{+\infty}$ and $\{\overset{(p_k)}{x}{}_n\}_{k=0}^{+\infty}$, $n \in Z_0^+$, converge coordinatewise to some $\tilde{\tilde{u}}$ and $\tilde{\tilde{x}}(n, x_0), n \in Z_0^+$, respectively.

Let $\overset{(p_k)}{x}{}_n^* = (\tilde{x}_n^1, \ldots, \tilde{x}_n^g, \overset{(p_k)}{x}{}_n^{g+1}, \overset{(p_k)}{x}{}_n^{g+2}, \ldots)$. Then

$$\|f_n(\tilde{\tilde{x}}_n) - f_n(\overset{(p_k)}{x}{}_n^*)\| \leq \delta(g)\|\tilde{\tilde{x}}_n - \overset{(p_k)}{x}{}_n\| \leq 2R\delta(g) \to 0 \quad \text{as} \quad g \to +\infty,$$

$$\|f_n(\overset{(p_k)}{x}{}_n) - f_n(\overset{(p_k)}{x}{}_n^*)\| \leq \delta(0) \sup_{1\leq i\leq g} \| \overset{(p_k)}{x}{}_n^i - \tilde{\tilde{x}}_n^i \| \to 0 \quad \text{as} \quad k \to +\infty.$$

This yields

$$\lim_{k\to+\infty} f_n(\overset{(p_k)}{x}{}_n) = f_n(\tilde{\tilde{x}}_n).$$

Let us perform the coordinatewise transition in p_k to the limit as $k \to +\infty$ in (3.26). We obtain the equality

$$\tilde{\tilde{x}}_{n+1} = \tilde{\tilde{x}}_n + \varepsilon f_n(\tilde{\tilde{x}}_n) - \tilde{\tilde{u}}, \quad n \in Z_0^+.$$

By virtue of Theorem 3.5, $\tilde{\tilde{x}}_n = \tilde{x}, \tilde{\tilde{u}} = u$ for all $n \in Z_0^+$.

Consider now a sequence of equalities of the form (3.26) for $p = 1, 2, 3, \ldots$ and its any subsequence

$$\overset{(r)}{x}{}_{n+1} = \overset{(r)}{x}{}_n + \varepsilon \overset{(r)}{f}{}_n(\overset{(r)}{x}{}_n) - \overset{(r)}{u}, \quad n \in Z_0^+, r = 1, 2, 3, \ldots.$$

For this subsequence, the whole above-presented analysis is valid. This means that there exist the subsequences $\{\overset{(l)}{u}\}$ and $\{\overset{(l)}{x}{}_n\}$ of the sequences $\{\overset{(r)}{u}\}_{r=1}^{+\infty}$ and $\{\overset{(r)}{x}{}_n\}_{r=1}^{+\infty}$, respectively, which converge coordinatewise as $l \to \infty$ to the same u and $\tilde{x}(n, x_0) \ \forall n \in Z_0^+$. By analogy with the proof of Theorem 22.1 in [121], this consideration yields obviously relations (3.27). $\square$

Let now $f_n(x_n) = A(n)x_n$, where $A(n)$ an infinite matrix N-periodic in $n \in Z_0^+$, $\|A(n)\| \le M_0 = const$, $|\varepsilon| < \frac{2}{M_0 N}$. Then Eq. (3.20) becomes linear and takes the form

$$x_{n+1} = x_n + \varepsilon A(n)x_n, \quad n \in Z_0^+. \tag{3.28}$$

For $M = M_0 R$ all conditions of Theorem 3.5 are satisfied. Hence, there exists the single control $u \in \mathfrak{M}$ such that the equation

$$x_{n+1} = x_n + \varepsilon A(n)x_n - u, \quad n \in Z_0^+, \tag{3.29}$$

has the unique N-periodic solution $\tilde{x}_n = \tilde{x}(n, x_0)$,

$$\tilde{x}_0 = x_0 \in D_f^0 = \{x \in D | \|x\| \le R - \frac{|\varepsilon| M_0 R N}{2}\},$$

related to the control u by the equality $u = \varepsilon \overline{A(\nu)\tilde{x}_\nu}$.

Equation (3.28) corresponds to the truncated equation

$$\overset{(p)}{x}_{n+1} = \overset{(p)}{x}_n + \varepsilon \overset{(p)}{A}(n) \overset{(p)}{x}_n, \quad n \in Z_0^+,$$

for which there exists the single control $\overset{(p)}{u}$ such that the solution $\overset{(p)}{x}_n = \overset{(p)}{x}(n, \overset{(p)}{x}_0)$ of the equation

$$\overset{(p)}{x}_{n+1} = \overset{(p)}{x}_n + \varepsilon \overset{(p)}{A}(n) \overset{(p)}{x}_n - \overset{(p)}{u}, \quad n \in Z_0^+,$$

is N-periodic.

In the following proposition, we formulate the conditions, under which relation (3.27) is satisfied in the meaning of the norm and indicate the estimate of the accuracy of an approximation of the solution $\tilde{x}_n$ by the function $\overset{(p)}{x}_n$.

Corollary 3.5. *Let* $\|A(n)\| \le M_0$, $|\varepsilon| < \frac{2}{M_0 N}$, *and let a matrix* $A(n)$ *be such that*

$$\|A(n) - \overset{(p)}{A}(n)\| \le \eta(p), \quad \{n, p\} \subset Z_0^+.$$

Then, for the N-periodic solution $\tilde{x}_n = \tilde{x}(n, x_0), x_0 \in D_f^0$, *of Eq. (3.29), whose initial value x_0 satisfies the inequality*

$$\|(0, 0, \ldots, x_0^{p+1}, x_0^{p+2}, \ldots)\| \le \zeta(p),$$

the limiting transition (3.27) holds in the meaning of the norm, if the relation $\gamma(p) = \max\{\eta(p), \zeta(p)\} \to 0$ *is satisfied as $p \to +\infty$. In this case, the following estimates hold:*

$$\|\tilde{x}_n - \overset{(p)}{x}_n\| \le \gamma(p)\frac{2 + |\varepsilon| N R}{2 - |\varepsilon| N M_0};$$

$$\|u - \overset{(p)}{u}\| \le \gamma(p)(|\varepsilon| M \frac{2 + |\varepsilon| N R}{2 - |\varepsilon| N M_0} + R).$$

Proof. Taking into account that, for $x_0 \in D_f$, both the functions $\tilde{x}_n = \tilde{x}(n, x_0)$ and $\overset{(p)}{x}_n = \overset{(p)}{x}(n, \overset{(p)}{x}_0)$ are bounded in the norm by a constant R for all $p \in Z^+$ and $n \in Z_0^+$, we write the inequalities

$$\|\tilde{x}_n - \overset{(p)}{x}_n\| \le \|x_0 - \overset{(p)}{x}_0\| +$$

$$+ |\varepsilon| \left\| \sum_{i=0}^{n-1} \left(A_i \tilde{x}_i - \overset{(p)}{A}_i \overset{(p)}{x}_i - \frac{1}{N} \sum_{\nu=0}^{N-1} (A_\nu \tilde{x}_\nu - \overset{(p)}{A}_\nu \overset{(p)}{x}_\nu) \right) \right\|$$

$$\le \zeta(p) + \frac{|\varepsilon| N}{2} \left(M_0 \sup_{i \in Z_0^+} \|\tilde{x}_i - \overset{(p)}{x}_i\| + \eta(p) R \right).$$

These inequalities yield the estimate

$$\|\tilde{x}_n - \overset{(p)}{x}_n\|_0 \le \frac{2\zeta(p) + |\varepsilon| N R \eta(p)}{2 - |\varepsilon| N M_0}.$$

For the control u, we have

$$\|u - \overset{(p)}{u}\| \le \frac{|\varepsilon|}{N} \sum_{\nu=0}^{N-1} \|A(\nu)\tilde{x}_\nu - \overset{(p)}{A}(\nu) \overset{(p)}{x}_\nu\|$$

$$\le \sum_{\nu=0}^{N-1} (M_0 \|\tilde{x}_n - \overset{(p)}{x}_n\|_0 + \eta(p) R),$$

which completes the proof. $\qquad \square$

We note that the assertion of Corollary 3.5 has no meaning, if there exists $n^0 \in [0, N-1]_Z$ such that the matrix $A(n^0)$ is invertible and $A^{-1}(n^0) \in \mathbf{\Gamma}$.

3^0. **Degenerate case.** If there exists at least one number $n^0 \in [0, N-1]_Z$, for which the matrix $A(n^0)$ is not invertible, or $A^{-1}(n^0)$ does not belong to the set $\mathbf{\Gamma}$, Eqs. (3.1), (3.3), and (3.9) are called degenerate. First, we consider the homogeneous degenerate equation

$$x_{n+1} = A(n)x_n, \quad n \in Z \tag{3.30}$$

corresponding to Eq. (3.1). It is clear that, for it, the different solutions with identical initial values x_l, l can exist. Since $A(n) \in \mathbf{\Gamma}$ for any integer n, these all solutions coincide on the semiaxis $[l, +\infty)$. We denote the set of solutions of Eq. (3.30) defined on the axis $(-\infty, +\infty)$ by G_1 and the set of all its other solutions by G_2.

It is obvious that every element from G_2 originates from one of the sets $X_{s+1} = Im A(s)$, where $s \in Z$, $A(s)$ is a noninvertible matrix, and

$ImA(s)$ is the image of the matrix $A(s)$, if it is considered as the operator of multiplication by elements from $\mathfrak{M}$. This solution on the semiaxis $[s+1, \infty)$ can be presented in the form

$$x(n, x^0_{s+1}) = \Omega^n_{s+1} x^0_{s+1}, \quad n \geq s+1, x^0_{s+1} \in X_{s+1},$$

where Ω^n_l is the matrix defined for $n \geq l \in Z$ by the equalities

$$\Omega^n_l = A(n-1)A(n-2)\cdots A(l), n > l, \quad \Omega^l_l = E.$$

The description of the set of initial values of elements from G_1 is a more complicated task. If, for example, the matrix $A(n)$ is invertible for $n < l$, then all solutions from G_1 are given by the formula $\Omega^n_l x^0_l$, where $x^0_l \in \mathfrak{M}$, and the matrix Ω^n_l is defined for all $n \in Z$, like the case of the nondegenerate equation (3.1), by equality (1.7).

We consider that the matrix $A(n)$ is periodic with period $N \in Z^+$. As an N-periodic solution of Eq. (3.30), we consider such element $x(n)$ from the set G_1, for which $x(n+N) = X(n)$ for all $n \in Z$.

Theorem 3.7. *In order that Eq. (3.30) have the unique N-periodic solution $x(n) = 0$, it is necessary and sufficient that the unity be a regular point of the operator of multiplication of the matrix Ω^N_0 on elements from $\mathfrak{M}$ or belong to its continuous spectrum. In the opposite case, this equation has the infinite number of N-periodic solutions.*

Proof.   It is easy to verify that the initial value $x_0 \in \mathfrak{M}$ determines the N-periodic solution $x = x(n, x_0)$ of Eq. (3.30) if and only if $x_0 \in Ker(\Omega^N_0 - E)$. For $n \geq 0$, we set $x(n, x_0) = \Omega^n_0 x_0$. For any $n < 0$, there exist $p(n) \in [0, N]_Z$ and $k(n) \in Z^+$ such that $p(n) = n + k(n)N$. This allows us to additionally define the solution for $n < 0$, by setting $x(n, x_0) = x(p(n), x_0)$. The constructed solution is N-periodic. Since Eq. (3.30) is degenerate, we cannot exclude the possibility for it to have other solutions with the same initial conditions. But those solutions are not N-periodic and coincide with the constructed N-periodic solution of the given equation for all $n \geq 0$.

Hence, there exist two possibilities corresponding to the conditions of Theorem 3.7:

1) $Ker(\Omega^N_0 - E) = \{0 \in \mathfrak{M}\}$. Equation (3.30) has the unique N-periodic solution $x_n = 0$;

2) $Ker(\Omega^N_0 - E) \neq \{0 \in \mathfrak{M}\}$. Then the unity belongs to the point spectrum of the operator Ω^N_0. It is clear that, in this case, there exist the infinite number of initial values x_0, every of which generates an N-periodic solution of Eq. (3.30). $\square$

We note that Lemma 3.1 for $\rho = 1$ is a partial case of Theorem 3.7, because the multiplier of Eq. (3.1) is simultaneously an eigenvalue of the operator Ω_0^N.

We now write the equation

$$\overset{(s)}{x}_{n+1} = \overset{(s)}{A}(n)\,\overset{(s)}{x}_n, \quad n \in Z, \tag{3.31}$$

which is obtained from Eq. (3.30) by the truncation with respect to x up to the s-th order.

We define the matrix $\overset{(s)}{\Omega}{}_l^n$ for $n \geq l \in Z$ by the equalities

$$\overset{(s)}{\Omega}{}_l^n = \overset{(s)}{A}(n-1)\,\overset{(s)}{A}(n-2)\cdots\overset{(s)}{A}(l), n > l, \quad \overset{(s)}{\Omega}{}_l^l = \overset{(s)}{E}$$

and denote the solution of Eq. (3.31) by $\overset{(s)}{x} = \overset{(s)}{x}(n, \overset{(s)}{x}_0)$.

Let $\|A(n)\| \leq P, \|x_0\| = L$, where P and L are positive constants.

Theorem 3.8. *If* $\overset{(s)}{x}_0 \in Ker(\overset{(s)}{\Omega}{}_0^N - \overset{(s)}{E})$ *for all* $s \in Z^+$, *then* $x_0 \in Ker(\Omega_0^N - E)$. *If, in this case,* $\|A(n) - \overset{(s)}{A}(n)\| \leq \alpha(s), \|x_0 - \overset{(s)}{x}_0\| \leq \beta(s)$ *and*

$$\max\{\alpha(s), \beta(s)\} = \gamma(s) \to 0 \tag{3.32}$$

as $s \to \infty$, *then, uniformly in* $n \in Z$,

$$\|x(n, x_0) - \overset{(s)}{x}(n, \overset{(s)}{x}_0)\| \leq \gamma(s) \max_{r \in \{0,1,\ldots,N-1\}} \{rLP^{r-1} + P^r\} \to 0 \tag{3.33}$$

as $s \to \infty$.

Proof. If $x_0 = 0 \in \mathfrak{M}$, then the assertions of Theorem 3.8 are obvious, and $\|x(n, x_0) - \overset{(s)}{x}(n, \overset{(s)}{x}_0)\| \to 0$ as $s \to \infty$ even if condition (3.32) is not satisfied.

Let $x_0 \neq 0 \in \mathfrak{M}$. To prove the first assertion of the theorem, it is sufficient to show that $\forall\, n \in [0, N-1]_Z$

$$\overset{(s)}{x}(n, \overset{(s)}{x}_0) \to x(n, x_0) \quad \text{as} \quad s \to \infty \tag{3.34}$$

in the coordinatewise meaning.

The proof of relation (3.34) can be made quite analogously to that of equality (1.18) in section 1.1.

To prove estimate (3.33), we write the equality

$$x(n+1, x_0) - \overset{(s)}{x}(n+1, \overset{(s)}{x}_0) = (A(n) - \overset{(s)}{A}(n))x(n, x_0)$$

$$+ \overset{(s)}{A}(n)(x(n, x_0) - \overset{(s)}{x}(n, \overset{(s)}{x}_0)), n \in Z.$$

Using it and the inductive reasoning, we obtain the estimate

$$\|x(r, x_0) - \overset{(s)}{x}(r, \overset{(s)}{x}_0)\| \le r\alpha(s)LP^{r-1} + \beta(s)P^r$$

for $r \in [0, N-1]_Z$. In view of inequality (3.32) and the periodicity of solutions, we pass from this estimate to (3.33), which completes the proof of the theorem. $\qquad\qquad\square$

We now consider the inhomogeneous degenerate equation

$$x_{n+1} = A(n)x_n + g_n, \quad n \in Z, \tag{3.35}$$

where $g_n = g(n)$ is an N-periodic function that, for all $n \in [0, N-1]_Z$, takes values from the space $\mathfrak{M}$.

We set

$$x(0, x_0) = x_0, \quad x(n, x_0) = \Omega_0^n x_0 + \sum_{i=1}^n \Omega_i^n g(i-1), n \ge i.$$

It is easy to verify that the constructed function $x = x(n, x_0)$ is the unique solution of Eq. (3.35) with the initial condition $x(0, x_0) = x_0$ on the semi-axis $[0, +\infty)$. In order that this solution be N-periodic on this set, it is necessary and sufficient that the following equality be satisfied:

$$x_0 = \Omega_0^N x_0 + \sum_{i=1}^N \Omega_i^N g(i-1), \quad N \ge i. \tag{3.36}$$

If the matrix $E - \Omega_0^N$ is invertible (nonresonance case), then equality (3.36) yields the single required initial value

$$x_0 = (E - \Omega_0^N)^{-1} \sum_{i=1}^N \Omega_i^N g(i-1).$$

Then the function

$$x_n = \Omega_0^n (E - \Omega_0^N)^{-1} \sum_{i=1}^N \Omega_i^N g(i-1) + \sum_{i=1}^n \Omega_i^n g(i-1), \quad n \in Z^+,$$

is the unique N-periodic solution of Eq. (3.35) with the initial value x_0 on the right semiaxis $[0, +\infty)$. By continuing it onto the left semiaxis $Z^- = (-\infty, 0)_Z$ by periodicity, we obtain an N-periodic solution of Eq. (3.35).

But if the matrix $E - \Omega_0^N$ is non invertible (resonance case), then the number of N-periodic solutions of Eq. (3.35) is unknown.

The following proposition extends Theorem 3.3 to the degenerate equation (3.35) both in the nonresonance and resonance cases.

Theorem 3.9. *If, for $x_0 \in \mathfrak{M}$, there exists a sequence of natural numbers $m_1 < m_2 < \cdots < m_\nu < \ldots$, where $m_\nu \to \infty$ as $\nu \to \infty$, for which the truncated equation*

$$\overset{(s)}{x}_{n+1} = \overset{(s)}{A}(n)\,\overset{(s)}{x}_n + \overset{(s)}{g}_n, \quad n \in Z,$$

has an N-periodic solution $\overset{(s)}{x} = \overset{(s)}{x}(n, \overset{(s)}{x}_0)\ \forall s = m_\nu, \nu \in Z^+$, then Eq. (3.35) has at least one N-periodic solution $x = x(n, x_0)$.

The proof of this assertion repeats the proof of Theorem 3.3 on the semiaxis $[0, +\infty)$ with the subsequent extension of the function x_n, which was obtained with the help of the limiting transition (3.13), to the left semiaxis.

Then we consider the quasilinear degenerate equation

$$x_{n+1} = A(n)x_n + g_n + \varepsilon F_n(x_n), \quad n \in Z, \tag{3.37}$$

where $F_n(x)$ and ε are the same as those in (3.9).

In the nonresonance case, we have an analog of Theorem 3.2 for Eq. (3.37).

We now introduce the following notation:

$$K_1 = \|(E - \Omega_0^N)^{-1}\| \max_{i \in [0, N-1]_Z} \|\Omega_i^N\|,$$

$$Q_n = \|A(n)\| + |\varepsilon| K, n \in \{0, 1, 2, \ldots\}.$$

Theorem 3.10. *If the function $F_n(x)$ satisfies the inequality*

$$\|F_n(x) - F_n(y)\| \le K\|x - y\|,$$

where $\{x, y\} \subset \mathfrak{M}, K = const > 0, n \in [0, N-1]_Z$, then, for any ε such that

$$|\varepsilon| < \frac{1}{KK_1(1 + \sum_{i=2}^N \prod_{\nu=0}^{i-2} Q_\nu)},$$

Eq. (3.37) has the unique N-periodic solution in the nonresonance case.

Proof. Proof is carried on by the complete analogy with the proof of Theorem 3.2. It is easy to verify that the set of solutions of the system of equations

$$x_0 = (E - \Omega_0^N)^{-1} \sum_{i=1}^N \Omega_i^N (g(i-1) + \varepsilon F_{i-1}(x_{i-1}));$$

$$x_n = \Omega_0^n (E - \Omega_0^N)^{-1} \sum_{i=1}^{N} \Omega_i^N (g(i-1) + \varepsilon F_{i-1}(x_{i-1}))$$

$$+ \sum_{i=1}^{n} \Omega_i^n (g(i-1) + \varepsilon F_{i-1}(x_{i-1}))$$

coincides with the set of N-periodic solutions of Eq. (3.37) on the semiaxis $[0, +\infty)$.

Under conditions of the theorem, the operator

$$Lx_0 = (E - \Omega_0^N)^{-1} \sum_{i=1}^{N} \Omega_i^N (g(i-1) + \varepsilon F_{i-1}(x_{i-1}))$$

has the single fixed point $x^* \in \mathfrak{M}$ in the space $\mathfrak{M}$. In other words, the function $x = x(n, x^*)$, $x_0 = x^*$, is the unique N-periodic solution of Eq. (3.37) on the semiaxis $[0, +\infty)$. It remains to extend it onto the whole axis by periodicity. $\qquad\square$

By completing the subsection, we consider the equation

$$x_{n+1} = (E + \varepsilon A(n))x_n + g(n) + \varepsilon F_n(x_n), \quad n \in Z, \qquad (3.38)$$

whose partial cases are Eqs. (3.20) and (3.28). We assume that the function $F_n(x)$ satisfies the conditions of Theorem 3.10, $\|F_n(x)\| \le M$ for all $x \in \mathfrak{M}, n \in Z$, where M is a positive constant.

We now show that, in all above-considered cases (nonresonance, resonance, nondegenerate, and degenerate ones), it is possible to apply the numerical-analytic method used in the study of Eq. (3.20) to the determination of periodic solutions of an equation of the form (3.38).

Lemma 3.4. *For any $\{a_0, a_1, \ldots, a_{N-1}\} \subset \mathfrak{M}$, $n \in \{0, 1, 2, \ldots, N-1\}$, the inequality*

$$\left\| \sum_{i=0}^{n} (a_i - \bar{a}_\nu) \right\| \le \frac{N}{2} \max_{0 \le i < N} \|a_i\|$$

holds.

Proof of the lemma follows from the relations

$$\left\| \sum_{i=0}^{n} (a_i - \bar{a}_\nu) \right\| = \left\| (1 - \frac{n+1}{N}) \sum_{i=0}^{n} a_i - \frac{n+1}{N} \sum_{i=n+1}^{N-1} a_i \right\|$$

$$\le \frac{2}{N}(N - n - 1)(n + 1) \max_{0 \le i < N} \|a_i\|.$$

On the set $[0, N-1]_Z$, we define a sequence of functions $x_n^{(s)} = x^{(s)}(n, x_0)$, $s \in Z^+$, by the recurrence relation

$$x_n^{(s)} = x_0 + \sum_{i=0}^{n-1} (g_i + \varepsilon(A(i)x_i^{(s-1)} + F_i(x_i^{(s-1)}))$$

$$\overline{- g_\nu + \varepsilon(A(\nu)x_\nu^{(s-1)} + F_\nu(x_\nu^{(s-1)})))},$$

$$x_n^{(0)} \equiv x_0, \tag{3.39}$$

and set $\|g(n)\| \leq G = const < \infty, n \in Z$.

The following assertion is an analog of Theorem 3.5.

Theorem 3.11. *For any ε such that $|\varepsilon| < \frac{2}{N(P+K)}$ and for any initial value $x_0 \in \mathfrak{M}$, there exists the single control $u \in \mathfrak{M}$ such that the equation*

$$x_{n+1} = (E + \varepsilon A(n)) x_n + g(n) + \varepsilon F_n(x_n) - u, \quad n \in Z, \tag{3.40}$$

has an N-periodic solution $\bar{x}_n = \bar{x}(n, x_0), \bar{x}(0, x_0) = x_0$.

In this case, $u = \overline{g_\nu + \varepsilon(A(\nu)\bar{x}_\nu + F_\nu(\bar{x}_\nu))}$, and the following estimates hold:

$$\|\bar{x}(n, x_0) - x^{(s)}(n, x_0)\| \leq \sigma(s),$$

$$\|u - u^{(s)}\| \leq |\varepsilon|(P + K)\sigma(s), \quad n \in [0, N-1]_Z,$$

where $x^{(s)}(n, x_0)$ is determined by relation (3.39),

$$u^{(s)} = \overline{g_\nu + \varepsilon(A(\nu)x_\nu^{(s)} + F_\nu(x_\nu^{(s)}))},$$

$$\sigma(s) = \frac{N(G + |\varepsilon|(P\|x_0\| + M))}{2 - |\varepsilon|N(P + K)} (\frac{|\varepsilon|N(P + K)}{2})^s \to 0, \quad s \to \infty.$$

Proof. We now show that, for all $n \in [0, N-1]_Z$, the sequence $\{x_n^{(s)}\}_{s=0}^{\infty}$ is convergent. Relation (3.39) and Lemma 3.4 yield the inequalities

$$\|x_n^{(s+1)} - x_n^{(s)}\| \leq |\varepsilon|\frac{N}{2}(P + K) \max_{i \in [0, N-1]_Z} \|x_i^{(s)} - x_i^{(s-1)}\|,$$

$$\|x_n^{(1)} - x_n^{(0)}\| \leq \frac{N}{2}(G + |\varepsilon|(P\|x_0\| + M)).$$

Whence we can easily obtain the estimate

$$\max_{i \in [0, N-1]_Z} \|x_i^{(s+1)} - x_i^{(s)}\|$$

$$\leq (|\varepsilon|\frac{N}{2}(P + K))^s (G + |\varepsilon|(P\|x_0\| + M))\frac{N}{2},$$

which testifies to the fundamentality and, hence, to the convergence of this sequence.

We set $\bar{x}_n = \lim_{s\to\infty} x_n^{(s)}$ and pass in (3.39) to the limit as $s \to \infty$. We obtain the equality

$$\bar{x}_n = x_0 + \sum_{i=0}^{n-1}(g_i + \varepsilon(A(i)\bar{x}_i + F_i(\bar{x}_i)) - \overline{g_\nu + \varepsilon(A(\nu)\bar{x}_\nu + F_\nu(\bar{x}_\nu))}).$$

Let us continue $\bar{x}_n$ N-periodically onto the whole axis. It is easy to verify that the construc-ted function $\bar{x}(n, x_0)$ is an N-periodic solution of Eq. (3.40) for $u = \overline{g_\nu + \varepsilon(A(\nu)\bar{x}_\nu + F_\nu(\bar{x}_\nu))}$ with initial values $0, x_0$.

Let now $\tilde{x}_n = \tilde{x}(n, x_0)$ be an N-periodic solution of Eq. (3.40) with initial values $0, x_0$.

Obviously, $\tilde{u} = \overline{g_\nu + \varepsilon(A(\nu)\tilde{x}_\nu + F_\nu(\tilde{x}_\nu))}$ in this case, and the inequality

$$\|\bar{x}_n - \tilde{x}_n\| \le \frac{|\varepsilon|N(P+K)}{2} \max_{i\in[0,N-1]_Z} \|\bar{x}_i - \tilde{x}_i\|, \quad n \in [0, N-1]_Z,$$

holds. This inequality yields $\bar{x}_n = \tilde{x}_n$ for all $n \in Z$. Hence, $\tilde{u} = u$.

Finally, the inequalities

$$\|\bar{x}_n - x_n^{(s)}\| \le \lim_{p\to\infty} \sum_{i=1}^{p} \|x_n^{(s+i)} - x_n^{(s+i-1)}\|$$

$$\le \frac{N(G + |\varepsilon|(P\|x_0\| + M))}{2}\left(\frac{|\varepsilon|N(P+K)}{2}\right)^s \sum_{i=0}^{\infty}\left(\frac{|\varepsilon|N(P+K)}{2}\right)^i,$$

$$\|u - u^{(s)}\| \le |\varepsilon|(P+K) \max_{i\in[0,N-1]_Z} \|\bar{x}_i - x_i^{(s)}\|$$

yield the estimates indicated in the statement of Theorem 3.11. $\qquad\square$

Corollary 3.6. *For any ε such that $|\varepsilon| < \frac{2}{N(P+K)}$, Eq. (3.38) has an N-periodic solution with initial values $0, x_0 \in \mathfrak{M}$ if and only if*

$$u = \overline{g_\nu + \varepsilon(A(\nu)\bar{x}_\nu + F_\nu(\bar{x}_\nu))} = 0,$$

where $\bar{x}(\nu, x_0)$ is defined by Theorem 3.11, and $x^{(s)}(\nu, x_0)$ is determined by relation (3.39).

We now write the truncated equation

$$\overset{(p)}{x}_{n+1} = (\overset{(p)}{E} + \varepsilon \overset{(p)}{A}(n))\overset{(p)}{x}_n + \overset{(p)}{g}(n) + \varepsilon \overset{(p)}{F}_n(\overset{(p)}{x}_n), \quad n \in Z,$$

corresponding to Eqs. (3.38). Here,

$$\overset{(p)}{F}(\overset{(p)}{x}) = \{F^1(x^1, x^2, \ldots, x^p, 0, 0, \ldots), \ldots, F^p(x^1, x^2, \ldots, x^p, 0, 0, \ldots)\}.$$

By Theorem 3.11 for each $p \in Z^+$, there exists the single control $\overset{(p)}{u} \in R^p$ such that the solution $\overset{(p)}{x}_n = \overset{(p)}{x}(n, \overset{(p)}{x}_0)$, $\overset{(p)}{x}(0, \overset{(p)}{x}_0) = \overset{(p)}{x}_0$ of the equation

$$\overset{(p)}{x}_{n+1} = (E + \varepsilon \overset{(p)}{A}(n)) \overset{(p)}{x}_n + \overset{(p)}{g}(n) + \varepsilon \overset{(p)}{F}_n(\overset{(p)}{x}_n) - \overset{(p)}{u}, \quad n \in Z,$$

with initial values 0, $\overset{(p)}{x}_0$ is N-periodic.

Below, we formulate an analog of Theorem 3.6.

Theorem 3.12. *Let the conditions of Theorem 3.11 be satisfied, and let there exist a function $\delta(s), s \in Z^+$, such that $\delta(s) \to 0$ as $s \to \infty$, and $\|F_n(x_1) - F_n(x_2)\| \leq \delta(s)\|x_1 - x_2\|$ for all $n \in Z, \{x_1, x_2\} \subset \mathfrak{M}$, if the first s corresponding coordinates x_1 and x_2 coincide. Then, coordinatewise,*

$$\overset{(p)}{x}(n, \overset{(p)}{x}_0) \to \bar{x}(n, x_0), n \in Z, \quad \overset{(p)}{u} \to u$$

as $p \to \infty$.

The proof is carried on analogously to that of Theorem 3.6.

Corollary 3.7. *Set $R > \frac{NG}{2}$. Then, in order that Theorem 3.11 be valid, it is sufficient to require the boundedness and the Lipschitz property of the function $F_n(x)$ on the set $D = \{x \in \mathfrak{M}|\|x\| \leq R\}$, to choose the initial value x_0 from the set $D_f^0 = \{x \in \mathfrak{M}|\|x\| \leq \frac{1}{2}(R(2 - |\varepsilon|NP) - N(G + |\varepsilon|M))\}$, and to subordinate the parameter ε to the condition*

$$|\varepsilon| < \min\{\frac{2}{N(P+K)}; \frac{2R - NG}{N(RP + M)}\}.$$

By Γ_D, we denote the boundary of the set D and present the sufficient conditions of existence of the N-periodic solution of Eq. (3.38) without any proof. The following proposition is an analog of Corollary 3.4.

Theorem 3.13. *Let the conditions of Corollary 3.7 be satisfied. If*

1) $u^{(s)}(x^0) = 0$ for some $x^0 \in D_f^0, s \in Z^+$;

2) there exists a closed set $D_1 \subset D_f^0$, which is topologically mapped by the function $u^{(s)}$ onto $u^{(s)}D_1$;

3) $\|u^{(s)}(x)\| \geq |\varepsilon|(P + K)\sigma(s)$ for all $x \in \Gamma_{D_1}$, where the function $\sigma(s)$ is defined in Theorem 3.11,

then there exists $x_0 \in D_1$ such that Eq. (3.37) has an N-periodic solution with initial values $0, x_0$.

3.2   Periodic solutions of nonlinear difference equations of the first order in an abstract Banach space

In a Banach space $\mathbf{W}$ with norm $\| * \|$, we consider the equation

$$x_{n+1} = x_n + f_n(x_n), \quad n \in Z, \tag{3.41}$$

where $f : (n, x) \to f_n(x)$ is a function, which is N-periodic in n and defined on the set $D_* = Z \times D$, and $D = \{x \in \mathbf{W} | \|x\| \le R\}$ is a closed ball in $\mathbf{W}$ with radius R.

We denote the set $\{0, 1, 2, \ldots, N - 1\}$ by Z_N and the average $\frac{1}{N} \sum_{\nu=0}^{N-1} a_\nu$, where $a_\nu \in \mathbf{W}$ $\forall \nu \in Z_N$, by $\bar{a}_\nu$. If $x(n)$ is a solution of Eq. (3.41), which is defined on the set $Z_N \bigcup \{N\}$ and satisfies the condition $x(N) = x(0)$, then it is the restriction of the N-periodic solution of the mentioned equation, which is defined on the set Z, on the set $Z_N \bigcup \{N\}$.

We set a sequence of functions $x_n^{(s)} = x^{(s)}(n, x_0)$ on the set Z_N by the recurrence relation

$$x_0^{(s)} = x_0, \quad x_n^{(0)} = x_0, \quad s \in Z_0^+,$$

$$x_n^{(s)} = x_0 + \sum_{i=0}^{n-1} (f_i(x_i^{(s-1)}) - \overline{f_\nu(x_\nu^{(s-1)})}), \quad s \in Z^+. \tag{3.42}$$

The following conditions are called conditions (A):

1) for all $\{x, \tilde{x}\} \subset D, n \in Z$, the inequalities

$$\|f_n(x)\| \le M, \quad \|f_n(x) - f_n(\tilde{x})\| \le \xi \|x - \tilde{x}\|,$$

where M and ξ are positive constants, are satisfied;

2) the set $\bar{D}_f = \{x \in D | \|x\| \le R - \frac{MN}{2}\}$ is nonempty.

Theorem 3.14. *Let conditions (A) and $\frac{\xi N}{2} < 1$ be satisfied. Then, $\forall x_0 \in \bar{D}_f$, there exists the single control $u \in \mathbf{W}$ such that the equation*

$$x_{n+1} = x_n + f_n(x_n) - u, \quad n \in Z, \tag{3.43}$$

has an N-periodic solution $x_n = \tilde{x}(n, x_0), \tilde{x}(0, x_0) = x_0$. In this case, $u = \overline{f_\nu(\tilde{x}_\nu)}$, and the following estimates hold:

$$\|\tilde{x}(n, x_0) - x^{(s)}(n, x_0)\| \le \sigma(s), \quad \|u - u^{(s)}\| \le \sigma(s)\xi,$$

where $x^{(s)}(n, x_0)$ is determined by the recurrence relation (3.42), and

$$u^{(s)} = \overline{f_\nu(x_\nu^{(s)})}, \quad \sigma(s) = \frac{\xi^s (\frac{N}{2})^{s+1} M}{1 - \xi \frac{N}{2}} \to 0 \tag{3.44}$$

as $s \to \infty$.

Proof.   For any $x_0 \in \bar{D}_f$, the functions $x_n^{(s)}(x_0)$ are defined properly, and their values do not leave the set D. Indeed, for $x_n^{(0)}$, this assertion is obvious. If it is valid for $x_n^{(s)}$, then it also holds for $x_n^{(s+1)}$, because the inequality

$$\|\sum_{i=0}^{n}(a_i - \bar{a}_\nu)\| \leq \frac{N}{2}\max_{i \in Z_N}\|a_i\|, \quad a_i \in \mathbf{W}, n \in Z_N,$$

ensures the validity of the estimate $\|x_n^{(s+1)} - x_0\| \leq \frac{MN}{2}$, and $\|x_0\| \leq R - \frac{MN}{2}$.

The inequalities

$$\|x_n^{(s+1)} - x_n^{(s)}\| \leq \frac{N}{2}\xi \max_{i \in Z_N}\|x_i^{(s)} - x_i^{(s-1)}\|,$$

$$\|x_n^{(1)} - x_n^{(0)}\| \leq \frac{NM}{2}$$

yield the estimate

$$\max_{i \in Z_N}\|x_i^{(s+1)} - x_i^{(s)}\| \leq (\xi\frac{N}{2})^s\frac{NM}{2},$$

which testifies to the fundamentality and, hence, to the convergence of the sequence $\{x_n^{(s)}\}_{s=0}^{\infty} \ \forall n \in Z_N$ in the space $\mathbf{W}$.

Setting $\tilde{x}_n = \lim_{s \to \infty} x_n^{(s)}$ and passing in (3.42) to the limit as $s \to \infty$, we obtain the equality

$$\tilde{x}_n = x_0 + \sum_{i=0}^{n-1}(f_i(\tilde{x}_i) - \overline{f_\nu(\tilde{x}_\nu)}). \tag{3.45}$$

Let us continue $\tilde{x}_n$ onto $Z\ N$-periodically. Then the last equality is satisfied for all $n \in Z$. This means that the constructed function $\tilde{x}(n, x_0)$ is an N-periodic solution of Eq. (3.43) for $u = \overline{f_\nu(\tilde{x}_\nu)}$ with initial values $0, x_0$.

The uniqueness of the control $u \in \mathbf{W}$, for which the solution of Eq. (3.43) with initial values $0, x_0$ is N-periodic, can be easily proved by contradiction.

Finally, the inequalities

$$\|\tilde{x}_n - x_n^{(s)}\| \leq \lim_{p \to \infty}\sum_{i=1}^{p}\|x_n^{(s+i)} - x_n^{(s+i-1)}\|$$

$$\leq \frac{MN}{2}(\frac{\xi N}{2})^s\sum_{i=0}^{\infty}(\frac{\xi N}{2})^i,$$

$$\|u - u^{(s)}\| \leq \xi \max_{i \in Z_N}\|\tilde{x}_i - x_i^{(s)}\|$$

yield the estimates indicated in the statement of Theorem 3.14. $\square$

We now set the mappings Δ and Δ_s on $\bar{D}_f$ as follows:
$\Delta(x_0) = \overline{f_\nu(\tilde{x}_\nu(x_0))}$, $\Delta_s(x_0) = \overline{f_\nu(x_\nu^{(s)}(x_0))}$, where $\tilde{x}_\nu(x_0)$ is defined by Theorem 3.14, and $x_\nu^{(s)}(x_0)$ is given by relation (3.42).

It is clear that Eq. (3.41) has an N-periodic solution $x_n = x(n)$, $x(0) = x_0 \in \bar{D}_f$ if and only if $\Delta(x_0) = 0$. In this case, $x_n = \tilde{x}(n, x_0)$.

The following assertion defines the necessary and sufficient conditions of existence of the solution of the equation $\Delta(x) = 0$ in some closed ball $S \subset \bar{D}_f$.

Theorem 3.15. *Under conditions of Theorem 3.14, the equation $\Delta(x) = 0$ has the solution $x^* \in S$ if and only if there exists a sequence $\{x_p\} \subset S$ convergent to x^* and such that $\Delta_p(x_p) \to 0$ as $p \to \infty$.*

Proof. First, we show that the mapping Δ is uniformly continuous on $\bar{D}_f$. With regard for (3.45) for any $\{x_1, x_2\} \subset \bar{D}_f$, we have

$$\|\tilde{x}(n, x_1) - \tilde{x}(n, x_2)\| \leq \|x_1 - x_2\| + \frac{\xi N}{2} \max_{0 \leq i < N} \|\tilde{x}(i, x_1) - \tilde{x}(i, x_2)\|,$$

whence

$$\|\tilde{x}(n, x_1) - \tilde{x}(n, x_2)\| \leq \frac{\|x_1 - x_2\|}{1 - \frac{\xi N}{2}},$$

$$\|\Delta(x_1) - \Delta(x_2)\| = \frac{1}{N} \| \sum_{\nu=0}^{N-1} (f_\nu(\tilde{x}(\nu, x_1)) - f_\nu(\tilde{x}(\nu, x_2)))\|$$

$$\leq \frac{\xi}{N} \sum_{\nu=0}^{N-1} \|\tilde{x}(\nu, x_1) - \tilde{x}(\nu, x_2)\| \leq \frac{\xi}{1 - \frac{\xi N}{2}} \|x_1 - x_2\|. \quad (3.46)$$

Let there exist a sequence $\{x_p\}$ indicated in the statement of the theorem. With regard for Theorem 3.14 and estimates (3.46), we write the inequalities

$$\|\Delta(x^*) - \Delta_p(x_p)\| \leq \|\Delta(x^*) - \Delta(x_p)\|$$

$$+ \|\Delta(x_p) - \Delta_p(x_p)\| \leq \frac{\xi}{1 - \frac{\xi N}{2}} \|x^* - x_p\| + \sigma(p)\xi,$$

which yield the equality $\Delta(x^*) = \lim_{p \to \infty} \Delta_p(x_p)$.

Necessity. Let $\Delta(x^*) = 0$ and $\{x_p\}$ be any sequence from S, which converges to $x^* \in S$. Then the continuity of the mapping Δ implies that $\|\Delta(x^*) - \Delta(x_p)\| \to 0$ as $p \to \infty$, i.e., $\|\Delta(x_p)\| \to 0$ as $p \to \infty$. But

$$\|\Delta_p(x_p)\| \leq \|\Delta(x_p)\| + \sigma(p)\xi,$$

where $\sigma(p) \to 0$ as $p \to \infty$, which completes the proof. $\qquad\square$

The following assertion can help us to seek the zeros of the mapping Δ.

Corollary 3.8. *In order that some subset $X \subset \bar{D}_f$ contain at least one zero of the mapping $\Delta(x)$, it is necessary that, for all natural s and any $x \in X$, the inequality*

$$\|\Delta_s(x)\| \leq \xi[\sigma(s) + \frac{d(X)}{1 - \frac{\xi N}{2}}]$$

be satisfied. Here, $d(X)$ is the diameter of the set X, and $\sigma(s)$ is determined by equality (3.44).

The proof of this corollary is easy.

We now formulate the sufficient conditions of existence of the solution of the equation $\Delta(x) = 0$ in some subset from $\bar{D}_f$ in the form of the following proposition.

Theorem 3.16. *Let the conditions of Theorem 3.14 be satisfied, let the mapping $\Delta(x)$ be Fréchet-differentiable in some domain $G_f \subset \bar{D}_f$, and let the derivative Δ' satisfy the Lipschitz condition*

$$\|\Delta'(x_1) - \Delta'(x_2)\| \leq L\|x_1 - x_2\|, \quad L = const > 0$$

in G_f. Let, in addition, there exist the inverse mapping $[\Delta'(x^0)]^{-1}$ at the point $x^0 \in G_f$, and

$$N_0 = \|[\Delta'(x^0)]^{-1}\|, \quad k = \|[\Delta'(x^0)]^{-1}\Delta(x^0)\|, \quad h = N_0 k L.$$

Then, if $h < 0.25$, and the closed ball $B(x^0, kt_0) \subset G_f$, where t_0 is the least root of the equation $ht^2 - t + 1 = 0$, then

1) the equation $\Delta(x) = 0$ has the unique solution x^* in this ball, and the sequence $\{x_n\}$, which is determined by the recurrence relation

$$x_{n+1} = x_n - [\Delta'(x^0)]^{-1}\Delta(x_n)$$

with zero approximation x^0, converges to this solution;

2) $\Delta_n(x_n) \to 0$ as $n \to \infty$;

3) for all $n \in Z$, the estimate

$$\|\tilde{x}(n, x^*) - x^{(p)}(n, x_p)\| \leq \frac{1}{1 - \frac{\xi N}{2}} \, \frac{q^p}{1 - q} \|[\Delta'(x^0)]^{-1}\Delta(x^0)\| + \sigma(p), \quad (3.47)$$

holds. Here, $\sigma(p)$ is determined by equality (3.44), and $q = \frac{1 - \sqrt{1 - 4h}}{2} < 0.5$.

Proof.   Item 1 of the assertion of Theorem 3.16 follows directly from [43], where a modified Newton–Kantorovich method of solution of operator equations is substantiated, and

$$\|x_n - x^*\| \leq \frac{q^n}{1 - q}\|[\Delta'(x^0)]^{-1}\Delta(x^0)\|. \tag{3.48}$$

Theorem 3.15 guarantees the existence of a sequence $\{y_n\}$, $n \in Z_0^+$, which is convergent in the norm to x^* as $n \to \infty$, in the ball $B(x^0, kt_0)$. Then, in this process, $\|x_n - y_n\| \to 0$. The relations

$$\|\Delta_n(x_n) - \Delta_n(y_n)\| \leq \|\Delta_n(x_n) - \Delta(x_n)\|$$
$$+ \|\Delta(y_n) - \Delta_n(y_n)\| + \|\Delta(x_n) - \Delta(y_n)\|$$
$$\leq 2\xi\sigma(n) + \frac{\xi}{1 - \frac{\xi N}{2}}\|x_n - y_n\| \to 0 \quad \text{for} \quad n \to \infty$$

prove item 2. Finally, inequalities (3.48) and

$$\|\tilde{x}(n, x^*) - x^{(m)}(n, x_p)\| \leq \frac{\|x^* - x_p\|}{1 - \frac{\xi N}{2}} + \sigma(m), \quad n \in Z, m \in Z^+, \tag{3.49}$$

give estimate (3.47), which completes the proof. $\qquad\qquad\square$

We denote the identity operator by E, the difference $f(x_1) - f(x_2)$ by $f(x)|_{x_2}^{x_1}$, and a ρ-neighborhood of the set D by D_ρ, i.e., $D_\rho = \{x \in \mathbf{W}\,|\,\|x\| < R + \rho\}$.

Lemma 3.5. *Let the conditions of Theorem 3.14 be satisfied, and let the function $f_n(x)$ be Fréchet-differentiable in a domain D_ρ $\forall n \in Z_N$. Moreover, $\forall\{x, x_1, x_2\} \subset D_\rho$ uniformly in $n \in Z_N$, the inequalities*

$$\left\|\frac{df_n(x)}{dx}\right\| \leq P = const > 0;$$

$$\left\|\frac{df_n(x)}{dx}\right|_{x_2}^{x_1}\right\| \leq L_0\|x_1 - x_2\|, \quad L_0 = const > 0;$$

$$PN < 2$$

hold. Then the functions $x_n^{(s)}(x)$, which are defined by relation (3.42), are Fréchet-differentiable in the domain

$$D_f = \{x \in \bar{D}_f\,|\,\|x\| < R - \frac{MN}{2}\}$$

for all $s \in Z_0^+$ and $n \in Z_N$. Moreover, $\forall\{x, x_1, x_2\} \subset D_f$ uniformly in $n \in Z_N$, the following estimates hold:

$$\left\|\frac{dx_n^{(s)}(x)}{dx}\right\| \leq \frac{2}{2 - NP}; \tag{3.50}$$

$$\left\|\frac{dx_n^{(s)}(x)}{dx}\Big|_{x_2}^{x_1}\right\| \leq \frac{NL_0}{2}\Big\{\frac{8}{(2-PN)^2(2-\xi N)} + (\frac{PN}{2})^{s-1}\Big\}\|x_1 - x_2\|. \quad (3.51)$$

Proof. For $s = 0$, relation (3.50) is obvious. By the method of complete mathematical induction, we prove that, for all $s \in Z^+$ and $n \in Z_N$,

$$\left\|\frac{dx_n^{(s)}(x)}{dx}\right\| \leq \sum_{i=0}^{s}(\frac{NP}{2})^i. \quad (3.52)$$

For $s = 1$, we have

$$\frac{dx_n^{(1)}(x)}{dx} = E + \sum_{i=0}^{n-1}(\frac{df_i(x)}{dx} - \frac{\overline{df_\nu(x)}}{dx}),$$

whence

$$\left\|\frac{dx_n^{(1)}(x)}{dx}\right\| \leq 1 + \frac{N}{2}\max_{i\in Z_N}\left\|\frac{df_i(x)}{dx}\right\| = 1 + \frac{NP}{2}.$$

Assuming that inequality (3.52) holds for all $1 < s \leq k$, we obtain the estimates

$$\left\|\frac{dx_n^{(k+1)}(x)}{dx}\right\| \leq 1 + \frac{NP}{2}\max_{i\in Z_N}\left\|\frac{dx_i^{(k)}(x)}{dx}\right\|$$

$$\leq 1 + \frac{NP}{2}\sum_{i=0}^{k}(\frac{NP}{2})^i = \sum_{i=0}^{k+1}(\frac{NP}{2})^i,$$

which prove inequality (3.52) for all natural s. This ensures the validity of estimate (3.50) for $PN < 2$.

Analogously, we will show that the inequality

$$\left\|\frac{dx_n^{(s)}(x)}{dx}\Big|_{x_2}^{x_1}\right\|$$

$$\leq \frac{NL_0}{2}\Big\{\frac{4}{(2-NP)(2-\xi N)}\sum_{i=0}^{s-2}(\frac{PN}{2})^i + (\frac{PN}{2})^{s-1}\Big\}\|x_1 - x_2\| \quad (3.53)$$

holds $\forall s \in \{2, 3, 4, \dots\}$ uniformly in $n \in Z_N$. Indeed, for $s = 1$,

$$\left\|\frac{dx_n^{(1)}(x)}{dx}\Big|_{x_2}^{x_1}\right\|$$

$$= \left\|\sum_{i=0}^{n-1}\{\frac{df_i(x)}{dx}\Big|_{x_2}^{x_1} - \frac{\overline{df_\nu(x)}}{dx}\Big|_{x_2}^{x_1}\}\right\| \leq \frac{L_0 N}{2}\|x_1 - x_2\|. \quad (3.54)$$

In view of the relation

$$\|x_n^{(s)}(x_1) - x_n^{(s)}(x_2)\| \leq \|x_1 - x_2\| + \frac{N}{2}\max_{i\in Z_n}\|f_i(x_i^{s-1}(x_1))$$

$$- f_i(x_i^{s-1}(x_2))\| \leq \sum_{i=0}^{s}(\frac{N\xi}{2})^i\|x_1 - x_2\| = \frac{\|x_1 - x_2\|}{1 - \frac{N\xi}{2}}$$

and inequality (3.54), we have, for $s = 2$,

$$\|\frac{dx_n^{(2)}(x)}{dx}|_{x_2}^{x_1}\| \leq \frac{N}{2}\max_{i\in Z_n}\|(\frac{df_i(x_i^1(x))}{dx}\frac{dx_i^{(1)}(x)}{dx})|_{x_2}^{x_1}\|$$

$$\leq \frac{N}{2}\max_{i\in Z_n}\{L_0\|x_i^{(1)}(x)|_{x_2}^{x_1}\|\|\frac{dx_i^{(1)}(x_1)}{dx_1}\| + \frac{PNL_0}{2}\|x_1 - x_2\|\}$$

$$\leq \frac{NL_0}{2}\{\frac{4}{(2 - NP)(2 - \xi N)} + \frac{PN}{2}\}\|x_1 - x_2\|.$$

By assuming that inequality (3.53) holds for $s \leq k$, we obtain the estimates

$$\|\frac{dx_n^{(k+1)}(x)}{dx}|_{x_2}^{x_1}\| \leq \frac{N}{2}\max_{i\in Z_n}\{L_0\|x_i^{(k)}(x)|_{x_2}^{x_1}\|\|\frac{dx_i^{(k)}(x_1)}{dx_1}\|$$

$$+ \|$$

$$fracdf_i(x_i^k(x_2))dx_i^{(k)}(x_2)\|\|\frac{dx_i^{(k)}(x)}{dx}|_{x_2}^{x_1}\|\}$$

$$\leq \frac{NL_0}{2}\{\frac{\frac{2}{2-NP}}{1 - \frac{\xi N}{2}}\sum_{i=0}^{k-1}(\frac{PN}{2})^i + (\frac{PN}{2})^k\}\|x_1 - x_2\|,$$

which prove inequality (3.51) $\forall s \in \{2, 3, 4, \ldots\}$. For $s = 0$ and $s = 1$, this inequality is proved directly. Lemma 3.5 is proved. $\qquad\square$

Lemma 3.6. *Under conditions of Lemma 3.5, the function $\tilde{x}_n(x)$ defined in Theorem 3.14 is Fréchet-differentiable on D_f for all $n \in Z_N$.*

Proof. It is sufficient to show that, $\forall n \in Z_N$, the sequence $\{\frac{dx_n^{(s)}(x)}{dx}\}_{s=1}^{\infty}$ converges as $s \to \infty$ uniformly with respect to $x \in D_f$. This sequence belongs to the space of linear bounded operators $\mathbf{L}(\mathbf{W}, \mathbf{W})$, which is complete. Therefore, it is sufficient to prove its fundamentality uniform with respect to $x \in D_f$.

We consider that $x \in D_f, n \in Z_N$ and introduce the following notation:

$$r_n^{(s)}(x) = \frac{dx_n^{(s)}(x)}{dx} - \frac{dx_n^{(s-1)}(x)}{dx};$$

$$p_n^{(s)}(x) = \frac{df_n(x_n^{(s)}(x))}{dx} - \frac{df_n(x_n^{(s-1)}(x))}{dx};$$

$$\tilde{\gamma} = \frac{L_0 N^2 M}{2(2 - NP)}.$$

Then

$$r_n^{(s)}(x) = \sum_{i=0}^{n-1} \{p_i^{(s-1)}(x) - \overline{p_\nu^{(s-1)}(x)}\},$$

whence

$$\|r_n^{(s)}(x)\| \le \frac{N}{2} \max_{i \in Z_N} \|p_i^{(s-1)}(x)\| \tag{3.55}$$

for any $s = 2, 3, 4, \ldots$. But

$$\|p_i^{(s-1)}(x)\| = \|\frac{df_i(x_i^{(s-1)}(x))}{dx_i^{(s-1)}(x)} \frac{dx_i^{(s-1)}(x)}{dx}$$

$$- \frac{df_i(x_i^{(s-2)}(x))}{dx_i^{(s-2)}(x)} \frac{dx_i^{(s-2)}(x)}{dx}\| \le P\|r_i^{(s-1)}(x)\| + \frac{L_0 NM}{2 - NP}(\frac{\xi N}{2})^{s-2}.$$

Whence, with regard for (3.55), we obtain the estimate

$$\|r_n^{(s)}(x)\| \le \frac{PN}{2} \max_{i \in Z_N} \|r_i^{(s-1)}(x)\| + \tilde{\gamma}(\frac{\xi N}{2})^{s-2}, \tag{3.56}$$

which holds for all $s = 2, 3, 4, \ldots$. For $s = 1$, the estimate $\|r_n^{(1)}\| \le \frac{NP}{2}$ can be directly verified.

For $s = 3, 4, 5, \ldots$, the inductive inequality (3.56) leads to the estimate

$$\|r_n^{(s)}(x)\| \le (\frac{PN}{2})^{s-2}[(\frac{PN}{2})^2 + \tilde{\gamma}] + \tilde{\gamma} \sum_{i=0}^{s-3} (\frac{PN}{2})^i (\frac{\xi N}{2})^{s-2-i},$$

which can be presented in the form

$$\|r_n^{(s)}(x)\| \le (\frac{PN}{2})^{s-2}[(\frac{PN}{2})^2 + \tilde{\gamma} + \tilde{\gamma} \sum_{i=0}^{s-3} (\frac{PN}{2})^{i-s+2} (\frac{\xi N}{2})^{s-2-i}].$$

In view of the convexity of the set D_ρ without any loss of generality, we can consider that $\xi < P$. Then

$$\sum_{i=0}^{s-3} (\frac{PN}{2})^{i-s+2} (\frac{\xi N}{2})^{s-2-i} = \sum_{i=0}^{s-3} (\frac{\xi}{P})^{s-i-2}$$

$$= \sum_{i=1}^{s-2} (\frac{\xi}{P})^i < \sum_{i=1}^{\infty} (\frac{\xi}{P})^i = \frac{\xi}{P - \xi}.$$

Thus, for $s = 3, 4, 5, \ldots,$

$$\|r_n^{(s)}(x)\| \leq \left(\frac{PN}{2}\right)^{s-2} K^*,$$

where $K^* = \left(\frac{PN}{2}\right)^2 + \frac{\tilde{\gamma}P}{P-\xi} = const.$

The proof of the lemma is completed, since $\frac{NP}{2} < 1.$ $\square$

Lemma 3.7. *Under conditions of Lemma 3.6, the mapping* $\Delta(x) = \overline{f_\nu(\tilde{x}_\nu(x))}$ *is Fréchet-differentiable on* D_f, *and,* $\forall\{x_1, x_2\} \subset D_f$, *the inequality*

$$\left\|\frac{d\Delta(x)}{dx}\big|_{x_2}^{x_1}\right\| \leq L\|x_1 - x_2\|$$

holds. Here,

$$L = \frac{4L_0}{(2 - NP)(2 - \xi N)}\left\{1 + \frac{PN}{2 - PN}\right\} = const.$$

Proof. Taking Lemmas 3.5 and 3.6 into account, it is easy to verify that, $\forall n \in Z_N$ and $\forall\{x_1, x_2\} \subset D_f$, the following relations hold:

$$\left\|\frac{d\tilde{x}_n(x)}{dx}\right\| \leq \frac{2}{2 - NP};$$

$$\left\|\frac{d\tilde{x}_n(x)}{dx}\big|_{x_2}^{x_1}\right\| \leq \frac{4NL_0}{(2 - PN)^2(2 - \xi N)}\|x_1 - x_2\|;$$

$$\frac{d\Delta(x)}{dx} = \frac{1}{N}\sum_{\nu=0}^{N-1}\frac{df_\nu(\tilde{x}_\nu(x))}{d\tilde{x}_\nu(x)}\frac{d\tilde{x}_\nu(x)}{dx}.$$

Using them, we obtain the inequalities

$$\left\|\frac{d\Delta(x)}{dx}\big|_{x_2}^{x_1}\right\| \leq \frac{1}{N}\sum_{\nu=0}^{N-1}\left\{\left\|\frac{df_\nu(x)}{dx}\big|_{\tilde{x}_\nu(x_2)}^{\tilde{x}_\nu(x_1)}\right\|\left\|\frac{d\tilde{x}_\nu(x_1)}{dx_1}\right\|\right.$$

$$\left. + \left\|\frac{df_\nu(\tilde{x}_\nu(x_2))}{d\tilde{x}_\nu(x_2)}\right\|\left\|\frac{d\tilde{x}_\nu(x)}{dx}\big|_{x_2}^{x_1}\right\|\right\}$$

$$\leq \frac{1}{N}\sum_{\nu=0}^{N-1}\left\{\frac{2L_0}{2 - NP}\frac{\|x_1 - x_2\|}{1 - \frac{\xi N}{2}} + \frac{4PNL_0}{(2 - PN)^2(2 - \xi N)}\|x_1 - x_2\|\right\},$$

which yield the assertion of Lemma 3.7. $\square$

Theorem 3.17. *Let us assume the validity of the assertions:*

1) the conditions of Lemma 3.5 are satisfied;

2) there exist a point $x^0 \in D_f$ and a sequence of indices $p_1 < p_2 < p_3 < \ldots$ such that, for $s \in Z^+$,

$$\left\| \frac{d\Delta_{p_s}(x^0)}{dx^0} - E \right\| \le l_0 = const < 1, \tag{3.57}$$

where E is the identity operator;

3) $h^* = LN_0^{*2}M < 0.25$, and the closed ball $B^*(x^0, k^*t^*)$, where $k^* = N_0^* M$, $N_0^* = \frac{1}{1-l_0}$, and t^* is the least root of the equation $h^*t^2 - t + 1 = 0$, belongs to D_f.

Then, in the ball B^*, there exists the single point x^* that defines the N-periodic solution $\tilde{x}(n, x^*), \tilde{x}(0, x^*) = x^*$ of Eq. (3.41). In this case,

$$\|\tilde{x}(n, x^*) - x^{(p)}(n, x_p)\| \le \frac{1}{1 - \frac{\xi N}{2}} \frac{q^{*p}}{1 - q^*} N_0^* M + \sigma(p), \tag{3.58}$$

where $q^* = \frac{1-\sqrt{1-4h^*}}{2} < 0.5$, the functions $x^{(p)}(n, x_p)$ are determined by relation (3.42), $\{x_p\}_{p=1}^{\infty}$, $x_0 = x^0$ is a sequence constructed by the method Newton–Kantorovich according to Theorem 3.16, and the function $\sigma(p)$ is defined in (3.44).

Proof. It is obvious that, under conditions of the formulated theorem, Lemmas 3.6 and 3.7 are valid. The continuity of the mapping $\frac{df_n(x)}{dx}$ $\forall n \in Z_N$ yields

$$\left\| \frac{d\Delta_m(x)}{dx} - \frac{d\Delta(x)}{dx} \right\| \to 0$$

as $m \to \infty$.

Condition (3.57) leads to the invertibility of the mapping $\frac{d\Delta(x^0)}{dx^0}$ at the point $x^0 \in D_f$. In this case,

$$\left\| \left[\frac{d_\Phi \Delta(x^0)}{dx^0} \right]^{-1} \right\| \le \frac{1}{1 - l_0} = N_0^*. \tag{3.59}$$

It is easy to verify that the ball B mentioned in Theorem 3.16 is contained in the ball B^*. Indeed, it follows from the inequality $k \le k^*$ that $h \le h^*$. This, in turn, ensures the validity of the estimate $t_0 \le t^*$. Thus, all conditions of Theorem 3.16 are satisfied. Estimate (3.58) follows from inequalities (3.47) and (3.59). $\qquad\square$

We introduce the notation

$$l_* = \sum_{i=1}^{\infty} i l_0^{i-1}, \quad \Gamma = 1 + \frac{2\xi N_0^*}{2 - \xi N}, \quad \Gamma_k = \sum_{i=1}^{k} \Gamma^{i-1},$$

$$L_s^* = \frac{4L_0}{(2-PN)(2-\xi N)}\{1 + \frac{PN}{2-PN}\} + L_0(\frac{PN}{2})^{p_s}$$

and prove the following significant assertion.

Theorem 3.18. *Let two first conditions of Theorem 3.17 be satisfied. We assume that, for some $s = s_0$, the inequality $h_{s_0}^* = N_0^{*^2} M L_{s_0}^* < 0.25$ holds, and the closed ball $B^{s_0}(x^0, k^* t_{s_0}^*)$, where $t_{s_0}^*$ is the least root of the equation $h_{s_0}^* t^2 - t + 1 = 0$, belongs to the set D_f.*

Then, in this ball, there exists the single point x^*, which determines an N-periodic solution $\tilde{x}(n, x^*)$ of Eq. (3.41) such that $\tilde{x}(0, x^*) = x^*$. In this case, the relation

$$\lim_{k \to \infty} \lim_{s \to \infty} x_k^{(p_s)} = x^* \tag{3.60}$$

holds, and, for all $\{k, n\} \subset Z^+$, $p_s \geq 3$, the inequality

$$\|\tilde{x}(n, x^*) - x^{(m)}(n, x_k^{(p_s)})\| \leq \frac{\|x^* - x_k^{(p_s)}\|}{1 - \frac{\xi N}{2}} + \sigma(m) \tag{3.61}$$

is satisfied. Here, $\{x_k^{(p_s)}\}_{k=0}^\infty$ is a sequence constructed by the Newton–Kantorovich method. As $k \to \infty$, the limit of the sequence is the solution of the equation $\Delta_{p_s}(x) = 0$,

$$\|x^* - x_k^{(p_s)}\| \leq \frac{q^{*k}}{1 - q^*} N_0^* M + \{Ml_*[\frac{2L_0}{2-NP}\sigma(p_s)$$
$$+ PK^* \frac{2(\frac{PN}{2})^{p_s-1}}{2-PN}] + N_0^* \xi \sigma(p_s)\}\Gamma_k. \tag{3.62}$$

Proof.   With regard for (3.49) – (3.51), the equality

$$\frac{d\Delta_{p_s}(x)}{dx} = \frac{1}{N}\sum_{\nu=0}^{N-1} \frac{df_\nu(x_\nu^{(p_s)}(x))}{dx_\nu^{(p_s)}(x)} \frac{dx_\nu^{(p_s)}(x)}{dx},$$

and the proof of Lemma 3.7, we obtain the inequality

$$\|\frac{d\Delta_{p_s}(x)}{dx}\big|_{x_2}^{x_1}\| \leq L_s^* \|x_1 - x_2\| \quad \forall\{x_1, x_2\} \subset D_f,$$

where $L_s^* \to L$ as $s \to \infty$.

It is easy to verify that, for all natural $s > s_0$, the estimate $L_{s+1}^* < L_s^*$ holds. This yields $h_{s+1}^* < h_s^*$. Hence, $t_{s+1}^* < t_s^* < t_{s_0}^*$, where $h_s^* = M_0^{*^2} M L_s^* < 0.25$, and t_s^* is the least root of the equation $h_s^* t^2 - t + 1 = 0$. In this case, the inclusion of balls

$$B^{s_0} \supset B^{s_0+1} \supset B^{s_0+2} \supset \cdots \supset B^*$$

is valid. Then, for all $s \geq s_0$, the equation $\Delta_{p_s}(x) = 0$ has a solution $x_{p_s} \in B^{s_0}$, and $x_{p_s} = \lim_{k \to \infty} x_k^{(p_s)}$, where the sequence $\{x_k^{(p_s)}\}_{k=0}^{\infty}$ is constructed by the Newton–Kantorovich method, and $x_0^{(p_s)} = x^0 \; \forall s \geq s_0$.

We now show that, for all $k \in Z^+$, there exists the limit $\lim_{s \to \infty} x_k^{(p_s)} = x_k$, and the sequence $\{x_k\}_{k=0}^{\infty}$ is a sequence of points, which determines the point x^* by the Newton–Kantorovich method, i.e.,

$$x_k = x_{k-1} - [\Delta'(x^0)]^{-1}\Delta(x_{k-1}), \quad x_0 = x^0.$$

Since

$$x_k^{(p_s)} = x_{k-1}^{(p_s)} - [\Delta'_{p_s}(x^0)]^{-1}\Delta_{p_s}(x_{k-1}^{(p_s)}), \quad x_0^{(p_s)} = x^0,$$

it is sufficient to prove

$$\|x_k - x_k^{(p_s)}\| \to 0 \quad \text{as} \quad s \to \infty. \tag{3.63}$$

Taking (3.57) into account, we obtain the relation

$$\|[\Delta'(x^0)]^{-1} - [\Delta'_{p_s}(x^0)]^{-1}\| \leq \left\| \sum_{i=0}^{\infty} \{(E - \Delta'(x^0))^i - (E - \Delta'_{p_s}(x^0))^i\} \right\|$$

$$\leq \|\Delta'(x^0) - \Delta'_{p_s}(x^0)\| \sum_{i=1}^{\infty} i l_0^{i-1} = \|\Delta'(x^0) - \Delta'_{p_s}(x^0)\| l_*,$$

which yields the estimate

$$\|x_1 - x_1^{(p_s)}\| \leq M l_* \|\Delta'(x^0) - \Delta'_{p_s}(x^0)\| + N_0^* \xi \sigma(p_s).$$

Analogously, we obtain

$$\|\Delta(x_{k-1}) - \Delta_{p_s}(x_{k-1}^{(p_s)})\| \leq \frac{2\xi}{2 - \xi N}\|x_{k-1} - x_{k-1}^{(p_s)}\| + \xi \sigma(p_s).$$

Now, we can write the chain of inequalities

$$\|x_k - x_k^{(p_s)}\| \leq \|x_{k-1} - x_{k-1}^{(p_s)}\| + \|[\Delta'(x^0)]^{-1} - [\Delta'_{p_s}(x^0)]^{-1}\|\|\Delta(x_{k-1})\|$$

$$+ \|[\Delta'_{p_s}(x_0)]^{-1}\|\|\Delta(x_{k-1}) - \Delta_{p_s}(x_{k-1}^{(p_s)})\|$$

$$\leq \|x_{k-1} - x_{k-1}^{(p_s)}\|(1 + \frac{2\xi N_0^*}{2 - \xi N}) + M l_* \|\Delta'(x^0) - \Delta'_{p_s}(x^0)\| + N_0^* \xi \sigma(p_s)$$

$$\leq \{\|x_{k-2} - x_{k-2}^{(p_s)}\|\Gamma$$

$$+ M l_* \|\Delta'(x^0) - \Delta'_{p_s}(x^0)\| + N_0^* \xi \sigma(p_s)\}\Gamma + M l_* \|\Delta'(x^0)$$

$$- \Delta'_{p_s}(x^0)\| + N_0^* \xi \sigma(p_s) \leq \|x_{k-2} - x_{k-2}^{(p_s)}\|\Gamma^3$$

$$+ (1 + \Gamma + \Gamma^2)M l_* \|\Delta'(x^0) - \Delta'_{p_s}(x^0)\| + (1 + \Gamma + \Gamma^2)N_0^* \xi \sigma(p_s)$$

$$\leq \cdots \leq \{M l_* \|\Delta'(x^0) - \Delta'_{p_s}(x^0)\| + N_0^* \xi \sigma(p_s)\}\Gamma_k. \tag{3.64}$$

Using the proof of Lemma 3.6, we obtain that, for $p_s \geq 3$,

$$\left\| \frac{d\tilde{x}_n(x^0)}{dx^0} - \frac{dx_n^{(p_s)}(x^0)}{dx^0} \right\| \leq \lim_{p \to \infty} \sum_{i=1}^{p} \left\| \frac{dx_n^{(p_s+i)}(x^0)}{dx^0} - \frac{dx_n^{(p_s+i-1)}(x^0)}{dx^0} \right\|$$

$$\leq \sum_{i=1}^{\infty} \| r_n^{p_s+i}(x^0) \| \leq K^* \sum_{i=1}^{\infty} \left(\frac{PN}{2} \right)^{p_s+i-2} \leq K^* \frac{2(\frac{PN}{2})^{p_s-1}}{2 - PN},$$

where $x_n^{(p_s)}(x^0)$ is obtained from relation (3.42).

Now, we can write the inequalities

$$\| \Delta'(x^0) - \Delta'_{p_s}(x^0) \| \leq \frac{1}{N} \sum_{\nu=0}^{N-1} \left\| \frac{df_\nu(\tilde{x}_\nu(x^0))}{dx^0} - \frac{df_\nu(x_\nu^{(p_s)}(x^0))}{dx^0} \right\|$$

$$\leq \frac{2L_0}{2 - NP} \max_{i \in Z_N} \| \tilde{x}_i(x^0) - x_i^{(p_s)}(x^0) \| + P \max_{i \in Z_N} \left\| \frac{d\tilde{x}_i(x^0)}{dx^0} - \frac{dx_i^{(p_s)}(x^0)}{dx^0} \right\|$$

$$\leq \frac{2L_0}{2 - NP} \sigma(p_s) + PK^* \frac{2(\frac{PN}{2})^{p_s-1}}{2 - PN}. \qquad (3.65)$$

It is obvious that inequalities (3.64) and (3.65) guarantee the validity of relation (3.63) and estimates (3.61), (3.62). Theorem 3.18 is proved. $\qquad \square$

We note that the above-performed proof does not yield the commutativity of the double limit in (3.60), since the limiting transition $\lim_{s \to \infty} x_k^{(p_s)} = x_k$ is not uniform with respect to $k \in Z^+$. Therefore, we cannot assert that the sequence $\{x_{p_s}\}$ converges to a point x^* as $s \to \infty$. In the case of a finite-dimensional space, it is possible to separate a subsequence convergent to x^* from this sequence due to the compactness of the closed ball B^{s_0}.

By completing this subsection, we consider Eq. (3.41) in the space $\mathbf{W} = \mathfrak{M}$ and formulate the assertion that is an analog of Corollary 3.4.

Corollary 3.9. *Let the conditions of Theorem 3.14 be satisfied. Moreover,*

1) for some natural s, there exists a point $x^0 \in D_f$ such that $\Delta_s(x^0) = 0 \in \mathfrak{M}$;

2) there exists a closed domain $D_1 \in D_f$ containing x^0 and such that Δ_s topologically maps D_1 onto $\Delta_s D_1$;

3) on the boundary Γ_{D_1} of the set D_1, the inequality

$$\inf_{\Gamma_{D_1}} \| \Delta_s(x) \| \geq \xi \sigma(s).$$

holds. Then Eq. (3.41) has an N-periodic solution $x_n = x(n)$, for which $x_0 = x(0) \in D_1$.

3.3 Periodic solutions of nonlinear difference equations of the second order

Consider the equation

$$\Delta^2 x_n = f_n(x_n, x_{n+1}), \quad n \in Z, \tag{3.66}$$

where $\Delta x_n = x_{n+1} - x_n$, $f_n(x, y) : \mathbf{W} \times \mathbf{W} \to \mathbf{W}$, $x_n \in \mathbf{W}$.

It is clear that this equation can be presented in the form

$$x_{n+2} = 2x_{n+1} - x_n + f_n(x_n, x_{n+1}), \quad n \in Z.$$

It is easy to verify that, for all $\{x_0^*, x_1^*\} \subset \mathbf{W}$, Eq. (3.66) on the set Z_0^+ has the unique solution $x_n = x(n, x_0^*, x_1^*)$ such that $x_0 = x(0, x_0^*, x_1^*) = x_0^*$, $x_1 = x(1, x_0^*, x_1^*) = x_1^*$. For all $n \in \{2, 3, 4, \ldots\}$, the solution is given by the relation

$$x_n = -(n-1)x_0^* + nx_1^* + \sum_{i=1}^{n-1}(n-i)f_{i-1}(x_{i-1}, x_i). \tag{3.67}$$

Nothing can be said about the continuability of this solution onto the set Z^-, if no additional conditions are imposed. If $f_n(x_n, x_{n+1}) = f_n^{(1)}(x_n) + f_n^{(2)}(x_{n+1})$ $\forall n \in Z$, the mappings $f_n^{(1)}(x)$ and $f_n^{(2)}(x)$ transfer the set $\mathbf{W}$ in itself, and the mapping $f_n^{(1)}(x) - x : \mathbf{W} \to \mathbf{W}$ is invertible, then the solution of Eq. (3.66) on the set Z exists for any initial values $\{x_0^*, x_1^*\} \subset \mathbf{W}$. For example, for an equation of the form $x_{n+2} = 2x_{n+1} - x_n + p_n(x_{n+1})$, $p_n(x) : \mathbf{W} \to \mathbf{W}$ $\forall n \in Z$, the solution x_n on Z_0^+ is given by relations (3.67), where the expression $f_{i-1}(x_{i-1}, x_i)$ is replaced by $p_{i-1}(x_i)$. Onto the set Z^-, the solution is continued by the formula

$$x_n = (n+3)x_0^* + nx_1^* - \sum_{i=1}^{-n}(n+i-1)p_{-i}(x_{1-i}).$$

In what follows, we consider the function $f_n(x, y)$ N-periodic in n on the set Z.

Using equality (3.67), it is easy to verify that the solution $x_n = x(n, x_0^*, x_1^*)$ of Eq. (3.66) is N-periodic on Z_0^+ if and only if its initial values x_0^*, x_1^* satisfy the system of equations

$$\sum_{k=0}^{N-1} f_k(x_k(x_0^*, x_1^*), x_{k+1}(x_0^*, x_1^*)) = 0,$$

$$x_1^* = x_0^* - \frac{1}{N}\sum_{i=1}^{N-1}(N-i)f_{i-1}(x_{i-1}, x_i). \tag{3.68}$$

Indeed, the conditions

$$x_0^* = x_N, \quad x_1^* = x_{N+1}$$

are necessary and sufficient for the existence of an N-periodic solution of Eq. (3.66) on the set Z_0^+, which is given by relations (3.67). In this case, we obtain the system of equalities

$$x_0^* = x_1^* + \frac{1}{N}\sum_{i=1}^{N-1}(N-i)f_{i-1}(x_{i-1}, x_i),$$

$$x_1^* = x_0^* - \frac{1}{N}\sum_{i=1}^{N}(N+1-i)f_{i-1}(x_{i-1}, x_i),$$

which can be obviously transformed in system (3.68).

If Eq. (3.66) has a solution N-periodic on Z_0^+, then it can be easily continued onto the whole set Z by periodicity. Thus, Eq. (3.66) has so many N-periodic solutions, how many different pairs of points $\{x_0^*, x_1^*\} \subset \mathbf{W}$ satisfy the system of equations (3.68).

We define the operator L acting on the sequence of points $\{a_0, a_1, a_2, \ldots\} \subset \mathbf{W}$ as follows:

$$La_0 = 0 \in \mathbf{W}, \quad La_n = \sum_{i=0}^{n-1}(a_i - \bar{a}_\nu), \quad \text{where} \quad \bar{a}_\nu = \frac{1}{N}\sum_{\nu=0}^{N-1}a_\nu, \quad n \in Z^+.$$

Then

$$L^2 a_0 = 0 \in \mathbf{W}, \quad L^2 a_1 = -\frac{1}{N}\sum_{i=1}^{N-1}\sum_{s=0}^{i-1}(a_s - \bar{a}_\nu),$$

$$L^2 a_n = \sum_{i=1}^{n-1}\sum_{k=0}^{i-1}(a_k - \bar{a}_\nu) - \frac{n}{N}\sum_{\nu=1}^{N-1}\sum_{k=0}^{\nu-1}(a_k - \bar{a}_\nu), \quad n \in Z^+\backslash\{1\}.$$

If $\{a_n\}$ is an N-periodic sequence, then the sequences $\{La_n\}$ and $\{L^2 a_n\}$ have also this property. It is easy to verify that, in this case, the estimates

$$\|La_n\| \leq \frac{N}{2}\max_{i \in Z_N}\|a_i\|, \quad \|L^2 a_n\| \leq \frac{N^2}{4}\max_{i \in Z_N}\|a_i\|$$

hold. We set $L^2 f_n(x, y) = g_n(x, y)$ and write the equation

$$\Delta^2 x_n = f_n(x_n, x_{n+1}), \quad n \in Z_0^+, \tag{3.69}$$

$$x_n = x_0 + g_n(x_n, x_{n+1}), \quad n \in Z_0^+, \tag{3.70}$$

$$\Delta^2 x_n = f_n(x_n, x_{n+1}) - \mu, \quad n \in Z_0^+, \tag{3.71}$$

where $\mu \in \mathbf{W}$.

Using relations (3.67) and (3.68), we can verify the validity of the following assertions:

1) any solution $x_n(x_0, x_1)$ of Eq. (3.69), which is N-periodic on Z_0^+, is a solution of Eq. (3.70), which is N-periodic on Z_0^+;

2) if $x_n(x_0, x_1)$ is the solution of Eq. (3.70), which is N-periodic on Z_0^+, then it is a solution of Eq. (3.71), which is N-periodic on Z_0^+, where we set $\mu = \overline{f_\nu(x_\nu, x_{\nu+1})}$;

3) any solution $x_n(x_0, x_1)$ of Eq. (3.71), which is N-periodic on Z_0^+, where we set $\mu = \overline{f_\nu(x_\nu, x_{\nu+1})}$, is a solution of Eq. (3.70), which is N-periodic on this set.

Indeed, assertion 1 follows at once from the equalities

$$x_n = -(n-1)x_0 + n\Big(x_0 - \frac{1}{N}\sum_{i=1}^{N-1}(N-i)f_{i-1}(x_{i-1}, x_i)\Big)$$

$$+ \sum_{i=1}^{n-1}(n-i)f_{i-1}(x_{i-1}, x_i)$$

$$= x_0 + \sum_{i=1}^{n-1}\sum_{k=0}^{i-1}f_k(x_k, x_{k+1}) - \frac{n}{N}\sum_{l=1}^{N-1}\sum_{s=0}^{l-1}f_s(x_s, x_{s+1})$$

$$= x_0 + \sum_{i=1}^{n-1}\sum_{k=0}^{i-1}f_k(x_k, x_{k+1}) - \frac{n(n-1)}{2N}\sum_{\nu=0}^{N-1}f_\nu(x_\nu, x_{\nu+1})$$

$$- \frac{n}{N}\sum_{l=1}^{N-1}\sum_{s=0}^{l-1}f_s(x_s, x_{s+1}) + \frac{n(N-1)}{2N}\sum_{r=0}^{N-1}f_r(x_r, x_{r+1})$$

$$= x_0 + \sum_{i=1}^{n-1}\sum_{k=0}^{i-1}\big(f_k(x_k, x_{k+1}) - \overline{f_\nu(x_\nu, x_{\nu+1})}\big)$$

$$- \frac{n}{N}\sum_{s=1}^{N-1}\sum_{k=0}^{s-1}\big(f_k(x_k, x_{k+1}) - \overline{f_\nu(x_\nu, x_{\nu+1})}\big)$$

$$= x_0 + g_n(x_n, x_{n+1}), \quad n \in \{2, 3, 4, \dots\};$$

$$x_1 = x_0 - \frac{1}{N}\sum_{i=1}^{N-1}(N-i)f_{i-1}(x_{i-1}, x_i) = x_0 + g_1(x_1, x_2),$$

and assertion 2 is a consequence of the equalities

$$x_1 = x_0 + g_1(x_1, x_2)$$

$$= x_0 - \frac{1}{N} \sum_{i=1}^{N-1} (N-i) f_{i-1}(x_{i-1}, x_i) + \frac{N-1}{2} \overline{f_n(x_n, x_{n+1})};$$

$$x_n = x_0 + g_n(x_n, x_{n+1}) = -(n-1)x_0 + nx_1$$

$$+ \sum_{i=1}^{n-1} (n-i) f_{i-1}(x_{i-1}, x_i) - \frac{n(n-1)}{2} \overline{f_n(x_n, x_{n+1})}, \quad n \in Z^+ \backslash \{1\}.$$

Assertion 3 is proved analogously.

We now set a sequence $\{x_n^{(m)}(x_0)\}_{m=0}^{\infty}$ on the set Z_0^+ by the recurrence relations

$$x_n^{(0)} \equiv x_0, \quad x_0^{(m)}(x_0) \equiv x_0,$$

$$x_n^{(m)}(x_0) = x_0 + g_n(x_n^{(m-1)}(x_0), x_{n+1}^{(m-1)}(x_0)), \quad m \in Z^+. \tag{3.72}$$

The following conditions are called conditions (V):

a) $f_n(x, y)$ is a function, which is N-periodic in $n \in Z$ and is defined on the set $D_1 = Z \times D \times D$, $D = \{x \in \mathbf{W} | \|x\| \leq R\}$. On this set,

$$\|f_n(x, y)\| \leq M,$$

$$\|f_n(x, y) - f_n(\bar{x}, \bar{y})\| \leq K_1 \|x - \bar{x}\| + K_2 \|y - \bar{y}\|,$$

where M, R, K_1, K_2 are positive constants independent of $n \in Z$, and $\{x, y, \bar{x}, \bar{y}\} \subset D$;

b) $\bar{D}^f = \{x \in D | \|x\| \leq R - \frac{N^2 M}{4}\} \neq \varnothing$, and $\hat{\gamma} = \frac{N^2}{4}(K_1 + K_2) < 1$.

Theorem 3.19. *Let conditions (V) be satisfied. Then, $\forall x_0 \in \bar{D}^f$, the sequence $\{x_n^{(m)}(x_0)\}_{m=0}^{\infty}$ converges uniformly in $n \in Z_0^+$ in the norm of the space $\mathbf{W}$ as $m \to \infty$ to the function $\tilde{x}_n(x_0)$, $\tilde{x}_0(x_0) = x_0$, which is the unique N-periodic solution of Eq. (3.70) on Z_0^+ with initial values $0, x_0$, and*

$$\|\tilde{x}_n(x_0) - x_n^{(m)}(x_0)\| \leq \sigma^*(m),$$

where

$$\sigma^*(m) = \frac{(\frac{N^2}{4})^{m+1} M (K_1 + K_2)^m}{1 - \frac{N^2}{4}(K_1 + K_2)} \to 0 \quad as \quad m \to \infty.$$

Proof. For any $x_0 \in \bar{D}^f$, the functions $x_n^{(m)}(x_0)$ are defined properly, and their values do not leave the set D. Indeed, this is true for $x_n^{(0)}(x_0)$. If this assertion is valid for $x_n^{(s)}(x_0)$, then it is also valid for $x_n^{(s+1)}(x_0)$, since $\|x_n^{(s+1)}(x_0) - x_0\| \leq MN^2/4$, and $\|x_0\| \leq R - MN^2/4$.

By the method of complete mathematical induction, it is easy to obtain the estimate

$$\|x_n^{(s+1)}(x_0) - x_n^{(s)}(x_0)\| \leq \frac{MN^2}{4}[\frac{N^2}{4}(K_1 + K_2)]^s, \quad n \in Z_0^+,$$

which testifies to the fundamentality and, hence, to the convergence of the sequence $\{x_n^{(m)}(x_0)\}_{m=0}^{\infty}$ to some function $\tilde{x}_n(x_0)$ as $m \to \infty$.

Performing the limiting transition in (3.72) as $m \to \infty$, we obtain the equality

$$\tilde{x}_n(x_0) = x_0 + g_n(\tilde{x}_n(x_0), \tilde{x}_{n+1}(x_0)), \quad n \in Z_0^+,$$

which implies that the constructed function is an N-periodic solution of Eq. (3.70) on Z_0^+, since the functions $x_n^{(m)}(x_0)$ on this set are N-periodic for all $m \in Z_0^+$.

The inequalities

$$\|\tilde{x}_n(x_0) - x_n^{(s)}(x_0)\| \leq \lim_{p \to \infty} \sum_{i=1}^{p} \|x_n^{(s+i)}(x_0) - x_n^{(s+i-1)}(x_0)\|$$

$$\leq \frac{MN^2}{4}[\frac{N^2}{4}(K_1 + K_2)]^s \sum_{i=0}^{\infty}[\frac{N^2}{4}(K_1 + K_2)]^i$$

yield the estimate indicated in the statement of Theorem 3.19.

Let us assume that there exists one more solution $y_n(x_0)$, $y_0(x_0) = x_0$, of Eq. (3.70), which is N-periodic on Z_0^+. By the inductive reasoning, we obtain the estimate

$$\|\tilde{x}_n(x_0) - y_n(x_0)\| \leq \xi_0^m \|\tilde{x}_{n+m}(x_0) - y_{n+m}(x_0)\|,$$

where $\xi_0^m = (\frac{N^2 K_2}{4 - N^2 K_1})^m \to 0$ as $m \to \infty$.

Taking the N-periodicity of solutions into account and passing to the limit as $m \to \infty$ in the last inequality, we obtain that $\tilde{x}_n(x_0)$ and $y_n(x_0)$ coincide on the set Z_0^+. This completes the proof of the theorem. $\square$

Assertions 1 and 2 and Theorem 3.19 yield directly the following proposition.

Corollary 3.10. *Let conditions (V) be satisfied, and let $x_0 \in \bar{D}^f$. In order that Eq. (3.66) have an N-periodic solution $x_n = x(n, x_0, x_1)$, it is necessary and sufficient that the equality*

$$\mu = \mu^* = \overline{f_\nu(\tilde{x}_\nu(x_0), \tilde{x}_{\nu+1}(x_0))} = 0 \in \mathbf{W}$$

hold. If the indicated solution $x_n = x(n, x_0, x_1)$ is N-periodic, and if $x_1 \neq x_1^*$, then the solution $x_n^* = x(n, x_0, x_1^*)$ of this equation cannot be N-periodic.

Thus, in order to find an N-periodic solution of Eq. (3.66), it is necessary to seek $x_0 \in \bar{D}^f$ such that it satisfies the equation $\mu^* = 0$. By continuing the function $\tilde{x}(x_0)$ onto Z^- by periodicity, we obtain the N-periodic solution of Eq. (3.66) on the set Z.

The conception of the subsequent reasoning is the same as that in the previous section, but the technique of its realization is significantly more complicated.

On $\bar{D}^f$, we define the mapping

$$\Delta(x) = \overline{f_n(\tilde{x}_n(x), \tilde{x}_{n+1}(x))}, \quad \Delta_s(x) = \overline{f_n(x_n^{(s)}(x), x_{n+1}^{(s)}(x))}$$

and formulate an analog of Corollary 3.8.

Corollary 3.11. *Let conditions (V) be satisfied. In order that some subset $X \subset \bar{D}^f$ contain at least one zero of the mapping $\Delta(x)$, it is necessary that, for all natural s and any $x \in X$, the inequality*

$$\|\Delta_s(x)\| \leq [\sigma^*(s) + \frac{d(X)}{1 - \frac{N^2}{4}(K_1 + K_2)}](K_1 + K_2),$$

where $d(X)$ is the diameter of the set X, hold.

Proof. Let us assume that there exists a point $x_0 \in X$ such that $\Delta(x_0) = 0$. We denote the difference $\tilde{x}_n(x_1) - \tilde{x}_n(x_2)$ by ω_n and the difference $f_n(\tilde{x}_n(x_1), \tilde{x}_{n+1}(x_1)) - f_n(\tilde{x}_n(x_2), \tilde{x}_{n+1}(x_2))$ by ω_n^*. Then, for any $\{x_1, x_2\} \subset \bar{D}^f$,

$$\|\omega_n\|_0 = \max_{i \in Z_N} \|\omega_i\| \leq \|x_1 - x_2\| + \frac{N^2}{4}\|\omega_n^*\|_0$$

$$\leq \|x_1 - x_2\| + \frac{N^2}{4}(K_1\|\omega_n\|_0 + K_2\|\omega_{n+1}\|_0),$$

whence

$$\|\omega_n\|_0 \leq \frac{\|x_1 - x_2\|}{1 - \frac{N^2}{4}K_1} + \frac{\frac{N^2}{4}K_2}{1 - \frac{N^2}{4}K_1}\|\omega_{n+1}\|_0.$$

By the method of complete mathematical induction, we obtain the estimate

$$\|\omega_n\|_0 \leq \frac{\|x_1 - x_2\|}{1 - \frac{N^2}{4}(K_1 + K_2)} + \left(\frac{\frac{N^2}{4}K_2}{1 - \frac{N^2}{4}K_1}\right)^p \|\omega_{n+p}\|_0,$$

which holds for any natural number p. Passing to the limit as $p \to \infty$ in the last inequality and taking into account that $\|\omega_{n+p}\|_0 \leq 2R$, we obtain the estimate

$$\|\tilde{x}_n(x_1) - \tilde{x}_n(x_2)\|_0 \leq \frac{\|x_1 - x_2\|}{1 - \frac{N^2}{4}(K_1 + K_2)}.$$

In this case,

$$\|\Delta(x_1) - \Delta(x_2)\| \leq \frac{1}{N} \sum_{\nu=0}^{N-1} (K_1 \|\tilde{x}_n(x_1) - \tilde{x}_n(x_2)\|_0$$

$$+ K_2 \|\tilde{x}_{n+1}(x_1) - \tilde{x}_{n+1}(x_2)\|_0) \leq \frac{K_1 + K_2}{1 - \frac{N^2}{4}(K_1 + K_2)} \|x_1 - x_2\|,$$

which means the equicontinuity of the mapping $\Delta(x)$ on $\bar{D}^f$ and, therefore, on X. We now estimate the difference $\Delta(x) - \Delta_s(x)$, $x \in \bar{D}^f$:

$$\|\Delta(x) - \Delta_s(x)\| \leq \frac{1}{N} \sum_{\nu=0}^{N-1} (K_1 \|\tilde{x}_n(x) - x_n^{(s)}(x)\|_0 + K_2 \|\tilde{x}_{n+1}(x)$$

$$- x_{n+1}^{(s)}(x)\|_0) \leq \frac{1}{N} \sum_{\nu=0}^{N-1} (K_1 \sigma^*(s) + K_2 \sigma^*(s)) = (K_1 + K_2)\sigma^*(s).$$

The above inequalities yield the estimate

$$\|\Delta_s(x)\| \leq (K_1 + K_2)\sigma^*(s) + \frac{K_1 + K_2}{1 - \frac{N^2}{4}(K_1 + K_2)} \|x - x_0\|.$$

The proof is completed. $\qquad\square$

We now prove several auxiliary propositions.

Lemma 3.8. *Let conditions (V) be satisfied, and, in this case, let a function $f_n(x, y)$ be defined on the set $Z \times D_\rho \times D_\rho$, where $D_\rho = \{x \in \mathbf{W} \,|\, \|x\| < R + \rho\}$, ρ is some positive constant, and let it be Fréchet-differentiable $\forall n \in Z_N$ in the domain $D_\rho^2 = D_\rho \times D_\rho$, and $\|\frac{df_n(x,y)}{d(x,y)}\| \leq P$, where P is a positive constant independent of $n \in Z_N$, and $(x, y) \in D_\rho^2$. If $\hat{\eta} = N^2 P/4 < 1$, then the functions $x_n^{(m)}(x)$ defined by relations (3.72) are Fréchet-differentiable in the domain $D^f = \{x \in \bar{D}^f \,|\, \|x\| < R - N^2 M/4\}$, and*

$$\left\|\frac{dx_n^{(m)}(x)}{dx}\right\| < \frac{1}{1 - \hat{\eta}} \tag{3.73}$$

for all $n \in Z_N$, $m \in Z_0^+$.

Proof. We write relation (3.72) in the form

$$x_n^{(0)}(x) = x, \quad x_0^{(m)}(x) = x, \quad n \in Z_N, m \in Z_0^+; \qquad (3.74)$$

$$x_1^{(m)}(x) = x - \frac{1}{N} \sum_{i=1}^{N-1} \sum_{s=0}^{i-1} (f_s(x_s^{(m-1)}(x), x_{s+1}^{(m-1)}(x))$$
$$- \overline{f_\nu(x_\nu^{(m-1)}(x), x_{\nu+1}^{(m-1)}(x)))}, \quad m \in Z^+; \quad (3.75)$$

$$x_n^{(m)}(x) = x + \sum_{i=1}^{n-1} \sum_{s=0}^{i-1} (f_s(x_s^{(m-1)}(x), x_{s+1}^{(m-1)}(x))$$
$$- \overline{f_\nu(x_\nu^{(m-1)}(x), x_{\nu+1}^{(m-1)}(x)))} - \frac{n}{N} \sum_{i=1}^{N-1} \sum_{s=0}^{i-1} (f_s(x_s^{(m-1)}(x), x_{s+1}^{(m-1)}(x))$$
$$- \overline{f_\nu(x_\nu^{(m-1)}(x), x_{\nu+1}^{(m-1)}(x)))}, \quad n \in Z_N\setminus\{0,1\}, \quad m \in Z^+. \quad (3.76)$$

As the norm $\|(x,y)\|$ of an element $(x,y) \in D_\rho^2$, we consider $\max\{\|x\|, \|y\|\}$, where $\|x\|$, $\|y\|$ are the norms in the domain D_ρ. It is obvious that the set D_ρ^2 is open in $\mathbf{W} \times \mathbf{W}$.

For $m = 0$, relation (3.74) implies that, $\forall n \in Z_N$,

$$\frac{dx_n^{(0)}(x)}{dx} = E, \quad \|\frac{dx_n^{(0)}(x)}{dx}\| = 1 < \frac{1}{1 - \hat\eta},$$

where E is the identity operator.

For $m \in Z^+$, we now prove inequality (3.73) by the method of complete mathematical induction. Let $m = 1$. Using (3.75), we write the equalities

$$\frac{dx_1^{(1)}(x)}{dx} = E - \frac{1}{N} \sum_{i=1}^{N-1} \sum_{s=0}^{i-1} (\frac{df_s(x,x)}{dx} - \frac{\overline{df_\nu(x,x)}}{dx})$$
$$= E - \frac{1}{N} \sum_{i=1}^{N-1} \sum_{s=0}^{i-1} (\frac{df_s(x,x)}{d(x,x)} \frac{dg(x)}{dx} - \frac{\overline{df_\nu(x,x)}}{d(x,x)} \frac{dg(x)}{dx}),$$

where $g(x)$ is a mapping of the open set D^f onto the set D_ρ^2, whose components $g^1(x)$ and $g^2(x)$, which map D^f onto D_ρ, are identity operators. Then

$$\|\frac{dg(x)}{dx}\| = \sup_{\|h\|=1} \|\frac{dg(x)}{dx} h\| = \sup_{\|h\|=1} \max\{\|\frac{dg^1(x)}{dx} h\|, \|\frac{dg^2(x)}{dx} h\|\}$$
$$\leq \max\{\|\frac{dg^1(x)}{dx}\|, \|\frac{dg^2(x)}{dx}\|\} = 1, \quad (h \in \mathbf{W}).$$

This yields

$$\|\frac{dx_1^{(1)}(x)}{dx}\| \le 1 + \frac{N-1}{2} \max_{s \in Z_N}\{\|\frac{df_s(x,x)}{d(x,x)}\|\|\frac{dg(x)}{dx}\|\}$$

$$\le 1 + \frac{P(N-1)}{2} \le 1 + \frac{PN^2}{4} < \frac{1}{1-\hat{\eta}},$$

because $N - 1 < \frac{N^2}{2}$ for all $N \in Z$.

We now estimate the derivative $\frac{dx_n^{(1)}(x)}{dx}$ in the norm for $n \in Z_N\backslash\{0,1\}$. Relation (3.76) yields

$$\|\frac{dx_n^{(1)}(x)}{dx}\| \le 1 + \frac{N^2}{4} \max_{s \in Z_N}\{\|\frac{df_s(x,x)}{d(x)}\|\} \le 1 + \frac{PN^2}{4} < \frac{1}{1-\hat{\eta}}.$$

Hence, estimate (3.73) for $m = 1$ is valid for all $n \in Z_N$. Let us assume that it holds for all $1 < m \le k$ uniformly in $n \in Z_N$. We now prove its validity for $m = k + 1$. For $n = 0$, estimate (3.73) is obvious.

For $n = 1$, we have

$$\|\frac{dx_1^{(k+1)}(x)}{dx}\| \le 1 + \frac{N-1}{2} \max_{s \in Z_N}\{\|\frac{df_s(x_s^{(k)}(x), x_{s+1}^{(k)}(x))}{d(x)}\|\}$$

$$\le 1 + \frac{N-1}{2} P \max_{s \in Z_N}\{\|\frac{dx_s^{(k)}(x)}{dx}\|, \|\frac{dx_{s+1}^{(k)}(x)}{dx}\|\}$$

$$\le 1 + \frac{\hat{\eta}}{1-\hat{\eta}} = \frac{1}{1-\hat{\eta}}.$$

For $n \in Z_N\backslash\{0,1\}$,

$$\|\frac{dx_n^{(k+1)}(x)}{dx}\| \le 1 + \frac{PN^2}{4} \max_{s \in Z_N}\{\|\frac{dx_s^{(k)}(x)}{dx}\|, \|\frac{dx_{s+1}^{(k)}(x)}{dx}\|\} \le \frac{1}{1-\hat{\eta}},$$

which completes the proof of Lemma 3.8. $\square$

Remark 3.2. The boundedness of the Fréchet derivative of the function $f_n(x,y)$ on D_ρ^2 implies that, under conditions (V), we may set $K_1 = K_2 = P$. Then we obtain the estimate $\hat{\eta} < 1$ a consequence of the inequality $\hat{\gamma} < 1$.

Lemma 3.9. *Let us assume that the conditions of Lemma 3.8 hold, and, for all $\{x_1, x_2, y_1, y_2\} \subset D_\rho$, the inequality*

$$\|\frac{df_n(x,y)}{d(x,y)}\Big|_{(x_2,y_2)}^{(x_1,y_1)}\| \le L_0 \max\{\|x_1 - x_2\|, \|y_1 - y_2\|\}$$

holds uniformly in $n \in Z_N$. Here, L_0 is a positive constant. Then the estimate

$$\left\| \frac{dx_n^{(m)}(x)}{dx} \Big|_{x_2}^{x_1} \right\| \leq \frac{L_0 \hat{\eta}}{P(1 - \hat{\eta})^2 (1 - \hat{\gamma})} \|x_1 - x_2\| \tag{3.77}$$

holds uniformly in $n \in Z_N$ and $m \in Z_0^+$.

Proof. We denote the difference $x_n^{(k)}(x_1) - x_n^{(k)}(x_2)$ by ω_n^k and the difference $f_n(x_n^{(k)}(x_1), x_{n+1}^{(k)}(x_1)) - f_n(x_n^{(k)}(x_2), x_{n+1}^{(k)}(x_2))$ by ω_n^{*k}. Then, for any $\{x_1, x_2\} \subset D^f$, the estimates

$$\|\omega_n^k\|_0 = \max_{i \in Z_N} \|\omega_i^k\| \leq \|x_1 - x_2\| + \frac{N^2}{4} \|\omega_n^{*k-1}\|_0$$

$$\leq \|x_1 - x_2\| + \frac{N^2}{4}(K_1 \|\omega_n^{k-1}\|_0 + K_2 \|\omega_{n+1}^{k-1}\|_0)$$

hold. They yield the inductive inequality

$$\|\omega_n^k\|_0 \leq \|x_1 - x_2\| + \hat{\gamma} \|\omega_n^{k-1}\|_0,$$

which leads to the relation

$$\|\omega_n^k\|_0 \leq \|x_1 - x_2\| \sum_{i=0}^{k-1} \hat{\gamma}^i + \hat{\gamma}^k \|\omega_n^0\|_0 = \|x_1 - x_2\| \sum_{i=0}^{k-1} \hat{\gamma}^i.$$

Thus, for all $n \in Z_N$ and $k \in Z_0^+$, the estimate

$$\|x_n^{(k)}(x_1) - x_n^{(k)}(x_2)\| \leq \frac{\|x_1 - x_2\|}{1 - \frac{N^2}{4}(K_1 + K_2)}$$

holds. For $m = 0$, the assertion of Lemma 3.9 is obvious. For $m \in Z^+$, we prove inequality (3.77) by the method of complete mathematical induction.

Let $m = 1$. For $m = 1$ and $n = 0$, this inequality is also obvious. Let us estimate $\frac{dx_1^{(1)}(x)}{dx} \big|_{x_2}^{x_1}$ in the norm. The inequalities

$$\left\| \frac{dx_1^{(1)}(x)}{dx} \Big|_{x_2}^{x_1} \right\| \leq \frac{1}{N} \sum_{i=1}^{N-1} \left\| \sum_{s=0}^{i-1} \left(\frac{df_s(x,x)}{dx} \Big|_{x_2}^{x_1} - \overline{\frac{df_s(x,x)}{dx} \Big|_{x_2}^{x_1}} \right) \right\|$$

$$\leq \frac{N-1}{2} \max_{s \in Z_N} \left\{ \left\| \frac{df_s(x,x)}{dx} \Big|_{x_2}^{x_1} \right\| \right\}$$

$$\leq \frac{N-1}{2} \max_{s \in Z_N} \left\{ \left\| \frac{df_s(x,x)}{d(x,x)} \right\| \left\| \frac{dg(x)}{dx} \Big|_{x_2}^{x_1} \right\| + \left\| \frac{df_s(x,x)}{d(x,x)} \Big|_{x_2}^{x_1} \right\| \left\| \frac{dg(x_2)}{dx} \right\| \right\}$$

$$\leq \frac{N-1}{2} L_0 \|x_1 - x_2\| \leq \frac{\hat{\eta} L_0}{P} \|x_1 - x_2\|$$

hold and yield estimate (3.77).

The validity of this estimate for $m = 1, n \in Z_N \backslash \{0, 1\}$ follows from the inequalities

$$\|\frac{dx_n^{(1)}(x)}{dx}|_{x_2}^{x_1}\| \le \frac{N^2}{4} \max_{s \in Z_N} \{\|\frac{df_s(x, x)}{dx}|_{x_2}^{x_1}\|\} \le \frac{\hat{\eta} L_0}{P} \|x_1 - x_2\|.$$

Hence, estimate (3.77) is valid for all $m = 1, n \in Z_N$.

Let us assume that, for all $1 < m \le k, n \in Z_N$, the inequality

$$\|\frac{dx_n^{(m)}(x)}{dx}|_{x_2}^{x_1}\| \le \frac{L_0 \hat{\eta}}{P(1 - \hat{\eta})(1 - \hat{\gamma})} \sum_{i=0}^{m-1} \hat{\eta}^i \|x_1 - x_2\| \qquad (3.78)$$

holds. It yields estimate (3.77). We now show that inequality (3.78) holds for $m = k + 1$.

For $x_0^{(k+1)}(x)$, this inequality is obvious. Taking the convexity of the set D^f into account and denoting the mapping $D^f \to D_\rho^2$ with components $g_s^{(k)1}(x) = x_s^{(k)}(x)$ and $g_s^{(k)2}(x) = x_{s+1}^{(k)}(x)$ by $g_s^{(k)}(x)$, we obtain the relation

$$\|\frac{dx_1^{(k+1)}(x)}{dx}|_{x_2}^{x_1}\|$$

$$\le \frac{N-1}{2} \max_{s \in Z_N} \{\|\frac{df_s(x_s^{(k)}(x_1), x_{s+1}^{(k)}(x_1))}{d(x_s^{(k)}, x_{s+1}^{(k)})}\|\|\frac{dg_s^{(k)}(x)}{dx}|_{x_2}^{x_1}\|$$

$$+ \|\frac{df_s(x_s^{(k)}(x), x_{s+1}^{(k)}(x))}{d(x_s^{(k)}, x_{s+1}^{(k)})}|_{x_2}^{x_1}\|\|\frac{dg_s^{(k)}(x_2)}{dx}\|\}$$

$$\le \frac{N-1}{2} \max_{s \in Z_N} \{P\|\frac{dg_s^{(k)}(x)}{dx}|_{x_2}^{x_1}\| + L_0 \max\{\|x_s^{(k)}(x)|_{x_2}^{x_1}\|, \|x_{s+1}^{(k)}(x)|_{x_2}^{x_1}\|\}$$

$$\times \max\{\|\frac{dx_s^{(k)}(x_2)}{dx}\|\|\frac{dx_{s+1}^{(k)}(x_2)}{dx}\|\}\}$$

$$\le \frac{N-1}{2} \max_{s \in Z_N} \{P \max\{\|\frac{dx_s^{(k)}(x)}{dx}|_{x_2}^{x_1}\|, \|\frac{dx_{s+1}^{(k)}(x)}{dx}|_{x_2}^{x_1}\|\} + \frac{L_0 \|x_1 - x_2\|}{(1 - \hat{\eta})(1 - \hat{\gamma})}\}$$

$$\le \frac{\hat{\eta}}{P} \{P \frac{L_0 \hat{\eta}}{P(1 - \hat{\eta})(1 - \hat{\gamma})} \sum_{i=0}^{k-1} \hat{\eta}^i + \frac{L_0}{(1 - \hat{\eta})(1 - \hat{\gamma})}\} \|x_1 - x_2\|$$

$$= \frac{L_0 \hat{\eta}}{P(1 - \hat{\eta})(1 - \hat{\gamma})} \sum_{i=0}^{k} \hat{\eta}^i \|x_1 - x_2\|.$$

It remains to estimate the difference

$$\frac{dx_n^{(k+1)}(x)}{dx}|_{x_2}^{x_1}$$

in the norm for $n \in Z_N \backslash \{0, 1\}$. Using (3.76) and performing the analogous consideration, we obtain the estimates

$$\|\frac{dx_n^{(k+1)}(x)}{dx}|_{x_2}^{x_1}\| \leq \frac{N^2}{4} \max_{s \in Z_N} \{\|\frac{df_s(x_s^{(k)}(x), x_{s+1}^{(k)}(x))}{dx}\|\}$$

$$\leq \frac{L_0 \hat{\eta}}{P(1 - \hat{\eta})(1 - \hat{\gamma})} \sum_{i=0}^{k} \hat{\eta}^i \|x_1 - x_2\|.$$

The proof of Lemma 3.9 is completed. $\qquad\square$

Lemma 3.10. *Under conditions of Lemma 3.9, the sequence* $\{\frac{dx_n^{(m)}(x)}{dx}\}_{m=0}^{\infty}$ *converges as* $m \to \infty$ *uniformly in* $x \in D^f$ *for all* $n \in Z_N$.

Proof. The indicated sequence belongs to the space $\mathbf{L}(\mathbf{W}, \mathbf{W})$, which is a complete one. Therefore, it is sufficient to prove its fundamentality. For $n = 0$, the assertion of Lemma 3.10 is obvious. We now show that, for all $m \in Z^+ \backslash \{1\}$ and $n \in Z_N \backslash \{0\}$, the inequality

$$\|\frac{dx_n^{(m)}(x)}{dx} - \frac{dx_n^{(m-1)}(x)}{dx}\| \leq \frac{\hat{\eta}}{P}\left(\sum_{i=0}^{m-2} \frac{\beta}{1 - \hat{\eta}} \hat{\gamma}^{m-2-i} \hat{\eta}^i + P\hat{\eta}^{m-1}\right), \quad (3.79)$$

where $\beta = \frac{L_0 N^2 M}{4}$, holds.

Let us prove estimate (3.79) for $m = 2$. Using equalities (3.75) and (3.76), we obtain the estimates

$$\|\frac{dx_1^{(1)}(x)}{dx} - \frac{dx_1^{(0)}(x)}{dx}\| \leq \frac{N-1}{2} \max_{s \in Z_N} \|\frac{df_s(x, x)}{dx}\| \leq \frac{N-1}{2} P \leq \hat{\eta};$$

$$\|\frac{dx_n^{(1)}(x)}{dx} - \frac{dx_n^{(0)}(x)}{dx}\| \leq \frac{N^2}{4} \max_{s \in Z_N} \|\frac{df_s(x, x)}{dx}\| \leq \frac{N^2}{4} P = \hat{\eta}, \quad n \in Z_N \backslash \{0, 1\}.$$

Then we write the chain of inequalities

$$\|\frac{dx_1^{(2)}(x)}{dx} - \frac{dx_1^{(1)}(x)}{dx}\|$$

$$\leq \frac{N-1}{2} \max_{s \in Z_N} \{\|\frac{df_s(x_s^{(1)}(x), x_{s+1}^{(1)}(x))}{dx} - \frac{df_s(x, x)}{dx}\|\}$$

$$\leq \frac{N-1}{2} \max_{s \in Z_N} \{\|\frac{df_s(x_s^{(1)}(x), x_{s+1}^{(1)}(x))}{d(x_s^{(1)}, x_{s+1}^{(1)})} \frac{dg_s^{(1)}(x)}{dx} - \frac{df_s(x, x)}{d(x, x)} \frac{dg(x)}{dx}\|\}$$

$$\leq \frac{N-1}{2} \max_{s\in Z_N}\{\|\frac{df_s(x_s^{(1)}(x), x_{s+1}^{(1)}(x))}{d(x_s^{(1)}, x_{s+1}^{(1)})} - \frac{df_s(x,x)}{d(x,x)}\|\|\frac{dg_s^{(1)}(x)}{dx}\|$$

$$+ \|\frac{df_s(x,x)}{d(x,x)}\|\|\frac{dg_s^{(1)}(x)}{dx} - \frac{dg(x)}{dx}\|\}$$

$$\leq \frac{N-1}{2} \max_{s\in Z_N}\{L_0 \max\{\|x_s^{(1)}(x) - x_s^{(0)}(x)\|, \|x_{s+1}^{(1)}(x) - x_{s+1}^{(0)}(x)\|\}$$

$$\times \max\{\|\frac{dx_s^{(1)}(x)}{dx}\|, \|\frac{dx_{s+1}^{(1)}(x)}{dx}\|\} + P\max\{\|\frac{dx_s^{(1)}(x)}{dx} - \frac{dx_s^{(0)}(x)}{dx}\|,$$

$$\|\frac{dx_{s+1}^{(1)}(x)}{dx} - \frac{dx_{s+1}^{(0)}(x)}{dx}\|\}\} \leq \frac{N-1}{2}(\frac{\beta}{1-\hat\eta} + P\hat\eta) \leq \frac{\hat\eta}{P}(\frac{\beta}{1-\hat\eta} + P\hat\eta),$$

which yield, for all $n \in Z_N\setminus\{0,1\}$, the inequality

$$\|\frac{dx_n^{(2)}(x)}{dx} - \frac{dx_n^{(1)}(x)}{dx}\|$$

$$\leq \frac{N^2}{4} \max_{s\in Z_N}\{\|\frac{df_s(x_s^{(1)}(x), x_{s+1}^{(1)}(x))}{dx} - \frac{df_s(x,x)}{dx}\|\} \leq \frac{\hat\eta}{P}(\frac{\beta}{1-\hat\eta} + P\hat\eta).$$

In other words, estimate (3.79) holds for $m = 2$ $\forall n \in Z_N\setminus\{0\}$. We assume that it is valid for all $2 < m \leq k$ and prove its validity for $m = k+1$. It is easy to verify that, $\forall n \in Z_N\setminus\{0\}$, the inequalities

$$\|\frac{dx_n^{(k+1)}(x)}{dx} - \frac{dx_n^{(k)}(x)}{dx}\|$$

$$\leq \frac{\hat\eta}{P} \max_{s\in Z_N}\{\|\frac{df_s(x_s^{(k)}(x), x_{s+1}^{(k)}(x))}{dx} - \frac{df_s(x_s^{(k-1)}(x), x_{s+1}^{(k-1)}(x))}{dx}\|\}$$

$$\leq \frac{\hat\eta}{P}\{\frac{\beta}{1-\hat\eta}\hat\gamma^{k-1} + P\max\{\|\frac{dx_s^{(k)}(x)}{dx}$$

$$- \frac{dx_s^{(k-1)}(x)}{dx}\|, \|\frac{dx_{s+1}^{(k)}(x)}{dx} - \frac{dx_{s+1}^{(k-1)}(x)}{dx}\|\}\}$$

$$\leq \frac{\hat\eta}{P}\{\frac{\beta}{1-\hat\eta}\hat\gamma^{k-1} + \hat\eta(\sum_{i=0}^{k-2}\frac{\beta}{1-\hat\eta}\hat\gamma^{k-2-i}\hat\eta^i + P\hat\eta^{k-1})\}$$

hold. The proof of estimate (3.79) is completed.

For $\hat\eta \neq \hat\gamma$, the right-hand side of estimate (3.79) is as follows:

$$\frac{\hat\eta\beta}{P(1-\hat\eta)(\hat\eta-\hat\gamma)}(\hat\eta^{m-1} - \hat\gamma^{m-1}) + \hat\eta^m.$$

Since $\hat\eta < 1$ and $\hat\gamma < 1$, this yields the assertion of Lemma 3.10.

But if $\hat{\eta} = \hat{\gamma}$, then the right-hand side of estimate (3.79) reads

$$\frac{\beta}{P(1-\hat{\gamma})}\hat{\gamma}^{m-1}(m-1) + \hat{\gamma}^m \leq (\frac{\beta}{P(1-\hat{\gamma})} + 1)\hat{\gamma}^{m-1}(m-1) \quad (m \geq 2),$$

which ensures the fundamentality of the sequence $\{\frac{dx_n^{(m)}(x)}{dx}\}_{m=0}^{\infty}$, because the series $\sum_{m=2}^{\infty}\hat{\gamma}^{m-1}(m-1)$ is convergent. Lemma 3.10 is proved. $\square$

This lemma yields at once the following proposition.

Corollary 3.12. *Under conditions of Lemma 3.9, the function $\tilde{x}_n(x)$ defined in Theorem 3.19 is Fréchet-differentiable on D^f for all $n \in Z_N$.*

The proof of Corollary 3.12 directly follows from Theorem 111 in [136].

Lemma 3.11. *Let the conditions of Lemma 3.9 be satisfied. Then the mapping $\Delta(x) = \overline{f_n(\tilde{x}_n(x), \tilde{x}_{n+1}(x))}$ is Fréchet-differentiable on the set D^f, and, for all $\{x_1, x_2\} \subset D^f$, the inequality*

$$\|\frac{d\Delta(x)}{dx}|_{x_2}^{x_1}\| \leq L\|x_1 - x_2\|,$$

where $L = const > 0$, holds.

Proof.   In view of Lemmas 3.8 $-$ 3.10, it is easy to verify that, for all $n \in Z_N$ and $\{x_1, x_2\} \subset D^f$, the following relations are valid:

$$\|\frac{d\tilde{x}_n(x)}{dx}\| \leq \frac{1}{1-\hat{\eta}}; \quad \|\frac{d\tilde{x}_n(x)}{dx}|_{x_2}^{x_1}\| \leq \frac{L_0\hat{\eta}}{P(1-\hat{\eta})^2(1-\hat{\gamma})}\|x_1 - x_2\|;$$

$$\frac{d\Delta(x)}{dx} = \frac{1}{N}\sum_{s=0}^{N-1}\frac{df_s(\tilde{x}_s(x), \tilde{x}_{s+1}(x))}{d(\tilde{x}_s, \tilde{x}_{s+1})}\frac{d\tilde{g}_s(x)}{dx},$$

where $\tilde{g}_s(x)$ is the mapping of the set D^f onto the set D_ρ^2 with components $\tilde{g}_s^1(x) = \tilde{x}_s(x)$, $\tilde{g}_s^2(x) = \tilde{x}_{s+1}(x)$.

Using these relations, we obtain the inequalities

$$\|\frac{d\Delta(x)}{dx}|_{x_2}^{x_1}\| \leq \frac{1}{N}\sum_{s=0}^{N-1}\{\|\frac{df_s(\tilde{x}_s(x), \tilde{x}_{s+1}(x))}{d(\tilde{x}_s, \tilde{x}_{s+1})}|_{x_2}^{x_1}\|\|\frac{d\tilde{g}_s(x_1)}{dx}\|$$
$$+ \|\frac{df_s(\tilde{x}_s(x_2), \tilde{x}_{s+1}(x_2))}{d(\tilde{x}_s, \tilde{x}_{s+1})}\|\|\frac{d\tilde{g}_s(x)}{dx}|_{x_2}^{x_1}\|\}$$

$$\leq \frac{1}{N} \sum_{s=0}^{N-1} \{ L_0 \max\{ \| \tilde{x}_s(x)|_{x_2}^{x_1} \|, \| \tilde{x}_{s+1}(x)|_{x_2}^{x_1} \| \}$$

$$\times \max\{ \| \frac{d\tilde{x}_s(x_1)}{dx} \|, \| \frac{d\tilde{x}_{s+1}(x_1)}{dx} \| \}$$

$$+ P \max\{ \| \frac{d\tilde{x}_s(x)}{dx}|_{x_2}^{x_1} \|, \| \frac{d\tilde{x}_{s+1}(x)}{dx}|_{x_2}^{x_1} \| \} \}$$

$$\leq \frac{L_0}{(1-\hat{\eta})(1-\hat{\gamma})} (1 + \frac{\hat{\eta}}{1-\hat{\eta}}) \|x_1 - x_2\|.$$

Denoting the constant factor $\frac{L_0}{(1-\hat{\eta})(1-\hat{\gamma})}(1 + \frac{\hat{\eta}}{1-\hat{\eta}})$ by L, we complete the proof of Lemma 3.11. $\square$

Lemma 3.12. *Let the conditions of Lemma 3.9 be satisfied, and let there exist a point $x^0 \in D^f$ and the sequence of indices $p_1 < p_2 < p_3 < \cdots < p_k < \ldots$ such that, $\forall s \in Z^+$,*

$$\| \frac{d\Delta_{p_s}(x^0)}{dx^0} - E \| \leq l_0 < 1.$$

Then the mapping $\frac{d\Delta(x^0)}{dx^0}$ is invertible, and

$$\| [\frac{d\Delta(x^0)}{dx^0}]^{-1} \| \leq N_0^* = \frac{1}{1-l_0}.$$

Proof. It is clear that the assertion of this lemma follows directly from the relation

$$\| \frac{d\Delta(x)}{dx} - \frac{d\Delta_m(x)}{dx} \| \to 0 \quad \text{as} \quad m \to \infty. \tag{3.80}$$

First, we estimate the difference

$$I_n^m(x) = \frac{d\tilde{x}_n(x)}{dx} - \frac{dx_n^{(m)}(x)}{dx} :$$

$$\|I_n^m(x)\| = \lim_{p \to +\infty} \| \frac{dx_n^{(m+p)}(x)}{dx} - \frac{dx_n^{(m)}(x)}{dx} \|$$

$$\leq \lim_{p \to +\infty} \sum_{i=1}^{p} \| \frac{dx_n^{(m+i)}(x)}{dx} - \frac{dx_n^{(m+i-1)}(x)}{dx} \|.$$

Using Lemma 3.10 for $m \geq 2$, $n \in Z_N \backslash \{0\}$, and $\hat{\eta} \neq \hat{\gamma}$ (e.g., $\hat{\eta} > \hat{\gamma}$), we obtain the inequalities

$$\|I_n^m(x)\| \leq \sum_{i=1}^{\infty} (\frac{\hat{\eta}\beta}{P(1-\hat{\eta})(\hat{\eta}-\hat{\gamma})} + 1)\hat{\eta}^{m+i-1} \leq \frac{\hat{\eta}^m}{1-\hat{\eta}} (\frac{\beta}{P(1-\hat{\eta})(\hat{\eta}-\hat{\gamma})} + 1).$$

They imply that $\|I_n^m(x)\| \to 0$ as $m \to \infty$, because $\hat{\eta} < 1$.

But if $\hat{\eta} = \hat{\gamma}$, then

$$\|I_n^m(x)\| \le \sum_{i=1}^{\infty} (\frac{\beta}{P(1-\hat{\gamma})} + 1)\hat{\gamma}^{m+i-1}(m+i-1)$$

$$= \hat{\gamma}^{m-1}(\frac{\beta}{P(1-\hat{\gamma})} + 1)(\sum_{i=1}^{\infty} i\hat{\gamma}^i + \frac{(m-1)\hat{\gamma}}{1-\hat{\gamma}}) \to 0$$

as $m \to \infty$, since the series $\sum_{i=1}^{\infty} i\hat{\gamma}^i$ converges to some number l^*.

Now, we have

$$\|\frac{d\Delta(x)}{dx} - \frac{d\Delta_m(x)}{dx}\|$$

$$\le \frac{1}{N} \sum_{s=0}^{N-1} \{\|\frac{df_s(\tilde{x}_s(x), \tilde{x}_{s+1}(x))}{d(\tilde{x}_s, \tilde{x}_{s+1})} - \frac{df_s(x_s^{(m)}(x), x_{s+1}^{(m)}(x))}{d(x_s^{(m)}, x_{s+1}^{(m)})}\|\|\frac{d\tilde{g}_s(x)}{dx}\|$$

$$+ \|\frac{df_s(x_s^{(m)}(x), x_{s+1}^{(m)}(x))}{d(x_s^{(m)}, x_{s+1}^{(m)})}\|\|\frac{d\tilde{g}_s(x)}{dx} - \frac{dg_s^{(m)}(x)}{dx}\|\}$$

$$\le \frac{1}{N} \sum_{s=0}^{N-1} \{L_0 \max\{\|\tilde{x}_s(x) - x_s^{(m)}(x)\|, \|\tilde{x}_{s+1}(x) - x_{s+1}^{(m)}(x)\|\}$$

$$\times \max\{\|\frac{d\tilde{x}_s(x)}{dx}\|, \|\frac{d\tilde{x}_{s+1}(x)}{dx}\|\} + P \max\{\|I_s^m(x)\|, \|I_{s+1}^m(x)\|\}\}$$

$$\le L_0\sigma^*(m)\frac{1}{1-\hat{\eta}} + P\sigma_*(m) \to 0 \quad \text{as} \quad m \to \infty,$$

because the factor $\sigma^*(m)$ defined in Theorem 3.19 and the factor $\sigma_*(m)$ defined for $m \ge 2$ by the relation

$$\sigma_*(m) = \begin{cases} \dfrac{\hat{\eta}^m}{1-\hat{\eta}}(\dfrac{\beta}{P(1-\hat{\eta})(\hat{\eta}-\hat{\gamma})} + 1), & \text{if} \quad \hat{\eta} > \hat{\gamma}; \\[2ex] \hat{\gamma}^{m-1}(\dfrac{\beta}{P(1-\hat{\gamma})} + 1)(l^* + \dfrac{(m-1)\hat{\gamma}}{1-\hat{\gamma}}), & \text{if} \quad \hat{\eta} = \hat{\gamma}; \\[2ex] \dfrac{\hat{\gamma}^m}{1-\hat{\gamma}}(\dfrac{\beta}{P(1-\hat{\eta})(\hat{\gamma}-\hat{\eta})} + 1) & \text{if} \quad \hat{\eta} < \hat{\gamma}, \end{cases}$$

tend to zero as $m \to \infty$. This completes the proof of estimate (3.80) and also Lemma 3.12. $\qquad\square$

We retain the formal notation

$$N_0 = \|[\Delta'(x^0)]^{-1}\|, \quad k = \|[\Delta'(x^0)]^{-1}\Delta(x^0)\|, \quad h = N_0 k L,$$

and t_0 (the least root of the equation $ht^2 - t + 1 = 0$) from the previous subsection and formulate the following proposition that establishes the sufficient conditions of existence of the N-periodic solution of Eq. (3.66).

Theorem 3.20. *Let conditions (V) and the following requirements be satisfied:*

1) $\forall n \in Z_N$, the function $f_n(x, y)$ is Fréchet-differentiable in the domain D_ρ^2, and, $\forall \{x, y, x_1, y_1\} \subset D_\rho$,

$$\|\frac{df_n(x, y)}{d(x, y)}\| \le P, \quad \|\frac{df_n(x, y)}{d(x, y)}|_{(x_2,y_2)}^{(x_1,y_1)}\| \le L_0 \max\{\|x_1 - x_2\|, \|y_1 - y_2\|\},$$

where P and L_0 are positive constants, and $N^2 P/4 < 1$;

2) there exist a point $x^0 \in D^f$ and the sequence of indices $p_1 < p_2 < p_3 < \cdots < p_k < \ldots$ such that, $\forall s \in Z^+$,

$$\|\frac{d\Delta_{p_s}(x^0)}{dx^0} - E\| \le l_0 < 1;$$

3) the constants L and N_0^* are such that $h^* = LN_0^{*2}M < 0.25$, and the closed ball $B^*(x^0, k^*t^*) \subset D^f$, where $k^* = N_0^*M$, and t^* is the least root of the equation $h^*t^2 - t + 1 = 0$.

Then there exists the single point x^* generating an N-periodic solution $\tilde{x}_n(x^*)$, $\tilde{x}_0(x^*) = x^*$ of Eq. (3.66) in the closed ball $B(x^0, kt_0) \subset B^*(x^0, k^*t^*)$. In this case, the inequality

$$\|\tilde{x}_n(x^*) - x_n^{(p)}(x_p)\| \le \frac{M}{1 - \hat{\gamma}}\left(\frac{N_0^* q^{*p}}{1 - q^*} + \frac{N^2 \hat{\gamma}^p}{4}\right) \tag{3.81}$$

holds. Here, $q^* = (1 - \sqrt{1 - 4h^*})/2 < 0.5$, the functions $x_n^{(p)}(x_p)$ are defined by relations (3.72), and $\{x_p\}_{p=1}^\infty$ is the sequence defined by the recurrence relation

$$x_0 = x^0, \quad x_{p+1} = x_p - [\frac{d\Delta(x^0)}{dx^0}]^{-1}\Delta(x_p), \quad p \in Z^+.$$

Proof. Since $k \le k^*$, $h \le h^*$, we have $t_0 \le t^*$, and the ball $B(x^0, kt_0)$ is embedded in the ball $B^*(x^0, k^*t^*)$ and, hence, in the set D^f. This allows us to obtain the estimate

$$\|x^* - x_p\| \le \frac{q^{*p}}{1 - q^*} N_0^* M,$$

which yields the inequalities

$$\|\tilde{x}_n(x^*) - x_n^{(p)}(x_p)\| \le \|\tilde{x}_n(x^*) - \tilde{x}_n(x_p)\| + \|\tilde{x}_n(x_p) - x_n^{(p)}(x_p)\|$$

$$\le \frac{\|x^* - x_p\|}{1 - \hat{\gamma}} + \sigma^*(p) \le \frac{q^{*p}}{1 - q^*} N_0^* M \frac{1}{1 - \hat{\gamma}} + \frac{(\frac{N^2}{4})^{p+1} M(K_1 + K_2)^p}{1 - \hat{\gamma}}.$$

$$\tag{3.82}$$

The last estimate yields inequality (3.81), which completes the proof of the theorem. $\qquad\square$

Let us consider the equation

$$\Delta_{p_s}(x) = \overline{f_n(x_n^{(p_s)}(x), x_{n+1}^{(p_s)}(x))} = 0. \tag{3.83}$$

By analogy with the proof of Lemma 3.11, it is easy to verify that, for all $\{x_1, x_2\} \subset D^f$ and $s \in Z^+$, the inequality

$$\left\| \frac{d\Delta_{p_s}(x)}{dx}\big|_{x_2}^{x_1} \right\| \le L\|x_1 - x_2\|$$

holds. In addition, it is obvious that

$$\left\| \left[\frac{d\Delta_{p_s}(x^0)}{dx^0} \right]^{-1} \right\| \le N_0^*.$$

Therefore, under conditions of Theorem 3.20, Eq. (3.83) for every $s \in Z^+$ in the closed ball $B^*(x^0, k^*t^*)$ has a solution x_{p_s}, which is the limit, as $k \to \infty$, of the sequence defined by the recurrence formula

$$x_{k+1}^{(p_s)} = x_k^{(p_s)} - \left[\frac{d\Delta_{p_s}(x^0)}{dx^0} \right]^{-1} \Delta_{p_s}(x_k^{(p_s)}).$$

For all $s \in Z^+$, $k \in Z_0^+$, the points $x_k^{(p_s)}$ and the point x^* belong to the ball $B^*(x^0, k^*t^*)$.

We introduce the notation

$$l_* = \sum_{i=1}^{\infty} i l_0^{i-1}; \quad G = 1 + \frac{N_0^*(K_1 + K_2)}{1 - \hat{\gamma}};$$

$$\varepsilon(p_s) = l_* M \{ L_0 \sigma^*(p_s) \frac{1}{1 - \hat{\eta}} + P\sigma_*(p_s) \} + N_0^*(K_1 + K_2)\sigma^*(p_s);$$

$$\delta(p_s) = N_0^*(K_1 + K_2)\sigma^*(p_s) + M \{ L_0 \sigma^*(p_s) \frac{1}{1 - \hat{\eta}} + P\sigma_*(p_s) \}.$$

Since $\sigma^*(p_s)$ and $\sigma_*(p_s)$ tend to zero as $s \to \infty$, $\varepsilon(p_s)$ and $\delta(p_s)$ have the same property.

We now formulate the assertion that allows one to approximate the function $\tilde{x}_n(x^*)$ with the function $x_n^{(m)}(x_k^{(p_s)})$ with arbitrary preassigned accuracy.

Theorem 3.21. *Under conditions of Theorem 3.20, the relation*

$$\lim_{k \to \infty} \lim_{s \to \infty} x_k^{(p_s)} = x^* \tag{3.84}$$

is valid, and, for $p_s > 2$ and $m > 2$, the inequalities

$$\|\tilde{x}_n(x^*) - x_n^{(m)}(x_k^{(p_s)})\| \leq \frac{\|x^* - x_k^{(p_s)}\|}{1 - \hat{\gamma}} + \sigma^*(m), \qquad (3.85)$$

$$\|x^* - x_k^{(p_s)}\| \leq \frac{q^{*k}}{1 - q^*} N_0^* M + \varepsilon(p_s) G^{k-1} + \delta(p_s) \sum_{i=0}^{k-2} G^i \qquad (3.86)$$

hold.

Proof. Estimate (3.85) is obtained analogously to estimate (3.82). We now show that, for $k \in Z_0^+$,

$$\lim_{s \to \infty} x_k^{(p_s)} = x_k, \qquad (3.87)$$

where $\{x_k\}$ is the sequence of points, which is obtained with the help of the modified Newton–Kantorovich method, which consists in the application of the recurrence formula indicated in Theorem 3.20 for $x_0 = x^0$. The inequality

$$\left\|\left[\frac{d\Delta(x^0)}{dx^0}\right]^{-1} - \left[\frac{d\Delta_{p_s}(x^0)}{dx^0}\right]^{-1}\right\| \leq \left\|\frac{d\Delta(x^0)}{dx^0} - \frac{d\Delta_{p_s}(x^0)}{dx^0}\right\| l_*$$

yields the estimates

$$\|x_1 - x_1^{(p_s)}\| \leq \left\|\frac{d\Delta(x^0)}{dx^0} - \frac{d\Delta_{p_s}(x^0)}{dx^0}\right\| l_* M + N_0^*(K_1 + K_2)\sigma^*(p_s) \leq \varepsilon(p_s).$$

We now write the recurrence relation

$$\|x_k - x_k^{(p_s)}\| \leq \|x_{k-1} - x_{k-1}^{(p_s)}\|$$

$$+ \left\{L_0\sigma^*(p_s)\frac{1}{1 - \hat{\eta}} + P\sigma_*(p_s)\right\}M + N_0^*\|\Delta(x_{k-1}) - \Delta_{p_s}(x_{k-1}^{(p_s)})\|$$

$$\leq \|x_{k-1} - x_{k-1}^{(p_s)}\|\left\{1 + \frac{N_0^*(K_1 + K_2)}{1 - \hat{\gamma}}\right\} + N_0^*(K_1 + K_2)\sigma^*(p_s)$$

$$+ \left\{L_0\sigma^*(p_s)\frac{1}{1 - \hat{\eta}} + P\sigma_*(p_s)\right\}M \leq \|x_{k-1} - x_{k-1}^{(p_s)}\|G + \delta(p_s).$$

Using it and the inductive reasoning, we obtain the estimate

$$\|x_k - x_k^{(p_s)}\| \leq \varepsilon(p_s) G^{k-1} + \delta(p_s) \sum_{i=0}^{k-2} G^i,$$

which yields relation (3.87), where the limiting transition is not uniform in $k \in Z_0^+$. From the last inequality, it is easy to obtain inequality (3.86), which guarantees the validity of relation (3.84). Theorem 3.21 is proved. We note that its proof does not yield the commutativity of the double limit in (3.84). $\qquad\square$

We now consider the equation

$$\Delta^2 x_n = f_n(x_n, x_{n+1}), \quad n \in Z, \tag{3.88}$$

where $x_n \in \mathfrak{M}$, $f_n(x, y) = \{f_n^{(1)}(x, y), f_n^{(2)}(x, y), f_n^{(3)}(x, y), \dots\} : \mathfrak{M} \times \mathfrak{M} \to \mathfrak{M}$ for each $n \in Z$, and write the corresponding truncated equation

$$\Delta^2 \overset{(p)}{x}_n = \overset{(p)}{f}_n(\overset{(p)}{x}_n, \overset{(p)}{x}_{n+1}), \quad n \in Z, p \in Z^+, \tag{3.89}$$

where

$$\overset{(p)}{x} = (x^1, x^2, \dots, x^p, 0, 0, 0, \dots),$$

$$\overset{(p)}{f}_n(x, y) = \{f_n^{(1)}(x, y), f_n^{(2)}(x, y), \dots, f_n^{(p)}(x, y), 0, 0, 0, \dots\}.$$

As was shown above under conditions (V) for Eq. (3.88), there exists the single control $\mu \in \mathfrak{M}$ such that the perturbed equation

$$\Delta^2 x_n = f_n(x_n, x_{n+1}) - \mu, \quad n \in Z_0^+, \tag{3.90}$$

has the unique N-periodic solution $x_n = x_n(x_0, x_1)$ taking a value $x_0 \in \bar{D}^f$ for $n = 0$. This solution coincides with the function $\tilde{x}_n(x_0)$ constructed by the above-indicated scheme, and the control $\mu = \overline{f_n(\tilde{x}_n, \tilde{x}_{n+1})} \in \mathfrak{M}$. The last assertion is also valid for Eq. (3.89). In other words, there exists the single control $\overset{(p)}{\mu}$ for every $p \in Z^+$ such that the equation

$$\Delta^2 \overset{(p)}{x}_n = \overset{(p)}{f}_n(\overset{(p)}{x}_n, \overset{(p)}{x}_{n+1}) - \overset{(p)}{\mu}, \quad n \in Z_0^+, \tag{3.91}$$

has the unique N-periodic solution $\widetilde{\overset{(p)}{x}}_n(\overset{(p)}{x}_0)$ taking a value $\overset{(p)}{x}_0$ for $n = 0$. In this case, the function $\overset{(p)}{x}_n(\overset{(p)}{x}_0)$ can be constructed analogously to the function $\tilde{x}_n(x_0)$, and $\overset{(p)}{\mu} = \widetilde{\overset{(p)}{f}_n(\overset{(p)}{x}_n, \overset{(p)}{x}_{n+1})}$.

We say that the function $f_n(x, y)$ satisfies the sharpened Cauchy–Lipschitz conditions on the set D_1, if the inequality

$$\|f_n(x, y) - f_n(x', y')\| \leq \varepsilon_1(m)\|x - x'\| + \varepsilon_2(m)\|y - y'\|,$$

where $\varepsilon_1(m) \to 0$ and $\varepsilon_2(m) \to 0$ as $m \to \infty$, holds $\forall n \in Z$. Here, $\{x, x', y, y'\} \subset D$, and m first corresponding coordinates of the points x and x' and m first corresponding coordinates of the points y and y' pairwise coincide.

Theorem 3.22. *Let conditions (V) be satisfied for Eq. (3.88), and let the function $f_n(x, y)$ satisfy the sharpened Cauchy–Lipschitz conditions on the set D_1. Then the relation*

$$\tilde{x}_n(x_0) = \lim_{p \to \infty} \lim_{s \to \infty} \overset{(p)}{x}{}_n^{(s)}(\overset{(p)}{x}{}_0), \quad n \in Z^+ \tag{3.92}$$

holds. Here, the internal and external limits are understood in the meaning of the norm and in the coordinatewise meaning, respectively.

Proof. It is obvious that, for any $x_0 \in \bar{D}^f$ the values of the functions $\overset{(p)}{x}{}_n^{(s)}(\overset{(p)}{x}{}_0)$ do not leave the set D for all $\{n, p, s\} \subset Z^+$, i.e., $\|\overset{(p)}{x}{}_n^{(s)}(\overset{(p)}{x}{}_0)\| \leq R$.

Since $\widetilde{\overset{(p)}{x}{}_n}(\overset{(p)}{x}{}_0) = \lim_{s \to \infty} \overset{(p)}{x}{}_n^{(s)}(\overset{(p)}{x}{}_0)$, we have $\forall \{n, p\} \subset Z^+$ $\widetilde{\overset{(p)}{x}{}_n}(\overset{(p)}{x}{}_0) \in D$, which yields the inequality

$$\|\overset{(p)}{\mu}\| \leq \frac{1}{N} \sum_{i=0}^{N-1} \|\overset{(p)}{f}{}_i(\widetilde{\overset{(p)}{x}{}_i}(\overset{(p)}{x}{}_0), \widetilde{\overset{(p)}{x}{}_{n+1}}(\overset{(p)}{x}{}_0))\| \leq M.$$

Using the method of diagonalization, we can choose a coordinatewise convergent subsequence from every bounded sequence in the space $\mathfrak{M}$. Therefore, there exists a sequence of indices $\{p_k\}_{k=1}^{\infty}$ such that the sequences $\{\widetilde{\overset{(p_k)}{\mu}}\}_{k=1}^{\infty}$ and $\{\widetilde{\overset{(p_k)}{x}{}_n}(\overset{(p_k)}{x}{}_0)\}_{k=1}^{\infty}$ for all $n \in Z^+$ converge to some $\bar{\mu}$ and $\bar{x}_n(x_0)$, respectively, as $k \to \infty$.

The following inequalities hold:

$$I_k = \|f_n(\bar{x}_n(x_0), \bar{x}_{n+1}(x_0)) - f_n(\widetilde{\overset{(p_k)}{x}{}_n}(\overset{(p_k)}{x}{}_0), \widetilde{\overset{(p_k)}{x}{}_{n+1}}(\overset{(p_k)}{x}{}_0))\|$$

$$\leq \|f_n(\bar{x}_n(x_0), \bar{x}_{n+1}(x_0)) - f_n((\bar{x}_n^1, \bar{x}_n^2, \ldots, \bar{x}_n^g, x_n^{g+1},$$

$$x_n^{g+2}, \ldots), (\bar{x}_{n+1}^1, \bar{x}_{n+1}^2, \ldots, \bar{x}_{n+1}^g, x_{n+1}^{g+1}, x_{n+1}^{g+2}, \ldots))\|$$

$$+ \|f_n((\bar{x}_n^1, \bar{x}_n^2, \ldots, \bar{x}_n^g, x_n^{g+1}, x_n^{g+2}, \ldots), (\bar{x}_{n+1}^1, \bar{x}_{n+1}^2, \ldots, \bar{x}_{n+1}^g,$$

$$x_{n+1}^{g+1}, x_{n+1}^{g+2}, \ldots)) - f_n(\widetilde{\overset{(p_k)}{x}{}_n}(\overset{(p_k)}{x}{}_0), \widetilde{\overset{(p_k)}{x}{}_{n+1}}(\overset{(p_k)}{x}{}_0))\|$$

$$\leq \varepsilon_1(g)\|\bar{x}_n - \widetilde{\overset{(p_k)}{x}{}_n}\| + \varepsilon_2(g)\|\bar{x}_{n+1} - \widetilde{\overset{(p_k)}{x}{}_{n+1}}\|$$

$$+ \varepsilon_1(0)\|(\bar{x}_n^1 - \widetilde{\overset{(p_k)}{x}{}_n^1}, \bar{x}_n^2 - \widetilde{\overset{(p_k)}{x}{}_n^2}, \ldots, \bar{x}_n^g - \widetilde{\overset{(p_k)}{x}{}_n^g}, 0, 0, \ldots)\|$$

$$+ \varepsilon_2(0)\|(\bar{x}^1_{n+1} - \widetilde{x^1_{n+1}}^{(p_k)}, \bar{x}^2_{n+1} - \widetilde{x^2_{n+1}}^{(p_k)}, \ldots, \bar{x}^g_{n+1} - \widetilde{x^g_{n+1}}^{(p_k)}, 0, 0, \ldots)\|$$

$$\leq 2R(\varepsilon_1(g) + \varepsilon_2(g)) + \varepsilon_1(0) \sup_{1 \leq i \leq g} |\bar{x}^i_n - \widetilde{x^i_n}^{(p_k)}| + \varepsilon_2(0) \sup_{1 \leq i \leq g} |\bar{x}^i_{n+1} - \widetilde{x^i_{n+1}}^{(p_k)}|.$$

Since $2R(\varepsilon_1(g) + \varepsilon_2(g)) \to 0$ as $g \to \infty$, there exists a number g^0 for any arbitrarily small real number $\nu > 0$ such that $2R(\varepsilon_1(g^0) + \varepsilon_2(g^0)) \leq \frac{\nu}{3}$. Then

$$I_k \leq \frac{\nu}{3} + \varepsilon_1(0) \sup_{1 \leq i \leq g^0} |\bar{x}^i_n - \widetilde{x^i_n}^{(p_k)}| + \varepsilon_2(0) \sup_{1 \leq i \leq g^0} |\bar{x}^i_{n+1} - \widetilde{x^i_{n+1}}^{(p_k)}|.$$

Let us fix a number $k = N$ such that, for $k \geq N$,

$$\sup_{1 \leq i \leq g^0} |\bar{x}^i_n - \widetilde{x^i_n}^{(p_k)}| \leq \frac{\nu}{3\varepsilon_1(0)}, \qquad \sup_{1 \leq i \leq g^0} |\bar{x}^i_{n+1} - \widetilde{x^i_{n+1}}^{(p_k)}| \leq \frac{\nu}{3\varepsilon_2(0)}.$$

Then we obtain the inequality $I_k \leq \nu$ for all $k \geq N$. In other words, for all $n \in Z^+$, the relation

$$\lim_{k \to \infty} f_n(\widetilde{x}_n^{(p_k)}(\widetilde{x}_0^{(p_k)}), \widetilde{x}_{n+1}^{(p_k)}(\widetilde{x}_0^{(p_k)})) = f_n(\bar{x}_n(x_0), \bar{x}_{n+1}(x_0))$$

holds.

In the equality

$$\Delta^2 \widetilde{x}_n^{(p_k)}(\widetilde{x}_0^{(p_k)}) = \widetilde{f}_n^{(p_k)}(\widetilde{x}_n^{(p_k)}(\widetilde{x}_0^{(p_k)}), \widetilde{x}_{n+1}^{(p_k)}(\widetilde{x}_0^{(p_k)})) - \widetilde{\mu}^{(p_k)}, \quad n \in Z^+,$$

we pass coordinatewise to the limit as $k \to \infty$ and obtain

$$\Delta^2 \bar{x}_n(x_0) = f_n(\bar{x}_n(x_0), \bar{x}_{n+1}(x_0)) - \bar{\mu}, \quad n \in Z^+.$$

The uniqueness of the control implies that $\bar{x}_n(x_0) = \tilde{x}_n(x_0)\ \forall n \in Z^+$ and $\bar{\mu} = \mu$. This means that the sequence $\left\{ \widetilde{x}_n^{(p_k)}(\widetilde{x}_0^{(p_k)}) \right\}_{k=1}^{\infty}$ converges $\forall n \in Z^+$ coordinatewise to $\tilde{x}_n(x_0)$ as $k \to \infty$.

Let us consider any subsequence of the sequence of equations (3.91):

$$\Delta^2 \overset{(r)}{x}_n = \overset{(r)}{f}_n(\overset{(r)}{x}_n, \overset{(r)}{x}_{n+1}) - \overset{(r)}{\mu}, \quad n \in Z_0^+, r \in Z^+.$$

It satisfies all the above-indicated reasoning, i.e., there exists a subsequence $\{\widetilde{\overset{(l)}{x}}_n(\overset{(l)}{x}_0)\}_{l=1}^{\infty}$ of the sequence $\{\overset{(r)}{x}_n(\overset{(r)}{x}_0)\}_{r=1}^{\infty}$, which converges coordinatewise to the function $\tilde{x}_n(x_0)$ as $l \to \infty$. In this case, $\overset{(l)}{\mu} \to \mu$ also in the coordinatewise meaning. Then the sequence $\{\overset{(p)}{x}_n(\overset{(p)}{x}_0)\}_{p=1}^{\infty}$ tends to $\tilde{x}_n(x_0)$ as $p \to \infty$ in the coordinatewise meaning $\forall n \in Z^+$. Obviously, $\lim_{p \to \infty} \overset{(p)}{\mu} = \mu$ in this case. This completes the proof of Theorem 3.22. $\qquad\square$

By concluding this subsection, we consider the equation

$$\Delta^2 x_n = A(n)x_n + B(n)x_{n+1}, \quad n \in Z, \tag{3.93}$$

in the space $\mathfrak{M}$. Here, $A(n) = [a_{ij}(n)]_{i,j=1}^{\infty}$ and $B(n) = [b_{ij}(n)]_{i,j=1}^{\infty}$ are infinite matrices N-periodic in n and such that $\|A(n)\| \leq K$, $\|B(n)\| \leq K$ $\forall n \in [0, N-1]_Z$, where $K = const < 2/N^2$. We also consider that the matrices $A(n)$ and $B(n)$ are not invertible $\forall n \in [0, N-1]_Z$ or the matrices inverse to them exist but do not belong to the set $\mathbf{\Gamma}$.

First, we note that conditions (V) are satisfied for Eq. (3.93). Indeed, for all $\{x, y\} \subset D$, the inequality

$$\|f_n(x, y)\| = \|A(n)x + B(n)y\| \leq 2KR$$

holds, the Lipschitz constants K_1 and K_2 are equal to K, and the set

$$\bar{D}^f = \{x \in D | \|x\| \leq R - \frac{KN^2 R}{2}\}$$

is not empty.

In this case, Eq. (3.89) corresponds to the equation

$$\Delta^2 \overset{(p)}{x}_n = \overset{(p)}{A}(n) \overset{(p)}{x}_n + \overset{(p)}{B}(n) \overset{(p)}{x}_{n+1}, \quad \{n, p\} \subset Z^+,$$

where p is the truncation order, Eq. (3.90) corresponds to the equation

$$\Delta^2 x_n = A(n)x_n + B(n)x_{n+1} - \mu, \quad n \in Z^+, \tag{3.94}$$

and Eq. (3.91) corresponds, in turn, to the equation

$$\Delta^2 \overset{(p)}{x}_n = \overset{(p)}{A}(n) \overset{(p)}{x}_n + \overset{(p)}{B}(n) \overset{(p)}{x}_{n+1} - \overset{(p)}{\mu}, \quad \{n, p\} \subset Z^+. \tag{3.95}$$

As above, we denote the N-periodic solution of Eq. (3.94) by $x_n = \tilde{x}_n(x_0)$ and the N-periodic solution of Eq. (3.95) by $\overset{(p)}{x}_n(\overset{(p)}{x}_0)$. These solutions can be constructed by the above-indicated procedure.

The following assertion is an analog of Corollary 3.5 formulated for Eq. (3.26).

Corollary 3.13. *If* $\|A(n) - \overset{(p)}{A}(n)\| \leq \eta_1(p)$, $\|B(n) - \overset{(p)}{B}(n)\| \leq \eta_2(p)$ $\forall n \in [0, N-1]_Z$, and $\eta_i(p) \to 0$ as $p \to \infty$ $(i = 1, 2)$, then, for any $x_0 \in \bar{D}^f$ such that $\|x_0 - \overset{(p)}{x}_0\| \to 0$ as $p \to \infty$, the limiting transitions in equality (3.92) are carried on in the meaning of the norm.*

Proof.   Let $\{x, x', y, y'\} \subset D$, let m first corresponding coordinates of the points x and x' coincide, and let this property occur also for the points y and y'. Then

$$\|A(n)x + B(n)y - (A(n)x' + B(n)y')\| \le \eta_1(m)\|x - x'\| + \eta_2(m)\|y - y'\|.$$

This inequality follows from the sharpened Cauchy–Lipschitz conditions for the function $A(n)x + B(n)y$, which ensures the validity of equality (3.92).

For all $n \in [0, N-1]_Z$, the following relations hold:

$$\|\tilde{x}_n(x_0) - \widetilde{\overset{(p)}{x}_n(\overset{(p)}{x}_0)}\| \le \|x_0 - \overset{(p)}{x}_0\| + \|g_n(\tilde{x}_n(x_0), \tilde{x}_{n+1}(x_0))$$

$$- g_n(\widetilde{\overset{(p)}{x}_n(\overset{(p)}{x}_0)}, \widetilde{\overset{(p)}{x}_{n+1}(\overset{(p)}{x}_0)})\| = \|x_0 - \overset{(p)}{x}_0\|$$

$$+ \|L^2(f_n(\tilde{x}_n(x_0), \tilde{x}_{n+1}(x_0)) - \overset{(p)}{f}_n(\widetilde{\overset{(p)}{x}_n(\overset{(p)}{x}_0)}, \widetilde{\overset{(p)}{x}_{n+1}(\overset{(p)}{x}_0)}))\|$$

$$\le \|x_0 - \overset{(p)}{x}_0\| + \frac{N^2}{4} \max_{0 \le i \le N-1} \|f_i(\tilde{x}_i(x_0), \tilde{x}_{i+1}(x_0)) - \overset{(p)}{f}_i(\widetilde{\overset{(p)}{x}_i(\overset{(p)}{x}_0)}, \widetilde{\overset{(p)}{x}_{i+1}(\overset{(p)}{x}_0)})\|$$

$$\le \|x_0 - \overset{(p)}{x}_0\| + \frac{N^2}{4} \max_{0 \le i \le N-1} \{R\|A(i) - \overset{(p)}{A}(i)\| + R\|B(i) - \overset{(p)}{B}(i)\|\}$$

$$+ \frac{KN^2}{4} \max_{0 \le i \le N-1} \|\tilde{x}_i(x_0) - \widetilde{\overset{(p)}{x}_i(\overset{(p)}{x}_0)}\| + \frac{KN^2}{4} \max_{0 \le i \le N-1} \|\tilde{x}_{i+1}(x_0) - \widetilde{\overset{(p)}{x}_{i+1}(\overset{(p)}{x}_0)}\|.$$

They yield the inequalities

$$\max_{0 \le i \le N-1} \|\tilde{x}_n(x_0) - \widetilde{\overset{(p)}{x}_n(\overset{(p)}{x}_0)}\| \le \frac{1}{1 - \frac{KN^2}{2}} \{\|x_0 - \overset{(p)}{x}_0\|$$

$$+ \frac{N^2 R}{4} \max_{0 \le i \le N-1} \{\|A(i) - \overset{(p)}{A}(i)\| + \|B(i) - \overset{(p)}{B}(i)\|\}\}$$

$$\le \frac{1}{1 - \frac{KN^2}{2}} \{\|x_0 - \overset{(p)}{x}_0\| + \frac{N^2 R}{4} (\eta_1(p) + \eta_2(p))\},$$

which complete the proof of Corollary 3.13.

Moreover, the estimate

$$\|\tilde{x}_n(x_0) - \overset{(p)}{x}_n^{(s)}(\overset{(p)}{x}_0)\| \le \frac{1}{1 - \frac{KN^2}{2}} \{\|x_0 - \overset{(p)}{x}_0\|$$

$$+ \frac{N^2 R}{4} (\eta_1(p) + \eta_2(p))\} + \frac{R(\frac{KN^2}{2})^{s+1}}{1 - \frac{KN^2}{2}},$$

holds $\forall n \in Z^+$, which allows us to approximate the function $x_n(x_0)$ by the function $\overset{(p)}{x}_n^{(s)}(\overset{(p)}{x}_0)$ with any preassigned accuracy. $\square$

Remark 3.3. Since the constant $K < 1$ for $N \geq 2$, the linear bounded operator $(E - A(n))x$ (E is the identity matrix), which transfers $\mathfrak{M}$ into $\mathfrak{M}$ with the help of the operation of multiplication of the matrix $E - A(n)$ by the vector $x \in \mathfrak{M}$, is invertible for all $n \in Z$. Then the solutions $x_n(x_0, x_1)$ of Eq. (3.93), including the N-periodic one, can be determined $\forall n \in Z^-$ by the recurrence relation

$$x_n = (E - A(n))^{-1}(2x_{n+1} + B(n)x_{n+1} - x_{n+2}).$$

3.4 Asymptotic periodicity of solutions of a linear equation in a complex Banach space

In the study of various discrete processes, one needs to consider the equation

$$x_{n+1} = Ax_n + b_n, \quad n \in Z_0^+, \tag{3.96}$$

in a complex Banach space $\mathfrak{B}$ with norm $\| * \|$, where A is a linear bounded operator in this space, $\{b_n\} = \{b_n\}_{n=0}^{\infty} = \{b_n, n \in Z_0^+\} \subset \mathfrak{B}$ is a T-periodic sequence, i.e., T is the least natural number such that $b_{n+T} = b_n$ for all $n \in Z_0^+$.

Let $\mathbf{L}(\mathfrak{B}, \mathfrak{B})$ be the Banach space of linear bounded operators acting from $\mathfrak{B}$ to $\mathfrak{B}$, $\sigma(A)$ is the spectrum of the operator $A \in \mathbf{L}(\mathfrak{B}, \mathfrak{B})$, and $\rho(A)$ is its resolvent set. By the symbol $\bigoplus$, we denote the algebraic direct sum of linear manifolds.

Theorem 3.23. *In order that the solutionxd*
$x = x(n, x_0), x_0 = x(0, x_0)$, of Eq. (3.96) be bounded on the set Z_0^+ for any $x_0 \in \mathfrak{B}$, and a T-periodic sequence $\{b_n\} \subset \mathfrak{B}$, it is necessary and sufficient that
1) $\sup_{n \geq 0} \|A^n\| < +\infty$;
2) $1 \in \rho(A^T)$.

First, we prove two auxiliary assertions.

Lemma 3.13. *Theorem 3.23 is valid for a 1-periodic sequence $\{b_n\} \subset \mathfrak{B}$ if and only if*
1) $\sup_{n \geq 0} \|A^n\| < +\infty$;
2) $1 \in \rho(A)$.

Proof. *Necessity.* We set $x_0 = 0 \in \mathfrak{B}$. If $b_n = b$ for all $n \in Z_0^+$, then, denoting the identity operator by A^0, we obtain the equality $x_{n+1} =$

$\sum_{k=0}^{n} A^k b$. By condition, there exists C_b such that, for all $n \in Z^+$, the inequality $\| \sum_{k=0}^{n} A^k b \| \le C_b$ holds.

Applying the principle of uniform boundedness to the family of operators $\{ \sum_{k=0}^{n} A^k, n \in Z^+ \}$, we see that there exists a constant $C > 0$ such that, for all $n \in Z^+$, the estimate $\| \sum_{k=0}^{n} A^k \| \le C$ holds.

If $1 \in \sigma(A)$, then

$$C \ge \| \sum_{k=0}^{n} A^k \| \ge \sup_{\lambda \in \sigma(A)} | \sum_{k=0}^{n} \lambda^k | \ge n + 1, \quad n \in Z^+.$$

We obtain the contradiction, which proves the necessity of condition 2 of Lemma 3.13.

Since

$$\| A^n \| = \| \sum_{k=0}^{n} A^k - \sum_{k=0}^{n-1} A^k \| \le 2C, \quad n \in Z^+,$$

the necessity of condition 1 of this lemma is also proved.

Sufficiency. Let $1 \in \rho(A)$, i.e., there exists $(A - I)^{-1} \in \mathbf{L}(\mathfrak{B}, \mathfrak{B})$, where I is the identity operator. From the identity

$$(\sum_{k=0}^{n} A^k)(A - I) = A^{n+1} - I,$$

we obtain

$$\sum_{k=0}^{n} A^k = (A^{n+1} - I)(A - I)^{-1}.$$

Whence, for all $n \in Z^+$, we have

$$\| \sum_{k=0}^{n} A^k \| \le (C + 1)\|(A - I)^{-1}\|,$$

where $C = \sup_{n \ge 0} \| A^n \|$. Thus, for all $n \in Z^+$,

$$\| x_{n+1} \| = \| A^{n+1} x_0 + \sum_{k=0}^{n} A^k b \| \le C \|x_0\| + (C + 1)\|b\|\|(A - I)^{-1}\|,$$

i.e., the solution $x = x(n, x_0)$ of Eq. (3.96) is bounded in $\mathfrak{B}$. $\qquad \square$

Lemma 3.14. *In order that the solution $x = x(n, x_0)$ of Eq. (3.96) be bounded on the set Z_0^+ for any $x_0 \in \mathfrak{B}$ and a T-periodic sequence $\{b_n\} \subset \mathfrak{B}$, it is necessary and sufficient that the sequence $\{x(kT, x_0), k \in Z_0^+\}$ be bounded for any $x_0 \in \mathfrak{B}$ and the T-periodic sequence $\{b_n\} \subset \mathfrak{B}$.*

Proof.   The *necessity* is obvious, since

$$\{x(kT, x_0), k \in Z_0^+\} \subset \{x(n, x_0), n \in Z_0^+\}.$$

Sufficiency. We set

$$M = \max\{ \max_{0 \le s \le T} \|A^s\|, \quad \max_{0 \le k \le T-1} \|\sum_{l=0}^{k} A^l b_{k-1}\|\}.$$

The T-periodicity of the sequence $\{b_n\}$ yields

$$x_{n+1} = A_{[n/T]T}^{n-[n/T]T+1} + \sum_{l=0}^{n-[n/T]T} A^l b_{n-l} =$$

$$= A_{[n/T]T}^{n-[n/T]T+1} + \sum_{l=0}^{n-[n/T]T} A^l b_{n-[n/T]T-l}.$$

Whence we have $\|x(n, x_0)\| \le M\|x([n/T]T, x_0)\|$, since $0 \le n - [n/T]T \le T-1$. By condition, the sequence $\{x(kT, x_0), k \in Z_0^+\}$ is bounded in $\mathfrak{B}$. Then the solution $x = x(n, x_0)$ of Eq. (3.96) has also this property.

$\square$

Proof of Theorem 3.23

Proof.   *Necessity.* We set

$$y_k = x(kT, x_0), \quad k \in Z_0^+$$

and show that the sequence $\{y_k\}$ can be presented in the form

$$y_{k+1} = A^T y_k + \sum_{l=0}^{T-1} A^l b_{T-l-1}. \quad k \in Z_0^+. \tag{3.97}$$

Indeed, for all $k \in Z_0^+$,

$$y_{k+1} = x((k+1)T, x_0) = A^{(k+1)T} x_0 + \sum_{l=0}^{(k+1)T-1} A^l b_{(k+1)T-l-1}$$

$$= A^{(k+1)T} x_0 + \sum_{l=T}^{(k+1)T-1} A^l b_{(k+1)T-l-1} + \sum_{l=0}^{T-1} A^l b_{T-l-1}$$

$$= A^T (A^{kT} x_0 + \sum_{l=0}^{kT-1} A^l b_{kT-l-1}) + \sum_{l=0}^{T-1} A^l b_{T-l-1}$$

$$= A^T y_k + \sum_{l=0}^{T-1} A^l b_{T-l-1}.$$

We set

$$b_n = \begin{cases} b, & \text{if } n = T - 1 (\text{mod } T), \\ 0 \in \mathfrak{B} & \text{in other cases.} \end{cases}$$

Then, in correspondence with (3.97),

$$y_{k+1} = A^T y_k + \sum_{l=1}^{T-1} A^l b_{T-l-1} + b_{T-1} = A^T y_k + b.$$

The arbitrariness of the choice of the element $b \in \mathfrak{B}$ and Lemma 3.13 yield the necessity of the conditions:

1) $\sup_{n \geq 0} \|A^{nT}\| < +\infty$;

2) $1 \in \rho(A^T)$.

The inequalities

$$\|A^n\| \leq \|A^{[n/T]T}\| \|A^{n-[n/T]T}\| \leq \sup_{0 \leq k \leq T-1} \|A^k\| \|A^{[n/T]T}\|$$

testify that the first condition is equivalent to the condition $\sup_{n \geq 0} \|A^n\| < +\infty$.

Sufficiency. Let us use equality (3.97). By Lemma 3.13 applied to the sequence $\{y_k\}$, this sequence is bounded, which leads, according to Lemma 3.14, to the boundedness of the solution $x = x(n, x_0)$ of Eq. (3.96). $\quad\square$

Definition 3.1. A solution $x = x_n = x(n, x_0)$ of Eq. (3.96) is called asymptotically periodic, if there exists a periodic sequence $\{c_n, n \in Z_0^+\} \subset \mathfrak{B}$ such that $\|x_n - c_n\| \to 0$ as $n \to \infty$.

We introduce the notation

$$\mathfrak{B}_a = \{x \in \mathfrak{B} | A^n x \to 0, n \to \infty\}, \quad \mathfrak{B}_c^T = \text{c.l.s.}\{x \in \mathfrak{B} | A^T x = \lambda x, |\lambda| = 1\},$$

$$\sigma_{per}(A^T) = \{\lambda \in \mathbb{C} | A^T x = \lambda x \text{ and there exists } s \in Z^+ : \lambda^s = 1\},$$

a closed linear shell (c.l.s.).

In what follows, we need the following assertion This allows us to obtain the estimate [166].

Proposition 3.1. *If, for any $x \in \mathfrak{B}$, the set $\overline{\{A^n x, n \geq 0\}}$ is compact in $\mathfrak{B}$, then $\mathfrak{B} = \mathfrak{B}_a \oplus \mathfrak{B}_c$, where $\mathfrak{B}_a$ and $\mathfrak{B}_c$ are invariant subspaces, which are defined by the above-indicated relations, and $\mathfrak{B}_c = \mathfrak{B}_c^1$.*

If A is the operator of contraction, then the operator $A_c = A|_{\mathfrak{B}_c}$, being the restriction of the operator A to the set $\mathfrak{B}_c$, is isometric, and $\mathfrak{B}_c$ is decomposed in an orthogonal topological direct sum of eigensubspaces. Moreover, the equality $\mathfrak{B} = \mathfrak{B}_a \oplus \mathfrak{B}_c$ implies that, $\forall x \in \mathfrak{B}$, the set $\overline{\{A^n x, n \geq 0\}}$ is a compact set in $\mathfrak{B}$.

Remark 3.4. Let A be the operator of contraction. If the number of eigensubspace A_c is finite, then their topological direct sum is closed by virtue of the orthogonality and, therefore, coincides with the algebraic one.

Theorem 3.24. *In order that, for any x_0 and a T-periodic sequence $\{b_n\} \subset \mathfrak{B}$, the solution $x = x(n, x_0)$ of Eq. (3.96) be asymptotically periodic, it is necessary and sufficient that the following conditions be satisfied:*

1) $1 \in \rho(A^T)$;

2) the set $\sigma_{per}(A^T)$ is finite, and

$$\{\lambda \in \mathbb{C} | A^T x = \lambda x, |\lambda| = 1\} = \sigma_{per}(A^T);$$

3) the set $\mathfrak{B}$ can be presented in the form of a direct sum $\mathfrak{B}_a \bigoplus \mathfrak{B}_c^T$.

Proof. *Necessity.* First, we consider the case where $T = 1$. According to Theorem 3.23, conditions 1 and 2 of Lemma 3.13 are necessarily satisfied.

In $\mathfrak{B}$, we introduce the norm by the formula

$$\|x\|_1 = \sup_{n \geq 0} \|A^n x\|, \quad x \in \mathfrak{B}.$$

Relative to this norm, which is equivalent to the initial norm, the operator A is a contraction one. Below, the norm $\| * \|_1$ will be denoted by $\| * \|$. We set $b_n = 0$ for $n \geq 0$. Then $x_n = A^n x_0$. We now prove that

$$\{\mu \in \mathbb{C} | Ax = \mu x, |\mu| = 1\} = \sigma_{per}(A),$$

i.e., that the relation $\lambda \in \{\mu \in \mathbb{C} | Ax = \mu x, |\mu| = 1\}$ yields the existence of a natural n such that $\lambda^n = 1$. Let us assume that, for any natural n, $\lambda^n \neq 1$. Hence, $\lambda = e^{i\pi\alpha}$, and α is an irrational number. In this case, for $x \in \mathfrak{B}$ such that $Ax = \lambda x, x \neq 0$, we have $A^n x = \lambda^n x = e^{in\pi\alpha} x$. Moreover, by virtue of the irrationality of α, we have

$$\overline{\{e^{in\pi\alpha}, n \geq 0\}} = \{z \in \mathbb{C} | |z| = 1\}$$

by the Kronecker theorem. This implies that the set of limiting points of the sequence $\{A^n x, n \geq 0\}$ is the set $\{zx | z \in \mathbb{C}, |z| = 1\}$. But this contradicts the asymptotic periodicity of the sequence $\{A^n x, n \geq 0\}$.

We now prove the finiteness of the set $\sigma_{per}(A)$. We consider the restriction A_c of the operator A to an invariant subspace $\mathfrak{B}_c$ and assume that $\sigma_{per}(A) = \sigma_{per}(A_c)$ is an infinite set. Let us choose a sequence $\{\lambda_n, n \geq 1\} \subset \sigma_{per}(A)$ such that $\lambda_m \neq \lambda_n$ for $n \neq m$. We consider the series $\sum_{n=1}^{\infty} y_n / n^2$, for which $A_c y_n = \lambda_m y_n, \|y_n\| = 1, n \geq 1$, and set

$$y = \sum_{n=1}^{\infty} y_n / n^2 \in \mathfrak{B}_c.$$

By assumption, the sequence $\{A_c^n y, n \geq 0\}$ is asymptotically periodic. Then there exists a natural number s such that $\|A_c^{n+s} y - A_c^n y\| \to 0$ as $n \to \infty$. But

$$\|A_c^{n+s} y - A_c^n y\| = \|A_c^s y - y\|$$

by virtue of the isometricity of the operator A_c. Whence we have $A_c^s y = y$, i.e.,

$$\sum_{n=1}^{\infty} \frac{\lambda_n^s y_n}{n^2} = \sum_{n=1}^{\infty} \frac{y_n}{n^2} \quad \Rightarrow \quad \sum_{n=1}^{\infty} \frac{(\lambda_n^s - 1)y_n}{n^2} = 0 \in \mathfrak{B}.$$

Since the sequence $\{y_n, n \geq 0\}$ is composed by construction from linearly independent elements, we obtain $\{\lambda_n, n \geq 1\} \subset \{e^{2\pi i l/s}, \quad l \in Z_0^+, 0 \leq l \leq s - 1\}$, which contradicts the assumption that $\lambda_m \neq \lambda_n$ for $n \neq m$. Thus, for $T = 1$, the necessity is proved.

The case where $T > 1$ is reduced to the previous one: if the sequence $\{x_n, n \geq 0\}$ is asymptotically periodic, then the sequence $\{z_k = x_{kT}, k \geq 0\}$ is asymptotically periodic and satisfies the equality

$$z_{k+1} = A^T z_k + \sum_{l=0}^{T-1} A^l b_{T-l-1}.$$

Choosing a T-periodic sequence $\{b_n, n \geq 0\}$ so that

$$b_{T-1} = -\sum_{l=0}^{T-1} A^l b_{T-l-1},$$

we obtain $z_{k+1} = A^T z_k, k \geq 0$. (It is clear that $A^{nT} x \to 0 \Leftrightarrow A^n x \to 0$ as $n \to \infty$).

Sufficiency. Let $T = 1$. Then $x_{n+1} = Ax_n + b, \quad n \geq 0, \quad b \in \mathfrak{B}$, which yields $x_n = A^n(x_0 + (A - I)^{-1}b) - (A - I)^{-1}b, \quad n \geq 1$. Therefore, it is sufficient to show that, for any $x \in \mathfrak{B}$, the sequence $\{x_n, n \geq 0\}$ given by the relation $x_{n+1} = Ax_n, n \geq 0, x_0 = x$ is asymptotically periodic. By the second condition of the theorem, $\sigma_{per}(A)$ is a finite set, and

$$\sigma_{per}(A) = \{\lambda \in \mathbb{C} | Ax = \lambda x, |\lambda| = 1\}.$$

Let

$$\sigma_{per}(A) = \{\lambda_i \in \mathbb{C} | 1 \leq i \leq m, \quad \forall i \quad \text{there exists} \quad n_i \in Z^+ : \lambda_i^{n_i} = 1\}.$$

By the third condition of the theorem, $\mathfrak{B} = \mathfrak{B}_a \oplus \mathfrak{B}_c$. In view of Proposition 3.1 and Remark 3.4, we obtain the representation

$$\mathfrak{B}_c = \bigoplus_{i=1}^{m} Ker(A - \lambda_i I),$$

because, by analogy with the proof of the necessity, we can consider the operator A to be a contraction one. Thus, for $x \in \mathfrak{B}$, the equality $x = x_a + x_c$ holds. Here, $x_a \in \mathfrak{B}_a$ and $x_c \in \mathfrak{B}_c$. Moreover, for $x_c \in \mathfrak{B}_c$, there exists

$$\{x_c^i | 1 \le i \le m, \forall i : Ax_c^i = \lambda_i x_c^i, \lambda_i^n = 1 \quad \text{and} \quad x_c = \sum_{i=1}^{m} x_c^i\}.$$

Then the sequence $\{A^k x_c, k \ge 0\}$ is periodic. Indeed, by introducing the notation $p = \mathrm{LCM}(n_1, \ldots, n_m)$, we obtain the equalities

$$A^p x_c = \sum_{i=1}^{m} \lambda_i^p x_c^i = \sum_{i=1}^{m} x_c^i = x_c.$$

On the other hand, the limiting transition $A^n x_a \to 0$ is valid $\forall x_a \in \mathfrak{B}_a$ as $n \to \infty$. Therefore, by taking the sequence $\{A^k x_c, k \ge 0\}$ as a sequence $\{c_k, k \ge 0\}$ in the definition of asymptotic periodicity for $\{x_k, k \ge 0\}$, we obtain the relation

$$\|x_k - c_k\| = \|A^k(x_a + x_c) - A^k x_c\| = \|A^k x_a\| \to 0, \quad k \to \infty.$$

Consider the case where $T > 1$. According to the proved sufficiency in the case where $T = 1$ and equality (3.97), we obtain that the sequence $\{x_{kT}, k \ge 0\}$ is asymptotically periodic. Then the sequence $\{x_{kT+r}, k \ge 0\}$ is also asymptotically periodic $\forall r : 1 \le r \le T - 1$. Indeed,

$$x_{kT+r} = A^r x_{kT} + \sum_{l=0}^{r-1} A^l b_{r-l-1}.$$

If $\{c_k, k \ge 0\} \subset \mathfrak{B}$ is a periodic sequence such that $\|x_{kT} - c_k\| \to 0$ as $k \to \infty$, then the periodic sequence

$$\{A^r c_k + \sum_{l=0}^{r-1} A^l b_{r-l-1}, k \ge 0\}$$

is asymptotically limiting for the sequence $\{x_{kT+r}, k \ge 0\}$. Since

$$\{x_n, n \ge 0\} = \bigcup_{r=0}^{T-1} \{x_{kT+r}, k \ge 0\},$$

the sequence $\{x_n, n \ge 0\}$ is asymptotically periodic, which completes the proof of Theorem 3.24. $\qquad\square$

3.5 Extension "to the left" of solutions of nonlinear degenerate difference equations

1^0. Equations of the first order. Consider the equation

$$x_{n+1} = x_n + f_n(x_n), \quad n \in Z_0^+, \tag{3.98}$$

where the mapping $f_n(x)$ is defined $\forall n \in Z_0^+$ on the set $\Omega_n \subset \mathbf{W}$ and takes values from $\mathbf{W}$, $\mathbf{W}$ is a Banach space, and the mapping $\Phi_n(x) = x + f_n(x)$ is defined on the set Ω_n and takes values from the set Ω_{n+1}, $x_0 \in \Omega_0$.

We now formulate the following problem: to find a solution $x_n(x_0)$ of Eq. (3.98) such that $x_k(x_0) = d \in \Omega_k$, where k is a fixed natural number.

We note that $x_k = \Phi_{k-1}\Phi_{k-2}...\Phi_0 x_0$. For this problem to be solvable, it is necessary and sufficient that the equation $\prod_{i=k-1}^{0} \Phi_i x_0 = d$ have at least one solution $x_0 \in \Omega_0$. If, $\forall n \in \{0, 1, 2, ..., k-1\}$, the mapping Φ_n is bijective, then such solution exists, is unique, and is determined by the formula $x_0 = \prod_{i=0}^{k-1} \Phi_i^{-1} d$. In the opposite case, Eq. (3.98) is called k-degenerate or simply degenerate. This equation can have no indicated solution or can have the infinite number of such solutions. It is obvious that the solution $x_n(x_0)$ of Eq. (3.98) satisfies the condition $x_k(x_0) = d$ only if x_0 is a solution of the equation $x_0 + \sum_{i=0}^{k-1} f_i(x_i) = d$.

The following conditions are called conditions (B):

1) for all $x \in D = \{x \in \mathbf{W}|\ \|x\| \le R = const > 0\}$ and $i \in \{0, 1, 2, ..., k-1\}$ $\|f_i(x)\| \le M = const > 0$;

2) the set $\overline{D_k} = \{x \in D|\ \|x\| \le R - kM\}$ is nonempty;

3) on the set $D_\rho = \{x \in \mathbf{W}|\ \|x\| < R + \rho\}$, the functions $f_i(x)$ ($i = \overline{1, k-1}$) are Fréchet-differentiable, and

$$\|f_i'(x)\| \le P = const > 0;$$

$$\|f_i'(x) - f_i'(\bar{x})\| \le L\|x - \bar{x}\|,$$

where $\{x, \bar{x}\} \subset D_\rho$, L and ρ are positive constants, and ρ is arbitrarily small.

We now introduce the notation $F(x) = x + \sum_{i=0}^{k-1} f_i(x_i(x)) - d$ and will consider the derivative only in the Fréchet meaning in this subsection.

Lemma 3.15. *If conditions (B) are satisfied, then the function $F(x)$ is differentiable on the set $D_k = \{x \in \bar{D}_k|\ \|x\| < R - kM\}$, and the derivative $F'(x)$ satisfies the Lipschitz condition*

$$\|F'(x) - F'(\bar{x})\| \le L_0\|x - \bar{x}\|, \quad \{x, \bar{x}\} \subset D_k,$$

where $L_0 = \frac{L}{P}(1+P)^{k-1}\{(1+P)^k - 1\}$.

Proof. With the help of the inductive reasoning, it is easy to show that, for each $i \in \{0, 1, 2, ..., k-1\}$, the point $x_i(x) \in D$, if $x \in \overline{D_k}$. Indeed, for $i = 0$, $x_0(x) = x \in \overline{D_k} \subset D$. The assumption that the point $x_i(x) \in D$ for $i = \overline{1, n}$, $(n \le k-2)$, yields the inequalities

$$\|x_{n+1}(x)\| \le \|x\| + \sum_{i=0}^{n} \|f_i(x_i(x))\| \le R - kM + (n+1)M \le R,$$

i.e., the point $x_{n+1}(x) \in D$.

We now prove that, for each $i \in \{0, 1, 2, ..., k-1\}$, $x_i(x)$ is the function differentiable on D_k, and

$$\|\frac{dx_i(x)}{dx}\| \le (1+P)^i. \tag{3.99}$$

This assertion is easily substantiated by the method of complete mathematical induction with regard for the equality

$$\frac{dx_{n+1}(x)}{dx} = \frac{dx_n(x)}{dx} + \frac{df_n(x_n(x))}{dx_n(x)} \cdot \frac{dx_n(x)}{dx}.$$

Inequality (3.99) yields the estimate

$$\|\frac{df_i(x_i(x))}{dx}\| \le P(1+P)^i, \quad i \in \{0, 1, 2, ..., k-1\}. \tag{3.100}$$

Moreover, the recurrence inequalities

$$\|x_i(x) - x_i(\bar{x})\| \le \|x - \bar{x}\| + \sum_{s=0}^{i-1} \|f_s(x_s(x)) - f_s(x_s(\bar{x}))\|$$

$$\le \|x - \bar{x}\| + P\sum_{s=0}^{i-1} \|x_s(x) - x_s(\bar{x})\|,$$

in which $i \in \{1, 2, 3, ..., k-1\}$, $\{x, \bar{x}\} \in D_k$, allow us to write the estimate

$$\|x_i(x) - x_i(\bar{x})\| \le \sum_{s=0}^{i} C_i^s P^s \|x - \bar{x}\|, \quad i = \overline{0, k-1}, \tag{3.101}$$

where C_i^s is the number of combinations of i elements from s ones. Indeed, estimate (3.101) is obvious for $i = 0$. By assuming that it is valid for $i = 1, 2, \ldots, p$ $(p \le k-2)$, we obtain that it holds for $i = p+1$:

$$\|x_{p+1}(x) - x_{p+1}(\bar{x})\| \le \|x - \bar{x}\|$$

$$+ P\sum_{s=0}^{p-1} \|x_s(x) - x_s(\bar{x})\| + P\|x_p(x) - x_p(\bar{x})\|$$

$$\le (1+P)\sum_{s=0}^{p} C_p^s P^s \|x - \bar{x}\| = \sum_{s=0}^{p+1} C_{p+1}^s P^s \|x - \bar{x}\|.$$

With regard for the relation $\left\|\frac{dx_0(x)}{dx} - \frac{dx_0(\bar{x})}{d\bar{x}}\right\| = 0$ and estimates (3.99) – (3.101), we write the following chain of inequalities for $i \in \{1, 2, ..., k-1\}$:

$$\left\|\frac{dx_i(x)}{dx} - \frac{dx_i(\bar{x})}{d\bar{x}}\right\|$$

$$\leq \left\|\frac{dx_{i-1}(x)}{dx} - \frac{dx_{i-1}(\bar{x})}{d\bar{x}}\right\|$$

$$+ \left\|\frac{df_{i-1}(x_{i-1}(x))}{dx_{i-1}(x)}\right\|\left\|\frac{dx_{i-1}(x)}{dx} - \frac{dx_{i-1}(\bar{x})}{d\bar{x}}\right\|$$

$$+ \left\|\frac{df_{i-1}(x_{i-1}(x))}{dx_{i-1}(x)} - \frac{df_{i-1}(x_{i-1}(\bar{x}))}{dx_{i-1}(\bar{x})}\right\|\left\|\frac{dx_{i-1}(\bar{x})}{d\bar{x}}\right\|$$

$$\leq (1+P)\left\|\frac{dx_{i-1}(x)}{dx} - \frac{dx_{i-1}(\bar{x})}{d\bar{x}}\right\|$$

$$+ L(1+P)^{i-1}\|x_{i-1}(x) - x_{i-1}(\bar{x})\|$$

$$\leq (1+P)^2\left\|\frac{dx_{i-2}(x)}{dx} - \frac{dx_{i-2}(\bar{x})}{d\bar{x}}\right\|$$

$$+ L(1+P)^{i-1}\|x_{i-2}(x) - x_{i-2}(\bar{x})\|$$

$$+ L(1+P)^{i-1}\|x_{i-1}(x) - x_{i-1}(\bar{x})\| \leq ...$$

$$\leq (1+P)^i\left\|\frac{dx_0(x)}{dx} - \frac{dx_0(\bar{x})}{d\bar{x}}\right\|$$

$$+ L(1+P)^{i-1}\sum_{s=0}^{i-1}\|x_s(x) - x_s(\bar{x})\|$$

$$\leq L(1+P)^{i-1}\sum_{s=0}^{i-1}\sum_{p=0}^{s}C_s^p P^p \|x - \bar{x}\|.$$

Now, we can estimate the norm of the difference of the operators $F'(x) - F'(\bar{x})$:

$$\|F'(x) - F'(\bar{x})\| \leq \sum_{i=0}^{k-1}\{\left\|\frac{df_i(x_i(x))}{dx_i(x)}\right\|\left\|\frac{dx_i(x)}{dx} - \frac{dx_i(\bar{x})}{d\bar{x}}\right\|$$

$$+ \left\|\frac{df_i(x_i(x))}{dx_i(x)} - \frac{df_i(x_i(\bar{x}))}{dx_i(\bar{x})}\right\|\left\|\frac{dx_i(\bar{x})}{d\bar{x}}\right\|\}$$

$$\leq \sum_{i=0}^{k-1}\{P\left\|\frac{dx_i(x)}{dx} - \frac{dx_i(\bar{x})}{d\bar{x}}\right\| + L(1+P)^i\|x_i(x) - x_i(\bar{x})\|\} \leq L\|x - \bar{x}\|$$

$$+ \sum_{i=1}^{k-1}\{PL(1+P)^{i-1}\sum_{s=0}^{i-1}\sum_{p=0}^{s}C_s^p P^p + L(1+P)^i\sum_{s=0}^{i}C_i^s P^s\}\|x - \bar{x}\|.$$

The last estimates prove Lemma 3.15. $\square$

Under conditions (B), the solution of the equation $F(x) = 0$ can be found with the help of the modified Newton–Kantorovich method. We now introduce the formal notation

$$N_0 = \frac{1}{2 - (1+P)^k}, \quad p = \|[F'(a)]^{-1}F(a)\|, \quad h = N_0 p L_0,$$

where point $a \in D_k$.

Theorem 3.25. *Let conditions (B) be satisfied, and let the inequalities* $(P+1)^k < 2$, $4h < 1$, *and* $p(1 - \sqrt{1 - 4h}) < 2h(R - kM - \|a\|)$ *hold. Then the closed ball*

$$\bar{S}\left(a, \frac{1 - \sqrt{1 - 4h}}{2h}p\right) \subset D_k$$

and it contains the unique solution x_0 *of the equation* $F(x) = 0$. *The sequence, which is defined by the recurrence relation*

$$x_{(n+1)} = x_{(n)} - [F'(a)]^{-1}F(x_{(n)}), \quad x_{(0)} = a,$$

converges to x_0, *and*

$$\|x_{(n)} - x_0\| \le \frac{g^n p}{1 - g} \le \frac{g^n(\|a\| + kM + \|d\|)}{(1 - g)(2 - (1+P)^k)},$$

where

$$g = \frac{1 - \sqrt{1 - 4h}}{2} < \frac{1}{2}.$$

Proof. By Lemma 3.15, the mapping $F(x)$ is differentiable on the set D_k, and the operator $F'(x)$ satisfies the Lipschitz condition with the coefficient L_0. Under conditions of the theorem, the derivative

$$F'(x) = E + \sum_{i=0}^{k-1} \frac{df_i(x_i(x))}{dx_i(x)} \frac{dx_i(x)}{dx},$$

where E is the identity operator.

Denoting $\sum_{i=0}^{k-1} \frac{df_i(x_i(x))}{dx_i(x)}\big|_{x=a}$ by B, we write the equality $F'(a) = E + B$. It is obvious that

$$\|B\| \le \sum_{i=0}^{k-1} \left\|\frac{df_i(x_i(x))}{dx_i(x)}\big|_{x=a}\right\| \le \sum_{i=0}^{k-1} P(1+P)^i$$

$$= P\frac{(1+P)^k - 1}{P} = (1+P)^k - 1 < 1.$$

In this case, the operator $F'(a)$ is invertible, and

$$\|[F'(a)]^{-1}\| = \|[E - (-B)]^{-1}\|$$

$$\leq \sum_{s=0}^{\infty} \|-B\|^s \leq \sum_{s=0}^{\infty} ((1+P)^k - 1)^s = \frac{1}{2 - (1+P)^k}.$$

Taking into account that the ball $\overline{S}$ belongs to D_k using [43], we complete the proof of Theorem 3.25. $\qquad\square$

We note that the method of contraction mappings cannot be directly applied to the solution of the equation $d - \sum_{i=0}^{k-1} f_i(x_i(x)) = x$, though estimate (3.100) and the inequality $(P+1)^k < 2$ imply that the norm of the derivative of the left-hand side of this equation with respect to x is strictly less than 1. The essence consists in that the mapping $d - \sum_{i=0}^{k-1} f_i(x_i(x))$ does transfer the ball $\overline{D_k}$ into itself under the condition that $\|d\| \leq R - 2kM$, whereas the second of conditions (B) does not ensure that the inequality $R - 2kM > 0$ holds.

We now present an illustrative example, which shows that the conditions of Theorem 3.25 are not contradictory.

Example 3.1. Consider the equation

$$x_{n+1} = x_n + E_n \sin x_n, \quad n \in Z_o^+, \tag{3.102}$$

where $x = colon\{x^{(1)}, x^{(2)}, x^{(3)}, ...\}$ belongs to the space $\mathfrak{M}$ of bounded number sequences with norm $\|x\| = \sup_i\{|x^{(i)}|, i = 1, 2, 3, ...\}$, and E_n is an infinite diagonal matrix with the diagonal

$$\{2^{-(n+1)}, 2^{-(n+2)}, 2^{-(n+3)}, ...\},$$

$$\sin x = colon\{\sin x^{(1)}, \sin x^{(2)}, \sin x^{(3)}, ...\}.$$

We set

$$k = 2, \quad d = \{\frac{1}{9}; \frac{1}{9}; \frac{1}{9}; ...\}.$$

It is clear that Eq. (3.102) can be presented in the form of the countable system of independent equations

$$x_{n+1}^{(s)} = x_n^{(s)} + \frac{1}{2^{n+s}} \sin x_n^{(s)}, \quad n \in Z_o^+, \quad s \in Z^+, \tag{3.103}$$

each of which is defined in the space R^1 .

We now verify conditions (B) for Eq. (3.103):

1) the inequalities

$$\left|\frac{1}{2^s}\sin x^{(s)}\right| \le \frac{1}{2^s}; \quad \left|\frac{1}{2^{s+1}}\sin x^{(s)}\right| \le \frac{1}{2^{s+1}}$$

hold $\forall x^{(s)} \in R^1$. Hence, we can set

$$M_s = \frac{1}{2^s} \le M = \frac{1}{2} \quad \forall s \in Z^+;$$

2) $kM_s = 1/2^{s-1}$, i.e., for $R_s > 1/2^{s-1}$, the set

$$\bar{D}_{k_s} = \{x^{(s)} \in R^1 \,||x^{(s)}| \le R_s - 2M_s\}$$

is not empty. It is obvious that, $\forall s \in Z^+$, we can take any number $R > 1$ as R_s;

3) since the Fréchet derivative of a scalar function coincides with its ordinary derivative,

$$\left|\left(\frac{1}{2^{s+1}}\sin x^{(s)}\right)'\right| = \left|\frac{1}{2^{s+1}}\cos x^{(s)}\right| \le \frac{1}{2^{s+1}}$$

on the whole set R^1, which allows one to set

$$P_s = \frac{1}{2^{s+1}} \le P = \frac{1}{4} \quad \forall s \in Z^+.$$

In addition, the inequality

$$\left|\frac{1}{2^{s+1}}\cos x^{(s)} - \frac{1}{2^{s+1}}\cos \bar{x}^{(s)}\right| \le \frac{1}{2^{s+1}}|x^{(s)} - \bar{x}^{(s)}|, \quad \{x^{(s)}, \bar{x}^{(s)}\} \subset R^1,$$

allows us to set

$$L_s = \frac{1}{2^{s+1}} \le L = \frac{1}{4} \quad \forall s \in Z^+.$$

Thus, conditions (B) are satisfied.

We write the equality

$$F_s(x^{(s)}) = x^{(s)} + \frac{1}{2^s}\sin x^{(s)} + \frac{1}{2^{s+1}}\sin(x^{(s)} + \frac{1}{2^s}\sin x^{(s)}) - \frac{1}{9}. \qquad (3.104)$$

Then

$$F'_s(x^{(s)}) = 1 + \frac{1}{2^s}\cos x^{(s)} + \frac{1}{2^{s+1}}\cos(x^{(s)} + \frac{1}{2^s}\sin x^{(s)})(1 + \frac{1}{2^s}\cos x^{(s)}).$$

It is obvious that

$$F_s(0) = -\frac{1}{9}; \quad F'_s(0) = 1 + \frac{1}{2^s} + \frac{1}{2^{s+1}}(1 + \frac{1}{2^s}).$$

By Lemma 3.15,

$$L_{0_s} = \frac{L_s}{P_s}(1 + P_s)\{(1 + P_s)^2 - 1\} = (1 + \frac{1}{2^{s+1}})(\frac{1}{2^s} + \frac{1}{2^{2(s+1)}}).$$

The derivative $F'_s(0)$ defines the mapping $R^1 \to R^1$, which is a homothety with coefficient

$$1 + \frac{1}{2^s} + \frac{1}{2^{s+1}}\left(1 + \frac{1}{2^s}\right).$$

The inverse mapping is a homothety with coefficient

$$\left\{1 + \frac{1}{2^s} + \frac{1}{2^{s+1}}\left(1 + \frac{1}{2^s}\right)\right\}^{-1}.$$

In this case,

$$p_s = \frac{1}{9} : \left\{1 + \frac{1}{2^s} + \frac{1}{2^{s+1}}\left(1 + \frac{1}{2^s}\right)\right\}, \quad N_{o_s} = \frac{1}{2 - \left(1 + \frac{1}{2^{s+1}}\right)^2};$$

$$h_s = N_{o_s} p_s L_{o_s} = \frac{1}{2 - \left(1 + \frac{1}{2^{s+1}}\right)^2}$$

$$\times \frac{\frac{1}{9}}{1 + \frac{1}{2^s} + \frac{1}{2^{s+1}}\left(1 + \frac{1}{2^s}\right)}\left(1 + \frac{1}{2^{s+1}}\right)\left(\frac{1}{2^s} + \frac{1}{2^{2(s+1)}}\right).$$

The condition $(P_s + 1)^2 < 2$ is satisfied $\forall s \in Z^+$, since

$$\left(\frac{1}{2^{s+1}} + 1\right)^2 \leq \left(\frac{1}{4} + 1\right)^2 = \frac{25}{16} < 2.$$

The condition $4h_s < 1$ is also satisfied $\forall s \in Z^+$, since

$$L_{o_s} \leq (1 + P)\{(1 + P)^2 - 1\} = \left(1 + \frac{1}{4}\right)\left(\frac{25}{16} - 1\right) = \frac{45}{64};$$

$$N_{o_s} \leq \frac{1}{2 - \frac{25}{16}} = \frac{16}{7}.$$

This implies that

$$4h_s \leq 4 \cdot \frac{45}{64} \cdot \frac{16}{7} \cdot \frac{1}{9} = \frac{5}{7} < 1.$$

The inequality $p(1 - \sqrt{1 - 4h}) < 2h(R - kM - \|a\|)$ from the condition of Theorem 3.25 takes the form $p_s(1 - \sqrt{1 - 4h_s}) < 2h_s(R - 2M_s)$ and is always satisfied $\forall s \in Z^+$, since $R > 1$ can be taken arbtrarily large.

Hence, for every natural s, there exists a number $x_o^{(s)}$ such that $x_2^{(s)}(x_o^{(s)}) = 1/9$. We need only to verify that the point $\{x_0^{(1)}, x_0^{(2)}, x_0^{(3)}, ...\}$ belongs to the space $\mathfrak{M}$. For this purpose, it is sufficient to show that the set of numbers $\rho_s = (1 - \sqrt{1 - 4h_s})p_s/2h_s$ is bounded from above $\forall s \in Z^+$.

The last assertion follows from the inequalities

$$\rho_s \leq \frac{1}{9}; \quad \frac{1 - \sqrt{1 - 4h_s}}{2h_s} = \frac{2}{1 + \sqrt{1 - 4h_s}} \leq 2 \quad \forall s \in Z^+.$$

The calculation indicates that, to within the seventh decimal point,

$$\rho_1 = 0.0663301, \quad \rho_2 = 0.0790594, \quad \rho_3 = 0.0943554,$$

$$\rho_4 = 0.1012378, \quad \rho_5 = 0.1064619, \quad \rho_6 = 0.1087466.$$

We note that, by equating the expression on the right-hand side of equality (3.104) to 0, we obtain the equation

$$\frac{1}{9} - \frac{1}{2^s} \sin x^{(s)} - \frac{1}{2^{s+1}} \sin(x^{(s)} + \frac{1}{2} \sin x^{(s)}) = x^{(s)}.$$

The modulus of the derivative of its left-hand side is strictly less than 1. This indicates the possibility to use the method of contraction mappings for the solution of the last equation. It is easy to see that Eq. (3.103) is not degenerate $\forall s \in Z^+$.

In the following proposition, we give the sufficient conditions of solvability of the above-formulated problem, under which the operator $F'(x)$ is the identity one.

Corollary 3.14. *Let conditions (B) be satisfied, and let there exist a point $a \in D_k$ such that, for all $n \in \{0, 1, 2, \ldots, k-1\}$ $f_n(a) = 0 \in \mathbf{W}$ and $f'_n(a)$ are zero operators. If, in this case, the inequality*

$$\frac{L}{P}(1+P)^{k-1}\{(1+P)^k - 1\} > \frac{1}{2(R - kM - \|a\|)}$$

is satisfied, then, for every $d \in \mathbf{W}$ such that $\|a - d\| < 1/4L_0$, there exists the unique solution x_0 of the equation $F(x) = 0$ in the closed ball

$$\bar{S}(a, \frac{1 - \sqrt{1 - 4\|a - d\|L_0}}{2L_0}).$$

Moreover, the sequence defined by the recurrence relation $x_{(n+1)} = x_{(n)} - F(x_{(n)})$ with the initial approximation $x_{(0)} = a$ converges to x_0, and

$$\|x_{(n)} - x_0\| \le \frac{g^n}{1 - g}\|a - d\|,$$

where

$$g = \frac{1 - \sqrt{1 - 4\|a - d\|L_0}}{2} < \frac{1}{2}.$$

Proof of this proposition repeats the proof of Theorem 3.25 with regard for the relations $x_i(a) = a$ and $\frac{dx_i(x)}{dx}\big|_{x=a} = E$ valid $\forall i \in \{0, 1, 2, ..., k-1\}$, and $\frac{df_i(x_i(x))}{dx_i(x)}\big|_{x=a}$ is the zero operator. Hence, $F'(x) = E$ at the point $x = a$.

Example 3.2. Let us assume that the mapping $f_i(x)$ is a linear bounded operator defined on $\mathbf{W}$ $\forall i \in \{0, 1, 2, ..., k-1\}$, and $\|f_i\| \leq M$. Then the derivative $f_i'(x)$ independent of x is defined $\forall x \in \mathbf{W}$, and $f_i'(x)h = f_i(h)$ $\forall h \in \mathbf{W}$, i.e., $\|f_i'\| = \|f_i\| \leq M$. The equalities

$$F'(x) = E + \frac{df_0(x)}{dx} + \sum_{i=1}^{k-1} \frac{df_i(x_i(x))}{dx_i(x)} \prod_{s=i-1}^{0} \left(E + \frac{df_s(x_s(x))}{dx_s(x)}\right)$$

$$= E + f_0 + \sum_{i=1}^{k-1} f_i \prod_{s=i-1}^{0} (E + f_s) = \prod_{i=k-1}^{0} (E + f_i) = \prod_{i=k-1}^{0} \Phi_i$$

hold. Hence, $F'(x)$ is independent of x and a stationary operator. Its invertibility means the invertibility of $\prod_{i=k-1}^{0} \Phi_i$, i.e., the posed problem have the unique solution.

If the condition $(M+1)^k < 2$ is satisfied, then $M < 1$, and the operator $\Phi_n = E + f_n$ is invertible $\forall n \in \{0, 1, 2, ..., k-1\}$, i.e., Eq. (3.98) is not degenerate. But if $f_i'(a)$ is the zero operator $\forall i \in \{0, 1, 2, ..., k-1\}$, then the operator $f_i(x)$ is also the zero one, and the problem becomes trivial: $x_0 = d$.

We now assume that the solution $x_n(x_0)$ of Eq. (3.98) for $n = k$ is not equal to d. We give some conditions and introduce a certain perturbation of Eq. (3.98), under which we obtain an equation, whose solution $z_n(x_0)$ for $n = k$ equals d.

Theorem 3.26. *Let*

$$\|f_n(x) - f_n(\bar{x})\| \leq K\|x - \bar{x}\| \tag{3.105}$$

$\forall n \in \{1, 2, 3, ..., k-1\}$ and $\forall \{x, \bar{x}\} \subset \mathbf{W}$, where the positive constant K satisfies the inequality $2kK < 1$. Then, $\forall x_0 \in \mathbf{W}$, there exists a control $\alpha(x_0) \in \mathbf{W}$ such that the solution $z_n(x_0)$ of the perturbed equation

$$z_{n+1} = z_n + f_n(z_n) + \alpha(x_0), \quad n \in Z_0^+, \tag{3.106}$$

satisfies the condition $z_k(x_0) = d$.

Proof. It is clear that, for $k = 1$, $\alpha(x_0) = d - x_0 - f_0(x_0)$. Let $k > 1$. We set the recurrence relation

$$z_0^{(s)} = x_0, \quad z_n^{(0)} = x_0, \quad s \in Z_0^+;$$

$$z_n^{(s)} = x_0 + \sum_{i=0}^{n-1} f_i(z_i^{(s-1)}) + n\alpha(s, x_0), \quad s \in Z^+,$$

where n takes values from the set $\{1, 2, 3, \ldots, k\}$.

It is easy to verify that, for all $s \in Z^+$ and

$$\alpha(s, x_0) = \frac{1}{k}(d - x_0 - \sum_{i=0}^{k-1} f_i(z_i^{(s-1)})),$$

the equality $z_k^{(s)} = d$ holds. In this case,

$$z_n^{(s)} = x_0 + \sum_{i=0}^{n-1} f_i(z_i^{(s-1)}) + \frac{n}{k}(d - x_0 - \sum_{i=0}^{k-1} f_i(z_i^{(s-1)})), \tag{3.107}$$

where $s \in Z^+$, $n \in \{1, 2, 3, \ldots, k\}$.

We now show that, for each n from this set, the sequence $\{z_n^{(s)}\}$ converges in the norm as $s \to \infty$. For $n = k$, it is obvious.

The equality

$$z_n^{(s+1)} - z_n^{(s)} = \sum_{i=0}^{n-1} (f_i(z_i^{(s)}) - f_i(z_i^{(s-1)})) + \frac{n}{k} \sum_{i=0}^{k-1} (f_i(z_i^{(s-1)}) - f_i(z_i^{(s)}))$$

yields the inequality

$$\|z_n^{(s+1)} - z_n^{(s)}\| \leq K \sum_{i=0}^{k-1} \|z_i^{(s)} - z_i^{(s-1)}\|$$

$$+ \frac{kK}{k} \sum_{i=0}^{k-1} \|z_i^{(s)} - z_i^{(s-1)}\| = 2K \sum_{i=0}^{k-1} \|z_i^{(s)} - z_i^{(s-1)}\|.$$

Whence we obtain the inductive estimate

$$\max_{n \in \overline{1,k-1}} \|z_n^{(s+1)} - z_n^{(s)}\| \leq 2kK \max_{n \in \overline{1,k-1}} \|z_n^{(s)} - z_n^{(s-1)}\|.$$

Since

$$z_n^{(1)} = x_0 + \sum_{i=0}^{n-1} f_i(x_0) + \frac{n}{k}(d - x_0 - \sum_{i=0}^{k-1} f_i(x_0)),$$

for $n \in \{1, 2, \ldots, k-1\}$, the last estimate yields the relation

$$\max_{n \in \overline{1,k-1}} \|z_n^{(s+1)} - z_n^{(s)}\| \le (2kK)^s \max_{n \in \overline{1,k-1}} \|z_n^{(1)} - x_0\| \le (2kK)^s K_1,$$

where

$$K_1 = \sum_{i=0}^{k-2} \|f_i(x_0)\| + \frac{k-1}{k}\|d - x_0 - \sum_{i=0}^{k-1} f_i(x_0)\| = const > 0.$$

Since $2Kk < 1$, the sequence $\{z_n^{(s)}\}_{s=1}^{\infty}$ is fundamental for each $n \in \{1, 2, 3, \ldots, k-1\}$. It follows from the completeness of the space $\mathbf{W}$ that it converges as $s \to \infty$, i.e., $\lim_{s \to \infty} z_n^{(s)} = \tilde{z}_n \in \mathbf{W}$.

In view of the continuity of the mappings $f_i(x)$ ($i = \{1, 2, \ldots, k-1\}$) and relation (3.107), we can verify that, for all $n \in \{1, 2, ..., k\}$, the equality

$$\tilde{z}_n = x_0 + \sum_{i=0}^{n-1} f_i(\tilde{z}_i) + \frac{n}{k}(d - x_0 - \sum_{i=0}^{k-1} f_i(\tilde{z}_i))$$

holds, and $\tilde{z}_k = d$.

It is obvious that

$$\tilde{z}_{n+1} = \tilde{z}_n + f_n(\tilde{z}_n) + \frac{1}{k}(d - x_0 - \sum_{i=0}^{k-1} f_i(\tilde{z}_i)), \quad n \in \{1, 2, \ldots, k-1\},$$

i.e., $\tilde{z}_n$ satisfies Eq. (3.106) $\forall n \in \{0, 1, 2, \ldots k-1\}$, if we set

$$\alpha(x_0) = \lim_{s \to \infty} \alpha(s, x_0) = \frac{1}{k}(d - x_0 - \sum_{i=0}^{k-1} f_i(\tilde{z}_i)).$$

Thus, the solution $z_n(x_0)$ of the perturbed equation

$$z_{n+1} = z_n + f_n(x_n) + \frac{1}{k}(d - x_0 - \sum_{i=0}^{k-1} f_i(\tilde{z}_i), \quad n \in Z_0^+,$$

equals d for $n = k$, and $z_n(x_0) = \tilde{z}_n$ on the set $\{0, 1, 2, \ldots, k\}$. The theorem is proved. $\quad\square$

Corollary 3.15. *Let $y_n = y_n(x_0)$ be a solution of the equation*

$$y_{n+1} = y_n + f_n(y_n) + \alpha(s, x_0), \quad n \in Z_0^+.$$

Then, under conditions of Theorem 3.26,

$$\|d - y_k\| \le \frac{K_1[(1+K)^k - 1]}{1 - 2kK}(2kK)^{s-1}. \tag{3.108}$$

Proof. For all $n \in \{1, 2, \ldots, k-1\}$, the relations

$$\|\tilde{z}_n - z_n^{(s)}\| = \lim_{p \to \infty} \|z_n^{(s+p)} - z_n^{(s)}\| \leq \lim_{p \to \infty} \sum_{i=1}^{p} \|z_n^{(s+i)} - z_n^{(s+i-1)}\|$$

$$\leq \sum_{i=1}^{\infty} \max_{n \in \overline{1, k-1}} \|z_n^{(s+i)} - z_n^{(s+i-1)}\| \leq \sum_{i=1}^{\infty} (2kK)^{s+i-1} K_1$$

$$= K_1 (2kK)^s \sum_{i=0}^{\infty} (2kK)^i = \frac{K_1}{1 - 2kK} (2kK)^s$$

hold.

Using them, we write the following inequalities:

$$\|\alpha(x_0) - \alpha(s+1, x_0)\| \leq \frac{1}{k} \sum_{i=0}^{k-1} \|f_i(\tilde{z}_i) - f_i(z_i^{(s)})\|$$

$$\leq \frac{K}{k} \sum_{i=0}^{k-1} \|\tilde{z}_i - z_i^{(s)}\| \leq \frac{KK_1}{1 - 2kK} (2kK)^s,$$

or

$$\|\alpha(x_0) - \alpha(s, x_0)\| \leq \frac{KK_1}{1 - 2kK} (2kK)^{s-1} \underset{s \to \infty}{\to} 0. \tag{3.109}$$

On the other hand, we have

$$\|z_k - y_k\| \leq \|z_{k-1} - y_{k-1}\| + K\|z_{k-1} - y_{k-1}\| + \|\alpha(x_0) - \alpha(s, x_0)\|$$

$$= (1 + K)\|z_{k-1} - y_{k-1}\| + \|\alpha(x_0) - \alpha(s, x_0)\|,$$

where $z_n = z_n(x_0)$ is a solution of Eq. (3.106).

In view of the inductive character of the last inequality, we obtain the estimate

$$\|z_k - y_k\| \leq \|\alpha(x_0) - \alpha(s, x_0)\| \sum_{i=0}^{k-1} (1 + K)^i.$$

With regard for (3.109), this estimate yields the relation

$$\|z_k - y_k\| \leq \frac{KK_1}{1 - 2kK} (2kK)^{s-1} \sum_{i=0}^{k-1} (1 + K)^i,$$

which results in inequality (3.108). Since $z_k = d$, Corollary 3.15 is proved.

$\square$

Corollary 3.16. *If, under conditions of Theorem 3.26, the inequality*

$$\max\{2kK; k[(1 + K)^k - 1]\} < 1,$$

holds, then the control $\alpha(x_0)$ is single.

Proof. Let us assume that the solution $\bar{z}_n(x_0)$ of the equation

$$z_{n+1} = z_n + f_n(z_n) + \beta(x_0), \quad n \in Z_0^+$$

is such that $\bar{z}_k(x_0) = d$. Then

$$\beta(x_0) = \frac{1}{k}\left(d - x_0 - \sum_{i=0}^{k-1} f_i(\bar{z}_i)\right),$$

which follows from the inductive equality $d = \bar{z}_{k-1} + f_{k-1}(\bar{z}_{k-1}) + \beta(x_0)$.

We write the following inequalities:

$$\|\alpha(x_0) - \beta(x_0)\| \leq \sum_{i=0}^{k-1} \|f_i(z_i) - f_i(\bar{z}_i)\| \leq Kk \max_{i \in \overline{1,k-1}} \|z_i - \bar{z}_i\|$$

$$\leq Kk\|\alpha(x_0) - \beta(x_0)\| \sum_{i=0}^{k-1} (1+K)^i \leq k((1+K)^k - 1)\|\alpha(x_0) - \beta(x_0)\|.$$

Since $k((1+K)^k - 1) < 1$, we have $\|\alpha(x_0) - \beta(x_0)\| = 0$, which completes the proof. $\square$

Under conditions of Theorem 3.26, it is required that inequality (3.105) hold on the whole space $\mathbf{W}$. This requirement can be weakened by the choice of the set D_k, which should depend on $d \in \mathbf{W}$.

Corollary 3.17. *Let the first of conditions (B) be satisfied, let inequality (3.105) be satisfied $\forall\{x, \bar{x}\} \subset D$, $2kK < 1$, and let the set*

$$D_f^o = \left\{x \in D \mid \|x\| \leq \frac{1}{2}\left(R - \frac{kM}{2} - \|d\|\right)\right\}$$

be nonempty. Then the assertion of Theorem 3.26 remains to hold $\forall x_0 \in D_f^o$.

Proof. It is sufficient to show that the points $z_n^{(s)}$ defined by equality (3.107) belong to the set D for all $n \in \{1, 2, 3, \ldots, k\}$, $s \in Z^+$, if only $x_0 \in D_f^o$. Indeed, it is obvious for $z_n^{(0)}$. But it is true for $z_n^{(s)}$, it is also true for $z_n^{(s+1)}$, since

$$\|z_n^{(s+1)} - x_0\| = \left\| \sum_{i=0}^{n-1} f_i(z_i^{(s)}) - \frac{n}{k}\sum_{i=0}^{k-1} f_i(z_i^{(s)}) + \frac{n}{k}(d - x_0)\right\|$$

$$\leq \left\| \sum_{i=0}^{n-1} \{f_i(z_i^{(s)}) - \frac{1}{k}\sum_{\nu=0}^{k-1} f_i(z_i^{(s)})\}\right\| + \|d - x_0\|$$

$$\leq \frac{k}{2} \max_{0 \leq i \leq k-1} \|f_i(z_i^{(s)})\| + \|d - x_0\| \leq \frac{kM}{2} + \|d\| + \|x_0\|.$$

This yields

$$\|z_n^{(s+1)}\| \le 2\|x_0\| + \frac{kM}{2} + \|d\| \le R. \qquad \square$$

Example 3.3. In the three-dimensional space, we consider the equation

$$x_{n+1} = x_n + A_n x_n, \quad n \in \{0, 1, 2\},$$

where

$$A_0 = \begin{pmatrix} -2 & 1 & 0 \\ 1 & -1 & 1 \\ 0 & 1 & 0 \end{pmatrix}, \quad A_1 = \begin{pmatrix} 0 & 0.1 & 0 \\ 0 & 0 & 0.1 \\ 0 & 0.1 & 0 \end{pmatrix}, \quad A_2 = \begin{pmatrix} 0.1 & 0 & 0 \\ 0.05 & 0 & 0.05 \\ 0.05 & 0.05 & 0 \end{pmatrix},$$

$x = colon(x^1, x^2, x^3)$, $\|x\| = \max\{|x^1|, |x^2|, |x^3|\}$, $\|A\| = \|[a_{ij}]_{i,j=1}^3\| = \max_i \sum_{j=1}^3 |a_{ij}|$.

Then $\|A_0\| = 3$, $\|A_1\| = \|A_2\| = 0.1$, the matrices $E + A_1$ and $E + A_2$ are inverse matrices, the matrix $E + A_0$ is not an inverse one, and E is the identity matrix. We set $d = colon\{1.21; 1.21; 1.21\}$. The check indicates that $(E + A_2)^{-1}d = colon\{1.1; 1.1; 1.1\}$, $(E + A_1)^{-1}colon\{1.1; 1.1; 1.1\} = colon\{1, 1, 1\}$, and the equation $colon\{1, 1, 1\} = x_0 + A_0 x_0$ has no solution x_0.

We choose, for example, $x_0 = colon\{2, 1, 2\}$ and find a perturbed equation, whose solution $y_n(x_0)$ is such that $\|y_3(x_0) - d\| \le 0.6$.

At once, we note that the conditions of Corollary 3.16 are satisfied. Since $k=3$ and $K=0.1$, we have $2kK < 1$ and $k[(1 + K)^2 - 1] < 1$.

We now calculate the constant K_1:

$$K_1 = \|A_0 x_0\| + \|A_1 x_0\| + \frac{2}{3}\|d - x_0 - A_0 x_0 - A_1 x_0 - A_2 x_0\|$$

$$= \left\| \begin{pmatrix} -2 & 1 & 0 \\ 1 & -1 & 1 \\ 0 & 1 & 0 \end{pmatrix} \begin{pmatrix} 2 \\ 1 \\ 2 \end{pmatrix} \right\| + \left\| \begin{pmatrix} 0 & 0.1 & 0 \\ 0 & 0 & 0.1 \\ 0 & 0.1 & 0 \end{pmatrix} \begin{pmatrix} 2 \\ 1 \\ 2 \end{pmatrix} \right\|$$

$$+ \frac{2}{3}\left\| \begin{pmatrix} 1.21 \\ 1.21 \\ 1.21 \end{pmatrix} - \begin{pmatrix} 2 \\ 1 \\ 2 \end{pmatrix} - \begin{pmatrix} -2 & 1 & 0 \\ 1 & -1 & 1 \\ 0 & 1 & 0 \end{pmatrix} \begin{pmatrix} 2 \\ 1 \\ 2 \end{pmatrix} - \begin{pmatrix} 0 & 0.1 & 0 \\ 0 & 0 & 0.1 \\ 0 & 0.1 & 0 \end{pmatrix} \begin{pmatrix} 2 \\ 1 \\ 2 \end{pmatrix} \right.$$

$$\left. - \begin{pmatrix} 0.1 & 0 & 0 \\ 0.05 & 0 & 0.05 \\ 0.05 & 0.05 & 0 \end{pmatrix} \begin{pmatrix} 2 \\ 1 \\ 2 \end{pmatrix} \right\| = \left\| \begin{pmatrix} -3 \\ 3 \\ 1 \end{pmatrix} \right\| + \left\| \begin{pmatrix} 0.1 \\ 0.2 \\ 0.1 \end{pmatrix} \right\|$$

$$+\frac{2}{3}\left\|\begin{pmatrix}1.21\\1.21\\1.21\end{pmatrix}-\begin{pmatrix}2\\1\\2\end{pmatrix}-\begin{pmatrix}-3\\3\\1\end{pmatrix}-\begin{pmatrix}0.1\\0.2\\0.1\end{pmatrix}-\begin{pmatrix}0.2\\0.2\\0.15\end{pmatrix}\right\|=3.2+\frac{2}{3}\left\|\begin{pmatrix}1.91\\-3.19\\-2.04\end{pmatrix}\right\|$$

$$=\tfrac{1598}{300}<5.32.$$

From the inequality

$$5.32\frac{0.331}{0.4}(0.6)^{s-1}\le 0.6,$$

we obtain $s\ge 5$. We choose the least value $s=5$.

To find $\alpha(5,x_0)$, we use the recurrence formula (3.107), i.e.,

$$\alpha(5,x_0)=\frac{1}{3}(d-x_0-A_0x_0-A_1z_1^{(4)}-A_2z_2^{(4)}).$$

In this case, we calculate $z_1^{(4)}$ and $z_2^{(4)}$ to within the third decimal point.

The following equalities hold:

$$z_1^{(1)}=\begin{pmatrix}2\\1\\2\end{pmatrix}+\begin{pmatrix}-3\\3\\1\end{pmatrix}+\frac{1}{3}\begin{pmatrix}1.91\\-3.19\\-2.04\end{pmatrix}=\begin{pmatrix}-0.363\\2.937\\2.320\end{pmatrix};$$

$$z_2^{(1)}=\begin{pmatrix}2\\1\\2\end{pmatrix}+\begin{pmatrix}-3\\3\\1\end{pmatrix}+\begin{pmatrix}0.1\\0.2\\0.1\end{pmatrix}+\frac{2}{3}\begin{pmatrix}1.91\\-3.19\\-2.04\end{pmatrix}=\begin{pmatrix}0.373\\2.073\\1.740\end{pmatrix};$$

$$z_1^{(2)}=\begin{pmatrix}2\\1\\2\end{pmatrix}-\begin{pmatrix}-3\\3\\1\end{pmatrix}+\frac{1}{3}\left[\begin{pmatrix}1.21\\1.21\\1.21\end{pmatrix}-\begin{pmatrix}2\\1\\2\end{pmatrix}-\begin{pmatrix}-3\\3\\1\end{pmatrix}-\begin{pmatrix}0&0.1&0\\0&0&0.1\\0&0.1&0\end{pmatrix}\begin{pmatrix}-0.363\\2.937\\2.320\end{pmatrix}\right.$$

$$\left.-\begin{pmatrix}0.1&0&0\\0.05&0&0.05\\0.05&0.05&0\end{pmatrix}\begin{pmatrix}0.373\\2.073\\1.740\end{pmatrix}\right]=\begin{pmatrix}-1\\4\\3\end{pmatrix}+\frac{1}{3}\begin{pmatrix}1.879\\-3.128\\-2.206\end{pmatrix}=\begin{pmatrix}-0.373\\2.957\\2.264\end{pmatrix};$$

$$z_2^{(2)}=\begin{pmatrix}-1\\4\\3\end{pmatrix}+\begin{pmatrix}0.293\\0.232\\0.293\end{pmatrix}+\begin{pmatrix}-0.746\\5.914\\4.528\end{pmatrix}=\begin{pmatrix}-1.453\\10.146\\7.821\end{pmatrix};$$

$$z_1^{(3)}=\begin{pmatrix}-1\\4\\3\end{pmatrix}+\frac{1}{3}\left[\begin{pmatrix}1.21\\1.21\\1.21\end{pmatrix}-\begin{pmatrix}2\\1\\2\end{pmatrix}-\begin{pmatrix}-3\\3\\1\end{pmatrix}-\begin{pmatrix}0&0.1&0\\0&0&0.1\\0&0.1&0\end{pmatrix}\begin{pmatrix}-0.373\\2.957\\2.264\end{pmatrix}\right.$$

$$-\begin{pmatrix} 0.1 & 0 & 0 \\ 0.05 & 0 & 0.05 \\ 0.05 & 0.05 & 0 \end{pmatrix}\begin{pmatrix} -1.453 \\ 10.146 \\ 7.821 \end{pmatrix}\Bigg] = \begin{pmatrix} -0.410 \\ 2.888 \\ 2.156 \end{pmatrix};$$

$$z_2^{(3)} = \begin{pmatrix} -1 \\ 4 \\ 3 \end{pmatrix} + \begin{pmatrix} 0.296 \\ 0.226 \\ 0.296 \end{pmatrix} + \tfrac{2}{3}\begin{pmatrix} 1.769 \\ -3.335 \\ -2.520 \end{pmatrix} = \begin{pmatrix} 0.475 \\ 2.003 \\ 1.616 \end{pmatrix};$$

$$z_1^{(4)} = \begin{pmatrix} -1 \\ 4 \\ 3 \end{pmatrix} + \tfrac{1}{3}\Bigg[\begin{pmatrix} 2.21 \\ -2.79 \\ -1.79 \end{pmatrix} - \begin{pmatrix} 0 & 0.1 & 0 \\ 0 & 0 & 0.1 \\ 0 & 0.1 & 0 \end{pmatrix}\begin{pmatrix} -0.410 \\ 2.888 \\ 2.156 \end{pmatrix}$$

$$-\begin{pmatrix} 0.1 & 0 & 0 \\ 0.05 & 0 & 0.05 \\ 0.05 & 0.05 & 0 \end{pmatrix}\begin{pmatrix} 0.475 \\ 2.003 \\ 1.616 \end{pmatrix}\Bigg] = \begin{pmatrix} -1 \\ 4 \\ 3 \end{pmatrix} + \begin{pmatrix} 0.625 \\ -1.036 \\ -0.734 \end{pmatrix} = \begin{pmatrix} -0.375 \\ 2.964 \\ 2.265 \end{pmatrix};$$

$$z_2^{(4)} = \begin{pmatrix} -1 \\ 4 \\ 3 \end{pmatrix} + \begin{pmatrix} 0.289 \\ 0.216 \\ 0.289 \end{pmatrix} + \begin{pmatrix} 1.250 \\ -2.072 \\ -1.468 \end{pmatrix} = \begin{pmatrix} 0.539 \\ 2.144 \\ 1.821 \end{pmatrix}.$$

Then

$$\alpha(5, x_0)$$

$$= \tfrac{1}{3}\Bigg[\begin{pmatrix} 2.21 \\ -2.79 \\ -1.79 \end{pmatrix} - \begin{pmatrix} 0 & 0.1 & 0 \\ 0 & 0 & 0.1 \\ 0 & 0.1 & 0 \end{pmatrix}\begin{pmatrix} -0.375 \\ 2.964 \\ 2.265 \end{pmatrix} - \begin{pmatrix} 0.1 & 0 & 0 \\ 0.05 & 0 & 0.05 \\ 0.05 & 0.05 & 0 \end{pmatrix}\begin{pmatrix} 0.539 \\ 2.144 \\ 1.821 \end{pmatrix}\Bigg]$$

$$= \tfrac{1}{3}\Bigg[\begin{pmatrix} 2.21 \\ -2.79 \\ -1.79 \end{pmatrix} - \begin{pmatrix} 0.296 \\ 0.226 \\ 0.296 \end{pmatrix} - \begin{pmatrix} 0.054 \\ 0.118 \\ 0.133 \end{pmatrix}\Bigg] = \begin{pmatrix} 0.620 \\ -1.044 \\ -0.740 \end{pmatrix}.$$

We now verify that the obtained perturbation satisfies the condition of the problem even if it is written to within the first decimal point. Indeed, we write the equation

$$y_{n+1} = y_n + A_n y_n + \begin{pmatrix} 0.6 \\ -1 \\ -0.7 \end{pmatrix}, \quad n \in \{0, 1, 2\}.$$

The calculations indicate that

$$y_1(x_0) = \begin{pmatrix} -0.4 \\ 3 \\ 2.3 \end{pmatrix}; \quad y_2(x_0) = \begin{pmatrix} 0.5 \\ 2.23 \\ 1.9 \end{pmatrix}; \quad y_3(x_0) = \begin{pmatrix} 1.15 \\ 1.35 \\ 1.3365 \end{pmatrix},$$

hence,

$$\|d - y_3(x_0)\| = \left\| \begin{pmatrix} 0.06 \\ -0.14 \\ -0.1265 \end{pmatrix} \right\| = 0.14 < 0.6,$$

which solves the posed problem.

We note that it is necessary to set s=9; s=13; s=18 in this example in order to attain the accuracy, at which $\| y_3(x_0) - d \|$ does not exceed 0.1; 0.01; 0.001, respectively. It is clear that the process of construction of the required perturbation can be realized with a computer.

2^0. **Equations of higher orders.** Consider now the equation

$$\Delta^m x_n = f_n(x_n, x_{n+1}, ..., x_{n+m-1}), \quad n \in Z_0^+, \tag{3.110}$$

where $x_n \in \mathbf{W}$, the function f_n maps $\mathbf{W}^m$ onto $\mathbf{W}$ $\forall n \in Z_0^+$, and m is a natural number larger than 1. We pose the problem: to find $X_0 = (x_0, x_1, ..., x_{m-1}) \in \mathbf{W}^m$, which determines the solution $x_n = x_n(X_0)$ of Eq. (3.110) on Z_0^+ such that $x_i(X_0) = x_i$ for all $i \in \{0, 1, ..., m-1\}$ and $x_{k+m}(X_0) = d$, where k and d are given elements from Z_0^+ and $\mathbf{W}$, respectively.

Lemma 3.16. *Equation (3.110) can be written in the form*

$$x_{n+m} = -\sum_{i=0}^{m-1} (-1)^{m-i} C_m^i x_{n+i} + f_n(x_n, x_{n+1}, ..., x_{n+m-1}), \quad n \in Z_0^+,$$

where C_m^i is the number of combinations from m elements in i.

Proof. Obviously, it is sufficient to show that, $\forall p \in Z^+$, th equality

$$\Delta^p x_n = \sum_{i=0}^{p} (-1)^{p-i} C_p^i x_{n+i} \tag{3.111}$$

holds. For $p = 1$, this is true. Let us assume that this is true for all $1 < p \le m$.

Then

$$\Delta^{m+1}x_n = \Delta(\Delta^m x_n) = \sum_{i=0}^{m}(-1)^{m-i}C_m^i x_{n+1+i} - \sum_{i=0}^{m}(-1)^{m-i}C_m^i x_{n+i}$$

$$= -(-1)^m x_n + \sum_{i=0}^{m-1}(-1)^{m-i}C_m^i x_{n+1+i} - \sum_{i=1}^{m}(-1)^{m-i}C_m^i x_{n+i} + x_{n+m+1}$$

$$= (-1)^{m+1}C_{m+1}^0 x_n + \sum_{i=1}^{m}(-1)^{m-i+1}\left(C_m^{i-1} + C_m^i\right)x_{n+i}$$

$$+ (-1)^{m+1-(m+1)}C_{m+1}^{m+1}x_{n+m+1} = \sum_{i=0}^{m+1}(-1)^{m+1-i}C_{m+1}^i x_{n+i},$$

i.e., equality (3.111) holds for $p = m + 1$, which completes the proof of the lemma. $\qquad\qquad\square$

We introduce the notation

$$\Phi_n(Y) = -\sum_{i=0}^{m-1}(-1)^{m-i}C_m^i y_i + f_n(Y),$$

where $Y = (y_0, y_1, ..., y_{m-1}) \in \mathbf{W}^m$.

If at least one of the mappings $\Phi_n : \mathbf{W}^m \to \mathbf{W}$ $(n = 0, 1, ..., k)$ is not invertible, then we call Eq. (3.110) degenerate.

It is clear that the above-posed problem for the degenerate equation (3.110) can have the infinity of solutions and can have none. Obviously, in order that at least one such solution exist, it is necessary and sufficient that the equation

$$\Psi(X_0) = -\sum_{i=0}^{m-1}(-1)^{m-i}C_m^i x_{k+i}(X_0)$$

$$+ f_k(x_k(X_0), x_{k+1}(X_0), ..., x_{k+m-1}(X_0)) - d = 0$$

have at least one solution $X_0 \in \mathbf{W}^m$.

Lemma 3.17. *If the mappings $f_n(n = \overline{0, k})$ are differentiable on $\mathbf{W}^m$, then the mapping $\Psi(X_0)$ has also this property.*

Proof. We denote the mapping

$$-\sum_{i=0}^{m-1}(-1)^{m-i}C_m^i x_{p+i} : \mathbf{W}^m \to \mathbf{W}$$

by $\bar{F}(X_p)$, where $X_p=(x_p, x_{p+1},\ldots, x_{p+m-1}) \in \mathbf{W}^m$, $p \in Z^+$. Since $\bar{F}(X_0)$ is a linear operator, it is differentiable with respect to X_0, and its derivative acts on any vector $h=(h_0, h_1,\ldots,h_{m-1}) \in \mathbf{W}^m$ in the following way:

$$\frac{d\bar{F}(X_0)}{dX_0} h = \bar{F}(h).$$

Then, obviously, there exists the derivative

$$\frac{dx_m(X_0)}{dX_0} = \frac{d\bar{F}(X_0)}{dX_0} + \frac{df_0(X_0)}{dX_0}.$$

Let us assume that, for all $n \in Z^+$ such that $n \leq p < k$, there exists the derivative

$$\frac{dx_{m+n}(X_0)}{dX_0}.$$

We write the equality

$$x_{m+p+1}(X_0) = \bar{F}(X_{p+1}(X_0)) + f_{p+1}(X_{p+1}(X_0)).$$

The mapping $X_{p+1}(X_0)$: $\mathbf{W}^m \to \mathbf{W}^m$ is composed from m components:

$$\gamma_i : \mathbf{W}^m \to \mathbf{W} \mid \gamma_i(X_0) = x_{p+i}(X_0), \quad (i = \overline{1,m}),$$

each of which is differentiable with respect to X_0 by assumption, since $p+i \leq m+p$ for $i = \overline{1,m}$. Then there exists the derivative $\frac{dX_{p+1}(X_0)}{dX_0}$, and

$$\frac{dx_{m+p+1}(X_0)}{dX_0} = \left[\frac{d\bar{F}(X_{p+1}(X_0))}{dX_{p+1}(X_0)} + \frac{df_{p+1}(X_{p+1}(X_0))}{dX_{p+1}(X_0)}\right]\frac{dX_{p+1}(X_0)}{dX_0}.$$

Using the principle of complete mathematical induction, we verify that the mapping $\Psi(_0)$ is differentiable with respect to $X_0 \in \mathbf{W}^m$. $\square$

Let us agree that the norm of an element $Y \in \mathbf{W}^m$ is the expression

$$\|Y\| = \max\{\|y_0\|, \|y_1\|, ..., \|y_{m-1}\|\}$$

where $\|y_i\|$ $(i = \overline{0, m - 1})$ is the norm in the space $\mathbf{W}$.

Let $Z=(z_0, z_1,\ldots, z_{m-1})$, and let $\overline{Z} = (\bar{z}_0, \bar{z}_1, ..., \bar{z}_{m-1})$ be any points from the space $\mathbf{W}^m$. Below, we give the following auxiliary proposition.

Lemma 3.18. *Let* $\forall Z \in \mathbf{W}^m$

$$\left\|\frac{df_i(Z)}{dZ}\right\| \leq P = const > 0 \quad (i = 0, 1, ..., k).$$

Then, $\forall\{Z, \overline{Z}\} \subset \mathbf{W}^m$, *the following relations hold:*

$$\left\|\frac{d\bar{F}(Z)}{dZ}\right\| \leq 2^m - 1; \tag{3.112}$$

$$\left\| \frac{d\bar{F}(Z)}{dZ} - \frac{d\bar{F}(\overline{Z})}{d\overline{Z}} \right\| = 0; \tag{3.113}$$

$$\left\| \frac{dX_p(Z)}{dZ} \right\| \le (2^m - 1 + P)^p \quad (0 \le p \le k); \tag{3.114}$$

$$\| X_p(Z) - X_p(\overline{Z}) \| \le (2^m - 1 + P)^p \| Z - \overline{Z} \| \quad (0 \le p \le k). \tag{3.115}$$

Proof. The relations

$$\left\| \frac{d\bar{F}(Z)}{dZ} \right\| = \sup_{\| h \| = 1} \left\| \frac{d\bar{F}(Z)}{dZ} h \right\| = \sup_{\| h \| = 1} \| \bar{F}(h) \|$$

$$= \sup_{\max\{ \| h_0 \|, \ldots, \| h_{m-1} \| \} = 1} \left\| - \sum_{i=0}^{m-1} (-1)^{m-i} C_m^i h_i \right\| \le 2^m - 1$$

prove inequality (3.112).

Equality (3.113) is proper, since $\left(\frac{d\varphi(Z)}{dZ} - \frac{d\varphi(\overline{Z})}{d\overline{Z}} \right)$ is the zero operator, which transfers $\forall h \in \mathbf{W}^m$ in $0 \in \mathbf{W}$.

We prove inequality (3.114) by the method of complete mathematical induction. Indeed, $X_0(Z) = E(Z)$, where E is the identity operator. Hence, $\left\| \frac{dX_0(Z)}{dZ} \right\| = 1$, and inequality (3.114) is satisfied for $p = 0$.

We assume that it holds for all $1 \le p \le n < k$ and will prove its validity for $p = n + 1$.

Let us write the equality

$$X_{n+1}(Z) = (x_{n+1}(Z), x_{n+2}(Z), \ldots, x_{n+m}(Z)).$$

Two cases can take place: $n + 1 < m$ or $n + 1 \ge m$.

In the first case,

$$X_{n+1}(Z) = (x_{n+1}(Z), \ldots, x_{m-1}(Z), x_m(Z), \ldots, x_{n+m}(Z)).$$

Then

$$\left\| \frac{dX_{n+1}(Z)}{dZ} \right\| \le \max \{ \left\| \frac{dx_{n+1}(Z)}{dZ} \right\|, \ldots, \left\| \frac{dx_{m-1}(Z)}{dZ} \right\|,$$

$$\left\| \frac{dx_m(Z)}{dZ} \right\|, \ldots, \left\| \frac{dx_{n+m}(Z)}{dZ} \right\| \}.$$

For $0 \le n \le m - 1$,

$$x_n(Z) = z_n = \sum_{i=0}^{n-1} 0 \cdot z_i + z_n + \sum_{i=n+1}^{m-1} 0 \cdot z_i$$

is a linear operator acting from $\mathbf{W}^m$ to $\mathbf{W}$. Hence,

$$\left\|\frac{dx_n(Z)}{dZ}\right\| = \sup_{\|h\|=1}\left\|\frac{dx_n(Z)}{dZ}h\right\| = \sup_{\|h\|=1}\|x_n(h)\|$$

$$= \sup_{\|h\|=1}\|h_n\| = \sup_{\max\{\|h_0\|,\dots,\|h_{m-1}\|\}=1}\|h_n\| = 1, \quad h \in \mathbf{W}^m.$$

Then

$$\left\|\frac{dX_{n+1}(Z)}{dZ}\right\| = \max\left\{1, \left\|\frac{dx_m(Z)}{dZ}\right\|, \dots, \left\|\frac{dx_{n+m}(Z)}{dZ}\right\|\right\}.$$

For all $1 \leq s \leq n$, we have

$$\left\|\frac{dx_{m+s}(Z)}{dZ}\right\| \leq \left\{\left\|\frac{d\varphi(X_s(Z))}{dX_s(Z)}\right\| + \left\|\frac{df_s(X_s(Z))}{dX_s(Z)}\right\|\right\}\left\|\frac{dX_s(Z)}{dZ}\right\|$$

$$\leq (2^m - 1 + P)(2^m - 1 + P)^s = (2^m - 1 + P)^{s+1},$$

$$\left\|\frac{dx_{m+s-1}(Z)}{dZ}\right\| \leq (2^m - 1 + P)^s.$$

With regard for the inequality $2^m - 1 + P > 1$ $(m \geq 1)$, we can conclude that

$$\left\|\frac{dX_{+1}(Z)}{dZ}\right\| = \left\|\frac{dx_{n+}(Z)}{dZ}\right\| \leq (2^m - 1 + P)^{n+1}.$$

It is easy to see that, for $n + 1 \geq m$, the last inequality holds. Hence, inequality (3.114) is proved.

Inequality (3.115) is a simple consequence of estimate (3.114), since $\mathbf{W}^m$ is a convex set. $\qquad\qquad\square$

To simplify the presentation, we denote the formal differences $f(Z) - f(\overline{Z})$ and $\frac{df(Z)}{dZ} - \frac{df(\overline{Z})}{d\overline{Z}}$ by $f|_{\overline{Z}}^{Z}$ and $df|_{\overline{Z}}^{Z}$, respectively. We now present the conditions, under which the derivative of the mapping $\Psi(X_0)$ has the Lipschitz property on $\mathbf{W}^m$.

Lemma 3.19. *Let the conditions of Lemmas 18.3 and 18.4 be satisfied, and let, for all $\{Z, \overline{Z}\} \subset \mathbf{W}^m$ and $n \in \{0, 1, 2, \dots, k\}$, the inequality*

$$\|df_n|_{\overline{Z}}^{Z}\| \leq L^*\|Z - \overline{Z}\|, \tag{3.116}$$

where L^* is a positive constant, hold. Then

$$\|d\Psi|_{\overline{Z}}^{Z}\| \leq L_0^*\|Z - \overline{Z}\|,$$

where

$$L_0^* = L^*\gamma^k\left(\gamma^k + \frac{\gamma^{k+1} - 1}{\gamma - 1}\right), \quad \gamma = 2^m - 1 + P.$$

Proof. The following equalities hold:

$$d\Psi|_{\overline{Z}}^{Z} = \frac{d\bar{F}(X_k(Z))}{dZ} - \frac{d\bar{F}(X_k(\overline{Z}))}{d\overline{Z}} + \frac{df_k(X_k(Z))}{dZ} - \frac{df_k(X_k(\overline{Z}))}{d\overline{Z}}$$

$$= d\bar{F}|_{X_k(\overline{Z})}^{X_k(Z)} \cdot \frac{dX_k(Z)}{dZ} + \frac{d\bar{F}(X_k(\overline{Z}))}{dX_k(\overline{Z})} \cdot dX_k|_{\overline{Z}}^{Z}$$

$$+ df_k|_{X_k(\overline{Z})}^{X_k(Z)} \cdot \frac{dX_k(Z)}{dZ} + \frac{df_k(X_k(\overline{Z}))}{dX_k(\overline{Z})} \cdot dX_k|_{\overline{Z}}^{Z}.$$

In view of $(3.112) - (3.116)$, we obtain the estimates:

$$\| d\Psi|_{\overline{Z}}^{Z} \| \le \gamma \| dX_k|_{\overline{Z}}^{Z} \| + L^* \| X_k|_{\overline{Z}}^{Z} \| \gamma^k$$

$$\le L^* \gamma^{2k} \| Z - \overline{Z} \| + \gamma \| dX_k|_{\overline{Z}}^{Z} \|. \qquad (3.117)$$

It is easy to see that

$$\| dX_k|_{\overline{Z}}^{Z} \| \le \max \{ \| dx_k|_{\overline{Z}}^{Z} \|, \| dx_{k+1}|_{\overline{Z}}^{Z} \|, ..., \| dx_{k+m-1}|_{\overline{Z}}^{Z} \| \}.$$

Let us assume that $k_{\mathfrak{i}} m$. Then

$$\| dX_k|_{\overline{Z}}^{Z} \| \le \max \{ \| dx_k|_{\overline{Z}}^{Z} \|, ..., \| dx_{m-1}|_{\overline{Z}}^{Z} \|, \| dx_m|_{\overline{Z}}^{Z} \|, ...$$

$$..., \| dx_{k+m-1}|_{\overline{Z}}^{Z} \| \} \le \max \{ \| dx_m|_{\overline{Z}}^{Z} \|, ..., \| dx_{m+k-1}|_{\overline{Z}}^{Z} \| \},$$

by taking into account that, for $k_{\mathfrak{i}} m$, the operator $dx_k|_{\overline{Z}}^{Z}$ transfers $\forall h \in \mathbf{W}^m$ to $0 \in \mathbf{W}$, i.e., it is a zero operator.

We now show that, for all $0 \le p \le k$,

$$\| dx_{m+p}|_{\overline{Z}}^{Z} \| \le L^* \gamma^p \frac{\gamma^{p+1} - 1}{\gamma - 1} \| Z - \overline{Z} \|. \qquad (3.118)$$

For $p = 0$, we have

$$\| dx_m|_{\overline{Z}}^{Z} \| \le \| d\bar{F}|_{\overline{Z}}^{Z} \| + \| df_0|_{\overline{Z}}^{Z} \| \le L^* \| Z - \overline{Z} \|,$$

i.e., estimate (3.118) is proper.

We assume that it is true for all $0 < p \le n < k$ and prove its validity for $p = n + 1$. In view of (3.117), we write the relations

$$\| dx_{m+n+1}|_{\overline{Z}}^{Z} \| \le \| \frac{d\bar{F}(X_{n+1}(Z))}{dZ} - \frac{d\bar{F}(X_{n+1}(\overline{Z}))}{d\overline{Z}} \|$$

$$+ \| \frac{df_{n+1}(X_{n+1}(Z))}{dZ} - \frac{df_{n+1}(X_{n+1}(\overline{Z}))}{d\overline{Z}} \|$$

$$\le L^* \gamma^{2(n+1)} \| Z - \overline{Z} \| + \gamma \| dX_{n+1}|_{\overline{Z}}^{Z} \| \le L^* \gamma^{2(n+1)} \| Z - \overline{Z} \| + \gamma \| dx_{m+n}|_{\overline{Z}}^{Z} \|$$

$$\le L^* \gamma^{n+1} \{ \gamma^{n+1} + \frac{\gamma^{n+1} - 1}{\gamma - 1} \} \| Z - \overline{Z} \| = L^* \gamma^{n+1} \frac{\gamma^{n+2} - 1}{\gamma - 1} \| Z - \overline{Z} \|,$$

which prove inequality (3.118).

With regard for estimates (3.117) and (3.118), we can verify that

$$\| d\Psi|_{\overline{Z}}^{Z}\| \leq L^*\gamma^{2k}\| Z - \overline{Z}\| + \gamma\| dx_{m+k-1}|_{\overline{Z}}^{Z}\| \leq L^*\gamma^{2k}\| Z - \overline{Z}\|$$

$$+ \gamma L^*\gamma^{k-1}\frac{\gamma^k - 1}{\gamma - 1}\| Z - \overline{Z}\| = L^*\gamma^k\{\gamma^k + \frac{\gamma^k - 1}{\gamma - 1}\}\| Z - \overline{Z}\|.$$

It is easy to see that the last estimate holds also in the case where $k \geq m$. Lemma 3.19 is proved. $\square$

The above-obtained results allow us to apply a modified Newton–Kantorovich method [38], [43] to the solution of the equation $\Psi(X_0)=0$. The following proposition is true.

Theorem 3.27. *Let the conditions of Lemma 3.19 be satisfied, let there exist a point $X_0^* = (x_0^*, x_1^*, ..., x_{m-1}^*) \in \mathbf{W}^m$, at which at least one of the partial derivatives $\frac{\partial\Psi(X_0^*)}{\partial x_p}$ $(0 \leq p \leq m - 1)$ is invertible, and let*

$$N_0 = \| [\frac{\partial\Psi(X_0^*)}{\partial x_p}]^{-1}\|, \quad J = \| [\frac{\partial\Psi(X_0^*)}{\partial x_p}]^{-1}\Psi(X_0^*)\|, \quad h = N_0 J L_0^*.$$

If

$$L^* < \frac{1}{4N_0 J\gamma^k(\frac{\gamma^{k+1}-1}{\gamma-1})},$$

then the ball $S : \| z - x_p^*\| \leq J t_0$ from $\mathbf{W}$, where t_0 is the least root of the equation $ht^2 - t + 1 = 0$, contains the single point x_p^0 such that

$$X^0 = (x_0^*, ..., x_{p-1}^*, x_p^0, x_{p+1}^*..., x_{m-1}^*) \in \mathbf{W}^m$$

is a solution of the equation $\Psi(X_0) = 0$.

Proof. First, we note that, under conditions of Theorem 3.27 for each $p \in\{0,1,2,...m\text{-}1\}$ at any point $X_0^* \in \mathbf{W}^m$, the partial derivative $\frac{\partial\Psi(X_0^*)}{\partial x_p}$ exists. It is equal to $\frac{dG(x_p^*)}{dx_p^*}$, where $G(x_p)$ means the mapping

$$\Psi(x_0^*, ..., x_{p-1}^*, x_p, x_{p+1}^*..., x_{m-1}^*) : \mathbf{W} \to \mathbf{W}$$

at fixed $\{x_0^*, ..., x_{p-1}^*, x_{p+1}^*..., x_{m-1}^*\} \subset \mathbf{W}$. It is obvious that, $\forall h_p \in \mathbf{W}$,

$$\frac{dG(x_p^*)}{dx_p^*}h_p = \frac{d\Psi(X_0^*)}{dX_0^*}(0, ..., 0, h_p, 0, ..., 0),$$

where $(0, ..., 0, h_p, 0, ..., 0) \in \mathbf{W}^m$.

This ensures the validity of the estimate

$$\| dG|_{\overline{x}_p}^{x_p}\| \leq L_0^*\|x_p - \overline{x}_p\|, \quad \{x_p, \overline{x}_p\} \subset \mathbf{W}.$$

Then the mapping $G(x_p) : \mathbf{W} \to \mathbf{W}$ satisfies the necessary conditions from [43], which completes the proof. We note that x_p^0 is the limit of the sequence defined by the recurrence formula

$$\xi_0 = x_p^*, \quad \xi_{n+1} = \xi_n - [\frac{dG(x_p^*)}{dx_p^*}]^{-1} G(\xi_n).$$

$\square$

Remark 3.5. The complexity of the application of Theorem 3.27 consists in the verification of the invertibility of the mapping $\frac{\partial \Psi(X_0^*)}{\partial x_p}$, which is a quite difficult problem for infinite-dimensional spaces.

We introduce the notation:

$$D^* = \{Z \in \mathbf{W}^m \,|\, \|Z\| < a = const > 0\};$$

$$D_k^* = \{Z \in \mathbf{W}^m \,|\, \|Z\| < b\},$$

where

$$b = (a - \tilde{N}\frac{(2^m - 1)^k - 1}{2^m - 2})(2^m - 1)^{-k}, \quad \tilde{N} = const > 0.$$

Corollary 3.18. *Let the conditions of Lemma 3.19 be satisfied in a domain* D^**, and let* $\|f_n(Z)\| \leq \tilde{N}$ *for all* $n \in \{0, 1, 2, \ldots, k\}$*. We assume that the set* D_k^* *is not empty,* $X_0^* \in D_k^*$*, and the ball* S *is contained in the set* $\{z \in \mathbf{W}|\, \|z\| < b\}$*. Then the assertion of Theorem 3.27 is valid.*

The proof of Corollary 3.18 is reduced to the verification of the estimate $\|x_{m+p}(X_0)\| < a$ for all $p \in \{0, 1, 2, \ldots, k - 1\}$ and $X_0 \in D_k^*$, which is not a difficult task.

Remark 3.6. The assertion of Theorem 3.27 remains true, if the partial derivative $\frac{\partial \Psi(X_0^*)}{\partial x_p}$ in its formulation is replaced by the total derivative $\frac{d\Psi(X_0^*)}{dX_0^*}$, the ball S is replaced by the ball $S_1 : \|Z - X_0^*\| \leq Jt_0$ from $\mathbf{W}^m$, and the point $x_p^0 \in S$ is replaced by the point $X^0 = (x_0^0, x_1^0, \ldots, x_{m-1}^0) \in S_1$, which ensures the validity of the equality $\Psi(X^0) = 0$. But, in this case, the operator $\frac{d\Psi(X_0^*)}{dX_0^*}$ cannot be invertible in a finite-dimensional space $\mathbf{W}$ even for $m=2$. Indeed, a homeomorphism that maps $\mathbf{W}^2$ onto $\mathbf{W}$ must exist. But these two spaces have different dimensionalities. In the case of an infinite-dimensional space $\mathbf{W}$, there exist the linear invertible operators, which map $\mathbf{W}^2$ onto $\mathbf{W}$.

We now give the example of such operator. Let $\mathbf{W} = \mathbf{H}$ be a separable Hilbert space, and let $\{e_n\}_{n=1}^{\infty}$ be its any orthonormalized basis. The space $\mathbf{H}^2 = \mathbf{H} \times \mathbf{H}$ becomes a Hilbert one, if we set $\langle(x_1, y_1), (x_2, y_2)\rangle = \langle x_1, x_2\rangle + \langle y_1, y_2\rangle$. In this case, $\|(x, y)\| = \sqrt{\langle(x, y), (x, y)\rangle} = \sqrt{\|x\|^2 + \|y\|^2}$, where $\langle \cdot, \cdot \rangle$ is a scalar product.

The mappings

$$F_1 : \mathbf{H}_1 = \{(x, 0\,)|\ x \in \mathbf{H}\} \to \mathbf{H}$$

and

$$F_2 : \mathbf{H}_2 = \{(0, y)|\ y \in \mathbf{H}\} \to \mathbf{H},$$

which are set by the formulas $F_1((x, 0)) = x$ and $F_2((0, y)) = y$, respectively, are isometric isomorphisms. By identifying the appropriate elements under these isomorphisms $(x, 0) = x, (0, y) = y$, we obtain $\mathbf{H}^2 = \mathbf{H} \oplus \mathbf{H}$, since $\langle(x, 0), (0, y)\rangle = 0$.

If $\{e_n'\}_{n=1}^{\infty}$ is one more orthonormalized basis in $\mathbf{H}$, then the system

$$a_1 = (e_1, 0), a_2 = (0, e_1'), a_3 = (e_2, 0), a_4 = (0, e_2'), ...,$$

$$a_{2n-1} = (e_n, 0), a_{2n} = (0, e_n'), ...$$

is the orthonormalized basis of the space $\mathbf{H}^2$.

We now introduce a mapping $A : \mathbf{H} \to \mathbf{H}^2$, by setting

$$A\left(x = \sum_{n=1}^{\infty} \alpha_n e_n\right) = \sum_{n=1}^{\infty} \alpha_n a_n.$$

It is obvious that A is a linear operator. By the Parseval equality, $\|Ax\|^2 = \sum_{n=1}^{\infty} |\alpha_n|^2 = \|x\|^2$. As a result, the operator A is isometric and, hence, bounded. Due to the isometricity, $Ax = 0 \Leftrightarrow x = 0$.

Therefore, A is an injective mapping, and there exists $A^{-1} : \mathbf{H}^2 \to \mathbf{H}$. In this case, A^{-1} is a linear homeomorphism.

We now assume that $\Psi(X_0) \neq 0$. It is of interest to find the conditions and to introduce a certain perturbation of Eq. (3.110), which will allow us to obtain an equation, whose solution $x_n(X_0)$ equals d for $n = m + k$.

If $k=0$, we can easily write the perturbed equation:

$$\Delta^m x_n = f_n(x_n, x_{n+1}, ..., x_{n+m-1}) + \alpha(X_0), \quad n \in Z_0^+,$$

where

$$\alpha(X_0) = d + \sum_{i=0}^{m-1} (-1)^{m-i} C_m^i x_i - f_0(X_0).$$

Let us solve this problem for $k=1$. The following proposition is valid.

Theorem 3.28. *Let,* $\forall \{Z, \overline{Z}\} \subset \mathbf{W}^m$,

$$\|f_1(Z) - f_1(\overline{Z})\| \le K\|Z - \overline{Z}\|, \tag{3.119}$$

where the positive constant K satisfies the inequality $K < m + 1$.

Then, $\forall X_0 \in \mathbf{W}^m$, *there exists the single perturbation $\alpha(X_0) \in \mathbf{W}$, under which the solution $z_n(X_0)$ of the equation*

$$\Delta^m z_n = f_n(z_n, z_{n+1}, ..., z_{n+m-1}) + \alpha(X_0), \quad n \in Z_0^+, \tag{3.120}$$

satisfies the condition $z_{m+1}(X_0) = d$.

Proof. We denote $(x_p^{(s)}, x_{p+1}^{(s)}, ..., x_{p+m-1}^{(s)}) \in \mathbf{W}^m$ by $X_p^{(s)}$, use the operator $\bar{F}$ defined in the proof of Lemma 3.17, and set the recurrence relations:

$$X_0^{(s)} = X_0 \quad (s \in Z_0^+);$$

$$x_m^{(0)} = \bar{F}(X_0) + f_0(X_0);$$

$$x_m^{(s)} = \bar{F}(X_0) + f_0(X_0) + \alpha(s, X_0) \quad (s \in Z^+);$$

$$x_{m+1}^{(s)} = \bar{F}(X_1^{(0)}) + f_1(X_1^{(s-1)}) + (m+1)\alpha(s, X_0) \quad (s \in Z^+).$$

For all $s \in Z^+$ and

$$\alpha(s, X_0) = \frac{1}{m+1}\{d - \bar{F}(X_1^{(0)}) - f_1(X_1^{(s-1)})\},$$

the equality $x_{m+1}^{(s)} = d$ holds. In this case,

$$x_m^{(s)} = \bar{F}(X_0) + f_0(X_0) + \frac{1}{m+1}\{d - \bar{F}(X_1^{(0)}) - f_1(X_1^{(s-1)})\}. \tag{3.121}$$

We now show that the sequence $\{x_m^{(s)}\}_{s=1}^{\infty}$ converges in the norm as $s \to \infty$. The equality

$$x_m^{(s+1)} - x_m^{(s)} = \frac{1}{m+1}\{f_1(X_1^{(s-1)}) - f_1(X_1^{(s)})\}$$

yields the inductive inequality

$$\|x_m^{(s+1)} - x_m^{(s)}\| \le \frac{K}{m+1}\|x_m^{(s)} - x_m^{(s-1)}\|,$$

from which we easily obtain the estimate

$$\|x_m^{(s+1)} - x_m^{(s)}\| \le \left(\frac{K}{m+1}\right)^s \|x_m^{(1)} - x_m^{(0)}\| = \left(\frac{K}{m+1}\right)^s \|\alpha(1, X_0)\|. \tag{3.122}$$

But

$$\| \alpha(1, X_0) \| = \frac{1}{m+1} \| d - \bar{F}(X_1^{(0)}) - f_1(X_1^{(0)}) \|$$

$$= \frac{1}{m+1} \| d + \sum_{i=0}^{m-2} (-1)^{m-i} C_m^i x_{i+1} + m \sum_{i=0}^{m-1} (-1)^{m-i} C_m^i x_i - m f_0(X_0)$$

$$- f_1(x_1, ..., x_{m-1}, - \sum_{i=0}^{m-1} (-1)^{m-i} C_m^i x_i + f_0(X_0)) \| = K^* = const \geq 0.$$

If $d \in \mathbf{W}$ is such that $K^*=0$, then it is sufficient to set $\alpha(X_0)=0$. We consider d such that $K^* > 0$. Since $K/(m+1) < 1$, we have $\|x_m^{(s+1)} - x_m^{(s)}\| \to 0$ as $s \to \infty$, and the sequence $\{ x_m^{(s)} \}_{s=1}^{\infty}$ is fundamental. The completeness of the space $\mathbf{W}$ yields its convergence in the norm of this space as $s \to \infty$ to a certain element $\tilde{x}_m \in \mathbf{W}$.

With regard for the continuity of the function f_1, we pass to the limit in (3.121) as $s \to \infty$ and obtain the equality

$$\tilde{x}_m = \bar{F}(X_0) + f_0(X_0) + \frac{1}{m+1}\{ d - \bar{F}(X_1^{(0)}) - f_1(x_1, ..., x_{m-1}, \tilde{x}_m)\}.$$

In Eq. (3.120), we set

$$\alpha(X_0) = \frac{1}{m+1}\{ d - \bar{F}(X_1^{(0)}) - f_1(x_1, ..., x_{m-1}, \tilde{x}_m)\}$$

and find $z_{m+1}(X_0)$. The equalities

$$z_m(X_0) = \bar{F}(X_0) + f_0(X_0) + \alpha(X_0) = \tilde{x}_m$$

guarantee the validity of the following transformations:

$$z_{m+1}(X_0) = - \sum_{i=0}^{m-2} (-1)^{m-i} C_m^i x_{i+1}$$

$$+ m\tilde{x}_m + f_1(x_1, ..., x_{m-1}, \tilde{x}_m) + \alpha(X_0)$$

$$= - \sum_{i=0}^{m-2} (-1)^{m-i} C_m^i x_{i+1} + m\bar{F}(X_0) + m f_0(X_0) + f_1(x_1, ..., x_{m-1}, \tilde{x}_m)$$

$$+ (m+1)\alpha(X_0) = \lim_{s \to \infty} x_{m+1}^{(s)} = d.$$

We now assume that there exists a perturbation $\beta(X_0) \in \mathbf{W}$ such that the solution $\bar{z}_n(X_0)$ of the equation

$$\Delta^m z_n = f_n(z_n, z_{n+1}, ..., z_{m+n-1}) + \beta(X_0), \quad n \in Z_0^+,$$

for $n = m + 1$ equals d. In this case,

$$\|\alpha(X_0) - \beta(X_0)\| = \frac{1}{m+1} \, \| \, f_1(x_1, ..., x_{m-1}, \bar{z}_m) - f_1(x_1, ..., x_{m-1}, \tilde{x}_m) \, \|$$

$$\leq \frac{K}{m+1} \| \, \tilde{x}_m - \bar{z}_m \| = \frac{K}{m+1} \|\alpha(X_0) - \beta(X_0)\|,$$

which yields $\alpha(X_0) = \beta(X_0)$. This completes the proof. $\qquad\square$

Corollary 3.19. *Under conditions of Theorem 3.28, the solution* $y_n(X_0)$
of the equation

$$\Delta^m y_n = f_n(y_n, y_{n+1}, ..., y_{n+m-1}) + \alpha(s, X_0), \quad n \in Z_0^+,$$

satisfies the inequality

$$\|y_{m+1} - d\| \leq (m + K + 1)\frac{KK^*}{m+1-K}(\frac{K}{m+1})^s. \tag{3.123}$$

Proof. The inequalities

$$\|\tilde{x}_m - x_m^{(s)}\| \leq \lim_{g \to \infty} \sum_{i=1}^{g} \|x_m^{(s+i)} - x_m^{(s+i-1)}\|$$

$$\leq \sum_{i=1}^{\infty}(\frac{K}{m+1})^{s+i-1}K^* = K^*(\frac{K}{m+1})^s : (1 - \frac{K}{m+1})$$

yield the estimates

$$\|\alpha(X_0) + \alpha(s, X_0)\| \leq \| \, f_1(x_1, ..., x_{m-1}, \tilde{x}_m) - f_1(x_1, ..., x_{m-1}, x_m^{(s)}) \, \|$$

$$\leq \frac{K}{m+1}\| \, \tilde{x}_m - x_m^{(s)}\| \leq \frac{KK^*(\frac{K}{m+1})^s}{m+1-K}. \tag{3.124}$$

The following relations hold:

$$\|z_{m+1} - y_{m+1}\| = \|- \sum_{i=0}^{m-2}(-1)^{m-i}C_m^i x_{i+1} + C_m^1(\bar{F}(X_0) + f(X_0) + \alpha(X_0))$$

$$+ f_1(x_1, ..., x_{m-1}, z_m) + \alpha(X_0) + \sum_{i=0}^{m-2}(-1)^{m-i}C_m^i x_{i+1} - C_m^1(\bar{F}(X_0) + f(X_0)$$

$$+ \alpha(s, X_0)) - f_1(x_1, ..., x_{m-1}, y_m) - \alpha(s, X_0)\| \leq (m+1)\|\alpha(X_0) - \alpha(s, X_0)\|$$

$$+ K\|z_m - y_m\| \leq (m+1+K)\|\alpha(X_0) - \alpha(s, X_0)\|.$$

In view of (3.124), these relations yield estimate (3.123). $\qquad\square$

We note that the conditions of Theorem 3.28 require that inequality (3.119) be satisfied in the whole space $\mathbf{W}^m$. This requirement can be weakened by the restriction of the set of all X_0 to a subset in $\mathbf{W}^m$ depending on d.

Corollary 3.20. *Let inequality (3.119) hold*

$$\forall \{ Z, \overline{Z} \} \subset D_1 = \{ Z \in \mathbf{W}^m \mid \|Z\| \leq R = const > 0 \},$$

and, in this ball, $\max\{\|f_0\|, \|f_1\|\} \leq M = const > 0$. *If the set*

$$D_k^1 = \{ Z \in D_1 \mid \|Z\| \leq \frac{R - M - \frac{K(d+M(m+1))}{(m+1-K)(m+1)}}{2^m - 1 + \frac{K(2^m + m2^m - 2m - 1)}{(m+1-K)(m+1)}} \} \tag{3.125}$$

is nonempty, then, $\forall X_0 \in D_k^1$, *the assertion of Theorem 3.28 remains proper.*

Proof. It is easy to see that it is sufficient to attain that the point $(x_1, x_2, \ldots x_{m-1}, x_m^{(s)})$ belongs to the set D_1 for all $s \in Z_0^+$.

With regard for the inequalities

$$\|x_m^{(0)}\| \leq (2^m - 1) \|X_0\| + \|f_0(X_0)\| \leq (2^m - 1) \|X_0\| + M,$$

relations (3.122) yield

$$\|x_m^{(s)}\| \leq \|x_m^{(0)}\| + K^* \sum_{i=1}^{s-1} (\frac{K}{m+1})^i < \|x_m^{(0)}\| + \frac{K^* K}{m+1-K}.$$

If $\|X_0\| \leq \frac{R-M}{2^m - 1}$, then

$$K^* \leq \frac{1}{m+1}(d + (2^m - m - 1) \|X_0\| + m (2^m - 1) \|X_0\| + (m+1)M).$$

The inequality

$$\|x_m^{(0)}\| + \frac{K(d + (2^m - m - 1) \|X_0\| + m (2^m - 1) \|X_0\| + (m+1)M)}{(m+1)(m+1-K)} \leq R$$

yields the estimate given by relation (3.125). $\square$

Chapter 4

Countable-point boundary-value problems for nonlinear differential equations

It is clear that the possibilities for a constructive study of various boundary-value problems for ordinary differential equations depend essentially on the dimensionality of the space, where a specific boundary-value problem is considered. Up to now, the boundary-value problems of various types in finite-dimensional spaces are most completely studied. The results of studies of these problems in abstract Banach spaces are presented in works, whose number is comparatively small (see, e.g., [10] – [12], [62] – [64], [27; 52; 87; 89–91; 101; 121]. Moreover, they concern mainly periodic boundary-value problems. The study of the boundary-value problems in the space $\mathfrak{M}$ was started in works [62] – [64], [90; 101], where the periodic boundary-value problem for equations of the first and second orders and the two-point boundary-value problem for a nonlinear equation of the first order with a linear boundary condition were considered.

Here, we will study the countably point boundary-value problems for ordinary nonlinear differential equations of the normal form and the equations unsolvable with respect to the derivative, which are defined in the space $\mathfrak{M}$. We will analyze the case where the countably point set of limiting moments belongs to a finite interval and the case where this set is not bounded and belongs to the positive semiaxis. A special attention is paid to the possibility to reduce these problems to the comprehensively investigated multipoint boundary-value problems in finite-dimensional spaces.

4.1 Boundary-value problem on the semiaxis

By $\mathfrak{M}^\infty$, we denote the space of sequences $\psi = (\psi_1, \psi_2, \psi_3, ...)$, $\psi_i \in \mathfrak{M}\ \forall i \in Z^+$, bounded in the norm $\|\psi\| = \sup_{i \in Z^+} \{\|\psi_i\|\}$, where $\|\psi_i\|$ is the norm in

289

the space $\mathfrak{M}$. The norm of the infinite matrix $A = [a_{ij}]_{i,j=1}^{\infty}$ consistent with the norm of a vector $x \in \mathfrak{M}$ is defined, as above, by the equality $\|A\| = \sup_i \sum_{j=1}^{\infty} |a_{ij}|$. Let $D = \{x| \ x \in \mathfrak{M}, \ \|x\| \leq M_0 = const > 0\}$ and $D^{\infty} = \{\psi \in \mathfrak{M}^{\infty}| \ \|\psi\| \leq M_0\}$. We pose the following problem: to find the solution of the equation

$$\frac{dx}{dt} = f(t, x), \tag{4.1}$$

which satisfies the condition

$$A_0 x(0) + \sum_{i=1}^{\infty} A_i x(t_i) = \varphi(x(0); x(t_1), x(t_2), ...), \tag{4.2}$$

where

$$0 < \ t_1 < t_2 < t_3 < ..., \quad \sup_i \{t_i\} = +\infty; \quad x = (x_1, x_2, x_3, ...) \in D;$$

$$f(t, x) = \{f_1(t, x), f_2(t, x), f_3(t, x), ...\} : [0, +\infty) \times D = \overline{D}_0 \to \mathfrak{M};$$

$A_i (i = 0, 1, 2, ...)$ are infinite matrices such that $\sum_{i=1}^{\infty} \|A_i\| < \infty$; the derivative $\frac{dx(t)}{dt}$ means the vector

$$\left(\frac{dx_1(t)}{dt}, \frac{dx_2(t)}{dt}, \frac{dx_3(t)}{dt}, ...\right);$$

the function

$$\varphi(\psi_1, \psi_2, ...) = \{\varphi_1(\psi_1, \psi_2, ...), \varphi_2(\psi_1, \psi_2, ...), ...\} : D^{\infty} \to \mathfrak{M}.$$

By $h(t)$, we denote the function $h : [0, +\infty) \to [0, T), \ T = const > 0$, which has the following properties:

1^*) $h(0) = 0, \ h(+\infty) = \lim_{t \to +\infty} h(t) = T;$

2^*)   on the semiaxis $[0, +\infty)$, there exists a continuous nonnegative derivative $h'(t)$ boun-ded by the constant $\overline{h'} = \sup_{t \geq 0} h'(t)$.

The following conditions are called conditions $(\mathbf{A})$:

1) $\forall \{\psi, \psi_*\} \subset D^{\infty}$, the inequalities

$$\|\varphi(\psi)\| \leq M_{\varphi}, \ \|\varphi(\psi) - \varphi(\psi_*)\| \leq K_{\varphi}\|\psi - \psi_*\|, \tag{4.3}$$

where M_{φ} and K_{φ} are positive constants, hold;

2) the function $f(t, x)$ is continuous on $\overline{D}_0$ in t, x, and there exists a function $h(t)$ with properties 1^* and 2^* such that

$$\|f(t, x)\| \leq M_h h'(t), \ \|f(t, x) - f(t, x')\| \leq K_h h'(t)\|x - x'\|, \tag{4.4}$$

where x, x' are any points from D, and M_h and K_h are positive constants independent of the points (t, x) and (t, x') from $\overline{D}_0$;

3) the matrix $\sum\limits_{i=1}^{\infty} h(t_i) A_i$ is invertible, and the inverse matrix $(\sum\limits_{i=1}^{\infty} h(t_i) A_i)^{-1}$ is bounded in the norm.

It is easy to see that then the matrix $\sum\limits_{i=1}^{\infty} \frac{h(t_i)}{T} A_i$ is invertible, and the inverse matrix

$$H_h = T(\sum_{i=1}^{\infty} h(t_i) A_i)^{-1}$$

is also bounded in the norm.

By $D_{\beta\varphi h}$, we denote a subset of D, every point x_0 of which is contained in the set D together with its $\beta_{\varphi h}$-neighborhood, where

$$\beta_{\varphi h}(x_0) = \frac{T}{2} M_h + \beta_{1\varphi h}(x_0),$$

$$\beta_{1\varphi h}(x_0) = \|H_h\| \|d - \sum_{i=0}^{\infty} A_i x_0\| + \sum_{i=1}^{\infty} \|H_h A_i\| \alpha_{1h}(t_i) M_h,$$

$$\alpha_{1h}(t) = 2h(t)(1 - \frac{h(t)}{T}),$$

and the vector $d = (d_1, d_2, ...)$ is defined so that $|d_i| = M_\varphi$, $sign\, d_i = -sign\, d_i^0$, and $d^0 = colon(d_1^0, d_2^0, ...) = \sum\limits_{j=0}^{\infty} A_j x_0$, $i \in Z^+$.

It is easy to verify that $0 \leq \alpha_{1h}(t) \leq \frac{T}{2}$.

The following conditions are called conditions **(B)**:

a) the set $D_{\beta\varphi h}$ is nonempty;

b) the inequality

$$Q_{\varphi h} = \frac{K_h T}{2}[1 + \sum_{i=1}^{\infty} \|H_h A_i\|] + K_\varphi \|H_h\| < 1$$

holds. Below, for the vector-function $f(\tau) = \{f_1(\tau), f_2(\tau), ...\}, \tau \in R^1$, we denote the vector $\{\int\limits_a^b f_1(\tau) d\tau, \int\limits_a^b f_2(\tau) d\tau, ...\}$ by $\int\limits_a^b f(\tau) d\tau$.

We now write formally the recurrence sequence of functions $\{x_m(t, x_0)\}_{m=0}^{\infty}$ as follows:

$$x_0(t, x_0) = x_0 \equiv (x_{01}, x_{02}, ...) \in D_{\beta\varphi h};$$

$$x_m(t, x_0) = x_0 + \int_0^t [f(\tau, x_{m-1}(\tau, x_0)) - \frac{h'(\tau)}{T} \int_0^\infty f(s, x_{m-1}(s, x_0))ds]d\tau$$

$$+ \frac{h(t)}{T} H_h\{\varphi(x_{m-1}(0); x_{m-1}(t_1), x_{m-1}(t_2), ...) - \sum_{i=0}^\infty A_i x_0$$

$$- \sum_{i=1}^\infty A_i \int_0^{t_i} [f(\tau, x_{m-1}(\tau, x_0)) - \frac{h'(\tau)}{T} \int_0^\infty f(s, x_{m-1}(s, x_0))ds]d\tau\},$$

$$m \in Z^+. \quad (4.5)$$

For such sequence to exist, it is sufficient that, for all $i \in Z^+$, the functions $x_i(t, x_0)$ be continuous in the norm in the variable t on the positive semiaxis and be bounded in the norm by a number M_0. Then the coordinatewise convergence of the integral $\int_0^\infty f(\tau, x_{m-1}(\tau, x_0))d\tau$ will be ensured by the first of conditions (4.4) and by properties 1^* and 2^* of the function $h(t)$.

Theorem 4.1. *Let us assume that conditions* **(A)** *and* **(B)** *are satisfied. Then*

1) the sequence $\{x_m(t, x_0)\}_{m=0}^\infty$ *defined by equalities (4.5) is uniformly convergent as* $m \to \infty$ *with respect to* $(t, x_0) \in [0, +\infty) \times D_{\beta\varphi h}$ *to the function* $x^*(t, x_0)$, *and,* $\forall m \in Z^+$,

$$\|x_m(t, x_0) - x^*(t, x_0)\| \le \frac{Q_{\varphi h}^m}{1 - Q_{\varphi h}} \beta_{\varphi h}(x_0); \quad (4.6)$$

2) the function $x^*(t, x_0)$ *satisfies the boundary condition (4.2) and is a solution of the equation*

$$\frac{dx}{dt} = f(t, x) + \mu \cdot h'(t), \quad (4.7)$$

where

$$\mu = \Delta_{\varphi h}(x_0) = \frac{1}{T} H_h\{\varphi(x^*(0); x^*(t_1), x^*(t_2), ...)$$

$$- \sum_{i=0}^\infty A_i x_0 - \sum_{i=1}^\infty A_i \int_0^{t_i} [f(\tau, x^*(\tau, x_0))$$

$$- \frac{h'(\tau)}{T} \int_0^\infty f(s, x^*(s, x_0))ds]d\tau\} - \frac{1}{T} \int_0^\infty f(\tau, x^*(\tau, x_0))d\tau; \quad (4.8)$$

3) if

$$\Delta_{\varphi h}(x_0) = 0, \qquad (4.9)$$

then the function $x^*(t, x_0)$ is a solution of the boundary-value problem (4.1), (4.2).

Proof.   For $m = 1$, relation (4.5) yields

$$x_1(t, x_0) = x_0 + \int_0^t [f(\tau, x_0) - \frac{h'(\tau)}{T} \int_0^\infty f(s, x_0)ds]d\tau$$

$$+ \frac{h(t)}{T} H_h\{\varphi(x_0; x_0, x_0, ...)$$

$$- \sum_{i=0}^\infty A_i x_0 - \sum_{i=1}^\infty A_i \int_0^{t_i} [f(\tau, x_0) - \frac{h'(\tau)}{T} \int_0^\infty f(s, x_0)ds]d\tau\}.$$

For all $i \in Z^+$, the integral $\int_0^\infty f_i(s, x_0)ds$ converges, since $f_i(s, x_0)$ is a function continuous in s on the interval $[0, +\infty)$, and

$$\int_0^\infty |f_i(s, x_0)|ds \leq \int_0^\infty M_h h'(s)ds = \lim_{A \to +\infty} M_h \int_0^A h'(s)ds = M_h T.$$

For any $\{t', t''\} \subset [0, +\infty)$, we have

$$\|x_1(t'', x_0) - x_1(t', x_0)\| \leq \| \int_{t'}^{t''} [f(\tau, x_0) - \frac{h'(\tau)}{T} \int_0^\infty f(s, x_0)ds]d\tau\|$$

$$+ \frac{|h(t'') - h(t')|}{T} \|H_h\| \|\varphi(x_0; x_0, x_0, ...) - \sum_{i=0}^\infty A_i x_0 - \sum_{i=1}^\infty A_i \int_0^{t_i} [f(\tau, x_0)$$

$$- \frac{h'(\tau)}{T} \int_0^\infty f(s, x_0)ds]d\tau\| \leq 2M_h \overline{h}' |t'' - t'| + \frac{|h(t'') - h(t')|}{T} \|H_h\|\gamma,$$

where $\gamma = M_\varphi + \|x_0\| \sum_{i=0}^\infty \|A_i\| + 2M_h T \sum_{i=1}^\infty \|A_i\| = const > 0$.

Since the function $h(t)$ is continuous on $[0, +\infty)$, the last inequalities guarantee the continuity of the function $x_1(t, x_0)$ on this interval.

We now estimate the difference $x_1(t, x_0) - x_0$ in the norm, by using the inequalities

$$\|\varphi(x_0, x_0, ...) - \sum_{i=0}^\infty A_i x_0\| \leq \|d - \sum_{i=0}^\infty A_i x_0\|,$$

$$\left\| \int_0^t [f(\tau, x_0) - \frac{h'(\tau)}{T} \int_0^\infty f(s, x_0)ds]d\tau \right\| \le M_h \alpha_{1h}(t).$$

The substantiation of the first inequality is easy. Let us prove the second one. For any positive real number A, the equalities

$$\int_0^t [f(\tau, x_0) - \frac{h'(\tau)}{T} \int_0^A f(s, x_0)ds]d\tau$$

$$= \int_0^t f(\tau, x_0)d\tau - \int_0^t \frac{h'(\tau)}{T} \left(\int_0^t f(s, x_0)ds + \int_t^A f(s, x_0)ds \right) d\tau$$

$$= \int_0^t f(\tau, x_0)d\tau - \frac{h(t)}{T} \int_0^t f(s, x_0)ds - \frac{h(t)}{T} \int_t^A f(s, x_0)ds$$

$$= (1 - \frac{h(t)}{T}) \int_0^t f(s, x_0)ds - \frac{h(t)}{T} \int_t^A f(s, x_0)ds$$

hold. With regard for the first of inequalities (4.4), we obtain the relation

$$\left\| \int_0^t [f(\tau, x_0) - \frac{h'(\tau)}{T} \int_0^A f(s, x_0)ds]d\tau \right\|$$

$$\le M_h[(1 - \frac{h(t)}{T})h(t) + \frac{h(t)}{T}(h(A) - h(t))].$$

Passing in it to the limit as $A \to +\infty$ and taking condition 1^* into account, we obtain the required inequality.

It is easy to verify now that the inequalities

$$\|x_1(t, x_0) - x_0)\| \le M_h \alpha_{1h}(t) + \|H_h\| \| d - \sum_{i=0}^\infty A_i x_0 \|$$

$$+ M_h \sum_{i=1}^\infty \|H_h A_i\| \alpha_{1h}(t_i) \le \frac{T}{2} M_h + \beta_{1\varphi h}(x_0) = \beta_{\varphi h}(x_0)$$

hold, i.e., $\forall x_0 \in D_{\beta\varphi h}$ $x_1(t, x_0) \in D$ $\forall t \in [0, +\infty)$.

We assume that, $\forall i \le k \in Z^+$, the functions $x_i(t, x_0)$ are continuous on $[0, +\infty)$, and their values do not leave the set D. By performing the analogous consideration, we verify that the function $x_{k+1}(t, x_0)$ possesses these properties for $x_0 \in D_{\beta\varphi h}$. By the principle of complete mathematical induction, the function $x_i(t, x_0)$ $\forall i \in Z^+$ has these properties, i.e., relations (4.5) define the functional sequence $\{x_m(t, x_0)\}_{m=0}^{\infty}$.

It is easy to verify by the direct substitution that the elements of this sequence satisfy the recurrence condition

$$A_0 x_m(0) + \sum_{i=1}^{\infty} A_i x_m(t_i) = \varphi(x_{m-1}(0); x_{m-1}(t_1), x_{m-1}(t_2), ...) \qquad (4.10)$$

for every $m \in Z^+$.

We now prove that the indicated sequence converges in the norm as $m \to \infty$. Since the space $\mathfrak{M}$ is complete, it is sufficient to prove its fundamentality. Using relations (4.3) – (4.5), we obtain

$$\|x_{m+1}(t, x_0) - x_m(t, x_0)\| \le$$

$$K_h[(1 - \frac{h(t)}{T}) \int_0^t h'(s)\|x_m(s, x_0) - x_{m-1}(s, x_0)\|ds$$

$$+ \frac{h(t)}{T} \int_t^{\infty} h'(s)\|x_m(s, x_0) - x_{m-1}(s, x_0)\|ds] + K_h \frac{h(t)}{T}$$

$$\times \sum_{i=1}^{\infty} \|H_h A_i\|[(1 - \frac{h(t_i)}{T}) \int_0^{t_i} h'(s)\|x_m(s, x_0) - x_{m-1}(s, x_0)\|ds$$

$$+ \frac{h(t_i)}{T} \int_{t_i}^{\infty} h'(s)\|x_m(s, x_0) - x_{m-1}(s, x_0)\|ds]$$

$$+ K_\varphi\|H_h\| \frac{h(t)}{T} \sup_i \{ \|x_m(t_i, x_0) - x_{m-1}(t_i, x_0)\|\}.$$

Let us introduce the notation

$$\bar{r}_{m+1} = \sup_{t \in [0, +\infty)} \|x_{m+1}(t, x_0) - x_m(t, x_0)\|.$$

Then the last inequality yields

$$\bar{r}_{m+1} \le \{K_h[(1 - \frac{h(t)}{T})h(t) + \frac{h(t)}{T}(T - h(t))] + K_h \frac{h(t)}{T} \sum_{i=1}^{\infty} \|H_h A_i\|$$

$$\times [(1 - \frac{h(t_i)}{T})h(t_i) + \frac{h(t_i)}{T}(T - h(t_i))] + K_\varphi\|H_h\|\}\bar{r}_m$$

$$\le \{\frac{K_h T}{2}[1 + \sum_{i=1}^{\infty} \|H_h A_i\|] + K_\varphi\|H_h\|\}\bar{r}_m = Q_{\varphi h}\bar{r}_m.$$

By the principle of complete mathematical induction, it is easy to verify that, for any natural number m,

$$\bar{r}_{m+1} \leq Q_{\varphi h}^{m} \bar{r}_1 \leq Q_{\varphi h}^{m} \beta_{\varphi h}(x_0).$$

For any natural m, $j > m$, the identity

$$x_{m+j}(t, x_0) - x_m(t, x_0) = \sum_{i=1}^{j} (x_{m+i}(t, x_0) - x_{m+i-1}(t, x_0))$$

holds. Estimating its right-hand side from above, we obtain the chain of inequalities

$$\|x_{m+j}(t, x_0) - x_m(t, x_0)\| \leq \sum_{i=1}^{j} \bar{r}_{m+i} \leq \sum_{i=0}^{j-1} Q_{\varphi h}^{m+i} \beta_{\varphi h}(x_0)$$

$$\leq Q_{\varphi h}^{m} \beta_{\varphi h}(x_0) \sum_{i=0}^{j-1} Q_{\varphi h}^{i} < \frac{Q_{\varphi h}^{m}}{1 - Q_{\varphi h}} \beta_{\varphi h}(x_0).$$

Since $\beta_{\varphi}(x_0) \leq M_0$ and $Q_{\varphi h} < 1$, $\lim\limits_{m \to \infty} x_m(t, x_0) = x^*(t, x_0)$ uniformly in $(t, x_0) \in [0, +\infty) \times D_{\beta \varphi h}$ in the norm of the space $\mathfrak{M}$ as $m \to \infty$, and inequality (4.6) is satisfied.

The sequence $\{f(\tau, x_m(\tau, x_0))\}_{m=1}^{\infty}$ converges to $f(\tau, x^*(\tau, x_0))$ uniformly in $\tau \in [0, +\infty)$ and $x_0 \in D_{\beta \varphi h}$. This follows obviously from the inequality

$$\|f(\tau, x_m(\tau, x_0)) - f(\tau, x^*(\tau, x_0))\| \leq K_h \cdot \overline{h'} \|x_m(\tau, x_0) - x^*(\tau, x_0)\|.$$

In addition, the sequence

$$\left\{ \frac{h'(\tau)}{T} \int_0^{\infty} f(s, x_m(s, x_0)) ds \right\}_{m=1}^{\infty}$$

converges uniformly in $\tau \in [0, +\infty)$ to

$$\frac{h'(\tau)}{T} \int_0^{\infty} f(s, x^*(s, x_0)) ds,$$

which follows from the relations

$$\left\|\frac{h'(\tau)}{T}\int_0^\infty f(s, x^*(s, x_0))ds - \frac{h'(\tau)}{T}\int_0^\infty f(s, x_m(s, x_0))ds\right\|$$

$$\leq \frac{h'(\tau)}{T}\int_0^\infty K_h h'(s)\|x^*(s, x_0) - x_m(s, x_0)\|ds$$

$$\leq K_h \frac{h'(\tau)}{T}\int_0^\infty \frac{Q_{\varphi h}^m}{1 - Q_{\varphi h}}\beta_{\varphi h}(x_0)h'(s)ds \leq \frac{K_h \overline{h'} Q_{\varphi h}^m}{1 - Q_{\varphi h}}(\frac{T}{2}M_h$$

$$+ \|H_h\|\|d - \sum_{i=0}^\infty A_i x_0\| + \sum_{i=1}^\infty \|H_h A_i\|\frac{T}{2}M_h) = Q_{\varphi h}^m \cdot \gamma_1 \to 0$$

as $m \to \infty$, where γ_1 is a positive constant for a fixed x_0.

This allows us to pass coordinatewise in (4.5) to the limit as $m \to \infty$ and to obtain the equality

$$x^*(t, x_0) = x_0 + \int_0^t [f(\tau, x^*(\tau, x_0)) - \frac{h'(\tau)}{T}\int_0^\infty f(s, x^*(s, x_0))ds]d\tau$$

$$+ \frac{h(t)}{T}H_h\{\varphi(x^*(0); x^*(t_1), x^*(t_2), ...) - \sum_{i=0}^\infty A_i x_0$$

$$- \sum_{i=1}^\infty A_i \int_0^{t_i}[f(\tau, x^*(\tau, x_0)) - \frac{h'(\tau)}{T}\int_0^\infty f(s, x^*(s, x_0))ds]d\tau\}.$$

It is easy to see that

$$\varphi(x_m(0); x_m(t_1), x_m(t_2), ...) \xrightarrow[m\to\infty]{} \varphi(x^*(0); x^*(t_1), x^*(t_2), ...)$$

in the norm of the space $\mathfrak{M}$. In addition, $\forall m \in Z^+$, we have

$$\left\|(A_0 x_m(0) + \sum_{i=1}^\infty A_i x_m(t_i)) - (A_0 x^*(0) + \sum_{i=1}^\infty A_i x^*(t_i))\right\|$$

$$\leq M_0\frac{Q_{\varphi h}^m}{1 - Q_{\varphi h}}\sum_{i=0}^\infty \|A_i\| \xrightarrow[m\to\infty]{} 0.$$

These relations imply that we can realize the limiting transition in equalities (4.10) in the norm as $m \to \infty$. This indicates that $x^*(t, x_0)$ satisfies the boundary condition (4.2), which completes the proof of two first assertions of Theorem 4.1. The proof of its third assertion is obvious. $\qquad\square$

Conditions $(\mathbf{B^0})$ are called conditions $(\mathbf{B})$, in which the inequality $Q_{\varphi h} < 1$ is replaced by the sharper estimate

$$K_h T\left[1 + \sum_{i=1}^{\infty} \|H_h A_i\|\right] + K_\varphi \|H_h\| < 1.$$

Corollary 4.1. *If the function $h(t)$ is chosen, then, under conditions $(\mathbf{A})$ and $(\mathbf{B^0})$, there exists no another value $\mu \in \mathfrak{M}$ such that the solution of Eq. (4.7) with the initial condition $x(0) = x_0 \in D_{\beta \varphi h}$ would satisfy the boundary condition (4.2).*

Proof.   Let us assume that there exist $\{\mu_1, \mu_2\} \subset \mathfrak{M}$ and the functions $x_1(t, x_0)$ and $x_2(t, x_0)$, which satisfy equality (4.2), and $x_1(0, x_0) = x_2(0, x_0) = x_0 \in D_{\beta \varphi h}$. Moreover, for all $t \in [0, \infty)$ and $i \in \{1, 2\}$, the relations

$$\|x_i(t, x_0)\| \leq M_0, \quad \frac{dx_i(t, x_0)}{dt} = f(t, x_i(t, x_0)) + \mu_i h'(t)$$

hold. Then

$$x_j(t, x_0) = x_0 + \int_0^t f(\tau, x_j(\tau, x_0))d\tau + \mu_j h(t), \ j \in \{1, \ 2\},$$

$$A_0 x(0) + \sum_{i=1}^{\infty} A_i \left\{ x_0 + \int_0^{t_i} f(\tau, x_j(\tau, x_0))d\tau + \mu_j h(t_i) \right\} = \tilde{\varphi}_j, \ j \in \{1, \ 2\},$$

where, by $\tilde{\varphi}_j$, we denote the expression

$$\varphi\left(x_0; x_0 + \int_0^{t_1} f(\tau, x_j(\tau, x_0))d\tau + \mu_j h(t_1), x_0\right.$$

$$\left. + \int_0^{t_2} f(\tau, x_j(\tau, x_0))d\tau + \mu_j h(t_2), ...\right).$$

We also denote $\sup_{t \in [0, +\infty)} \|x_1(t, x_0) - x_2(t, x_0)\|$ by r. Then the above equalities yield the relations

$$\|x_1(t, x_0) - x_2(t, x_0)\| \leq \int_0^t K_h h'(\tau) r \, d\tau + \|\mu_1 - \mu_2\| h(t);$$

$$r \leq K_h r T + \|\mu_1 - \mu_2\| T;$$

$$\sum_{i=1}^{\infty} A_i\{ \int_0^{t_i} (f(\tau, x_1(\tau)) - f(\tau, x_2(\tau)))d\tau + h(t_i)(\mu_1 - \mu_2)\} = \tilde{\varphi}_1 - \tilde{\varphi}_2;$$

$$\|\mu_1 - \mu_2\| \le \frac{1}{T}(\sum_{i=1}^{\infty} \|H_h A_i\| \int_0^{t_i} K_h h'(\tau) r d\tau$$

$$+ \|H_h\| K_\varphi \sup_{i \in N}\{ \int_0^{t_i} K_h h'(\tau) r d\tau + \|\mu_1 - \mu_2\| h(t_i)\})$$

$$\le \frac{1}{T}(K_h r T \sum_{i=1}^{\infty} \|H_h A_i\| + \|H_h\| K_\varphi K_h r T + \|H_h\| K_\varphi T \|\mu_1 - \mu_2\|).$$

They, in turn, result in the estimates

$$\|\mu_1 - \mu_2\| \le \frac{K_h(\sum_{i=1}^{\infty} \|H_h A_i\| + K_\varphi \|H_h\|)}{1 - K_\varphi \|H_h\|} r,$$

$$r \le (K_h T + \frac{T K_h(\sum_{i=1}^{\infty} \|H_h A_i\| + K_\varphi \|H_h\|)}{1 - K_\varphi \|H_h\|}) r = \frac{T K_h(1 + \sum_{i=1}^{\infty} \|H_h A_i\|)}{1 - K_\varphi \|H_h\|} r,$$

where the coefficient of r on the right-hand side of the last equality is strictly less than 1.

This implies that $r = 0$, i.e., $x_1(t, x_0) \equiv x_2(t, x_0)\ \forall\ t \in [0, +\infty)$. In this case, of course, $\mu_1 = \mu_2$, which completes the proof. $\square$

In what follows, we agree to call a function $\Delta_{\varphi h}(x_0)$ of the form (4.8) as an exact defining function, its values μ for a fixed x_0 as a controlling parameter or the control, and Eq. (4.9) as an exact defining equation. In addition to the exact defining equation, we will consider the approximate defining equation $\Delta_{\varphi h m}(x_0) = 0$, where

$$\Delta_{\varphi h m}(x_0) = \frac{1}{T} H_h\{\varphi(x_m(0); x_m(t_1), x_m(t_2), ...)$$

$$- \sum_{i=0}^{\infty} A_i x_0 - \sum_{i=1}^{\infty} A_i \int_0^{t_i} [f(\tau, x_m(\tau, x_0))$$

$$- \frac{h'(\tau)}{T} \int_0^{\infty} f(s, x_m(s, x_0)) ds] d\tau\} - \frac{1}{T} \int_0^{\infty} f(\tau, x_m(\tau, x_0)) d\tau \quad (4.11)$$

is an approximate defining function.

A topological mapping $\Im$ of the closed set $D \subset \mathfrak{M}$ onto $\mathfrak{M}$ means a homeomorphism $\Im : D \mapsto \Im(D)$, which transfers the boundary of the set D to the boundary of its image $\Im(D)$.

Corollary 4.2. *The following assertions are true:*

1) if, under conditions $(\mathbf{A})$ *and* $()$, *there exists a closed subset* $D_1 \subset D_{\beta\varphi h}$ *such that, for some* $m \in Z^+$, *the function* $\Delta_{\varphi h m}$ *topologically maps* D_1 *onto* $\Delta_{\varphi h m} D_1$, *the equation* $\Delta_{\varphi h m}(x_0) = 0$ *has the unique solution* x^0 *in* D_1, *and, on the boundary* Γ_{D_1} *of the set* D_1, *the inequality*

$$\inf_{x \in \Gamma_{D_1}} \|\Delta_{\varphi h m}(x)\| \geq \sigma(m)$$

$$= (\frac{1}{T} K_\varphi \|H_h\| + \frac{1}{2} K_h \sum_{i=1}^{\infty} \|H_h A_i\| + K_h) \frac{Q_{\varphi h}^m}{1 - Q_{\varphi h}} M_0 \quad (4.12)$$

is satisfied, then the boundary-value problem (4.1), (4.2) has a solution $x = x^*(t)$ *with the initial condition* $x^*(0) = x_0^* \in D_1$;

2) for any function $h(t)$ *such that conditions* $(\mathbf{A})$ *i* $(\mathbf{B^0})$, *are satisfied, in order that some subset* $D_2 \subset D_{\beta\varphi h}$ *contain the initial value* $x^*(0) = x_0^*$ *of the solution of this boundary-value problem, it is necessary that,* $\forall m \in Z^+$ *and* $\forall \overline{x}_0 \in D_2$, *the inequality*

$$\|\Delta_{\varphi h m}(\overline{x}_0)\| \leq \sup_{x_0 \in D_2} \{\frac{1}{T}\|R_h\| + (\frac{1}{T} K_\varphi \|H_h\|$$

$$+ K_h(\frac{1}{2} \sum_{i=1}^{\infty} \|H_h A_i\| + 1))\} \frac{1 + \|R_h\|}{1 - Q_{\varphi h}} \|\overline{x}_0 - x_0\| + \sigma(m, \overline{x}_0), \quad (4.13)$$

be satisfied. Here R_h *stands for* $H_h \sum_{i=0}^{\infty} A_i$, *and* $\sigma(m, \overline{x}_0)$ *means the expression*

$$(\frac{1}{T} K_\varphi \|H_h\| + \frac{1}{2} K_h \sum_{i=1}^{\infty} \|H_h A_i\| + K_h) \frac{Q_{\varphi h}^m}{1 - Q_{\varphi h}} \beta_{\varphi h}(\overline{x}_0).$$

Proof.   In view of conditions (4.3) and (4.4) and inequality (4.6) $\forall m \in$

Z^+, relations (4.8) and (4.11) yield

$$\|\Delta_{\varphi h}(x_0) - \Delta_{\varphi h m}(x_0)\| \leq \frac{1}{T}\|H_h\|\|\varphi(x^*(0); x^*(t_1), x^*(t_2), ...)$$

$$- \varphi(x_m(0); x_m(t_1), x_m(t_2), ...)\| + \frac{1}{T}\sum_{i=1}^{\infty}\|H_h A_i\|$$

$$\times \{(1 - \frac{h(t_i)}{T})\int_0^{t_i}\|f(\tau, x^*(\tau, x_0)) - f(\tau, x_m(\tau, x_0))\|d\tau$$

$$+ \frac{h(t_i)}{T}\int_{t_i}^{\infty}\|f(\tau, x^*(\tau, x_0)) - f(\tau, x_m(\tau, x_0))\|d\tau\}$$

$$+ \frac{1}{T}\int_0^{\infty}\|f(\tau, x^*(\tau, x_0)) - f(\tau, x_m(\tau, x_0))\|d\tau$$

$$\leq \frac{1}{T}\|H_h\|K_\varphi \frac{Q_{\varphi h}^m}{1 - Q_{\varphi h}}\beta_{\varphi h}(x_0) + \frac{1}{T}\sum_{i=1}^{\infty}\|H_h A_i\|K_h \alpha_{1h}(t_i)$$

$$\times \frac{Q_{\varphi h}^m}{1 - Q_{\varphi h}}\beta_{\varphi h}(x_0) + \frac{1}{T}K_h \int_0^{\infty} h'(\tau)d\tau \frac{Q_{\varphi h}^m}{1 - Q_{\varphi h}}\beta_{\varphi h}(x_0)$$

$$\leq (\frac{1}{T}K_\varphi\|H_h\| + \frac{1}{2}K_h \sum_{i=1}^{\infty}\|H_h A_i\| + K_h)\frac{Q_{\varphi h}^m}{1 - Q_{\varphi h}}\beta_{\varphi h}(x_0) = \sigma(m, x_0),$$

$$(4.14)$$

where $\sigma(m, x_0) \to 0$ as $m \to \infty$, $x_0 \in D_{\beta \varphi h}$, and $\beta_{\varphi h}(x_0) \leq M_0$.

In view of (4.8), $\forall\{x_0', x_0''\} \subset D_{\beta \varphi h}$, we write the estimate

$$\|\Delta_{\varphi h}(x_0') - \Delta_{\varphi h}(x_0'')\| \leq \{\frac{1}{T}\|R_h\| + (\frac{1}{T}K_\varphi\|H_h\|$$

$$+ K_h(\frac{1}{2}\sum_{i=1}^{\infty}\|H_h A_i\| + 1))\frac{1 + \|R_h\|}{1 - Q_{\varphi h}}\}\|x_0' - x_0''\|, \quad (4.15)$$

which yields the equicontinuity of the mapping $\Delta_{\varphi h}$ on the set D_1.

According to Lemma 13.1 in [114], we set

$$\Delta_{\varphi h m}D_1 - \sigma(m) = \{x \in \Delta_{\varphi h m}D_1|\ \|x - y\|$$

$$\leq \sigma(m) \Rightarrow y \in \Delta_{\varphi h m}D_1\} \subset \Delta_{\varphi h}D_1.$$

We note that $0 \in \Delta_{\varphi h m}D_1$ by condition. Therefore, for the inclusion $0 \in \Delta_{\varphi h m}D_1 - \sigma(m)$, it is sufficient that the zero enters the set $\Delta_{\varphi h m}D_1$ together with its $\sigma(m)$-neighborhood. This is ensured by inequality (4.12).

Hence,

$$0 \in \Delta_{\varphi hm} D_1 - \sigma(m) \subset \Delta_{\varphi h} D_1,$$

which proves the first assertion of the corollary.

We now substantiate its second assertion. For all $m \in Z^+$ and $\bar{x}_0 \in D_2$, relations (4.14) and (4.15) yield the inequality

$$\|\Delta_{\varphi hm}(\bar{x}_0)\| \leq \{\frac{1}{T}\|R_h\| + (\frac{1}{T}K_\varphi\|H_h\| + K_h(\frac{1}{2}\sum_{i=1}^{\infty}\|H_h A_i\| + 1))$$

$$\times \frac{1 + \|R_h\|}{1 - Q_{\varphi h}}\}\|\bar{x}_0 - x_0^*\| + \sigma(m, \bar{x}_0),$$

which ensures the validity of inequality (4.13). This completes the proof of Corollary 4.2. $\qquad\square$

For the conditions of Theorem 4.1 to be satisfied, it is essential that a function $h(t)$ with the above-indicated properties exist. Consider this question in more details. We denote the set of points $\{x \in \mathfrak{M} \mid \|x\| < M_0 + \rho\}$, where ρ is an arbitrarily small positive constant, by D_ρ and the Cartesian product $[0, +\infty) \times D_\rho$ by $\bar{D}_\rho$.

The following proposition is true.

Theorem 4.2. *Let, $\forall t \in [0, +\infty)$, the function $f(t, x)$ be differentiable in the Fréchet meaning with respect to $x \in D_\rho$ uniformly in $x \in D_\rho$ and is continuous in $t \in [0, +\infty)$, and let*

$$\|f(t, x)\| \leq M^*, \ \|\frac{\partial f(t, x)}{\partial x}\| \leq P^*(t),$$

where $M^* = const > 0$, and the function $P^*(t)$ takes nonnegative values and is independent of $x \in D_\rho$. If, in this case, $\int\limits_0^{\infty} \sup\limits_{x \in D} \|f(t, x)\|dt$ is convergent, and

$$\sup_{t \geq 0}\{\frac{\sup\limits_{x \in D_\rho} \|\frac{\partial f(t,x)}{\partial x}\|}{\sup\limits_{x \in D} \|f(t, x)\|}\} = \xi < \infty, \tag{4.16}$$

then, the function $\tilde{h}(t) = \int\limits_0^t \sup\limits_{x \in D} \|f(s, x)\|ds$

1) satisfies conditions 1^*, 2^* for $T = \int\limits_0^{\infty} \sup\limits_{x \in D} \|f(t, x)\|dt$;

2) satisfies inequalities (4.4) for $M_h = 1$, $K_h \geq \xi$.

Proof. We denote $\sup\limits_{x\in D}\|f(t,x)\|$ by $\tilde{f}(t)$ and show that $\tilde{f}(t)$ is continuous in t on $[0,+\infty)$. It is obvious that, $\forall i \in Z^+$, $\{t_1, t_2\} \subset [0,+\infty)$, $x \in D_\rho$, the inequality

$$\big|\,|f_i(t_1,x)| - |f_i(t_2,x)|\,\big| \le |f_i(t_1,x) - f_i(t_2,x)|,$$

is satisfied. It yields the relations

$$\sup_{i\in Z^+}\big|\,|f_i(t_1,x)| - |f_i(t_2,x)|\,\big|$$
$$\le \sup_{i\in Z^+}|f_i(t_1,x) - f_i(t_2,x)| = \|f(t_1,x) - f(t_2,x)\|.$$

But

$$\sup_{i\in Z^+}\big|\,|f_i(t_1,x)| - |f_i(t_2,x)|\,\big| \ge \big|\,\sup_{i\in Z^+}|f_i(t_1,x)| - \sup_{i\in Z^+}|f_i(t_2,x)|\,\big|,$$

i.e.,

$$\big|\,\|f(t_1,x)\| - \|f(t_2,x)\|\,\big| \le \|f(t_1,x) - f(t_2,x)\|.$$

Then

$$\sup_{x\in D}\big|\,\|f(t_1,x)\| - \|f(t_2,x)\|\,\big| \le \sup_{x\in D}\|f(t_1,x) - f(t_2,x)\|.$$

Whence we obtain

$$|\tilde{f}(t_1) - \tilde{f}(t_2)| = \big|\sup_{x\in D}\|f(t_1,x)\| - \sup_{x\in D}\|f(t_2,x)\|\,\big|$$
$$\le \sup_{x\in D}\big|\,\|f(t_1,x)\| - \|f(t_2,x)\|\,\big| \le \sup_{x\in D}\|f(t_1,x) - f(t_2,x)\|.$$

Let t_0 be any point from $[0,+\infty)$. By the condition of the theorem, $\forall\varepsilon > 0$, $\exists\delta > 0$ such that the inequality $|t - t_0| < \delta$ yields the inequality $\|f(t,x) - f(t_0,x)\| < \varepsilon$ $\quad\forall x \in D$. It is clear that, for the indicated t, $\sup\limits_{x\in D}\|f(t,x) - f(t_0,x)\| \le \varepsilon$, i.e., $|\tilde{f}(t) - \tilde{f}(t_0)| \le \varepsilon$ for $|t - t_0\| < \delta$, which proves the continuity of the function $\tilde{f}(t)$ at the point t_0 and, hence, on the interval $[0,+\infty)$.

In this case, for any finite $t \ge 0$, the integral $\tilde{h}(t)$ and $\tilde{h}'(t) = \tilde{f}(t)$ exists. Then

$$\|f(t,x)\| \le \sup_{x\in D}\|f(t,x)\| = \tilde{f}(t) = \tilde{h}'(t) = 1 \cdot \tilde{h}'(t),$$

i.e., the first of inequalities (4.4) with $M_h = 1$ holds.

Since the set D_ρ is convex and open, we have

$$\|f(t,x') - f(t,x'')\| \le \sup_{x\in D_\rho}\Big\|\frac{\partial_\Phi f(t,x)}{\partial_\Phi x}\Big\| \cdot \|x' - x''\|$$

$\forall \{(t, x'), (t, x'')\} \subset \overline{D}_\rho$. Hence, the constant K_h in the second inequality in (4.4) must be chosen from the condition that the inequality

$$K_h \geq \frac{\sup\limits_{x \in D_\rho} \|\frac{\partial f(t,x)}{\partial x}\|}{\sup\limits_{x \in D} \|f(t,x)\|}$$

holds $\forall t \geq 0$. Therefore, it is sufficient to set $K_h \geq \xi$.

The validity of conditions 1^* and 2^* for the function $\tilde{h}(t)$ is obvious. The theorem is proved. $\qquad\qquad\square$

Remark 4.1. Condition (4.16) can be replaced by the condition of separability of the function $\tilde{f}(t)$ from zero, i.e., by the inequality $\inf\limits_{t \geq 0} \tilde{f}(t) = \ell = const > 0$, by assuming the function $P^*(t)$ to be a positive constant P^*. Then K_h can be chosen from the inequality $K_h \geq \frac{P^*}{\inf\limits_{t \geq 0} \tilde{f}(t)}$. The last estimate is more coarse, than that indicated in the theorem.

Example 4.1. We consider the function $f(t, x) = (f_1(t, x), f_2(t, x), ...)$, where $f_i(t, x) = e^{-t}(\sin x_i + \cos x_{i+1})$, $i \in Z^+$, and set $D = \{x \in \mathfrak{M} \mid \|x\| \leq \frac{\pi}{2}\}$. We show that it satisfies the conditions of Theorem 4.2.

It is obvious that

$$\|f(t, x)\| = \sup_i \{|e^{-t}(\sin x_i + \cos x_{i+1})|\} = e^{-t} \sup_i \{|\sin x_i + \cos x_{i+1}|\},$$

whence we have $\tilde{f}(t) = e^{-t} \sup\limits_{x \in D} \{\sup\limits_i \{ |\sin x_i + \cos x_{i+1}| \}\} = e^{-t} \cdot 2$.

The integral

$$\int\limits_0^\infty \sup\limits_{x \in D} \|f(t, x)\| dt = \int\limits_0^\infty 2e^{-t} dt = -2e^{-t}|_0^\infty = 2,$$

and, hence, it converges.

We show that the Fréchet derivative of $f(t, x)$ with respect to x for a fixed $t \geq 0$ i sset by the infinite matrix $L_{t,x} = [\ell_{ij}]_{i,j=1}^\infty$, whose elements are defined by the relation

$$\ell_{ij} = \begin{cases} e^{-t} \cos x_i, & j = i, \\ -e^{-t} \sin x_{i+1}, & j = i+1, \\ 0, & j \neq i, j \neq i+1, \end{cases} \quad .$$

The matrix acts on an element from $\mathfrak{M}$ by means of the operation of multiplication of a matrix by a vector, by realizing the mapping $\mathfrak{M} \to \mathfrak{M}$. By definition, it is sufficient to establish that

$$\lim_{\|\Delta x\| \to 0} \frac{\|\alpha(t, x, \Delta x)\|}{\|\Delta x\|} = 0 \quad \forall t \in [0, +\infty), \{x, x + \Delta x\} \subset D_\rho,$$

where

$$\alpha(t, x, \Delta x) = f(t, x + \Delta x) - f(t, x) - L_{t,x} \cdot \Delta x,$$

$$\alpha(t, x, \Delta x) = \{\alpha_1(t, x, \Delta x), \alpha_2(t, x, \Delta x), ...\}.$$

For the indicated x and Δx, $\|\Delta x\| \neq 0$, we have

$$\alpha_i(t, x, \Delta x) = f_i(t, x + \Delta x) - f_i(t, x) - \sum_{j=1}^{\infty} \ell_{ij} \Delta x_j$$

$$= f_i(t, x + \Delta x) - f_i(t, x) - \sum_{j=i}^{i+1} \ell_{ij} \Delta x_j$$

$$= e^{-t}\{(\sin(x_i + \Delta x_i) + \cos(x_{i+1} + \Delta x_{i+1})) - (\sin x_i + \cos x_{i+1})$$
$$-(\cos x_i \cdot \Delta x_i - \sin x_{i+1} \cdot \Delta x_{i+1})\}$$

$$= e^{-t}\{2 \sin \frac{\Delta x_i}{2} \cos(x_i + \frac{\Delta x_i}{2}) - 2 \sin(x_{i+1} + \frac{\Delta x_{i+1}}{2})$$
$$\times \sin \frac{\Delta x_{i+1}}{2} - \cos x_i \cdot \Delta x_i + \sin x_{i+1} \cdot \Delta x_{i+1}\}$$

$$= e^{-t}\{\sin x_i \cos \Delta x_i + \cos x_i \sin \Delta x_i - \sin \Delta x_i - \cos x_i \cdot \Delta x_i$$
$$-(\cos x_{i+1} \cos \Delta x_{i+1} - \sin x_{i+1} \sin \Delta x_{i+1} - \cos x_{i+1} + \sin x_{i+1} \cdot \Delta x_{i+1})\}.$$

This yields

$$|\alpha_i(t, x, \Delta x)| \leq e^{-t}\{|\sin x_i| \cdot |1 - \cos \Delta x_i| + |\cos x_i| \cdot |\sin \Delta x_i - \Delta x_i|$$
$$+ |\cos x_{i+1}| \cdot |1 - \cos \Delta x_{i+1}| + |\sin x_{i+1}| \cdot |\sin \Delta x_{i+1} - \Delta x_{i+1}|\}$$
$$\leq e^{-t}\{(|1 - \cos \Delta x_i| + |\sin \Delta x_i - \Delta x_i|) + (|1 - \cos \Delta x_{i+1}|+$$
$$+ |\sin \Delta x_{i+1} - \Delta x_{i+1}|)\}.$$

Since, for $\Delta x_i \in (-\frac{\pi}{2}; \frac{\pi}{2},)$ $|\sin \Delta x_i| \leq |\Delta x_i| \leq |tg\ \Delta x_i|$, we have

$$|1 - \cos \Delta x_i| + |\sin \Delta x_i - \Delta x_i| \leq |1 - \cos \Delta x_i| + |\sin \Delta x_i - tg\Delta x_i|$$
$$= |1 - \cos \Delta x_i| \cdot (1 + |tg\Delta x_i|)$$
$$= 2 \sin^2 \frac{\Delta x_i}{2}(1 + |tg\ \Delta x_i|) \leq \frac{(\Delta x_i)^2}{2}(1 + |tg\Delta x_i|).$$

The previous inequality takes the form

$$|\alpha_i(t, x, \Delta x)| \leq \frac{e^{-t}}{2}\{(\Delta x_i)^2(1 + |tg\Delta x_i|) + (\Delta x_{i+1})^2(1 + |tg\Delta x_{i+1}|)\}.$$

By considering the behavior of $\alpha_i(t, x, \Delta x)$ as $\|\Delta x\| \to 0$, it is sufficient to restrict ourselves by such values Δx, for which $\|\Delta x\| \leq \frac{\pi}{4}$; then $|tg\Delta x_i| \leq 1$, $|tg\Delta x_{i+1}| \leq 1$ and

$$\frac{\|\alpha(t, x, \Delta x)\|}{\|\Delta x\|} = \frac{\sup\limits_{i}\{|\alpha_i(t, x, \Delta x)|\}}{\|\Delta x\|} \leq \frac{\sup\limits_{i}\{\frac{e^{-t}}{2}(2(\Delta x_i)^2 + 2(\Delta x_{i+1})^2)\}}{\|\Delta x\|}$$

$$\leq \frac{2e^{-t}\|\Delta x\|^2}{\|\Delta x\|} = 2e^{-t}\|\Delta x\| \to 0$$

for $\|\Delta x\| \to 0$.

Therefore, the derivative $\frac{\partial f(t,x)}{\partial x}$ is, indeed, defined by the matrix $L_{t,x}$. Then we have

$$\|\frac{\partial f(t, x)}{\partial x}\| = \sup\limits_{i}\{|\ell_{ii}| + |\ell_{i,i+1}|\} = \sup\limits_{i}\{e^{-t}|\cos x_i| + e^{-t}|\sin x_i\} \leq 2e^{-t},$$

which yields

$$\xi = \sup\limits_{t \geq 0} \frac{2e^{-t}}{2e^{-t}} = 1.$$

In this case, the conditions of Remark 4.1 are not satisfied, since $\inf\limits_{t \geq 0} \tilde{f}(t) = \inf\limits_{t \geq 0} 2e^{-t} = 0$.

It is easy to see that, in the presented example for a fixed t, the matrix $L_{t,x}$ for the function $f(t, x)$ is constructed analogously to the construction of the Jacobian of the function $z : R^m \to R^n$. The following question arises: Under which conditions does this analogy conserve in the general case?

Consider the function $f(x) = \{f_1(x), f_2(x), ...\}$, which is defined in an open ball $S = S(0, \delta) = \{x \in \mathfrak{M}|\|x\| < \delta\}$ and takes values from the space $\mathfrak{M}$, i.e., $f : S \to \mathfrak{M}$. It is obvious that, $\forall i \in Z^+$, $f_i : S \to R^1$. Let us assume that the number function $f_i(x)$ has the partial Fréchet derivative with respect to x_j, $j \in Z^+$ at the point $x^0 = \{x_1^0, x_2^0, ...\} \in S$. It is the derivative of the mapping

$$f_j^i : (-\delta, +\delta) \subset R^1 \to f_i(x_1^0, \ldots, x_{j-1}^0, x_j, x_{j+1}^0, \ldots) \subset R^1$$

at the point x_j^0 and defines a homothety $R^1 \to R^1$ with constant coefficient denoted by $\frac{\partial f_i(x)}{\partial x_j}(x^0)$.

We consider that the function $f(x)$ belongs to the set $\hat{C}^1_{Lip}(S)$, if
1) $\forall x \in S$, $\|f(x)\| \leq P = const > 0$, and the inequality

$$\|f(x') - f(x'')\| \leq K\varepsilon(m)\|x' - x''\|$$

is satisfied (here, x', x'' are any points of the ball S, whose m first corresponding coordinates coincide, $K = const > 0$, and $\varepsilon(m) \to 0$ as $m \to \infty$);

2) $\forall \{i, j\} \subset Z^+$, there exist $\frac{\partial f_i(x)}{\partial x_j}$ at every point $x \in S$, and the matrix $[\frac{\partial f_i(x)}{\partial x_j}]^\infty_{i,j=1}$ is bounded in the norm by a constant $M' > 0$, which is independent of $x \in S$;

3) $\forall j \in Z^+$, the vector-function $\frac{\partial f(x)}{\partial x_j} = (\frac{\partial f_1(x)}{\partial x_j}, \frac{\partial f_2(x)}{\partial x_j}, ...)$ satisfies a sharpened Hölder condition, i.e.,

$$\left\| \frac{\partial f(x')}{\partial x_j} - \frac{\partial f(x'')}{\partial x_j} \right\| \le K' \varepsilon'(m) \|x' - x''\|^\alpha,$$

where x', x'' are any points from S, whose m first corresponding coordinates coincide, K', α are positive constants, and the series $\sum\limits_{m=0}^{\infty} \varepsilon'(m)$ is convergent.

The following proposition is valid.

Theorem 4.3. *If the function $f(x) \in \hat{C}^1_{Lip}(S)$, then it has the Fréchet derivative with respect to x at every point $x \in S$, which acts by means of the multiplication of the matrix*

$$\frac{\partial f(x)}{\partial x} = [\frac{\partial f_i(x)}{\partial x_j}]^\infty_{i,j=1}$$

by a vector $\forall h \in \mathfrak{M}$.

Proof.   We take any element $x \in S$ and introduce its increment $h = (h_1, h_2, ...)$ so that, simultaneously, $x + h \in S$. It is necessary to prove that

$$f(x + h) - f(x) = [\frac{\partial f_i(x)}{\partial x_j}]^\infty_{i,j=1} \cdot h + \alpha(x, h), \qquad (4.17)$$

where $\frac{\|\alpha(x,h)\|}{\|h\|} \to 0$ as $\|h\| \to 0$.

For any $n \in Z^+$, the n-th coordinate $\alpha_n(x, h)$ of a vector-function $\alpha(x, h)$ takes the form

$$\alpha_n(x, h) = f_n(x + h) - f_n(x) - \sum\limits_{m=1}^{\infty} \frac{\partial f_n(x)}{\partial x_m} \cdot h_m.$$

First, we prove the equality

$$f_n(x + h) - f_n(x)$$
$$= f_n(x_1 + h_1, x_2 + h_2, ...) - f_n(x_1, x_2 + h_2, x_3 + h_3, ...)$$
$$+ \sum\limits_{m=2}^{\infty} (f_n(x_1, ..., x_{m-1}, x_m + h_m, x_{m+1} + h_{m+1}, ...)$$
$$- f_n(x_1, ..., x_{m-1}, x_m, x_{m+1} + h_{m+1}, ...)). \qquad (4.18)$$

Consider the p-th partial sum S_p of the series on the right-hand side of (4.18):

$$S_p = f_n(x_1 + h_1, x_2 + h_2, ...) - f_n(x_1, x_2 + h_2, x_3 + h_3, ...)$$

$$+ \sum_{m=2}^{p} (f_n(x_1, ..., x_{m-1}, x_m + h_m, x_{m+1} + h_{m+1}, ...) -$$

$$- f_n(x_1, ..., x_{m-1}, x_m, x_{m+1} + h_{m+1}, ...))$$

$$= f_n(x_1 + h_1, x_2 + h_2, ...) - f_n(x_1, ..., x_{p-1}, x_p, x_{p+1} + h_{p+1}, x_{p+2} + h_{p+2}, ...).$$

It is necessary to show that, $\forall \varepsilon > 0$, $\exists p(\varepsilon) \in Z^+$ such that, $\forall p > p(\varepsilon)$, the inequality

$$\|f_n(x + h) - f_n(x) - S_p\| < \varepsilon \tag{4.19}$$

holds. It is easy to see that

$$\|f_n(x + h) - f_n(x) - S_p\|$$

$$= \|f_n(x_1, x_2, ..., x_p, x_{p+1} + h_{p+1}, x_{p+2} + h_{p+2}, ...)$$

$$- f_n(x_1, x_2, ..., x_p, x_{p+1}, x_{p+2}, ...)\| \le K\varepsilon(p) \sup_{i \ge p+1} \{ |h_i| \} \le K\|h\|\varepsilon(p).$$

Since $\varepsilon(p) \to 0$ as $p \to \infty$, we can indicate, for any $\varepsilon > 0$, $p(\varepsilon)$ such that, for all $p > p(\varepsilon)$, $\varepsilon(p) < \frac{\varepsilon}{K\|h\|}$, and, hence, inequality (4.19) and equality (4.18) are satisfied.

We apply the formula of finite increments in x_1 to the first difference on the right-hand side of (4.18) and in the appropriate argument x_m to the rest of differences. We obtain the equality

$$\alpha_n(x + h) = \frac{\partial f_n}{\partial x_1}(x_1 + \theta_1 h_1, x_2 + h_2, x_3 + h_3, ...)h_1$$

$$+ \sum_{m=2}^{\infty} \frac{\partial f_n}{\partial x_m}(x_1, ..., x_{m-1}, x_m + \theta_m h_m, x_{m+1} + h_{m+1}, ...)h_m$$

$$- \sum_{m=1}^{\infty} \frac{\partial f_n}{\partial x_m}(x_1, x_2, ...x_m, ...)h_m.$$

Taking into account that, for convergent series, the difference of the limits of sequences of partial sums equals the limit of the differences, we can replace the right-hand side of the last equality by the series composed from the differences of corresponding terms of the given series:

$$\alpha_n(x+h) = (\frac{\partial f_n}{\partial x_1}(x_1+\theta_1 h_1, x_2+h_2, x_3+h_3, \ldots)$$

$$-\frac{\partial f_n}{\partial x_1}(x_1, x_2, \ldots x_m, \ldots))h_1$$

$$+\sum_{m=2}^{\infty}(\frac{\partial f_n}{\partial x_m}(x_1, \ldots, x_{m-1}, x_m+\theta_m h_m, x_{m+1}+h_{m+1}, \ldots)$$

$$-\frac{\partial f_n}{\partial x_m}(x_1, x_2, \ldots x_m, \ldots))h_m.$$

By applying the sharpened Hölder inequality to each term on the right-hand side of the equality, we have

$$|\alpha_n(x,h)| \le \sum_{m=0}^{\infty} K'\varepsilon'(m)\|(\theta_{m+1}h_{m+1}, h_{m+2}, h_{m+3}, \ldots)\|^{\alpha} \cdot |h_{m+1}|$$

$$\le K'\|h\|^{1+\alpha}\sum_{m=0}^{\infty}\varepsilon'(m),$$

which yields (4.17). Hence, the theorem is proved. $\qquad\square$

We now assume that the equality $t = h_*(s)$ defines a diffeomorphism $h_* : [0, T) \to [0, +\infty)$, $h_*(0) = 0$, $\lim_{s \to T} h_*(s) = +\infty$, and the derivatives $h'_*(s)$ and $(h_*^{-1}(t))'$ do not become zero on the corresponding sets.

It is easy to see that, for any $T > 0$, such diffeomorphism can be set by the equality $t = tg(\frac{\pi}{2T}s)$.

Then the change of variables $y = x(h_*(s))$ reduces the boundary-value problem (4.1), (4.2) to a problem of the form

$$\frac{dy}{ds} = g(s, y), \tag{4.20}$$

$$A_0 y(0) + \sum_{i=1}^{\infty} A_i y(s_i) = \varphi(y(0); y(s_1), y(s_2), \ldots), \quad s_i \in [0, T) \; \forall i \in N, \tag{4.21}$$

where $g(s, y) = f(h_*(s), y) \cdot h'_*(s)$, $s_i = h_*^{-1}(t_i)$.

If the function $y = \psi(s)$ is a solution of the boundary-value problem (4.20), (4.21), then the function $x = \psi(h_*^{-1}(t))$ gives a solution of problem (4.1), (4.2).

Let us assume that the function $g(s, y)$ can be additionally defined at the point $s = T$ so that Eq. (4.20) with the boundary condition

$$A_0 y(0) + \sum_{i=1}^{\infty} A_i y(s_i) + C y(T) = \varphi_1(y(0), y(T); y(s_1), y(s_2), ...), \quad (4.22)$$

where $\varphi_1(y(0), y(T); y(s_1), y(s_2), ...) = \varphi(y(0); y(s_1), y(s_2), ...) + C y(T)$, $C = [c_{ij}]_{i,j=1}^{\infty}$ is some constant infinite matrix, would have a solution $y = \psi(s)$ on the interval $[0, T]$. Then the function $\psi(s)$ defines simultaneously a solution of problem (4.20), (4.21) on the interval $[0, T)$. To the solution of problem (4.20), (4.22), we may apply the method of truncation by K.P. Persidskii [84] and reduce it to a multipoint boundary-value problem in a finite-dimensional space.

4.2 Boundary-value problems on an interval

Let the solutions of Eq. (4.1) are subordinated to the boundary condition

$$A_0 x(0) + \sum_{i=1}^{\infty} A_i x(t_i) + C x(T) = \varphi(x(0), x(T); x(t_1), x(t_2), ...), \quad (4.23)$$

$$0 < t_i < t_{i+1} < T, \ i \in Z^+.$$

Here, $x \in D$, $f(t, x) : [0, T] \times D = D_0 \to \mathfrak{M}$, A_i and C are the infinite matrices bounded in the norm, and $\sum_{i=1}^{\infty} \|A_i\| < \infty$, $\varphi(\psi_1, \psi_2, ...) : D^{\infty} \to \mathfrak{M}$.

We consider that the function $\varphi(\psi)$ satisfies condition (4.3), the function $f(t, x)$ is continuous in the totality of variables on D_0, and, $\forall \ \{x, x'\} \subset D$,

$$\|f(t, x)\| \leq M = const > 0, \ \|f(t, x) - f(t, x')\| \leq K\|x - x'\|, \quad (4.24)$$

where $K = const > 0$.

Let, in addition, the following conditions be satisfied:

a_1) the matrix $\sum_{i=1}^{\infty} \frac{t_i}{T} A_i + C$ is invertible, and the matrix H inverse to it is bounded in the norm;

b_1) the set $D_{\beta\varphi}$ of points $x_0 \in \mathfrak{M}$, which belong to the domain D together with their β_φ-neighborhoods, is nonempty, and

$$\beta_\varphi(x_0) = \frac{T}{2} M + \beta_{1\varphi}(x_0),$$

$$\beta_{1\varphi}(x_0) = \|H\| \cdot \|d - (\sum_{i=0}^{\infty} A_i + C)x_0\| + \sum_{i=1}^{\infty} \|HA_i\|\alpha_1(t_i)M,$$

$$0 \le \alpha_1(t) = 2t(1 - \frac{t}{T}) \le \frac{T}{2};$$

the vector $d \in \mathfrak{M}$ is chosen so that $|d_i| = M_\varphi$, $sign\ d_i = -sign\ d_i^0$, $i \in Z^+$,
and $d^0 = colon(d_1^0, d_2^0, ...) = (\sum_{i=0}^{\infty} A_i + C)x_0;$

$c_1)$ $\quad Q_\varphi = \frac{KT}{2}[1 + \|H\| \sum_{i=1}^{\infty} \|A_i\|] + K_\varphi\|H\| < 1.$

We now write formally the recurrence sequence of functions

$$x_m(t, x_0) = x_0 + \int_0^t [f(\tau, x_{m-1}(\tau, x_0)) - \frac{1}{T}\int_0^T f(s, x_{m-1}(s, x_0))ds]d\tau$$

$$+ \frac{t}{T}H\{\varphi(x_{m-1}(0), x_{m-1}(T); x_{m-1}(t_1), x_{m-1}(t_2), ...) - (\sum_{i=0}^{\infty} A_i + C)x_0$$

$$- \sum_{i=1}^{\infty} A_i \int_0^{t_i} [f(\tau, x_{m-1}(\tau, x_0)) - \frac{1}{T}\int_0^T f(s, x_{m-1}(s, x_0))ds]d\tau\}, \quad (4.25)$$

$$m = 1, 2, ..., \ x_0(t, x_0) = (x_{01}, x_{02}, ...) \equiv x_0, \ x_0 \in D_{\beta\varphi},$$

which satisfies the recurrence boundary conditions

$$A_0 x_m(0) + \sum_{i=1}^{\infty} A_i x_m(t_i) + C x_m(T)$$

$$= \varphi(x_{m-1}(0), x_{m-1}(T); x_{m-1}(t_1), x_{m-1}(t_2), ...).$$

The following proposition testifies on the existence of sequence (4.25)
and its convergence to the function $x^*(t, x_0)$, which satisfies the equality

$$x^*(t, x_0) = x_0 + \int_0^t [f(\tau, x^*(\tau, x_0)) - \frac{1}{T}\int_0^T f(s, x^*(s, x_0))ds]d\tau$$

$$+ \frac{t}{T}H\{\varphi(x^*(0), x^*(T); x^*(t_1), x^*(t_2), ...) - (\sum_{i=0}^{\infty} A_i + C)x_0$$

$$- \sum_{i=1}^{\infty} A_i \int_0^{t_i} [f(\tau, x^*(\tau, x_0)) - \frac{1}{T}\int_0^T f(s, x^*(s, x_0))ds]d\tau\}.$$

Theorem 4.4. *Let conditions (4.3), (4.24), and $a_1 - c_1$ be satisfied. Then the sequence $\{x_m(t, x_0)\}_{m=0}^{\infty}$ defined by equalities (4.25) is uniformly convergent in $(t, x_0) \in [0, T] \times D_{\beta\varphi}$ as $m \to \infty$ to the function $x^*(t, x_0)$, and, for all natural m,*

$$\|x_m(t, x_0) - x^*(t, x_0)\| \leq \frac{Q_\varphi^m}{1 - Q_\varphi} \beta_\varphi(x_0).$$

The function $x^*(t, x_0)$ satisfies the boundary condition (4.23) and is a solution of the pertur-bed equation

$$\frac{dx}{dt} = f(t, x) + \mu,$$

where μ is determined by the relation

$$\mu = \Delta_\varphi(x_0) = \frac{1}{T} H\{\varphi(x^*(0), x^*(T); x^*(t_1), x^*(t_2), ...) - (\sum_{i=0}^{\infty} A_i + C)x_0$$

$$- \sum_{i=1}^{\infty} A_i \int_0^{t_i} [f(\tau, x^*(t, x_0)) - \frac{1}{T} \int_0^T f(s, x^*(s, x_0))ds] d\tau\}$$

$$- \frac{1}{T} \int_0^T f(\tau, x^*(\tau, x_0)) d\tau.$$

In this case, if $\Delta_\varphi(x_0) = 0$, then $x^*(t, x_0)$ is a solution of problem (4.1), (4.23).

The proof of the theorem is quite analogous to that of Theorem 4.1. Therefore, we omit it here.

Remark 4.2. If, under conditions of Theorem 4.4, inequality c_1 is replaced by the stronger condition

$$c_1^0) \qquad KT\{1 + \|H\| \sum_{i=1}^{\infty} \|A_i\| + \|H\| \cdot \|C\|\} + K_\varphi \|H\| < 1,$$

then there exists no another value of μ such that the solution of the equation $\frac{dx}{dt} = f(t, x) + \mu$ with the initial condition $x(0) = x_0$ would satisfy the boundary condition (4.23).

The proof of this proposition is analogous to that of Corollary 4.1.

Analogous results can be obtained for Eq. (4.1), whose solutions must satisfy the multipoint boundary condition

$$A_0 x(0) + \sum_{i=1}^{p} A_i x(t_i) + C x(T) = \varphi_p(x(0), x(T); x(t_1), ..., x(t_p)), \qquad (4.26)$$

where the function $\varphi_p(\psi) = \varphi_p(\psi_1, \psi_2, ..., \psi_{p+2}) : D^{p+2} \to \mathfrak{M}$, $f(t, x) :$ $D_0 \to \mathfrak{M}$, $x \in D$, and the matrices A_i $(i = \overline{0, p})$ and C are bounded in the norm.

We consider that the function $f(t, x)$ satisfies conditions (4.24), and the function $\varphi_p(\psi)$ is such that, $\forall \{ \psi, \psi_* \} \subset D^{p+2}$,

$$\|\varphi_p(\psi)\| \le M_{\varphi p}, \quad \|\varphi_p(\psi) - \varphi_p(\psi_*)\| \le K_{\varphi p}\|\psi - \psi_*\|, \tag{4.27}$$

where $M_{\varphi p}$ and $K_{\varphi p}$ are positive constants.

We now impose the following conditions on the boundary-value problem (4.1), (4.26):

a_2) the matrix $\sum_{i=1}^{p} \frac{t_i}{T} A_i + C$ is invertible, and the matrix H_p inverse to it is bounded in the norm;

b_2) the set $D_{\beta \varphi p}$ of points $x_0 \in \mathfrak{M}$, which enter the domain D together with their $\beta_{\varphi p}$-neighborhoods, is nonempty. Here,

$$\beta_{\varphi p}(x_0) = \frac{T}{2}M + \beta_{1\varphi p}(x_0),$$

$$\beta_{1\varphi p}(x_0) = \|H_p\|\|d_p - (\sum_{i=0}^{p} A_i + C)x_0\| + \sum_{i=1}^{p} \|H_p A_i\|\alpha_1(t_i)M,$$

$$d_p = (d_{1p}, d_{2p}, ...) \in \mathfrak{M}, \quad |d_{ip}| = M_{\varphi p}, \ i \in Z^+,$$

$$sign \ d_{ip} = -sign \ d_{ip}^0, \quad \text{and} \quad d_p^0 = colon(d_{1p}^0, d_{2p}^0, ...) = (\sum_{i=0}^{p} A_i + C)x_0;$$

c_2) $\quad Q_{\varphi p} = \frac{KT}{2}[1 + \|H_p\| \sum_{i=1}^{p} \|A_i\|] + K_{\varphi p}\|H_p\| < 1.$

We can establish that all terms of the sequence

$$x_{pm}(t, x_0) = x_0 + \int_0^t [f(\tau, x_{pm-1}(\tau, x_0)) - \frac{1}{T}\int_0^T f(s, x_{pm-1}(s, x_0))ds]d\tau$$

$$+ \frac{t}{T}H_p\{\varphi_p(x_{pm-1}(0), x_{pm-1}(T); x_{pm-1}(t_1), ..., x_{pm-1}(t_p)) - (\sum_{i=0}^{p} A_i + C)x_0$$

$$- \sum_{i=1}^{p} A_i \int_0^{t_i} [f(\tau, x_{pm-1}(\tau, x_0)) - \frac{1}{T}\int_0^T f(s, x_{pm-1}(s, x_0))ds]d\tau\}, \tag{4.28}$$

$$m = 1, 2, ..., \quad x_{p0}(t, x_0) = (x_{01}, x_{02}, ...) \equiv x_0, \ x_0 \in D_{\beta \varphi p},$$

satisfy the recurrence boundary conditions

$$A_0 x_{p_m}(0) + \sum_{i=1}^{p} A_i x_{p_m}(t_i) + C x_{p_m}(T)$$

$$= \varphi_p(x_{p_{m-1}}(0), x_{p_{m-1}}(T); x_{p_{m-1}}(t_1), ..., x_{p_{m-1}}(t_p))$$

for any $x_0 \in D_{\beta\varphi p}$. Then we will formulate an analog of Theorem 4.1, which follows directly from Theorem 4.4, for the boundary-value problem (4.1), (4.26).

Corollary 4.3. *Let us assume that conditions (4.24), (4.27), and $a_2 - c_2$ are satisfied. Then*

1) the sequence of functions $\{x_{p_m}(t, x_0)\}_{m=0}^{\infty}$ defined by equality (4.28) is uniformly conver-gent in $(t, x_0) \in [0, T] \times D_{\beta\varphi p}$ as $m \to \infty$ to the limiting function $x_p(t, x_0)$, and

$$\|x_{p_m}(t, x_0) - x_p(t, x_0)\| \leq \frac{Q_{\varphi p}^m}{1 - Q_{\varphi p}} \beta_{\varphi p}(x_0);$$

2) the function $x_p(t, x_0)$ is a solution of the equation $\frac{dx}{dt} = f(t, x) + \mu_p$ with the boundary condition (4.26), where μ_p is defined by the relation

$$\mu_p = \Delta_{\varphi p}(x_0) = \frac{1}{T} H_p \{\varphi_p(x_p(0), x_p(T); x_p(t_1), ..., x_p(t_p))$$

$$- (\sum_{i=0}^{p} A_i + C)x_0 - \sum_{i=1}^{p} A_i \int_{0}^{t_i} [f(\tau, x_p(t, x_0))$$

$$- \frac{1}{T} \int_{0}^{T} f(s, x_p(s, x_0))ds]d\tau\} - \frac{1}{T} \int_{0}^{T} f(\tau, x_p(\tau, x_0))d\tau.$$

In this case, if $\Delta_{\varphi_p}(x_0) = 0$, then $x_p(t, x_0)$ is a solution of problem (4.1), (4.26).

Remark 4.3. If, under conditions of Corollary 4.3, condition c_2 is replaced by the condition

$$c_2^0) \qquad KT\{1 + \|H_p\| \sum_{i=1}^{p} \|A_i\| + \|H_p\|\|C\|\} + K_{\varphi p}\|H_p\| < 1,$$

then there is no another value of μ_p, for which the solution of the equation $\frac{dx}{dt} = f(t, x) + \mu_p$ with the initial condition $x(0) = x_0$ would satisfy the boundary condition (4.26).

In what follows, we use the condition

$$c_2^*) \qquad KT\{1 + \|H_p\| \sum_{i=1}^{p} \|A_i\| + \|H_p\|\|C\|\} + K_{\varphi p}\|H_p\| \leq q = const < 1,$$

which makes the previous inequality sharper.

It is obvious that, for the boundary-value problems (4.1), (4.23) and (4.1), (4.26), we can easily formulate propositions analogous to Corollary 4.2.

4.3   Reduction to a finite-dimensional multipoint case

We introduce the notation

$$\overset{(n)}{x} = (x_1, x_2, ..., x_n), \quad \overset{(n)}{x_0} = (x_{01}, x_{02}, ..., x_{0n}); \quad \overset{(n)}{f} = (f_1, f_2, ..., f_n),$$

$$\overset{(n)}{f}(t, \overset{(n)}{x}) = (f_1(t, x_1, ..., x_n, 0, 0, ...), ..., f_n(t, x_1, ..., x_n, 0, 0, ...)),$$

$$\psi = (\psi_1, \psi_2, ...), \quad \psi_i = (\psi_{1i}, \psi_{2i}, ...), \quad \overset{(n)}{\psi}_i = (\psi_{1i}, ..., \psi_{ni}), \quad \{n, i\} \subset Z^+;$$

$$\varphi_p(\psi_1, ..., \psi_{p+2}) = \{\varphi_{1p}(\psi_1, ..., \psi_{p+2}), \varphi_{2p}(\psi_1, ..., \psi_{p+2}), ...\}$$
$$= \varphi(\psi_1, ..., \psi_{p+2}, 0, 0, ...)$$
$$= \{\varphi_1(\psi_1, ..., \psi_{p+2}, 0, 0, ...), \varphi_2(\psi_1, ..., \psi_{p+2}, 0, 0, ...) ... \};$$

$$\overset{(n)}{\varphi}_p(\overset{(n)}{\psi}_1, ..., \overset{(n)}{\psi}_{p+2}) = \{\varphi_{1p}(\overset{(n)}{\psi}_1, ..., \overset{(n)}{\psi}_{p+2}), ..., \varphi_{np}(\overset{(n)}{\psi}_1, ..., \overset{(n)}{\psi}_{p+2})\}$$
$$= \{\varphi_1(\overset{(n)}{\psi}_1, 0, 0, ...; ...; \overset{(n)}{\psi}_{p+2}, 0, 0, ...; 0, 0, ...), ...$$
$$..., \varphi_n(\overset{(n)}{\psi}_1, 0, 0, ...; ...; \overset{(n)}{\psi}_{p+2}, 0, 0, ...; 0, 0, ...)\} \; ;$$

$\overset{(n)}{A}_i = [a_{jk}^{(i)}]_{j,k=1}^{n}$ $(i \in Z^+)$ and $\overset{(n)}{C} = [c_{jk}]_{j,k=1}^{n}$ are the $n \times n$ matrices, which are obtained by the truncation of the matrices $A_i = [a_{jk}^{(i)}]_{j,k=1}^{\infty}$ and $C = [c_{jk}]_{j,k=1}^{\infty}$, respectively.

In addition to the boundary-value problem (4.1), (4.26), we consider a boundary-value problem for the equation

$$\frac{d\overset{(n)}{x}}{dt} = \overset{(n)}{f}(t, \overset{(n)}{x}) \tag{4.29}$$

with the boundary condition

$$\overset{(n)}{A_0}\,\overset{(n)}{x}(0) + \sum_{i=1}^{p}\overset{(n)}{A_i}\,\overset{(n)}{x}(t_i) + \overset{(n)}{C}\,\overset{(n)}{x}(T)$$

$$= \overset{(n)}{\varphi}_p(\overset{(n)}{x}(0,\overset{(n)}{x_0}),\ \overset{(n)}{x}(T,\overset{(n)}{x_0});\ \overset{(n)}{x}(t_1,\overset{(n)}{x_0}),...,\overset{(n)}{x}(t_p,\overset{(n)}{x_0})). \quad (4.30)$$

The last problem is a multipoint boundary-value problem in the space R^n and is quite well studied. We will seek the conditions, which allow one to reduce the boundary-value problem (4.1), (4.26) to the boundary-value problem (4.29), (4.30).

We consider that the functions $f(t,x)$ and $\varphi_p(\psi)$ satisfy conditions (4.3) and (4.27), respectively, like above. We also consider that $y \in R^n$ belongs to the set $\tilde{D}^{(n)}$, if $(y,0,0,0,...) \in D$. It is clear that, $\forall\{\overset{(n)}{x},\overset{(n)}{x'}\} \subset \tilde{D}^{(n)}$ and $t \in [0,T]$, the inequalities

$$\|\overset{(n)}{f}(t,\overset{(n)}{x})\| \le M,\ \|\overset{(n)}{f}(t,\overset{(n)}{x}) - \overset{(n)}{f}(t,\overset{(n)}{x'})\| \le K\|\overset{(n)}{x} - \overset{(n)}{x'}\| \quad (4.31)$$

hold.

If $y_i \in \tilde{D}^{(n)}$, $i \in \{1,2,...,p+2\}$, then $Y = (y_1,y_2,...y_{p+2}) \in \tilde{D}^{(n)^{p+2}}$. It is obvious that, $\forall\{Y_1,Y_2\} \subset \tilde{D}^{(n)^{p+2}}$, the inequalities

$$\|\overset{(n)}{\varphi}_p(Y_1)\| \le M_{\varphi p},\ \|\overset{(n)}{\varphi}_p(Y_1) - \overset{(n)}{\varphi}_p(Y_2)\| \le K_{\varphi p}\|Y_1 - Y_2\| \quad (4.32)$$

are satisfied.

We now impose the following conditions on problem (4.29), (4.30):

a_3) there exists the matrix $\overset{(n)}{H_p}$ inverse to the matrix $\sum_{i=1}^{p}\frac{t_i}{T}\overset{(n)}{A_i} + \overset{(n)}{C}$;

b_3) the set $\tilde{D}^{(n)}_{\overset{(n)}{\beta}\varphi p}$ of points $\overset{(n)}{x_0} = (x_{01},x_{02},...,x_{0n}) \in R^n$ such that the corresponding elements $(x_{01},...,x_{0n},0,0,...)$ belong to the domain D together with their $\overset{(n)}{\beta}_{\varphi p}$-neighborhoods is nonempty. Here,

$$\overset{(n)}{\beta}_{\varphi p}(x_0) = \frac{T}{2}M + \overset{(n)}{\beta}_{1\varphi p}(x_0),$$

$$\overset{(n)}{\beta}_{1\varphi p}(x_0) = \|\overset{(n)}{H_p}\|\|\overset{(n)}{d}_p - (\sum_{i=0}^{p}\overset{(n)}{A_i} + \overset{(n)}{C})\overset{(n)}{x_0}\| + \sum_{i=1}^{p}\|\overset{(n)}{H_p}\overset{(n)}{A_i}\|\alpha_1(t_i)M,$$

the vector $\overset{(n)}{d}_p = \{\overset{(n)}{d}_{1p},\overset{(n)}{d}_{2p},...,\overset{(n)}{d}_{np}\} \in R^n$ is chosen so that $|\overset{(n)}{d}_{ip}| = M_{\varphi p}$, $sign\ \overset{(n)}{d}_{ip} = -sign\ \overset{(n)}{d}^{0}_{ip}$, $\overset{(n)}{d}^{0}_{p} = colon\{\overset{(n)}{d}^{0}_{1p},\overset{(n)}{d}^{0}_{2p},...,\overset{(n)}{d}^{0}_{np}\} = (\sum_{i=0}^{p}\overset{(n)}{A_i} + \overset{(n)}{C})\overset{(n)}{x}_0;$

$\mathrm{c_3})\quad \overset{(n)}{Q}_{\varphi p} = \frac{KT}{2}[1 + \|\overset{(n)}{H}_p\| \overset{p}{\underset{i=1}{\sum}} \|\overset{(n)}{A}_i\|] + K_{\varphi p}\|\overset{(n)}{H}_p\| < 1.$

Thus, it follows from conditions (4.3), (4.27) and $\mathrm{a_3 - c_3}$ that the sequence given by the relation

$$\overset{(n)}{x}_{pm}(t, \overset{(n)}{x}_0) = \overset{(n)}{x}_0 + \int_0^t [\overset{(n)}{f}(\tau, \overset{(n)}{x}_{pm-1}(\tau, \overset{(n)}{x}_0))$$

$$-\frac{1}{T}\int_0^T \overset{(n)}{f}(s, \overset{(n)}{x}_{pm-1}(s, \overset{(n)}{x}_0))ds]d\tau$$

$$+\frac{t}{T}\overset{(n)}{H}_p\{\overset{(n)}{\varphi}_p(\overset{(n)}{x}_{pm-1}(0, \overset{(n)}{x}_0), \overset{(n)}{x}_{pm-1}(T, \overset{(n)}{x}_0); \overset{(n)}{x}_{pm-1}(t_1, \overset{(n)}{x}_0),$$

$$..., \overset{(n)}{x}_{pm-1}(t_p, \overset{(n)}{x}_0)) - (\overset{p}{\underset{i=0}{\sum}}\overset{(n)}{A}_i + \overset{(n)}{C})\overset{(n)}{x}_0$$

$$-\overset{p}{\underset{i=1}{\sum}}\overset{(n)}{A}_i \int_0^{t_i}[\overset{(n)}{f}(\tau, \overset{(n)}{x}_{pm-1}(\tau, \overset{(n)}{x}_0))$$

$$-\frac{1}{T}\int_0^T \overset{(n)}{f}(s, \overset{(n)}{x}_{pm-1}(s, \overset{(n)}{x}_0))ds]d\tau\}, \quad m = 1, 2, ..., \quad \overset{(n)}{x}_{p0}(t, \overset{(n)}{x}_0) \equiv \overset{(n)}{x}_0,$$

converges uniformly in $(t, \overset{(n)}{x}_0) \in [0, T] \times \tilde{D}^{(n)}_{\underset{\beta\,\varphi p}{(n)}}$ to the function $\overset{(n)}{x}_p(t, \overset{(n)}{x}_0)$.

This function satisfies the equality

$$\overset{(n)}{x}_p(t, \overset{(n)}{x}_0) = \overset{(n)}{x}_0 + \int_0^t [\overset{(n)}{f}(\tau, \overset{(n)}{x}_p(\tau, \overset{(n)}{x}_0)) - \frac{1}{T}\int_0^T \overset{(n)}{f}(s, \overset{(n)}{x}_p(s, \overset{(n)}{x}_0))ds]d\tau$$

$$+\frac{t}{T}\overset{(n)}{H}_p\{\overset{(n)}{\varphi}_p(\overset{(n)}{x}_p(0, \overset{(n)}{x}_0), \overset{(n)}{x}_p(T, \overset{(n)}{x}_0); \overset{(n)}{x}_p(t_1, \overset{(n)}{x}_0), ..., \overset{(n)}{x}_p(t_p, \overset{(n)}{x}_0))$$

$$-(\overset{p}{\underset{i=0}{\sum}}\overset{(n)}{A}_i + \overset{(n)}{C})\overset{(n)}{x}_0 - \overset{p}{\underset{i=1}{\sum}}\overset{(n)}{A}_i \int_0^{t_i}[\overset{(n)}{f}(\tau, \overset{(n)}{x}_p(\tau, \overset{(n)}{x}_0))$$

$$-\frac{1}{T}\int_0^T \overset{(n)}{f}(s, \overset{(n)}{x}_p(s, \overset{(n)}{x}_0))ds]d\tau\} \quad (4.33)$$

and is a solution of the perturbed equation

$$\frac{d\overset{(n)}{x}}{dt} = \overset{(n)}{f}(t, \overset{(n)}{x}) + \overset{(n)}{\mu_p} \tag{4.34}$$

with the boundary condition (4.30), where

$$\overset{(n)}{\mu_p} = \overset{(n)}{\Delta}_{\varphi p}(\overset{(n)}{x}_0)$$

$$= \frac{1}{T}\overset{(n)}{H}_p\{\overset{(n)}{\varphi}_p(\overset{(n)}{x}_p(0,\overset{(n)}{x_0}n),\ \overset{(n)}{x}_p(T,\overset{(n)}{x_0});\ \overset{(n)}{x}_p(t_1,\overset{(n)}{x_0}),...,\overset{(n)}{x}_p(t_p,\overset{(n)}{x_0}))$$

$$- (\sum_{i=0}^{p}\overset{(n)}{A_i} + \overset{(n)}{C})\overset{(n)}{x_0} - \sum_{i=1}^{p}\overset{(n)}{A_i}\int_{0}^{t_i}[\overset{(n)}{f}(\tau,\overset{(n)}{x}_p(\tau,\overset{(n)}{x_0}))$$

$$- \frac{1}{T}\int_{0}^{T}\overset{(n)}{f}(s,\overset{(n)}{x}_p(s,\overset{(n)}{x_0}))ds]d\tau\} - \frac{1}{T}\int_{0}^{T}\overset{(n)}{f}(\tau,\overset{(n)}{x}_p(\tau,\overset{(n)}{x_0}))d\tau.$$

If $\overset{(n)}{\Delta}_{\varphi p}(\overset{(n)}{x}_0) = 0$, then the function $\overset{(n)}{x}_p(t,\overset{(n)}{x_0})$ is a solution of the boundary-value problem (4.29), (4.30).

We consider that if the function $f(t,x) \in \hat{C}_{Lip}(x)$ is continuous in the domain D_0, it is bounded by a constant M and satisfies the sharpened Cauchy–Lipschitz condition with respect to x, i.e., the inequality

$$\|f(t,x') - f(t,x'')\| \le \alpha(t)\varepsilon(m)\|x' - x''\|$$

holds for any points x', x'' from the domain D, whose m first corresponding coordinates coincide, $\alpha(t) \ge 0$ is a function continuous on $[0,T]$, and $\varepsilon(m)$ tends to 0 as $m \to \infty$.

We set $K = \max\limits_{t\in[0,T]} \alpha(t) \cdot \varepsilon(0)$ and introduce the notation

$$\beta^*_{\varphi p}(x,n_0) = \frac{\frac{2}{KT} - 1}{\sum\limits_{i=1}^{p}\|\overset{(n_0)}{A}_i\| + \frac{2K_\varphi}{KT}}(M_\varphi + (\sum_{i=0}^{p}\|A_i\| + \|C\|)\|x\|) + \frac{M}{K}, \quad n_0 \in Z^+;$$

$D^*_{\beta\varphi p}$ is a set, every point of which belongs to the set D together with its $\beta^*_{\varphi p}$-neighborhood;

$$\{\psi_i,\tilde{\psi}_i\} \subset D\ \forall i \in Z^+,\ \psi = (\psi_1,\psi_2,\psi_3,...) \in D^\infty,$$

$$\tilde{\psi} = (\psi_1,\psi_2...,\psi_n,\tilde{\psi}_{n+1},\tilde{\psi}_{n+2},...) \in D^\infty,$$

$$\psi_i^g = (\psi_{1i},...,\psi_{gi},\bar{\psi}_{g+1i},\bar{\psi}_{g+2i},...) \subset D,\ i \in Z^+,\ \psi^g = \{\psi_1^g,\psi_2^g,...\} \subset D^\infty.$$

We say that $\varphi(\psi) \in \hat{C}_{Lip}(\psi)$, if the function $\varphi(\psi)$ is bounded on D^∞ by a constant M_φ, and, for all $\{\,\psi, \tilde{\psi}, \psi^g\} \subset D^\infty$, the inequalities

$$\|\varphi(\psi) - \varphi(\tilde{\psi})\| \le \delta_0(n)\|\psi - \tilde{\psi}\|, \tag{4.35}$$

$$\|\varphi(\psi) - \varphi(\psi^g)\| \le \delta(g)\|\psi - \psi^g\|, \tag{4.36}$$

hold. In this case, $\delta_0(n) \to 0$ as $n \to \infty$, and $\delta(g) \to 0$ as $g \to \infty$.

We now present the example of a function from the set $\hat{C}_{Lip}(\psi)$. We set

$$\varphi(\psi) = \{\varphi_1(\psi), \varphi_2(\psi), \varphi_3(\psi), ...\}, \quad D = [0,1]^\infty,$$

and, by $trig\ \psi_{ij}$ $(i, j = 1, 2, ...)$, denote the functions $\sin \psi_{ij}$ or $\cos \psi_{ij}$. Let

$$\varphi_i(\psi) \;=\; \frac{1}{2}\,trig\ \psi_{11} + \frac{1}{4}\,trig\ \psi_{21} + \frac{1}{8}\,trig\ \psi_{31} + ...$$

$$+\frac{1}{2}\Big(\frac{1}{2}\,trig\ \psi_{12} + \frac{1}{4}\,trig\ \psi_{22} + \frac{1}{8}\,trig\ \psi_{32} + ...\Big)$$

$$+\frac{1}{4}\Big(\frac{1}{2}\,trig\ \psi_{13} + \frac{1}{4}\,trig\ \psi_{23} + \frac{1}{8}\,trig\ \psi_{33} + ...\Big) + ...,$$

$i = 1, 2, 3, ...$ It is obvious that $|\varphi_i(\psi)| \le 1 + \frac{1}{2} + \frac{1}{4} + ... = 2$, and, hence, $\|\varphi(\psi)\| \le 2$. We now estimate the modulus of the difference $\varphi_i(\psi) - \varphi_i(\tilde{\psi})$:

$$|\varphi_i(\psi) - \varphi_i(\tilde{\psi})| \le \frac{1}{2^n}\Big\{\frac{1}{2}|trig\ \psi_{1n+1} - trig\ \tilde{\psi}_{1n+1}|$$

$$+\frac{1}{4}|trig\ \psi_{2n+1} - trig\ \tilde{\psi}_{2n+1}| + ...$$

$$+\frac{1}{2^{n+1}}\Big\{\frac{1}{2}|trig\ \psi_{1n+2} - trig\ \tilde{\psi}_{1n+2}| + \frac{1}{4}|trig\ \psi_{2n+2} - trig\ \tilde{\psi}_{2n+2}| + ...\Big\} + ...$$

$$\le \frac{1}{2^n}\Big\{\frac{1}{2}|\psi_{1n+1} - \tilde{\psi}_{1n+1}| + \frac{1}{4}|\psi_{2n+1} - \tilde{\psi}_{2n+1}| + ...\Big\} + \frac{1}{2^{n+1}}\Big\{\frac{1}{2}|\psi_{1n+2} - \tilde{\psi}_{1n+2}|$$

$$+\frac{1}{4}|\psi_{2n+2} - \tilde{\psi}_{2n+2}| + ...\Big\} + ... \le \frac{1}{2^n}\|\psi_{n+1} - \tilde{\psi}_{n+1}\|$$

$$+\frac{1}{2^{n+1}}\|\psi_{n+2} - \tilde{\psi}_{n+2}\| + ... \le \frac{1}{2^{n-1}}\|\psi - \tilde{\psi}\|.$$

This implies that

$$\|\varphi(\psi) - \varphi(\tilde{\psi})\| \le \frac{1}{2^{n-1}}\|\psi - \tilde{\psi}\|,$$

and $\frac{1}{2^{n-1}} \to 0$ as $n \to \infty$.

We now estimate the modulus of the difference $\varphi_i(\psi) - \varphi_i(\psi^g)$:

$$|\varphi_i(\psi) - \varphi_i(\psi^g)| \le \frac{1}{2^{g+1}} |trig\ \psi_{g+11} - trig\ \bar\psi_{g+11}|$$

$$+ \frac{1}{2^{g+2}} |trig\ \psi_{g+21} - trig\ \bar\psi_{g+21}| + \ldots$$

$$+ \frac{1}{2} \Big(\frac{1}{2^{g+1}} |trig\ \psi_{g+12} - trig\ \bar\psi_{g+12}| + \frac{1}{2^{g+2}} |trig\ \psi_{g+22} - trig\ \bar\psi_{g+22}| + \ldots \Big) + \ldots$$

$$\le \frac{1}{2^g} \|\psi_1 - \psi_1^g\| + \frac{1}{2} \cdot \frac{1}{2^g} \|\psi_2 - \psi_2^g\| + \frac{1}{4} \cdot \frac{1}{2^g} \|\psi_3 - \psi_3^g\| + \ldots \le$$

$$\le \frac{1}{2^g} \Big(1 + \frac{1}{2} + \frac{1}{4} + \ldots \Big) \|\psi - \psi^g\| = \frac{1}{2^{g-1}} \|\psi - \psi^g\|.$$

This implies that

$$\|\varphi(\psi) - \varphi(\psi^g)\| \le \frac{1}{2^{g-1}} \|\psi - \psi^g\|,$$

i.e., the constructed function belongs to $\hat{C}_{Lip}(\psi)$, since $\frac{1}{2^{g-1}} \to 0$ as $g \to \infty$.

In what follows, we set $\delta(0) = \delta_0(0) = K_\varphi$.

We recall that, in the case where we consider the elementwise convergence of the sequence of matrices $\{ \overset{(n)}{A} = [a_{ij}^{(n)}]_{i,j=1}^{n} \}_{n=1}^{\infty}$ or the coordinatewise convergence of the sequence of vectors $\{ \overset{(n)}{x} = (x_1^{(n)}, \ldots, x_n^{(n)}) \}_{n=1}^{\infty}$ as $n \to \infty$ in the previous designations, the quantity $\overset{(n)}{A}$ means the matrix $[a_{ij}]_{i,j=1}^{\infty}$, whose elements

$$a_{ij} = \begin{cases} a_{ij}^{(n)} & \text{for} \quad i \le n \ \text{ and } \ j \le n, \\ 0 & \text{for} \quad i > n \ \text{ or } \quad j > n, \end{cases}$$

By $\overset{(n)}{x}$ we denote the vector $(x_1^{(n)}, \ldots, x_n^{(n)}, 0, 0, 0, \ldots)$.

Lemma 4.1. *Let $f(t,x) \in \hat{C}_{lip}(x)$, let the set $D_{\beta\varphi}^*$ be nonempty, and let conditions a_2, c_2^0, (4.3) and (4.36) be satisfied. If, $\forall\, n \ge n_0$, conditions a_3, c_3 are satisfied, then, $\forall x_0 \in D_{\beta\varphi p}^*$,*

$$\lim_{n\to\infty} \overset{(n)}{x_p}(t, \overset{(n)}{x_0}) = x_p(t, x_0), \quad \lim_{n\to\infty} \overset{(n)}{\mu_p} = \mu_p \tag{4.37}$$

in the meaning of the coordinatewise convergence.

Proof.   It is obvious that condition (4.3) yields condition (4.27), since we may consider that, for any natural p, the constants $M_{\varphi p}$ and $K_{\varphi p}$ are equal to the constants M_φ and K_φ, respectively. The inclusion $f(t,x) \in \hat{C}_{Lip}(x)$ on D_0 yields inequalities (4.24) and inequalities (4.31) for all $n \in Z^+$. Moreover, relation (4.27) yields inequalities (4.32) $\forall\, n \in Z^+$. In this case, the constants M, K, $M_{\varphi p}$ and $K_{\varphi p}$ are independent of n.

If $x_0 \in D^*_{\beta\varphi p}$, then $x_0 \in D_{\beta\varphi p}$ and, $\forall n \geq n_0$, $\overset{(n)}{x_0} \in \tilde{D}^{(n)}_{\overset{(n)}{\beta}\varphi p}$. Indeed, it is easy to verify that the sufficient conditions for two last inclusions are the inequalities

$$\beta_{\varphi p} < \beta^*_{\varphi p}, \quad \overset{(n)}{\beta}_{\varphi p} < \beta^*_{\varphi p} \ (n \geq n_0),$$

respectively.

We prove only the first inequality, since the second can be proved analogously. Condition c_2 yields the inequalities

$$\frac{KT}{2}[1 + \|H_p\| \sum_{i=1}^{p} \|A_1\|] < 1,$$

$$\|H_p\| < \frac{\frac{2}{KT} - 1}{\sum\limits_{i=1}^{p} \|A_i\| + \frac{2K_\varphi}{KT}} \leq \frac{\frac{2}{KT} - 1}{\sum\limits_{i=1}^{p} \|\overset{(n_0)}{A_i}\| + \frac{2K_\varphi}{KT}}.$$

In view of condition b_2, we obtain

$$\beta_{\varphi p}(x) \leq \frac{T}{2}M + \|H_p\|(\|d_p\| + (\sum_{i=0}^{p} \|A_i\| + \|C\|)\|x\|)$$

$$+ \sum_{i=1}^{p} \|H_p A_i\| \frac{T}{2}M \leq \frac{T}{2}M(1 + \sum_{i=1}^{p} \|H_p A_i\|)$$

$$+ \frac{\frac{2}{KT} - 1}{\sum\limits_{i=1}^{p} \|\overset{(n_0)}{A_i}\| + \frac{2K_\varphi}{KT}} (M_\varphi + (\sum_{i=0}^{p} \|A_i\| + \|C\|)\|x\|) < \beta^*_{\varphi p}(x, n_0).$$

Hence, for any natural $n \geq n_0$, there exists a controlling parameter $\overset{(n)}{\mu_p} = \overset{(n)}{\Delta}_{\varphi p}(\overset{(n)}{x_0})$. With regard for condition c_3, we arrive at the inequalities

$$\|\overset{(n)}{\Delta}_{\varphi p}(\overset{(n)}{x_0})\|$$

$$\leq \frac{1}{T}(\|\overset{(n)}{H_p}\|M_\varphi + \|\overset{(n)}{H_p}(\sum_{i=0}^{p} \overset{(n)}{A_i} + \overset{(n)}{C})\overset{(n)}{x_0}\|) + M(\frac{1}{2}\sum_{i=1}^{p} \|\overset{(n)}{H_p}\overset{(n)}{A_i}\| + 1)$$

$$\leq \frac{1}{T}\|\overset{(n)}{H}_p\|M_\varphi + \frac{1}{T}\|\overset{(n)}{H}_p\|(\sum_{i=0}^{p}\|\overset{(n)}{A}_i\| + \|\overset{(n)}{C}\|)\|\overset{(n)}{x}_0\|$$

$$+ \frac{M}{2}\|\overset{(n)}{H}_p\|\sum_{i=1}^{p}\|\overset{(n)}{A}_i\| + M$$

$$\leq \frac{1}{T}\|\overset{(n)}{H}_p\|M_\varphi + \frac{1}{T}\|\overset{(n)}{H}_p\|(\sum_{i=0}^{p}\|A_i\| + \|C\|)\|x_0\|$$

$$+ \frac{M}{2}\|\overset{(n)}{H}_p\|\sum_{i=1}^{p}\|A_i\| + M$$

$$= \frac{\|\overset{(n)}{H}_p\|}{T}(M_\varphi + (\sum_{i=0}^{p}\|A_i\| + \|C\|)\|x_0\| + \frac{MT}{2}\sum_{i=1}^{p}\|A_i\|) + M$$

$$\leq \frac{\frac{2}{KT} - 1}{T\sum_{i=1}^{p}\|\overset{(n_0)}{A}_i\| + \frac{2K_\varphi}{K}}$$

$$\times (M_\varphi + (\sum_{i=0}^{p}\|A_i\| + \|C\|)\|x_0\| + \frac{MT}{2}\sum_{i=1}^{p}\|A_i\|) + M \leq M',$$

where $n \geq n_0$, $x_0 \in D^*_{\beta\varphi p}$, $0 < M' = const < \infty$. Thus, the sequence $\{\overset{(n)}{\mu}_p\}_{n=n_0}^{\infty}$ is uniformly bounded in the norm of the space $\mathfrak{M}$ according to the above agreement. With the help of the well-known method of diagonalization, we can separate a subsequence $\{\overset{(s_i)}{\mu}_p\}_{i=1}^{\infty}$ from it. This subsequence is coordinatewise convergent as $i \to \infty$.

Consider the sequence of equations

$$\frac{d\overset{(s_i)}{x}}{dt} = \overset{(s_i)}{f}(t, \overset{(s_i)}{x}) + \overset{(s_i)}{\mu}_p, \ i \in Z^+, \tag{4.38}$$

by replacing the index n by the index s_i in (4.34). Each of these equations corresponds to the boundary condition (4.30), where the same replacement of the index n is made. The solution $\overset{(s_i)}{x}_p(t, \overset{(s_i)}{x}_0)$ of the appropriate boundary-value problem satisfies equality (4.33), in which n is replaced by s_i. Since, $\forall \ i \in Z^+, t \in [0, T]$, $\|\overset{(s_i)}{x}_p(t, \overset{(s_i)}{x}_0)\| \leq M_0$, the sequence $\{\overset{(s_i)}{x}_p(t, \overset{(s_i)}{x}_0)\}_{i=1}^{\infty}$ is uniformly bounded on this interval. We show that it is equicontinuous.

Let $t^{(1)} < t^{(2)}$ be any points of the indicated interval. Then, by designating s_i as k for convenience, we have

$$\overset{(k)}{x}_p(t^{(2)}, \overset{(k)}{x}_0) - \overset{(k)}{x}_p(t^{(1)}, \overset{(k)}{x}_0)$$

$$= \int_{t^{(1)}}^{t^{(2)}} [\overset{(k)}{f}(\tau, \overset{(k)}{x}_p(\tau, \overset{(k)}{x}_0)) - \frac{1}{T}\int_0^T \overset{(k)}{f}(s, \overset{(k)}{x}_p(s, \overset{(k)}{x}_0))ds]d\tau$$

$$+ \frac{t^{(2)} - t^{(1)}}{T} \overset{(k)}{H}_p\{\overset{(k)}{\varphi}_p(\overset{(k)}{x}_p(0, \overset{(k)}{x}_0), \overset{(k)}{x}_p(T, \overset{(k)}{x}_0); \overset{(k)}{x}_p(t_1, \overset{(k)}{x}_0), ..., \overset{(k)}{x}_p(t_p, \overset{(k)}{x}_0))$$

$$- (\sum_{i=0}^{p} \overset{(k)}{A}_i + \overset{(k)}{C}) \overset{(k)}{x}_0$$

$$- \sum_{i=1}^{p} \overset{(k)}{A}_i \int_0^{t_i} [\overset{(k)}{f}(\tau, \overset{(k)}{x}_p(\tau, \overset{(k)}{x}_0)) - \frac{1}{T}\int_0^T \overset{(k)}{f}(s, \overset{(k)}{x}_p(s, \overset{(k)}{x}_0))ds]d\tau\}.$$

Since

$$\|\int_{t^{(1)}}^{t^{(2)}} [\overset{(k)}{f}(\tau, \overset{(k)}{x}_p(\tau, \overset{(k)}{x}_0)) - \frac{1}{T}\int_0^T \overset{(k)}{f}(s, \overset{(k)}{x}_p(s, \overset{(k)}{x}_0))ds]d\tau\| \leq 2M(t^{(2)} - t^{(1)}),$$

the previous inequality yields the estimate

$$\|\overset{(k)}{x}_p(t^{(2)}, \overset{(k)}{x}_0) - \overset{(k)}{x}_p(t^{(1)}, \overset{(k)}{x}_0)\| \leq 2M(t^{(2)} - t^{(1)}) + \frac{t^{(2)} - t^{(1)}}{T}\|\overset{(k)}{H}_p$$

$$\times\{\|\overset{(k)}{\varphi}_p(\overset{(k)}{x}_p(0), \overset{(k)}{x}_p(T); \overset{(k)}{x}_p(t_1), ..., \overset{(k)}{x}_p(t_p))\| + (\sum_{i=0}^{p} \|\overset{(k)}{A}_i\| + \|\overset{(k)}{C}\|)\|\overset{(k)}{x}_0\|$$

$$+ \sum_{i=1}^{p} \|\overset{(k)}{A}_i\|M\alpha_1(t_i)\} \leq (t^{(2)} - t^{(1)})(2M + \frac{\|\overset{(k)}{H}_p\|}{T}\{M_\varphi +$$

$$(\sum_{i=0}^{\infty} \|A_i\| + \|C\|)\|x_0\| + \sum_{i=1}^{\infty} \|A_i\|\frac{MT}{2}\}) \leq (t^{(2)} - t^{(1)})(2M$$

$$+ \frac{\frac{2}{KT} - 1}{T\sum_{i=1}^{p} \|\overset{(n_0)}{A}_i\| + \frac{2K_\varphi}{K}}(M_\varphi + (\sum_{i=0}^{\infty} \|A_i\| + \|C\|)\|x_0\|) + \frac{MT}{2}\sum_{i=1}^{\infty} \|A_i\|)$$

$$\leq (t^{(2)} - t^{(1)})M'', \quad 0 < M'' = const < \infty,$$

which guarantees the equicontinuity of the sequence $\{\overset{(s_i)}{x}_p(t,\overset{(s_i)}{x_0})\}_{i=1}^{\infty}$ on the segment $[0,T]$, since the constant M'' is independent of s_i.

Using the Arzela theorem and the method of diagonalization one more, we choose a subsequ-ence $\{\overset{(k_i)}{x}_p(t,\overset{(k_i)}{x_0})\}_{i=1}^{\infty}$ from this sequence, which is uniformly convergent in $t \in [0,T]$ in the coordinatewise meaning.

From the sequence of equations (4.38), we separate the subsequence

$$\frac{d\overset{(k_i)}{x}}{dt} = \overset{(k_i)}{f}(t,\overset{(k_i)}{x}) + \overset{(k_i)}{\mu_p}, \quad i \in Z^+,$$

where each equation corresponds to the boundary condition obtained from (4.30) by the replacement of the index n by k_i.

We denote the limit of the sequence $\{\overset{(s_i)}{\mu_p}\}_{i=1}^{\infty}$ as $i \to \infty$ by $\bar{\mu}_p$ and the limit of the sequence $\{\overset{(k_i)}{x}_p(t,\overset{(k_i)}{x_0})\}_{i=1}^{\infty}$ as $i \to \infty$ by $\bar{x}_p(t,x_0)$. It is clear that $\lim\limits_{i\to\infty}\overset{(k_i)}{\mu_p} = \bar{\mu}_p$. All these limits are considered, of course, in the coordinatewise meaning.

We now show that

$$\bar{x}_p(t,x_0) = x_p(t,x_0), \quad \bar{\mu}_p = \mu_p. \tag{4.39}$$

We denote $\overset{(k_i)}{x}_p(t,\overset{(k_i)}{x_0})$ by $\overset{(k_i)}{x}_p = (\overset{(k_i)}{x_{1p}},...,\overset{(k_i)}{x_{k_ip}})$ and $\overset{(k_i)}{f}(t,\overset{(k_i)}{x}_p)$ by $\overset{(k_i)}{f} = (\overset{(k_i)}{f_1},\overset{(k_i)}{f_2},...,\overset{(k_i)}{f_{k_i}})$, $i \in Z^+$. Consider the sequence $\{\overset{(k_i)}{f_\ell}\}_{i=1}^{\infty}$, where ℓ is a fixed natural number. In this sequence, a finite number of elements corresponding to values $i \in \{1,2,...,m\}$ are equal to zero, if $k_m < \ell \leq k_{m+1}$. It is obvious that $\overset{(k_i)}{f_\ell} = f_\ell(t,\overset{(k_i)}{x_{1p}},\overset{(k_i)}{x_{2p}},...,\overset{(k_i)}{x_{k_ip}},0,0,0,...) \; \forall i \geq m+1$. Hence, the sequence $\{\overset{(k_i)}{f_\ell}\}_{i=1}^{\infty}$ reads

$$\underbrace{0,\, 0,\, 0,\, ...,\, 0,}_{m} \; f_\ell(t,\overset{(k_{m+1})}{x_{1p}},\overset{(k_{m+1})}{x_{2p}},...,\overset{(k_{m+1})}{x_{k_{m+1}p}},0,0,...\,,$$

$$f_\ell(t,\overset{(k_{m+2})}{x_{1p}},\overset{(k_{m+2})}{x_{2p}},...,\overset{(k_{m+2})}{x_{k_{m+2}p}},0,0,...)\,,\,...\,.$$

We now prove that $\{\overset{(k_i)}{f_\ell}\}_{i=1}^{\infty} \to f_\ell(t,\bar{x}_p(t,x_0))$ as $i \to \infty$. For this purpose, we estimate the modulus of the difference

$$I_\ell^{(k_i)} = |f_\ell(t,\overset{(k_i)}{x_{1p}},\overset{(k_i)}{x_{2p}},\overset{(k_i)}{x_{3p}},...) - f_\ell(t,\bar{x}_{1p},\bar{x}_{2p},\bar{x}_{3p},...)|\,, \; i \geq m+1,$$

where $\overset{(k_i)}{x_{k_i+1}},\overset{(k_i)}{x_{k_i+2}},...$ are zeros.

We will carry out the consideration analogous to that in [121]. We preassign an arbitrarily small real number $\nu > 0$ and write the inequality

$$I_\ell^{(k_i)} \leq |f_\ell(t, \overset{(k_i)}{x}_{1p}(t), \overset{(k_i)}{x}_{2p}(t), ...)$$

$$- f_\ell(t, \bar{x}_{1p}(t), ..., \bar{x}_{gp}(t), \overset{(k_i)}{x}_{g+1p}(t), \overset{(k_i)}{x}_{g+2p}(t), ...)|$$

$$+ |f_\ell(t, \bar{x}_{1p}(t), ..., \bar{x}_{gp}(t), \overset{(k_i)}{x}_{g+1p}(t), \overset{(k_i)}{x}_{g+2p}(t), ...) - f_\ell(t, \bar{x}_{1p}(t), \bar{x}_{2p}(t), ...)|.$$

We denote the first modulus on its right-hand side by $A(\ell, g)$ and the second one by $B(\ell, g)$. Since $f(t, x) \in \hat{C}_{Lip}(x)$, we have

$$B(\ell, g) \leq \alpha(t) 2 M_0 \varepsilon(g) \leq \max_{t \in [0,T]} \alpha(t) 2 M_0 \varepsilon(g),$$

where $\varepsilon(g) \to 0$ as $g \to \infty$. Choosing g^0 so that $\varepsilon(g^0) < \nu$, we obtain the inequality $B(\ell, g^0) < 2 M_0 \nu \max_{t \in [0,T]} \alpha(t)$.

For $A(\ell, g^0)$, the inequality

$$A(\ell, g^0) \leq \alpha(t) \varepsilon(0) \sup\{|\overset{(k_i)}{x}_{1p} - \bar{x}_{1p}|, ..., |\overset{(k_i)}{x}_{g^0 p} - \bar{x}_{g^0 p}|\}$$

holds.

Since $\overset{(k_i)}{x}_p$ tends to $\bar{x}_p$ as $i \to \infty$ uniformly in $t \in [0, T]$ in the coordinatewise meaning, there exists a number $N(\ell, \nu)$ such that, $\forall\, k_i \geq N(\ell, \nu)$,

$$\sup\{|\overset{(k_i)}{x}_{1p} - \bar{x}_{1p}|, ..., |\overset{(k_i)}{x}_{g^0 p} - \bar{x}_{g^0 p}|\} < \nu.$$

Obviously, the inequality $k_i \geq N(\ell, \nu)$ is satisfied $\forall\, i \geq i_0 \in Z^+$. Thus, $\forall i \geq i_0$,

$$I_\ell^{(k_i)} \leq \nu(2 M_0 + \varepsilon(0)) \max_{t \in [0,T]} \alpha(t) \; \forall\, t \in [0, T].$$

This means that the sequence $\{\overset{(k_i)}{f_\ell}\}_{i=1}^\infty$ tends to $f_\ell(t, \bar{x}_p)$ as $i \to \infty$ uniformly in $t \in [0, T]$. Since $\ell \in Z^+$ is chosen arbitrarily, the sequence $\{f^{(k_i)}(t, \overset{(k_i)}{x}_p)\}_{i=1}^\infty$ tends to $f(t, \bar{x}_p)$ as $i \to \infty$ uniformly in $t \in [0, T]$ in the coordinatewise meaning. But

$$\frac{d\,\overset{(k_i)}{x}_p(t, \overset{(k_i)}{x}_0)}{dt} = \overset{(k_i)}{f}(t, \overset{(k_i)}{x}_p(t, \overset{(k_i)}{x}_0)) + \overset{(k_i)}{\mu}_p, \quad \overset{(k_i)}{x}_p(0, \overset{(k_i)}{x}_0) = \overset{(k_i)}{x}_0 \; \forall i \in Z^+.$$

Passing coordinatewise in the last equality to the limit as $i \to \infty$ and applying the theorem of existence of the derivative of the limit of the sequence of scalar differentiable functions, we obtain

$$\frac{d\bar{x}_p(t, x_0)}{dt} = f(t, \bar{x}_p(t, x_0)) + \bar{\mu}_p, \quad \bar{x}_p(t, x_0) = x_0. \tag{4.40}$$

In addition, $\forall i \in Z^+$, the equality

$$\overset{(k_i)}{A_0}\,\overset{(k_i)}{x_p}(0,\,\overset{(k_i)}{x_0}) + \sum_{j=1}^{p}\overset{(k_i)}{A_j}\,\overset{(k_i)}{x_p}(t_j,\,\overset{(k_i)}{x_0}) + \overset{(k_i)}{C}\,\overset{(k_i)}{x_p}(T,\,\overset{(k_i)}{x_0})$$

$$= \overset{(k_i)}{\varphi_p}(\overset{(k_i)}{x_p}(0,\,\overset{(k_i)}{x_0})\,,\,\overset{(k_i)}{x_p}(T,\,\overset{(k_i)}{x_0})\,;\,\overset{(k_i)}{x_p}(t_1,\,\overset{(k_i)}{x_0}),...,\overset{(k_i)}{x_p}(t_p,\,\overset{(k_i)}{x_0})) \quad (4.41)$$

holds.

In it, let us pass coordinatewise to the limit as $i \to \infty$. For simplicity, we choose the first coordinate. Since the left-hand side of this equality contains a finite sum, we restrict ourselves by the transition to the limit along the first coordinate of the product $\overset{(k_i)}{A_0}\,\overset{(k_i)}{x_p}(0,\,\overset{(k_i)}{x_0})$. We have

$$L_1 = \lim_{i\to\infty}\sum_{j=1}^{k_i} a_{1j}^{(0)}\,\overset{(k_i)}{x_{jp}}(0,\,\overset{(k_i)}{x_0}) = \lim_{i\to\infty}\sum_{j=1}^{\infty} a_{1j}^{(0)}\,\overset{(k_i)}{x_{jp}}(0,\,\overset{(k_i)}{x_0}),$$

where we set $\overset{(k_i)}{x_{jp}}(0,\,\overset{(k_i)}{x_0}) = 0$ for $j > k_i$. The boundedness of the matrix A_0 in the norm implies that the last series converges uniformly in i, since the inequality $|\overset{(k_i)}{x_{jp}}(0,\,\overset{(k_i)}{x_0})| \le M_0$ is satisfied $\forall\{i,j\} \subset Z^+$. Then

$$L_1 = \sum_{j=1}^{\infty} a_{1j}^{(0)}\,\lim_{i\to\infty}\overset{(k_i)}{x_{jp}}(0,\,\overset{(k_i)}{x_0}) = \sum_{j=1}^{\infty} a_{1j}^{(0)}\,\bar{x}_{jp}(0,x_0).$$

We now estimate the modulus of the difference

$$G_1^{(k_i)} = |\,\overset{(k_i)}{\varphi_{1p}}(\overset{(k_i)}{x_p}(0),\,\overset{(k_i)}{x_p}(T);\,\overset{(k_i)}{x_p}(t_1),...,\overset{(k_i)}{x_p}(t_p))$$

$$- \varphi_{1p}(\bar{x}_p(0),\bar{x}_p(T);\bar{x}_p(t_1),...,\bar{x}_p(t_p))|$$

$$= |\varphi_1(\overset{(k_i)}{x_p}(0),0,0,...;\,\overset{(k_i)}{x_p}(T),0,0,...;\,\overset{(k_i)}{x_p}(t_1),$$

$$0,0,...;...;\,\overset{(k_i)}{x_p}(t_p),0,0,...;0,0,0,...)$$

$$- \varphi_1(\bar{x}_p(0),\bar{x}_p(T);\bar{x}_p(t_1),...,\bar{x}_p(t_p),0,0,0,...)|.$$

We denote $(\overset{(k_i)}{x_p}(t),0,0,...)$ by $\overset{(k_i)}{x_\infty}(t) = (\overset{(k_i)}{x_p}(t), x_{k_i+1}(t), \overset{(k_i)}{x_{k_i+2}}(t),...),$

where $\overset{(k_i)}{x}_r(t)$ for $r > k_i$ are identical zeros. We have

$$
\begin{aligned}
G_1^{(k_i)} &= |\varphi_1(\overset{(k_i)}{x}_\infty(0), \overset{(k_i)}{x}_\infty(T); \overset{(k_i)}{x}_\infty(t_1), ..., \overset{(k_i)}{x}_\infty(t_p), 0, 0, 0, ...) \\
&\quad - \varphi_1(\bar{x}_p(0), \bar{x}_p(T); \bar{x}_p(t_1), ..., \bar{x}_p(t_p), 0, 0, 0, ...)| \\
&\leq |\varphi_1(\overset{(k_i)}{x}_\infty(0), \overset{(k_i)}{x}_\infty(T); \overset{(k_i)}{x}_\infty(t_1), ..., \overset{(k_i)}{x}_\infty(t_p), 0, 0, 0, ...) \\
&\quad - \varphi_1[(\bar{x}_{1p}(0), ..., \bar{x}_{gp}(0), \overset{(k_i)}{x}_{g+1p}(0), \overset{(k_i)}{x}_{g+2p}(0), ...), \\
&\qquad (\bar{x}_{1p}(T), ..., \bar{x}_{gp}(T), \overset{(k_i)}{x}_{g+1p}(T), \overset{(k_i)}{x}_{g+2p}(T), ...); \\
&\qquad (\bar{x}_{1p}(t_1), ..., \bar{x}_{gp}(t_1), \overset{(k_i)}{x}_{g+1p}(t_1), \overset{(k_i)}{x}_{g+2p}(t_1), ...), ... \\
&\qquad ..., (\bar{x}_{1p}(t_p), ..., \bar{x}_{gp}(t_p), \overset{(k_i)}{x}_{g+1p}(t_p), \overset{(k_i)}{x}_{g+2p}(t_p), ...), \\
&\qquad 0, 0, 0, ...]| + |\varphi_1[*] - \varphi_1(\bar{x}_p(0), \bar{x}_p(T); \bar{x}_p(t_1), ..., \bar{x}_p(t_p), 0, 0, 0, ...)|,
\end{aligned}
$$

where $\varphi_1[*]$ is equal to the subtrahend, which is positioned under the sign of modulus of the first difference on the right-hand side of the last inequality.

Thus,

$$
\begin{aligned}
G_1^{(k_i)} &\leq \delta(g) \sup\{\sup\{|\overset{(k_i)}{x}_{g+1p}(0) - \bar{x}_{g+1p}(0)|, |\overset{(k_i)}{x}_{g+2p}(0) - \bar{x}_{g+2p}(0)|, ...\}, \\
&\qquad ..., \sup\{|\overset{(k_i)}{x}_{g+1p}(t_p) - \bar{x}_{g+1p}(t_p)|, |\overset{(k_i)}{x}_{g+2p}(t_p) - \bar{x}_{g+2p}(t_p)|, ...\}\} \\
&\quad + \delta(0) \sup\{\sup\{|\overset{(k_i)}{x}_{1p}(0) - \bar{x}_{1p}(0)|, ..., |\overset{(k_i)}{x}_{gp}(0) - \bar{x}_{gp}(0)|\}, \\
&\qquad ..., \sup\{|\overset{(k_i)}{x}_{1p}(t_p) - \bar{x}_{1p}(t_p)|, ..., |\overset{(k_i)}{x}_{gp}(t_p) - \bar{x}_{gp}(t_p)|\}\},
\end{aligned}
$$

and the first term on the right-hand side of this inequality does not exceed $2\delta(g)M_0$.

Let ν be an arbitrarily small positive number. Since $\delta(g) \to 0$ as $g \to \infty$, we may choose $g = g^0$ so that $2\delta(g^0)M_0 < \nu$. The uniform in $t \in [0, T]$ coordinatewise convergence of $\overset{(k_i)}{x}_p(t)$ to $\bar{x}_p(t)$ as $i \to \infty$ yields the existence of $i_0 \in Z^+$ such that the inequality

$$
|\overset{(k_i)}{x}_{rp}(t) - \bar{x}_{rp}(t)| \leq \frac{\nu}{\delta(0)},
$$

where $r \in \{1, 2, ..., g^0\}$, holds $\forall i \geq i_0, t \in [0, T]$. Then, $\forall i \geq i_0$,

$$
G_1^{(k_i)} \leq \nu + \nu = 2\nu.
$$

This means that

$$
\lim_{i \to \infty} \overset{(k_i)}{\varphi}_{1p}(\overset{(k_i)}{x}_p(0), \overset{(k_i)}{x}_p(T); \overset{(k_i)}{x}_p(t_1), ..., \overset{(k_i)}{x}_p(t_p))
$$

$$
= \varphi_{1p}(\bar{x}_p(0), \bar{x}_p(T); \bar{x}_p(t_1), ..., \bar{x}_p(t_p)).
$$

The analogous analysis can be performed for the other coordinates of the vector $\overset{(k_i)}{\varphi_p}$ in (4.41).

Thus, the function $\bar{x}_p(t, x_0)$ satisfies equality (4.26), i.e., it is a solution of the perturbed equation $\frac{dx}{dt} = f(t, x) + \bar{\mu}_p$ with the boundary condition (4.26), since equalities (4.40) are satisfied. The uniqueness of the control μ_p for a fixed x_0 yields equalities (4.39).

Consider any subsequence $\frac{d\overset{(r)}{x}}{dt} = \overset{(r)}{f}(t, \overset{(r)}{x})$ of the sequence of equations (4.29) with the boundary condition (4.30), in which the index n is replaced by the index r. This subsequence satisfies all the above-presented require-ments, i.e., there exists the subsequence $\{\overset{(\ell)}{x}_p(t, \overset{(\ell)}{x}_0)\}_{\ell=1}^{\infty}$ of the sequence $\{\overset{(r)}{x}_p(t, \overset{(r)}{x}_0)\}_{r=1}^{\infty}$, which is coordinatewise convergent again to the function $x_p(t, x_0)$, and $\overset{(\ell)}{\mu}_p \to \mu_p$ as $\ell \to \infty$ in the coordinatewise meaning. As is known [97, p. 276], in this case, the sequences $\{\overset{(n)}{x}_p(t, \overset{(n)}{x}_0)\}_{n=1}^{\infty}$ and $\{\overset{(n)}{\mu}_p\}_{n=1}^{\infty}$ are coordinatewise convergent to $x_p(t, x_0)$ and μ_p, respectively, as $n \to \infty$, which completes the proof of the lemma. $\qquad\square$

We introduce the notation

$$\beta_{\varphi}^*(x, n_0, p_0) = \frac{\frac{2}{KT} - 1}{\sum\limits_{i=1}^{p_0} \| \overset{(n_0)}{A_i} \| + \frac{2K_{\varphi}}{KT}} (M_{\varphi} + (\sum_{i=0}^{\infty} \|A_i\| + \|C\|)\|x\|) + \frac{M}{K},$$

$\{n_0, p_0\} \subset Z^+$; $D_{\beta\varphi}^*$ is a set, every point of which belongs to the set D together with its $\beta_{\varphi}^*(x, n_0, p_0)$-neighborhood.

Theorem 4.5. *Let $f(t, x) \in \hat{C}_{lip}(x)$, $\varphi(\psi) \in \hat{C}_{lip}(\psi)$, let the set $D_{\beta\varphi}^*$ be nonempty, and let, $\forall n \geq n_0$, $p \geq p_0$, conditions a_i $(i = 1, 2, 3)$, c_2^0, and c_3, where $K_{\varphi p} = K_{\varphi}$, be satisfied. Then, $\forall x_0 \in D_{\beta\varphi}^*$,*

$$x^*(t, x_0) = \lim_{p\to\infty} (\lim_{n\to\infty} \overset{(n)}{x}_p(t, \overset{(n)}{x}_0)), \tag{4.42}$$

$$\mu = \lim_{p\to\infty} (\lim_{n\to\infty} \overset{(n)}{\mu}_p), \tag{4.43}$$

where the convergence in n is coordinatewise, the convergence in p holds in the norm of the space $\mathfrak{M}$, and $x^*(t, x_0)$ and μ are defined in Theorem 4.4. If condition c_2^0 is replaced by condition c_2^*, there exists no other value of $\mu \in \mathfrak{M}$ such that the solution of the equation $\frac{dx}{dt} = f(t, x) + \mu$ with the initial condition $x(0) = x_0$ would satisfy the boundary condition (4.23).

Proof.   First, we note that the inclusion $x_0 \in D^*_{\beta\varphi}$ yields the inclusions $x_0 \in D_{\beta\varphi}$, $x_0 \in D_{\beta\varphi p}$, and $\overset{(n)}{x}_0 \in \tilde{D}^{(n)}_{\overset{(n)}{\beta}_{\varphi p}}$ $\forall n \geq n_0$, $p \geq p_0$. This can be easily verified analogously to the reasoning in Lemma 4.1. We note that the matrices

$$H^{-1} = \sum_{i=1}^{\infty} \frac{t_i}{T} A_i + C \quad \text{and} \quad H_p^{-1} = \sum_{i=1}^{p} \frac{t_i}{T} A_i + C, \ p \geq p_0,$$

satisfy the relations $\lim\limits_{p \to \infty} H_p^{-1} = H^{-1}$ in the meaning of a matrix norm. Moreover, the matrix H_p is bounded in the norm uniformly in $p \geq p_0$ by the constant

$$\frac{\frac{2}{KT} - 1}{\sum\limits_{i=1}^{p_0} \| \overset{(n_0)}{A_i} \| + \frac{2K_\varphi}{KT}}.$$

Therefore, the inequality

$$\|H - H_p\| \leq \|H\| \|H^{-1} - H_p^{-1}\| \|H_p\|$$

yields

$$\|H - H_p\| \leq \nu(p), \ p \geq p_0, \tag{4.44}$$

where $\nu(p) \to 0$ as $p \to \infty$. It is clear that condition c_1 is satisfied.

Since the function $x_p(t, x_0)$ satisfies the equality

$$x_p(t, x_0) = x_0 + \int_0^t [f(\tau, x_p(\tau, x_0)) - \frac{1}{T} \int_0^T f(s, x_p(s, x_0))ds] d\tau$$

$$+ \frac{t}{T} H_p \{ \varphi_p(x_p(0), x_p(T); x_p(t_1), ..., x_p(t_p)) - (\sum_{i=0}^{p} A_i + C) x_0$$

$$- \sum_{i=1}^{p} A_i \int_0^{t_i} [f(\tau, x_p(\tau, x_0)) - \frac{1}{T} \int_0^T f(s, x_p(s, x_0))ds] d\tau \}, \ p \geq p_0,$$

and the function $x^*(t, x_0)$ satisfies the equality written prior to the formulation of Theorem 4.4, the inequality

$$\|x_p(t, x_0) - x^*(t, x_0)\| \leq \Gamma_1 + \Gamma_2 + \Gamma_3$$

holds. Here,

$$\Gamma_1 = \| \int_0^t [f(\tau, x^*(\tau, x_0)) - f(\tau, x_p(\tau, x_0))$$

$$- \frac{1}{T} \int_0^T (f(s, x^*(s, x_0)) - f(s, x_p(s, x_0)))ds]d\tau\|$$

$$\leq K\alpha_1(t) \sup_{t\in[0,T]} \|x^*(t, x_0) - x_p(t, x_0)\|;$$

$$\Gamma_2 = \|H\{\varphi(x^*(0), x^*(T); x^*(t_1), x^*(t_2), ...) - (\sum_{i=0}^{\infty} A_i + C)x_0\}$$

$$-H_p\{\varphi_p(x_p(0), x_p(T); x_p(t_1), ..., x_p(t_p)) - (\sum_{i=0}^{p} A_i + C)x_0\}\|$$

$$\leq \|H\varphi(x^*(0), x^*(T); x^*(t_1), ...) - H_p\varphi_p(x_p(0), x_p(T); x_p(t_1), ..., x_p(t_p))\|$$

$$+ \|H(\sum_{i=0}^{\infty} A_i + C)x_0 - H_p(\sum_{i=0}^{p} A_i + C)x_0\|$$

$$\leq \|H\|\|\varphi(x^*(0), x^*(T); x^*(t_1), ...)$$

$$-\varphi_p(x_p(0), x_p(T); ..., x_p(t_p))\| + \|H - H_p\|\|\varphi_p(x_p(0), x_p(T); ..., x_p(t_p))\|$$

$$+ M_0\|H\| \sum_{i=p+1}^{\infty} \|A_i\| + M_0\|H - H_p\|(\|C\| + \sum_{i=0}^{p} \|A_i\|)$$

$$\leq \|H - H_p\|\{M_\varphi + M_0(\|C\| + \sum_{i=0}^{\infty} \|A_i\|)\} + M_0\|H\| \sum_{i=p+1}^{\infty} \|A_i\|$$

$$+ \|H\|\|\varphi(x^*(0), x^*(T); x^*(t_1), ...) - \varphi(x_p(0), x_p(T); x_p(t_1), ...)\|$$

$$+ \|H\|\|\varphi(x_p(0), x_p(T); x_p(t_1), ..., x_p(t_p), 0, 0, 0, ...)$$

$$- \varphi(x_p(0), x_p(T); ..., x_p(t_p), x_p(t_{p+1}), ...)\| \leq K^{(1)}\|H - H_p\|$$

$$+\|H\|K_\varphi \sup_{t\in[0,T]} \|x^*(t, x_0) - x_p(t, x_0)\| + \|H\|M_0\{\sum_{i=p+1}^{\infty} \|A_i\| + \delta_0(p+2)\};$$

$$\Gamma_3 = \|H \sum_{i=1}^{\infty} A_i \int_0^{t_i} [f(\tau, x^*(\tau, x_0)) - \frac{1}{T} \int_0^T f(s, x^*(s, x_0))ds]d\tau$$

$$-H_p \sum_{i=1}^{p} A_i \int_0^{t_i} f(\tau, x_p(\tau, x_0)) - \frac{1}{T} \int_0^T f(s, x_p(s, x_0))ds]d\tau\|$$

$$\leq \|H\|\|\sum_{i=1}^{p} A_i \int_0^{t_i} [f(\tau, x^*(\tau, x_0)) - f(\tau, x_p(\tau, x_0))$$

$$- \frac{1}{T} \int_0^{T} (f(s, x^*(s, x_0)) - f(s, x_p(s, x_0)))ds]d\tau\|$$

$$+ \|H\|\|\sum_{i=p+1}^{\infty} A_i \int_0^{t_i} [f(\tau, x^*(\tau, x_0)) - \frac{1}{T} \int_0^{T} f(s, x^*(s, x_0))ds]d\tau\|$$

$$+ \|H - H_p\|\|\sum_{i=1}^{p} A_i \int_0^{t_i} [f(\tau, x_p(\tau, x_0)) - \frac{1}{T} \int_0^{T} f(s, x_p(s, x_0))ds]d\tau\|$$

$$\leq \|H\|\frac{TK}{2} \sum_{i=1}^{\infty} \|A_i\| \sup_{t\in[0,T]} \|x^*(t, x_0)) - x_p(t, x_0))\|$$

$$+ \|H\|\frac{TM}{2} \sum_{i=p+1}^{\infty} \|A_i\| + \|H - H_p\|\frac{TM}{2} \sum_{i=1}^{\infty} \|A_i\|.$$

Setting $\sup_{t\in[0,T]} \|x^*(t, x_0) - x_p(t, x_0)\| = \xi$, we obtain the estimate

$$\xi \leq \frac{KT}{2}\xi + K^{(1)}\|H - H_p\| + \|H\|K_\varphi\xi + \|H\|M_0\{ \sum_{i=p+1}^{\infty} \|A_i\| + \delta_0(p+2)\}$$

$$+ \frac{K\|H\|T}{2}\xi \sum_{i=1}^{\infty} \|A_i\| + \|H\|\frac{TM}{2} \sum_{i=p+1}^{\infty} \|A_i\| + \|H - H_p\|\frac{TM}{2} \sum_{i=1}^{\infty} \|A_i\|.$$

Taking into account that the constant $K^{(1)}$ is independent of p, the series $\sum_{i=1}^{\infty} \|A_i\|$ converges, $\delta_0(p) \to 0$ as $p \to \infty$, as well as condition c_1 and relation (4.44), the last estimate yields the inequality $\xi \leq \frac{\eta(p)}{1-Q_\varphi}$, where $\eta(p) \to 0$ as $p \to \infty$, i.e., the sequence $\{x_p(t, x_0)\}_{p=p_0}^{\infty}$ is convergent in the norm of the space $\mathfrak{M}$ as $p \to \infty$ to $x^*(t, x_0) \ \forall x_0 \in D^*_{\beta\varphi}$. Using relation (4.37), we obtain (4.42), which completes the proof of the theorem, since the proof of equality (4.43) is obvious, and condition c_2^* yields condition c_1^0, i.e., Remark 4.2 is valid. $\square$

Corollary 4.4. *Under conditions of Theorem 4.5, the estimate*

$$\|x^*(t, x_0) - \overset{(n)}{x}_p(t, \overset{(n)}{x}_0)\| \leq L_1(p) + L_2(n), \quad n \geq n_0, \ p \geq p_0, \qquad (4.45)$$

where the sum $L_1(p) + L_2(n)$ is bounded uniformly in $p \geq p_0$, $n \geq n_0$, and $L_1(p) \to 0$ as $p \to \infty$.

Proof. Let us estimate the difference $x^*(t, x_0) - \overset{(n)}{x}_p(t, \overset{(n)}{x_0})$ in the norm, by supplementing the vector $\overset{(n)}{x}_p$ by zeros, as was mentioned earlier. We obtain the inequality

$$\|x^*(t, x_0) - \overset{(n)}{x}_p(t, \overset{(n)}{x_0})\| \leq \|x^*(t, x_0) - x_p(t, x_0)\| + \|x_p(t, x_0) - \overset{(n)}{x}_p(t, \overset{(n)}{x_0})\|$$

$$\leq \frac{\eta(p)}{1 - Q_\varphi} + \|x_p(t, x_0) - \overset{(n)}{x}_p(t, \overset{(n)}{x_0})\| \ \forall \, t \in [0, T] \, , \ x_0 \in D^*_{\beta\varphi}.$$

We now write the estimate

$$\|x_p(t, x_0) - \overset{(n)}{x}_p(t, \overset{(n)}{x_0})\| \leq \Gamma^0_1 + \Gamma^0_2 + \Gamma^0_3,$$

where

$$\Gamma^0_1 = \| \int_0^t [f(\tau, x_p(\tau, x_0)) - \overset{(n)}{f}(\tau, \overset{(n)}{x}_p(\tau, \overset{(n)}{x_0})) - \frac{1}{T} \int_0^T (f(s, x_p(s, x_0)) -$$

$$- \overset{(n)}{f}(s, \overset{(n)}{x}_p(s, \overset{(n)}{x_0})))ds]d\tau \| \leq \alpha_1(t) \sup_{t \in [0,T]} \|f(t, x_p(t, x_0)) - \overset{(n)}{f}(t, \overset{(n)}{x}_p(t, \overset{(n)}{x_0}))\|;$$

$$\Gamma^0_2 = \|H_p\{\varphi_p(x_p(0), x_p(T); x_p(t_1), ..., x_p(t_p)) - (\sum_{i=0}^{p} A_i + C)x_0\}$$

$$- \overset{(n)}{H}_p\{\overset{(n)}{\varphi}_p(\overset{(n)}{x}_p(0), \overset{(n)}{x}_p(T); \overset{(n)}{x}_p(t_1), ..., \overset{(n)}{x}_p(t_p)) - (\sum_{i=0}^{p} \overset{(n)}{A}_i + \overset{(n)}{C}) \overset{(n)}{x_0}\}\|;$$

$$\Gamma^0_3 = \|H_p \sum_{i=1}^{p} A_i \int_0^{t_i} [f(\tau, x_p(\tau, x_0)) - \frac{1}{T} \int_0^T f(s, x_p(s, x_0))ds]d\tau$$

$$- \overset{(n)}{H}_p \sum_{i=1}^{p} \overset{(n)}{A}_i \int_0^{t_i} [\overset{(n)}{f}(\tau, \overset{(n)}{x}_p(\tau, \overset{(n)}{x_0})) - \frac{1}{T} \int_0^T \overset{(n)}{f}(s, \overset{(n)}{x}_p(s, \overset{(n)}{x_0}))ds]d\tau\|;$$

The following inequalities hold:

$$\|f(t, x_p(t, x_0)) - \overset{(n)}{f}(t, \overset{(n)}{x}_p(t, \overset{(n)}{x_0}))\| \leq \|f(t, x_p(t, x_0)) - f(t, \overset{(n)}{x}_p(t, \overset{(n)}{x_0}))\|$$

$$+ \|f(t, \overset{(n)}{x}_p(t, \overset{(n)}{x_0})) - \overset{(n)}{f}(t, \overset{(n)}{x}_p(t, \overset{(n)}{x_0}))\| \leq K\|x_p(t, x_0) - \overset{(n)}{x}_p(t, \overset{(n)}{x_0})\|$$

$$+ \sup\{|f_{n+1}(t, \overset{(n)}{x}_p(t, \overset{(n)}{x_0}))|, |f_{n+2}(t, \overset{(n)}{x}_p(t, \overset{(n)}{x_0}))|, ...\};$$

$$\Gamma_2^0 \leq \|H_p \varphi_p(x_p(0), x_p(T); x_p(t_1), ..., x_p(t_p))$$

$$\overset{(n)}{H_p}\ \overset{(n)}{\varphi_p}(\overset{(n)}{x_p}(0), \overset{(n)}{x_p}(T); \overset{(n)}{x_p}(t_1), ..., \overset{(n)}{x_p}(t_p))\|$$

$$+ \|H_p(\sum_{i=0}^{p} A_i + C)x_0 - \overset{(n)}{H_p}(\sum_{i=0}^{p} \overset{(n)}{A_i} + \overset{(n)}{C})\,\overset{(n)}{x_0}\|$$

$$\leq \|H_p\|\|\varphi_p(x_p(0), x_p(T); x_p(t_1), ..., x_p(t_p))$$

$$- \overset{(n)}{\varphi_p}(\overset{(n)}{x_p}(0), \overset{(n)}{x_p}(T); \overset{(n)}{x_p}(t_1), ..., \overset{(n)}{x_p}(t_p))\| + \|H_p - \overset{(n)}{H_p}\|$$

$$\times \|\overset{(n)}{\varphi_p}(\overset{(n)}{x_p}(0), \overset{(n)}{x_p}(T); \overset{(n)}{x_p}(t_1), ..., \overset{(n)}{x_p}(t_p))\| + \|G_p\|\|x_0 - \overset{(n)}{x_0}\|$$

$$+ \|G_p - \overset{(n)}{G_p}\| \cdot \|\overset{(n)}{x_0}\| \leq \|H_p\| \, K_{\varphi p} \sup_{t\in[0,T]} \left\| x_p(t, x_0) - \overset{(n)}{x_p}(t, \overset{(n)}{x_0}) \right\|$$

$$+ \|H_p\| \sup_{\psi_i \in D} \{|\varphi_{n+1}(\psi_1, \psi_2, ..., \psi_{p+2}, 0, 0, ...)|,$$

$$|\varphi_{n+2}(\psi_1, \psi_2, ..., \psi_{p+2}, 0, 0, ...)|, ...\} + \|G_p\|\|x_0 - \overset{(n)}{x_0}\| + M_0\|G_p - \overset{(n)}{G_p}\|,$$

where

$$G_p = H_p(\sum_{i=0}^{p} A_i + C)\ ,\ \overset{(n)}{G_p} = \overset{(n)}{H_p}(\sum_{i=0}^{p} \overset{(n)}{A_i} + \overset{(n)}{C});$$

$$\Gamma_3^0 \leq \|H_p\| \sum_{i=1}^{p} \{\|A_i\|\| \int_{0}^{t_i} [f(\tau, x_p(\tau, x_0)) - \overset{(n)}{f}(\tau, \overset{(n)}{x_p}(\tau, \overset{(n)}{x_0}))$$

$$- \frac{1}{T} \int_{0}^{T} (f(s, x_p(s, x_0)) - \overset{(n)}{f}(s, \overset{(n)}{x_p}(s, \overset{(n)}{x_0})))ds]d\tau\|$$

$$+ \|A_i - \overset{(n)}{A_i}\|\| \int_{0}^{t_i} [\overset{(n)}{f}(\tau, \overset{(n)}{x_p}(\tau, \overset{(n)}{x_0})) - \frac{1}{T} \int_{0}^{T} \overset{(n)}{f}(s, \overset{(n)}{x_p}(s, \overset{(n)}{x_0}))ds]d\tau\|\}$$

$$+ \|H_p - \overset{(n)}{H_p}\| \sum_{i=1}^{p} \|\overset{(n)}{A_i}\|\| \int_{0}^{t_i} [\overset{(n)}{f}(\tau, \overset{(n)}{x_p}(\tau, \overset{(n)}{x_0})) - \frac{1}{T} \int_{0}^{T} \overset{(n)}{f}(s, \overset{(n)}{x_p}(s, \overset{(n)}{x_0}))ds]d\tau\|$$

$$\leq \|H_p\| \sum_{i=1}^{p} \{\|A_i\|\frac{T}{2} \sup_{t\in[0,T]} \|f(t, x_p(t, x_0)) - \overset{(n)}{f}(t, \overset{(n)}{x_p}(t, \overset{(n)}{x_0}))\|$$

$$+ \frac{TM}{2}\|A_i - \overset{(n)}{A_i}\|\} + \|H_p - \overset{(n)}{H_p}\| \sum_{i=1}^{p} \|\overset{(n)}{A_i}\|\frac{TM}{2};$$

$$\|G_p - \overset{(n)}{G_p}\| \le \|H_p\|(\sum_{i=0}^{p} \|A_i - \overset{(n)}{A_i}\| + \|C - \overset{(n)}{C}\|)$$

$$+ \|H_p - \overset{(n)}{H_p}\|(\sum_{i=0}^{p} \|\overset{(n)}{A_i}\| + \|\overset{(n)}{C}\|).$$

We introduce the notation

$$\sup_{\psi_i \in D} \{|\varphi_{n+1}(\psi_1, \psi_2, ..., \psi_{p+2}, 0, 0, ...)|, |\varphi_{n+2}(\psi_1, \psi_2, ..., \psi_{p+2},$$

$$0, 0, ...)|, ...\} = \xi_*(p, n),$$

$$\sup_{t \in [0,T]} \|x_p(t, x_0) - \overset{(n)}{x_p}(t, \overset{(n)}{x_0}))\| = \xi^0(p, n), \quad \|x_0 - \overset{(n)}{x_0}\| = \eta^0(n),$$

$$\|\{\underbrace{0, ..., 0}_{n}, f_{n+1}(t, x), f_{n+2}(t, x), ...\}\| = \mu(n, t, x),$$

$$\sup_{(t,x) \in D_0} \mu(n, t, x) = \mu^0(n),$$

$$\|A_i - \overset{(n)}{A_i}\| = \overline{A_i}^{\,n}, \quad \|H_p - \overset{(n)}{H_p}\| = \overline{H}_p^{\,n}, \quad \|C - \overset{(n)}{C}\| = \overline{C}^n.$$

In view of the previous estimates, we can write the inequality

$$\xi^0(p, n) \le \frac{T}{2}(K\xi^0(p, n) + \mu^0(n)) + \|H_p\|K_{\varphi p}\xi^0(p, n) + \|G_p\|\eta^0(n)$$

$$+ \|H_p\|\xi_*(p, n) + M_0\|H_p\|(\sum_{i=0}^{p} \overline{A_i}^{\,n} + \overline{C}^n) + M_0\overline{H}_p^{\,n}(\sum_{i=0}^{p} \|\overset{(n)}{A_i}\| + \|\overset{(n)}{C}\|)$$

$$+\|H_p\| \sum_{i=1}^{p} \|A_i\|\frac{T}{2}(K\xi^0(p, n) + \mu^0(n))+\|H_p\|\frac{TM}{2}\overline{A_i}^{\,n}+\overline{H}_p^{\,n}\frac{TM}{2} \sum_{i=1}^{p} \|A_i\|,$$

or

$$(1 - Q_{\varphi p})\xi^0(p, n) \le \frac{T\mu^0(n)}{2}$$

$$+ \|G_p\|\eta^0(n) + M_0\|H_p\|(\sum_{i=0}^{p} \overline{A_i}^{\,n} + \overline{C}^n + \xi_*(p, n))$$

$$+ M_0\overline{H}_p^{\,n}(\sum_{i=0}^{p} \|\overset{(n)}{A_i}\| + \|\overset{(n)}{C}\|) + \frac{\|H_p\|T\mu^0(n)}{2} \sum_{i=1}^{p} \|A_i\|$$

$$+ \|H_p\|\frac{TM}{2}\overline{A_i}^{\,n} + \overline{H}_p^{\,n}\frac{TM}{2} \sum_{i=1}^{p} \|A_i\|.$$

We denote the right-hand side of the last inequality by $\Phi(p,n)$ and obtain the estimate

$$\|x_p(t,x_0) - \overset{(n)}{x}_p(t,\overset{(n)}{x_0}))\| \leq \frac{\Phi(p,n)}{1-Q_{\varphi p}}. \tag{4.46}$$

We sharpen the last inequality, by replacing the signs of summation $\sum\limits_{i=0}^{p}$, $\sum\limits_{i=1}^{p}$ by $\sum\limits_{i=0}^{\infty}$, $\sum\limits_{i=1}^{\infty}$, respectively, in the formula for $\Phi(p,n)$. We replace $\|H_p\|$ by the constant

$$\frac{\frac{2}{KT} - 1}{\sum\limits_{i=0}^{p_0} \| \overset{(n_0)}{A}_i \| + \frac{2K_\varphi}{KT}},$$

which is present in Theorem 4.5. Moreover, we replace the constants $K_{\varphi p}$ by the constant K_φ, $Q_{\varphi p}$ by a constant q, and, finally, $\xi_*(p,n)$ by

$$\xi_*^0(n) = \sup_{\psi_i \in D} \{|\varphi_{n+1}(\psi_1,\psi_2,...)|, |\varphi_{n+2}(\psi_1,\psi_2,...)|, ...\}.$$

As a result, we obtain an inequality, which is "cruder" than estimate (4.46):

$$\|x_p(t,x_0) - \overset{(n)}{x}_p(t,\overset{(n)}{x_0}))\| \leq \frac{\Phi^*(n)}{1-q}.$$

In other words, $\forall t \in [0,T]$, $p \geq p_0$, $n \geq n_0$,

$$\|x^*(t,x_0) - \overset{(n)}{x}_p(t,\overset{(n)}{x_0}))\| \leq \frac{\eta(p)}{1-Q_\varphi} + \frac{\Phi^*(n)}{1-q},$$

which proves inequality (4.45) and also the formulated corollary. $\square$

Remark 4.4. The conditions of Theorem 4.5 do not imply that $L_2(n) \to 0$ as $n \to \infty$.

4.4   Another means of the reduction. Conditions of commutativity of the limiting transitions (4.42) and (4.43)

Together with the boundary-value problems (4.1), (4.23) and (4.29), (4.30), we consider the boundary-value problem (4.29) with the boundary condition

$$\overset{(n)}{A}_0 \overset{(n)}{x}(0) + \sum_{i=1}^{\infty} \overset{(n)}{A}_i \overset{(n)}{x}(t_i) + \overset{(n)}{C} \overset{(n)}{x}(t)$$

$$= \overset{(n)}{\varphi}(\overset{(n)}{x}(0,\overset{(n)}{x_0}), \overset{(n)}{x}(T,\overset{(n)}{x_0}); \overset{(n)}{x}(t_1,\overset{(n)}{x_0}), \overset{(n)}{x}(t_2,\overset{(n)}{x_0}),...). \tag{4.47}$$

We consider that conditions (4.3) and (4.24) are satisfied. Moreover, the following restrictions hold as well:

a_4) the matrix $\sum\limits_{i=0}^{\infty} \frac{t_i}{T} \overset{(n)}{A_i} + \overset{(n)}{C}$ is invertible and the matrix $\overset{(n)}{H}$ inverse to it is bounded in the norm;

b_4) the set $\tilde{D}_{\underset{\overset{(n)}{\beta}\varphi}{(n)}}^{(n)}$ is nonempty. It is composed from points $\overset{(n)}{x_0} \in R^n$ such that the points $(x_{01}, ..., x_{0n}, 0, 0, ...)$ enter the set D together with their $\overset{(n)}{\beta}\varphi$-neighborhoods, where

$$\overset{(n)}{\beta}\varphi(\overset{(n)}{x_0}) = \frac{T}{2}M + \overset{(n)}{\beta}_{1\varphi}(\overset{(n)}{x_0}), \quad \overset{(n)}{\beta}_{1\varphi}(\overset{(n)}{x_0})$$

$$= \|\overset{(n)}{H}\|\|[\overset{(n)}{d^*} - (\sum_{i=0}^{\infty} \overset{(n)}{A_i} + \overset{(n)}{C})\overset{(n)}{x_0}]\| + \sum_{i=1}^{\infty} \|\overset{(n)}{H}\overset{(n)}{A_i}\|\alpha_1(t_i)M,$$

and $\overset{(n)}{d^*} \in R^n$ is selected so that $|\overset{(n)}{d_i^*}| = M_\varphi$, $\operatorname{sign} \overset{(n)}{d_i^*} = -\operatorname{sign} \overset{(n)}{d_i^0}$. Moreover,

$$\overset{(n)}{d_i^0} = colon(\overset{(n)}{d_1^0}, \overset{(n)}{d_2^0}, ..., \overset{(n)}{d_n^0}) = (\sum_{i=0}^{\infty} \overset{(n)}{A_i} + \overset{(n)}{C})\overset{(n)}{x_0};$$

c_4) $\quad \overset{(n)}{Q_\varphi} = \frac{KT}{2}[1 + \|\overset{(n)}{H}\| \sum\limits_{i=1}^{\infty} \|\overset{(n)}{A_i}\|] + K_\varphi\|\overset{(n)}{H}\| < 1.$

Repeating the reasoning in the proof of Theorem 4.1, we can conclude that the sequence

$$\overset{(n)}{x}_m(t, \overset{(n)}{x}_0) = \overset{(n)}{x_0} + \int\limits_0^t [\overset{(n)}{f}(\tau, \overset{(n)}{x}_{m-1}(\tau, \overset{(n)}{x_0}))$$

$$- \frac{1}{T}\int\limits_0^T \overset{(n)}{f}(s, \overset{(n)}{x}_{m-1}(s, \overset{(n)}{x_0}))ds]d\tau$$

$$+ \frac{t}{T}\overset{(n)}{H}\{\overset{(n)}{\varphi}(\overset{(n)}{x}_{m-1}(0, \overset{(n)}{x_0}), \overset{(n)}{x}_{m-1}(T, \overset{(n)}{x_0}); \overset{(n)}{x}_{m-1}(t_1, \overset{(n)}{x_0}), \overset{(n)}{x}_{m-1}(t_2, \overset{(n)}{x_0}), ...)$$

$$- (\sum_{i=0}^{\infty} \overset{(n)}{A_i} + \overset{(n)}{C})\overset{(n)}{x_0} - \sum_{i=1}^{\infty} \overset{(n)}{A_i} \int\limits_0^{t_i} [\overset{(n)}{f}(\tau, \overset{(n)}{x}_{m-1}(\tau, \overset{(n)}{x_0}))$$

$$- \frac{1}{T}\int\limits_0^T \overset{(n)}{f}(s, \overset{(n)}{x}_{m-1}(s, \overset{(n)}{x_0}))ds]d\tau\}, \quad m \in Z^+, \quad \overset{(n)}{x}_0(t, \overset{(n)}{x}_0) \equiv \overset{(n)}{x_0} \in \tilde{D}_{\underset{\overset{(n)}{\beta}\varphi}{(n)}}^{(n)},$$

converges uniformly in $\overset{(n)}{x_0} \in \tilde{D}^{(n)}_{\underset{\beta\ \varphi}{(n)}}$, $t \in [0, T]$ to the function $\overset{(n)}{x}(t, \overset{(n)}{x_0})$, which satisfies the equality

$$
\overset{(n)}{x}(t, \overset{(n)}{x_0}) = \overset{(n)}{x_0} + \int_0^t [\overset{(n)}{f}(\tau, \overset{(n)}{x}(\tau, \overset{(n)}{x_0})) - \frac{1}{T}\int_0^T \overset{(n)}{f}(s, \overset{(n)}{x}(s, \overset{(n)}{x_0}))ds]d\tau
$$

$$
+ \frac{t}{T} \overset{(n)}{H}\{\overset{(n)}{\varphi}(\overset{(n)}{x}(0, \overset{(n)}{x_0}), \overset{(n)}{x}(T, \overset{(n)}{x_0}); \overset{(n)}{x}(t_1, \overset{(n)}{x_0}), \overset{(n)}{x}(t_2, \overset{(n)}{x_0}), ...)
$$

$$
- (\sum_{i=0}^{\infty} \overset{(n)}{A_i} + \overset{(n)}{C})\overset{(n)}{x_0} - \sum_{i=1}^{\infty} \overset{(n)}{A_i} \int_0^{t_i} [\overset{(n)}{f}(\tau, \overset{(n)}{x}(\tau, \overset{(n)}{x_0}))
$$

$$
- \frac{1}{T}\int_0^T \overset{(n)}{f}(s, \overset{(n)}{x}(s, \overset{(n)}{x_0}))ds]d\tau\}.
$$

For any natural m, we have

$$
\|\overset{(n)}{x}_m(t, \overset{(n)}{x}_0) - \overset{(n)}{x}(t, \overset{(n)}{x}_0)\| \le \frac{(\overset{(n)}{Q}_\varphi)^m}{1 - \overset{(n)}{Q}_\varphi} \overset{(n)}{\beta_\varphi}(\overset{(n)}{x_0}).
$$

In this case, $\overset{(n)}{x}(t, \overset{(n)}{x}_0)$ is a solution of the equation

$$
\frac{d\overset{(n)}{x}}{dt} = \overset{(n)}{f}(t, \overset{(n)}{x}) + \overset{(n)}{\mu}, \tag{4.48}
$$

where

$$
\overset{(n)}{\mu} = \overset{(n)}{\Delta}_\varphi(\overset{(n)}{x_0}) = \frac{1}{T}\overset{(n)}{H}\{\overset{(n)}{\varphi}(\overset{(n)}{x}(0, \overset{(n)}{x_0}), \overset{(n)}{x}(T, \overset{(n)}{x_0}); \overset{(n)}{x}(t_1, \overset{(n)}{x_0}), \overset{(n)}{x}(t_2, \overset{(n)}{x_0}), ...)
$$

$$
- (\sum_{i=0}^{\infty} \overset{(n)}{A_i} + \overset{(n)}{C})\overset{(n)}{x_0} - \sum_{i=1}^{\infty} \overset{(n)}{A_i} \int_0^{t_i} [\overset{(n)}{f}(\tau, \overset{(n)}{x}(\tau, \overset{(n)}{x_0}))
$$

$$
- \frac{1}{T}\int_0^T \overset{(n)}{f}(s, \overset{(n)}{x}(s, \overset{(n)}{x_0}))ds]d\tau\} - \frac{1}{T}\int_0^T \overset{(n)}{f}(\tau, \overset{(n)}{x}(\tau, \overset{(n)}{x_0}))d\tau,
$$

satisfies the boundary conditions (4.47) and, if $\overset{(n)}{\Delta}_\varphi(\overset{(n)}{x}_0) = 0$, is a solution of the boundary-value problem (4.39), (4.47).

We sharpen inequality c_3, by replacing its condition c_3^0:

$$
c_3^0) \qquad \frac{KT}{2}[1 + \|\overset{(n)}{H_p}\| \sum_{i=1}^{p} \|\overset{(n)}{A_i}\|] + K_{\varphi p}\|\overset{(n)}{H_p}\| \le q_0 < 1.
$$

Lemma 4.2. *Let conditions (4.3), (4.24), and (4.35) be satisfied, and let the set $D^*_{\beta\varphi}$ be nonempty. In addition, let conditions a_3 and c_3^0 be satisfied $\forall n \geq n_0, p \geq p_0$. Then conditions a_4, b_4, c_4 hold, and, $\forall x_0 \in D^*_{\beta\varphi}, n \geq n_0$,*

$$\lim_{p\to\infty} \overset{(n)}{x}_p(t, \overset{(n)}{x_0}) = \overset{(n)}{x}(t, \overset{(n)}{x_0}), \quad \lim_{p\to\infty} \overset{(n)}{\mu}_p = \overset{(n)}{\mu}$$

in the meaning of the norm.

Proof. It is easy to verify that the inclusion $x_0 \in D^*_{\beta\varphi}$ yields the inclusion $\overset{(n)}{x_0} \in \tilde{D}^{(n)}_{\overset{(n)}{\beta}\overset{(n)}{\varphi}} \ \forall n \geq n_0$. In addition, the inequalities

$$\left\| \sum_{i=1}^{\infty} \frac{t_i}{T} \overset{(n)}{A_i} + \overset{(n)}{C} - \overset{(n)}{H}_p^{-1} \right\| \leq \sum_{i=p+1}^{\infty} \frac{t_i}{T} \|\overset{(n)}{A_i}\| \leq \sum_{i=p+1}^{\infty} \|A_i\|$$

and the estimate

$$\|\overset{(n)}{H}_p\| \leq \frac{\frac{2}{KT} - 1}{\sum_{i=1}^{p_0} \| \overset{(n_0)}{A_i} \| + \frac{2K_\varphi}{KT}} \quad (n \geq n_0, \ p \geq p_0)$$

guarantee the existence of the matrix $\overset{(n)}{H} \ (n \geq n_0)$ and the boundedness of its norm by the same constant. Hence, $\| \overset{(n)}{H} - \overset{(n)}{H}_p \| \leq \sigma(p) \to 0$ as $p \to \infty$ $(p \geq p_0)$, and condition a_4 is satisfied. Condition c_3^0 guarantees the validity of condition c_4.

Denoting $\sup_{t\in[0,T]} \| \overset{(n)}{x}(t, \overset{(n)}{x_0})) - \overset{(n)}{x}_p(t, \overset{(n)}{x_0})) \|$ by $\xi^*(p, n)$ and executing the analysis analogous to the proof of Theorem 4.5, we obtain the estimates

$$\xi^*(p, n) \leq \Gamma_1^* + \Gamma_2^* + \Gamma_3^*;$$

$$\Gamma_1^* \leq \frac{KT}{2} \xi^*(p, n);$$

$$\Gamma_2^* \leq K^{(1)} \| \overset{(n)}{H} - \overset{(n)}{H}_p \| + \| \overset{(n)}{H} \| K_\varphi \xi^*(p, n)$$

$$+ M_0 \| \overset{(n)}{H} \| \{ \sum_{i=p+1}^{\infty} \|A_i\| + \delta_0(p + 2) \};$$

$$\Gamma_3^* \leq \| \overset{(n)}{H} \| \frac{TK}{2} \sum_{i=1}^{\infty} \|A_i\| \xi^*(p, n) + \| \overset{(n)}{H} \| \frac{TM}{2} \sum_{i=p+1}^{\infty} \|A_i\|$$

$$+ \| \overset{(n)}{H} - \overset{(n)}{H}_p \| \frac{TM}{2} \sum_{i=1}^{\infty} \|A_i\|,$$

which yields

$$\xi^*(p,n)(1 - \overset{(n)}{Q_\varphi}) \le K^{(1)}\|\overset{(n)}{H} - \overset{(n)}{H_p}\| + M_0\|\overset{(n)}{H}\|\{\sum_{i=p+1}^{\infty}\|A_i\| + \delta_0(p+2)\}$$

$$+ \|\overset{(n)}{H}\|\frac{TM}{2}\sum_{i=p+1}^{\infty}\|A_i\| + \|\overset{(n)}{H} - \overset{(n)}{H_p}\|\frac{TM}{2}\sum_{i=1}^{\infty}\|A_i\|$$

or

$$\xi^*(p,n) \le \frac{\eta^*(p,n)}{1 - \overset{(n)}{Q_\varphi}}, \quad \eta^*(p,n) \to 0 \quad \text{as} \quad p \to \infty.$$

Hence, $\forall n \ge n_0$, the sequence $\{\overset{(n)}{x_p}(t, \overset{(n)}{x_0})\}_{p=p_0}^{\infty}$ converges to the function $\overset{(n)}{x}(t, \overset{(n)}{x_0})$ as $p \to \infty$. The relation

$$\|\overset{(n)}{\mu} - \overset{(n)}{\mu_p}\| \le K\xi^*(p,n) + \Gamma_2^* + \Gamma_3^* \to 0 \quad \text{as} \quad p \to \infty$$

completes the proof of the lemma. $\qquad\qquad\qquad\qquad\qquad\qquad\qquad\square$

Theorem 4.6. *Let $f(t,x) \in \hat{C}_{lip}(x)$, $\varphi(\psi) \in \hat{C}_{lip}(\psi)$, let the set $D_{\beta\varphi}^*$ be nonempty, and let, $\forall\, n \ge n_0, p \ge p_0$, conditions a_1, a_3, c_1^0, and c_3^0, in which $K_{\varphi p} = K_\varphi$, be satisfied. Then, $\forall x_0 \in D_{\beta\varphi}^*$,*

$$x^*(t, x_0) = \lim_{n\to\infty}\left(\lim_{p\to\infty}\overset{(n)}{x_p}(t, \overset{(n)}{x_0})\right), \ \mu = \lim_{n\to\infty}\left(\lim_{p\to\infty}\overset{(n)}{\mu_p}\right),$$

where the convergence in n is coordinatewise, and the convergence in p is in the norm of the space $\mathfrak{M}$. In this case, there exists no other value of $\mu \in \mathfrak{M}$ such that the solution of the equation $\frac{dx}{dt} = f(t,x) + \mu$ with the initial condition $x(0) = x_0$ would satisfy the boundary condition (4.23).

Proof. Let us estimate the controlling parameter $\overset{(n)}{\mu}$ in Eq. (4.48) for $n \ge n_0$:

$$\overset{(n)}{\mu} \le \frac{\|\overset{(n)}{H}\|}{T}\left(M_\varphi + (\sum_{i=0}^{\infty}\|\overset{(n)}{A_i}\| + \|\overset{(n)}{C}\|)\|\overset{(n)}{x_0}\| + \frac{MT}{2}\sum_{i=1}^{\infty}\|\overset{(n)}{A_i}\|\right) + M$$

$$\le \frac{\frac{2}{KT} - 1}{\sum_{i=1}^{p_0}\|\overset{(n_0)}{A_i}\| + \frac{2K_\varphi}{KT}}(M_\varphi + (\sum_{i=0}^{\infty}\|A_i\| + \|C\|)\|x_0\|$$

$$+ \frac{MT}{2}\sum_{i=1}^{\infty}\|A_i\|) + M = const < \infty, \quad x_0 \in D_{\beta\varphi}^*.$$

Hence, the sequence $\{\overset{(n)}{\mu}\}_{n=n_0}^{\infty}$ is uniformly bounded in the norm of the space $\mathfrak{M}$. The sequence $\{\overset{(n)}{x}(t,\overset{(n)}{x_0})\}_{n=n_0}^{\infty}$ is uniformly bounded by a constant M_0 and is equicontinuous on the interval $[0,T]$, which follows from the inequality

$$\|\overset{(n)}{x}(t^{(2)},\overset{(n)}{x_0}) - \overset{(n)}{x}(t^{(1)},\overset{(n)}{x_0})\| \leq (t^{(2)} - t^{(1)})M_0'',$$

where $\{t^{(1)}, t^{(2)}\} \subset [0,T]$, $t^{(2)} > t^{(1)}$, and

$$M_0'' = 2M + \frac{\frac{2}{KT} - 1}{T\sum_{i=1}^{p_0} \|\overset{(n_0)}{A_i}\| + \frac{2K_\varphi}{K}}$$

$$\times (M_\varphi + (\sum_{i=0}^{\infty} \|A_i\| + \|C\|)\|x_0\| + \frac{MT}{2}\sum_{i=1}^{\infty} \|A_i\|).$$

This allows us to perform the consideration analogous to the proof of Lemma 4.1 and to substantiate the validity of the relations

$$\lim_{n\to\infty} \overset{(n)}{x}(t,\overset{(n)}{x_0}) = x^*(t,x_0) , \quad \lim_{n\to\infty} \overset{(n)}{\mu} = \mu ,$$

where the limiting transitions are considered in the coordinatewise meaning. In view of Lemma 4.2, we complete the proof of the theorem. $\qquad\square$

Corollary 4.5 (of Theorems 4.5 and 4.6). *Let $f(t,x) \in \hat{C}_{lip}(x)$, $\varphi(\psi) \in \hat{C}_{lip}(\psi)$, let the set $D_{\beta\varphi}^*$ be nonempty, and let, $\forall n \geq n_0, p \geq p_0$, conditions a_1, a_2, a_3, c_2^*, and c_3^0, in which $K_{\varphi p} = K_\varphi$ be satisfied. Then Theorems 4.5 and 4.6 hold, and the limiting transitions (4.42) and (4.43) are commutative with respect to n and p.*

The proof of the last proposition is easy.

Corollary 4.6. *Let $f(t,x) \in \hat{C}_{lip}(x)$, $\varphi(\psi) \in \hat{C}_{lip}(\psi)$, let the set $D_{\beta\varphi}^*$ be nonempty, and let, $\forall p \geq p_0$, the inequality*

$$\|\sum_{i=1}^{p} \frac{t_i}{T}A_i + C - E\| \leq q^* = const < 1, \tag{4.49}$$

where E is an infinite identity matrix, hold. If, in this case,

$$KT(1 + \frac{1}{1-q^*}(\sum_{i=1}^{\infty} \|A_i\| + \|C\|)) + \frac{K_\varphi}{1-q^*} \leq q_* = const < 1, \tag{4.50}$$

then all conditions of Theorems 4.5 and 4.6 are satisfied, and the limiting transitions (4.42) and (4.43) are commutative with respect to n and p.

Proof.   Estimate (4.49) yields the inequalities

$$\left\|\sum_{i=1}^{\infty}\frac{t_i}{T}A_i + C - E\right\| \le q^*, \quad \left\|\sum_{i=1}^{p}\frac{t_i}{T}\overset{(n)}{A}_i + \overset{(n)}{C} - \overset{(n)}{E}\right\| \le q^*,$$

which implies that, $\forall n \in Z^+, p \ge p_0$, the matrices

$$H = \sum_{k=0}^{\infty}\left(E - \sum_{i=1}^{\infty}\frac{t_i}{T}A_i - C\right)^k, \quad H_p = \sum_{k=0}^{\infty}\left(E - \sum_{i=1}^{p}\frac{t_i}{T}A_i - C\right)^k,$$

$$\overset{(n)}{H}_p = \sum_{k=0}^{\infty}\left(\overset{(n)}{E} - \sum_{i=1}^{p}\frac{t_i}{T}\overset{(n)}{A}_i - \overset{(n)}{C}\right)^k,$$

exist, and their norms do not exceed $1/(1 - q^*)$. Then inequality (4.50) yields conditions c_2^* and c_3^0, where we set $K_{\varphi p} = K_\varphi$, which completes the proof. $\qquad\square$

Example 4.2. Consider the system of equations

$$\frac{dx_n}{dt} = \frac{1 - 2t}{2^{n+2}}x_{n+1}^2, \ n \in Z^+, \tag{4.51}$$

and the boundary condition

$$A_0 x(0) + \sum_{i=1}^{\infty}A_i x(t_i) + C x(T) = d, \tag{4.52}$$

where $T = 1$, $D_0 = [0, 1] \times D = [0, 1] \times \{x \in \mathfrak{M}|\ \|x\| \le 2\}$, $d = \{d_1, d_2, d_3, ...\} \in \mathfrak{M}$ is a constant vector, $A_0 = -E$, $C = E$, E is the infinite identity matrix,

$$d_i = \frac{1}{2\cdot 4^{i+1}}e^{\frac{1}{2\cdot 4^{i+1}}}, \ t_i = \frac{1}{2} + \sqrt{\frac{1}{4} - \frac{1}{2^{i+1}}}, \ i = 1, 2, ...,$$

and the elements of the matrices $A_i = [a_{jk}^{(i)}]_{j,k=1}^{\infty}$ $(i = 1, 2, ...)$ are defined by the relation

$$a_{jk}^{(i)} = \begin{cases} \frac{1}{2\cdot 4^{i+1}}, & \text{if } j = k = i, \\ 0, & \text{if } (j - i)^2 + (k - i)^2 \ne 0. \end{cases}$$

It is easy to verify that:

1) inequalities (4.3), in which we set $M_\varphi = \|d\| = \frac{e^{\frac{1}{32}}}{32}$ and $K_\varphi = 0$, are satisfied;

2) on the set D_0, inequalities (4.24), in which we set $K = \frac{1}{2}$ and $M = \frac{1}{2}$, are satisfied;

3) $\|C\| = 1$, $\|A_i\| \le \frac{1}{32}$ $(i = 1, 2, ...)$;

4) the series $\sum\limits_{i=1}^{\infty} \|A_i\|$ converges to the number $\frac{1}{2}\sum\limits_{i=1}^{\infty}\frac{1}{4^{i+1}} = \frac{1}{24}$;

5) the matrix $\sum\limits_{i=1}^{\infty}\frac{t_i}{T}A_i + C$ is invertible, and the elements of the matrix $H = [h_{jk}]_{j,k=1}^{\infty}$ inverse to it take the form

$$h_{jk} = \begin{cases} (1 + \frac{1}{2\cdot 4^{j+1}}(\frac{1}{2} + \sqrt{\frac{1}{4} - \frac{1}{2^{j+1}}}))^{-1} & \text{for} \quad j = k, \\ 0 & \text{for} \quad j \neq k, \end{cases}$$

and $\|H\| \leq 1$;

6) $Q_\varphi = \frac{KT}{2}[1 + \|H\|\sum\limits_{i=1}^{\infty}\|A_i\|] = \frac{1}{4}[1 + \frac{1}{24}] = \frac{25}{96} < 1.$

We now show that the set $D_{\beta\varphi}$ is not empty. For this purpose, it is sufficient that there exist at least one point $x_0 \in D$, for which $\|x_0\| + \beta_\varphi(x_0) \leq 2$. The inequality

$$\|x_0\| + \beta_\varphi(x_0) \leq \|x_0\| + \frac{T}{2}M + \|H\|(\|d\|$$

$$+ \|\sum_{i=0}^{\infty}A_i + C\|\|x_0\|) + \|H\|\sum_{i=1}^{\infty}\|A_i\|\alpha_1(t_i)M$$

$$\leq \|x_0\| + \frac{1}{4} + (\frac{1}{32}e^{\frac{1}{32}} + \frac{1}{32}\|x_0\|) + 1\cdot\frac{1}{24}\cdot\frac{1}{2}\cdot\frac{1}{2} = \frac{33}{32}\|x_0\| + \frac{25}{96} + \frac{1}{32}e^{\frac{1}{32}} \leq 2$$

implies that any point $x_0 \in D$ such that

$$\frac{33}{32}\|x_0\| + \frac{25}{96} + \frac{1}{32}e^{\frac{1}{32}} \leq 2$$

belongs to the set D together with its β_φ-neighborhood. The simple calculation indicates that every point of the set

$$D_\gamma = \{x_0 \in D \mid \|x_0\| \leq 1.6556\}$$

has such property. Thus, on the Cartesian product $[0, 1] \times D_\gamma$, all conditions of Theorem 4.4 are satisfied.

By a simple substitution in Eq. (4.51), it is easy to verify that the function $x^*(t) = \{x_1^*(t), x_2^*(t), ...\}$, where

$$x_n^*(t) = \exp(\frac{t - t^2}{2^{n+2}}), \quad n = 1, 2, ..., \tag{4.53}$$

is its solution defined on the whole number axis. At $t = 0$, this solution takes a value $x_0 = \{1, 1, 1, ...\} \in D_\gamma$. In this case, it satisfies the boundary condition (4.52), i.e., is a solution of the boundary-value problem (4.51), (4.52).

However, condition c_1^0 is not satisfied. Therefore, we cannot assert that the obtained solution coincides with the function $x^*(t, x_0)$ found as the limit of successive approximations, whose existence is guaranteed by Theorem 4.4.

It is easy to verify that function (4.53) is a solution of the system of equations

$$\frac{dx_n}{dt} = \frac{1 - 2t}{2^{n+2}} x_n, \quad n = 1, 2, \ldots, \tag{4.54}$$

which is defined on the whole number axis, i.e., a solution of the boundary-value problem (4.54), (4.52). We consider this problem in more details, by setting $D_1 = \{x \in \mathfrak{M} \mid \|x\| \le 4\}$, $D_0^1 = [0, 1] \times D_1$. In this case, conditions $1 - 5$, where we set $K = \frac{1}{4}$ and $M = \frac{1}{2}$, are satisfied, the inequality $Q_\varphi < 1$ holds, and every point from the set $D_\gamma^* = \{x_0 \in D_1 \mid \|x_0\| \le 3.5949\}$ belongs to the set D_1 together with its β_φ-neighborhood. Thus, on the Cartesian product $[0, 1] \times D_\gamma^*$, all conditions of Theorem 4.4 are satisfied, and the point $x_0 = \{1, 1, 1, \ldots\} \in D_\gamma^*$. As distinct from the previous problem, condition c_1^0 is satisfied in this case:

$$KT\{1 + \sum_{i=1}^{\infty} \|H\|\|A_i\| + \|H\|\|C\|\} = \frac{1}{4}\{1 + \frac{1}{24} + 1\} = \frac{49}{96} < 1.$$

Hence, we may assert that function (4.53) on the interval $[0, 1]$ coincides with the function $x^*(t, x_0)$, $x_0 = \{1, 1, 1, \ldots\}$, which is obtained as the limit of successive approximations constructed in Theorem 4.4 for the boundary-value problem (4.54), (4.52).

We now verify whether, in this case, the limiting transitions (4.42) and (4.43) are satisfied. For this purpose, it is simple to verify whether the conditions of Corollary 4.6 are satisfied.

The function $f(t, x)$ entering the right-hand side of equality (4.54) belongs to the space $\hat{C}_{Lip}(x)$ on D_0^1, since the inequalities

$$\|f(t, x) - f(t, \bar{x})\| \le \sup_{i \ge m+1} \{\frac{|1 - 2t|}{2^{i+2}}|x_i - \bar{x}_i|\} \le |1 - 2t|\frac{1}{2^{m+3}}\|x - \bar{x}\| \quad ,$$

where $\varepsilon(m) = \frac{1}{2^{m+3}} \to 0$ as $m \to \infty$, hold $\forall t \in [0, 1]$ for any $\{x, \bar{x}\} \subset D_1$, whose m first coordinates coincide.

For all natural p, inequality (4.49)

$$\|\sum_{i=1}^{p} \frac{t_i}{T} A_i + C - E\| \le \sum_{i=1}^{p} \|\frac{t_i}{T} A_i\| < \sum_{i=1}^{\infty} \|A_i\| = \frac{1}{24} = q^* < 1$$

holds.

Inequality (4.50), where we set $K_\varphi = 0$, is also satisfied:

$$KT(1 + \frac{1}{1-q^*}(\sum_{i=1}^{\infty} \|A_i\| + \|C\|)) = \frac{1}{4}(1 + \frac{24}{23}(\frac{1}{24} + 1)) = \frac{12}{23} = q_* < 1.$$

But a simple calculation shows that the set $D^*_{\beta\varphi}$ is empty for the problem under consideration. Hence, we can say nothing about the possibility to present function (4.53) in the form (4.42). This question requires the additional study.

Finally, we consider the system of equations (4.54) with the boundary condition (4.52), where we set $d = 0 \in \mathfrak{M}$. For this boundary-value problem, the set $D^*_{\beta\varphi}$ is not empty. It is easy to verify that every point from the set $D^1_\gamma = \{x \in D_1| \ \|x\| \le 0.00436\}$ belongs to the set D_1 together with its $\beta^*_\varphi(x, 1, 1)$- neighborhood. It is also easy to verify that, $\forall\{n, p\} \subset Z^+$, there exist the matrices

$$\overset{(n)}{H_p} = [\overset{(n)}{h_{pjk}}]^n_{j,k=1} = (\sum_{i=1}^{p} \frac{t_i}{T} \overset{(n)}{A_i} + \overset{(n)}{C})^{-1},$$

whose elements take the form

$$\overset{(n)}{h_{pjk}} = \begin{cases} (1 + \frac{1}{2\cdot 4^{j+1}}(\frac{1}{2} + \sqrt{\frac{1}{4} - \frac{1}{2^{j+1}}}))^{-1}, & \text{if } \ j = k; \\ 0, & \text{if } \ j \ne k \end{cases}$$

for $p \ge n$ and

$$\overset{(n)}{h_{pjk}} = \begin{cases} (1 + \frac{1}{2\cdot 4^{j+1}}(\frac{1}{2} + \sqrt{\frac{1}{4} - \frac{1}{2^{j+1}}}))^{-1}, & \text{if } \ j = k \le p; \\ 1, & \text{if } \ p < j = k \le n \end{cases}$$

for $p < n$. All these matrices are bounded in the norm by 1. Thus, all conditions of Corollary 4.6 and, hence, all conditions of Theorems 4.5 and 4.6 are satisfied.

4.5　Boundary-value problems for differential equations unsolvable with respect to the derivative

Consider the equation

$$\frac{dx}{dt} = f(t, x, \frac{dx}{dt}), \tag{4.55}$$

whose right-hand side is defined on the set $(t, x, x_1) \in \overline{D}_1 = [0, +\infty) \times D \times D_1$, where $D_1 = \{x_1 \ |x_1 \in \mathfrak{M}, \ \|x_1\| \le M_1 = const > 0\}$, and the set D is defined in Subsection 4.1.

The following conditions are called conditions $(\mathbf{A}_1)$:

1) $\forall \{\psi, \psi_*\} \subset D^\infty$, inequalities (4.3) are satisfied;

2) the function $f(t, x, x_1)$ is continuous on $\overline{D}_1$ in (t, x, x_1), and there exists a function $h(t)$ with properties 1*, 2* (see Subsection 4.1) such that, for all $\{x, x'\} \subset D$, $\{x_1, x_1'\} \subset D_1$, the inequalities

$$\|f(t, x, x_1)\| \le M_h h'(t),$$

$$\|f(t, x, x_1) - f(t, x', x_1')\| \le [K_h \|x - x'\| + K_{1h}\|x_1 - x_1'\|]h'(t) \qquad (4.56)$$

hold. Here, M_h, K_h, and K_{1h} are positive constants, which are independent of the choice of the points (t, x, x_1) and (t, x', x_1') from the set $\overline{D}_1$;

3) the matrix $\sum\limits_{i=1}^{\infty} h(t_i) A_i$ satisfies the third of conditions $(\mathbf{A})$.

Retaining the content of the designations H_h, $\beta_{1\varphi h}(x_0)$, $\beta_{\varphi h}(x_0)$, $D_{\beta\varphi h}$ from section 4.1, we denote the expression $2M_h\overline{h'} + \frac{h'}{T}\beta_{1\varphi h}(x_0)$ by $\gamma_{\varphi h}(x_0)$ and the set of elements $x_0 \in D_1$, which enter D_1 together with their $\gamma_{\varphi h}(x_0)$-neighborhoods, by $D_{1\gamma\varphi h}$.

The following conditions are called conditions $(\mathbf{B}_1)$:

a) the intersection $D_{\beta\gamma\varphi h}$ of the sets $D_{\beta\varphi h}$ and $D_{1\gamma\varphi h}$ is nonempty;

b) the matrix norm

$$Q_0 =$$

$$\begin{bmatrix} \frac{K_h T}{2}\left(1 + \sum\limits_{i=1}^{\infty} \|H_h A_i\|\right) + K_\varphi\|H_h\| & \frac{K_{1h}T}{2}\left(1 + \sum\limits_{i=1}^{\infty} \|H_h A_i\|\right) \\ K_h\overline{h'}\left(2 + \frac{1}{2}\sum\limits_{i=1}^{\infty} \|H_h A_i\|\right) + \frac{h'}{T}K_\varphi\|H_h\| & K_{1h}\overline{h'}\left(2 + \frac{1}{2}\sum\limits_{i=1}^{\infty} \|H_h A_i\|\right) \end{bmatrix}$$

is less than 1.

It is easy to see that condition a) guarantees that the zero element of the set $\mathfrak{M}$ belongs to set $D_{\beta\gamma\varphi h}$.

Consider the recurrence sequence of vector-functions $\{x_m(t, x_0)\}_{m=0}^{\infty}$ formally denoted as

$$x_0(t, x_0) = x_0 = (x_{01}, x_{02}, ...) \in D_{\beta\gamma\varphi h};$$

$$x_m(t, x_0) = x_0 + \int_0^t [f(\tau, x_{m-1}(\tau, x_0), \frac{dx_{m-1}(\tau, x_0)}{d\tau})$$

$$- \frac{h'(\tau)}{T} \int_0^\infty f(s, x_{m-1}(s, x_0), \frac{dx_{m-1}(s, x_0)}{ds})ds]d\tau$$

$$+ \frac{h(t)}{T} H_h \{\varphi(x_{m-1}(0); x_{m-1}(t_1), x_{m-1}(t_2), ...) - \sum_{i=0}^{\infty} A_i x_0$$

$$- \sum_{i=1}^{\infty} A_i \int_0^{t_i} [f(\tau, x_{m-1}(\tau, x_0), \frac{dx_{m-1}(\tau, x_0)}{d}\tau)$$

$$- \frac{h'(\tau)}{T} \int_0^{\infty} f(s, x_{m-1}(s, x_0), \frac{dx_{m-1}(s, x_0)}{d}s)ds]d\tau\}, \quad m \in Z^+. \quad (4.57)$$

The sufficient conditions for the existence of such sequence are the continuity of the functions $x_i(t, x_0)$ and $\frac{dx_i(t, x_0)}{dt}$ in $t \geq 0$ in the norm and their membership to the sets D and D_1, respectively, for all $i \in Z^+$. In this case, the first of inequalities (4.56) and properties 1^*, 2^* of the function $h(t)$ ensure the coordinatewise convergence of the improper integral $\int_0^{\infty} f(\tau, x_{m-1}(\tau, x_0), \frac{dx_{m-1}(\tau, x_0)}{d\tau})d\tau$.

Theorem 4.7. *Let conditions* $(\mathbf{A_1})$ *and* $(\mathbf{B_1})$ *be satisfied. Then*

1) as $m \to \infty$, *the sequence* $\{x_m(t, x_0)\}_{m=0}^{\infty}$ *defined by equalities (4.57) converges uniformly in* $(t, x_0) \in [0, +\infty) \times D_{\beta\gamma\varphi h}$ *to a function* $x^*(t, x_0)$ *differentiable with respect to* t, *and the sequence* $\{\frac{dx_m(t, x_0)}{dt}\}_{m=0}^{\infty}$ *converges to a function* $\frac{dx^*(t, x_0)}{dt}$, *and,* $\forall m \in Z^+$,

$$\sup_{t \in [0, +\infty)} \{\|x_m(t, x_0) - x^*(t, x_0)\| , \|\frac{dx_m(t, x_0)}{dt} - \frac{dx^*(t, x_0)}{dt}\|\}$$

$$\leq \|Q_0^m (E_2 - Q_0)^{-1} Z_1^0\|, \quad (4.58)$$

where E_2 *is the* 2×2 *identity matrix,* $Z_1^0 = \begin{bmatrix} \beta_{\varphi h}(x_0) \\ \gamma_{\varphi h}(x_0) \end{bmatrix}$;

2) the function $x^*(t, x_0)$ *satisfies the boundary condition (4.2) and is a solution of the equation*

$$\frac{dx}{dt} = f(t, x, \frac{dx}{dt}) + \mu \cdot h'(t), \quad (4.59)$$

where

$$\mu = \Delta_{\varphi h}^0(x_0) = \frac{1}{T} H_h \{\varphi(x^*(0); x^*(t_1), x^*(t_2), ...) - \sum_{i=0}^{\infty} A_i x_0$$

$$- \sum_{i=1}^{\infty} A_i \int_0^{t_i} [f(\tau, x^*(\tau, x_0), \frac{dx^*(\tau, x_0)}{d\tau})$$

$$-\frac{h'(\tau)}{T}\int_0^\infty f(s,x^*(s,x_0),\frac{dx^*(s,x_0)}{ds})ds]d\tau\}$$

$$-\frac{1}{T}\int_0^\infty f(\tau,x^*(\tau,x_0),\frac{dx^*(\tau,x_0)}{d\tau})d\tau; \quad (4.60)$$

3) if

$$\Delta^0_{\varphi h}(x_0) = 0, \qquad (4.61)$$

then the function $x^*(t,x_0)$ *is a solution of the boundary-value problem* *(4.55), (4.2).*

Proof. Proof of this theorem is analogous to that of Theorem 4.1. For $m = 1$, relation (4.57) yields the equality

$$x_1(t,x_0) = x_0 + \int_0^t [f(\tau,x_0,0) - \frac{h'(\tau)}{T}\int_0^\infty f(s,x_0,0)ds]d\tau$$

$$+ \frac{h(t)}{T}H_h\{\varphi(x_0;x_0,x_0,...)\}$$

$$- \sum_{i=0}^\infty A_i x_0 - \sum_{i=1}^\infty A_i \int_0^{t_i} [f(\tau,x_0,0) - \frac{h'(\tau)}{T}\int_0^\infty f(s,x_0,0)ds]d\tau\}.$$

For any $i \in Z^+$, the function $f_i(s,x_0,0)$ is continuous in s on the interval $[0,+\infty)$, and the first estimate in (4.56) ensures the convergence of the improper integral $\int_0^\infty f_i(s,x_0,0)ds$. For the difference $x_1(t'',x_0) - x_1(t',x_0)$, the estimate obtained in the proof of Theorem 4.1 is valid, i.e., the function $x_1(t,x_0)$ is continuous in the norm on the interval $[0,+\infty)$. We now prove that it is continuously differentiable on this interval. Indeed, executing the coordinatewise differentiation of this function, we obtain the equality

$$\frac{dx_1(t,x_0)}{dt}$$

$$= f(t,x_0,0) - \frac{h'(t)}{T}\int_0^\infty f(s,x_0,0)ds + \frac{h'(t)}{T}H_h\{\varphi(x_0;x_0,x_0,...)\}$$

$$- \sum_{i=0}^\infty A_i x_0 - \sum_{i=1}^\infty A_i \int_0^{t_i} [f(\tau,x_0,0) - \frac{h'(\tau)}{T}\int_0^\infty f(s,x_0,0)ds]d\tau\}.$$

Using this equality, it is easy to verify the continuity of $\frac{dx_1(t,x_0)}{dt}$ in t in the meaning of the norm on the interval $[0, +\infty)$. It is also easy to prove the validity of the inequalities

$$\|x_1(t, x_0) - x_0\| \le \beta_{\varphi h}(x_0), \quad \|\frac{dx_1(t, x_0)}{dt}\| \le \gamma_{\varphi h}(x_0).$$

In other words, the function $x_1(t, x_0)$ takes values from the set $D \ \forall \ t \ge 0$ and $x_0 \in D_{1\beta\gamma\varphi h}$, and the values of its derivative $\frac{dx_1(t,x_0)}{dt}$ belong to the set D_1. The first inequality is substantiated in the proof of Theorem 4.1, and the second follows from the estimates

$$\|\frac{dx_1(t, x_0)}{dt}\| \le \|f(t, x_0, 0)\| + \frac{\bar{h}'}{T} \int_0^\infty \|f(s, x_0, 0)\| ds$$

$$+ \frac{h'(t)}{T}\{\|H_h\|\|\varphi(x_0; x_0, x_0, ...) - \sum_{i=0}^\infty A_i x_0\|$$

$$+ \sum_{i=1}^\infty \|H_h A_i\|\|\int_0^{t_i} [f(\tau, x_0, 0) - \frac{h'(\tau)}{T} \int_0^\infty f(s, x_0, 0)ds]d\tau\|\}$$

$$\le 2M_h\bar{h}' + \frac{\bar{h}'}{T}\{\|H_h\|\|d - \sum_{i=0}^\infty A_i x_0\|$$

$$+ M_h \sum_{i=1}^\infty \|H_h A_i\|\alpha_{1h}(t_i) \le 2M_h\bar{h}' + \frac{\bar{h}'}{T}\beta_{1\varphi h}(x_0) = \gamma_{\varphi h}(x_0),$$

where d is the same vector, as that in Subsection 4.1.

Hence, $\forall \ t \ge 0$, $x_0 \in D_{\beta\gamma\varphi h}$, the function $x_1(t, x_0)$ and its derivative take values from the sets D and D_1, respectively.

Let us assume that, for $t \ge 0 \ \forall m \le k \in Z^+$, the functions $x_m(t, x_0)$ are continuous in the norm and take values in set D, and let their derivatives with respect to t take the form

$$\frac{dx_m(t, x_0)}{dt} = f(t, x_{m-1}(t, x_0), \frac{dx_{m-1}(t, x_0)}{dt})$$

$$- \frac{h'(t)}{T} \int_0^\infty f(s, x_{m-1}(s, x_0), \frac{dx_{m-1}(s, x_0)}{ds})ds$$

$$+ \frac{h'(t)}{T} H_h\{\varphi(x_{m-1}(0); x_{m-1}(t_1), x_{m-1}(t_2), ...) - \sum_{i=0}^\infty A_i x_0$$

$$-\sum_{i=1}^{\infty} A_i \int_0^{t_i} [f(\tau, x_{m-1}(\tau, x_0), \frac{dx_{m-1}(\tau, x_0)}{d\tau})$$

$$-\frac{h'(\tau)}{T} \int_0^{\infty} f(s, x_{m-1}(s, x_0), \frac{dx_{m-1}(s, x_0)}{ds})ds]d\tau\}, \quad (4.62)$$

be continuous in the norm, and take values in D_1. By performing the analogous reasoning, it is easy to establish that the function $x_{k+1}(t, x_0)$ has the same properties. By the principle of complete mathematical induction, the same properties are inherent in an arbitrary function $x_i(t, x_0)$, $i \in Z^+$.

Hence, equalities (4.57) define, indeed, the sequence $\{x_m(t, x_0)\}_{m=0}^{\infty}$, whose elements satisfy, as it can be verified with the help of the direct substitution, the recurrence boundary condition (4.10) for every $m \in Z^+$. We will show that this sequence and the sequence of derivatives $\{\frac{dx_m(t, x_0)}{dt}\}_{m=0}^{\infty}$ are convergent as $m \to \infty$ in the norm. For this purpose, it is sufficient to verify their fundamentality by virtue of the completeness of the space $\mathfrak{M}$.

Denoting the difference

$$f(t, x_m(t, x_0), \frac{dx_m(t, x_0)}{dt}) - f(t, x_{m-1}(t, x_0), \frac{dx_{m-1}(t, x_0)}{dt})$$

by $\overline{f}(t, x_0)$ and taking (4.3) and (4.57) into account, we obtain, $\forall\, m \in Z^+$, the inequalities

$$\|x_{m+1}(t, x_0) - x_m(t, x_0)\| \le \|\int_0^t [\overline{f}(\tau, x_0) - \frac{h'(\tau)}{T} \int_0^{\infty} \overline{f}(s, x_0)ds]d\tau\|$$

$$+ \frac{h(t)}{T}\|H_h\|\|\varphi(x_m(0); x_m(t_1), ...) - \varphi(x_{m-1}(0); x_{m-1}(t_1), ...)\|$$

$$+ \frac{h(t)}{T}\sum_{i=1}^{\infty}\|H_h A_i\|\|\int_0^{t_i} \overline{f}(\tau, x_0) - \frac{h'(\tau)}{T}\int_0^{\infty}\overline{f}(s, x_0)ds]d\tau\|$$

$$\le [K_h \sup_{t\in[0,+\infty)} \|x_m(t, x_0) - x_{m-1}(t, x_0)\|$$

$$+ K_{1h} \sup_{t\in[0,+\infty)} \|\frac{dx_m(t, x_0)}{dt} - \frac{dx_{m-1}(t, x_0)}{dt}\|]\alpha_{1h}(t)$$

$$+ \|H_h\|K_\varphi \sup_i\{ \|x_m(t_i, x_0) - x_{m-1}(t_i, x_0)\| \}$$

$$+ \sum_{i=1}^{\infty} \|H_h A_i\| \alpha_{1h}(t_i) [K_h \sup_{t \in [0,+\infty)} \|x_m(t_i, x_0) - x_{m-1}(t, x_0)\|$$

$$+ K_{1h} \sup_{t \in [0,+\infty)} \|\frac{dx_m(t, x_0)}{dt} - \frac{dx_{m-1}(t, x_0)}{dt}\|]. \quad (4.63)$$

Analogously, equality (4.62) yields the estimate

$$\|\frac{dx_{m+1}(t, x_0)}{dt} - \frac{dx_m(t, x_0)}{dt}\| \le [K_h \|x_m(t, x_0) - x_{m-1}(t, x_0)\|$$

$$+ K_{1h} \|\frac{dx_m(t, x_0)}{dt} - \frac{dx_{m-1}(t, x_0)}{dt}\|]h'(t)$$

$$+ \frac{h'(t)}{T}[K_h \sup_{t \in [0,+\infty)} \|x_m(t, x_0) - x_{m-1}(t, x_0)\|$$

$$+ K_{1h} \sup_{t \in [0,+\infty)} \|\frac{dx_m(t, x_0)}{dt} - \frac{dx_{m-1}(t, x_0)}{dt}\|]T$$

$$+ \frac{h'(t)}{T} K_\varphi \|H_h\| \sup_i \{ \|x_m(t_i, x_0) - x_{m-1}(t_i, x_0)\| \}$$

$$+ \frac{h'(t)}{T} \sum_{i=1}^{\infty} \|H_h A_i\| \alpha_{1h}(t_i) [K_h \sup_{t \in [0,+\infty)} \|x_m(t_i, x_0) - x_{m-1}(t_i, x_0)\|$$

$$+ K_{1h} \sup_{t \in [0,+\infty)} \|\frac{dx_m(t_i, x_0)}{dt} - \frac{dx_{m-1}(t_i, x_0)}{dt}\|]. \quad (4.64)$$

In view of the notation

$$\bar{r}_{m+1} = \sup_{t \in [0,+\infty)} \|x_{m+1}(t, x_0) - x_m(t, x_0)\|,$$

$$\bar{r}_{1m+1} = = \sup_{t \in [0,+\infty)} \|\frac{dx_{m+1}(t, x_0)}{dt} - \frac{dx_m(t, x_0)}{dt}\|,$$

inequalities (4.63) and (4.64) yield, $\forall m \in Z^+$, the relations

$$\bar{r}_{m+1} \le [K_h(\frac{T}{2} + \sum_{i=1}^{\infty} \|H_h A_i\| \alpha_{1h}(t_i))$$

$$+ K_\varphi \|H_h\|]\bar{r}_m + K_{1h}(\frac{T}{2} + \sum_{i=1}^{\infty} \|H_h A_i\| \alpha_{1h}(t_i))\bar{r}_{1m},$$

$$\bar{r}_{1m+1} \le [\overline{h'} K_h(2 + \frac{1}{T} \sum_{i=1}^{\infty} \|H_h A_i\| \alpha_{1h}(t_i)) + \frac{\overline{h'}}{T} K_\varphi \|H_h\|]\bar{r}_m$$

$$+ \overline{h'} K_{1h}(2 + \frac{1}{T} \sum_{i=1}^{\infty} \|H_h A_i\| \alpha_{1h}(t_i))\bar{r}_{1m},$$

which result in the coordinatewise inequality

$$\begin{bmatrix} \overline{r}_{m+1} \\ \overline{r}_{1m+1} \end{bmatrix} \leq Q_0 \begin{bmatrix} \overline{r}_m \\ \overline{r}_{1m} \end{bmatrix}, \quad m \in Z^+, \tag{4.65}$$

since $\alpha_{1h}(t_i) \leq \frac{T}{2} \ \forall i \in Z^+$.

We note that, in the coordinatewise meaning, $\begin{bmatrix} \overline{r}_1 \\ \overline{r}_{11} \end{bmatrix} \leq Z_1^0$. Taking (4.65) into account and performing the inductive transformations, we obtain the estimate

$$\sup_{t\in[0,+\infty)} \{\|x_{m+j}(t,x_0) - x_m(t,x_0)\|, \| \frac{dx_{m+j}(t,x_0)}{dt}$$

$$- \frac{dx_m(t,x_0)}{dt}\|\} \leq \|Q_0^m \sum_{i=0}^{j-1} Q_0^i Z_1^0\|. \tag{4.66}$$

Whence, by virtue of the second condition $(\mathbf{B}_1)$ uniformly in $t \in [0,+\infty)$, we obtain

$$\lim_{m\to\infty} x_m(t,x_0) = x^*(t,x_0), \quad \lim_{m\to\infty} \frac{dx_m(t,x_0)}{dt} = x_1^*(t,x_0),$$

where the limiting transitions are understood in the meaning of the norm of the space $\mathfrak{M}$.

Since $\|Q_0\| < 1$, there exists the matrix $(E_2 - Q_0)^{-1}$ inverse to the matrix $E_2 - Q_0$, and $(E_2 - Q_0)^{-1} = \sum_{i=0}^{\infty} Q_0^i$. Then it follows from (4.66) that inequality (4.58) holds.

The obvious inequality

$$\|f(t, x_m(t,x_0), \frac{dx_m(t,x_0)}{dt}) - f(t, x^*(t,x_0), x_1^*(t,x_0))\|$$

$$\leq K_h \overline{h'}\|x_m(t,x_0) - x^*(t,x_0)\| + K_{1h}\overline{h'}\|\frac{dx_m(t,x_0)}{dt} - x_1^*(t,x_0)\|$$

yields the uniform in $t \in [0,+\infty)$ convergence of the sequence

$$\{f(t, x_m(t,x_0), \frac{dx_m(t,x_0)}{dt})\}_{m=1}^{\infty}$$

in the norm to the function $f(t, x^*(t,x_0), x_1^*(t,x_0))$. In addition, the sequence

$$\{\frac{h'(t)}{T} \int_0^{\infty} f(t, x_m(t,x_0), \frac{dx_m(t,x_0)}{dt})dt\}_{m=1}^{\infty}$$

converges uniformly in $t \in [0, +\infty)$ to the function

$$\frac{h'(t)}{T} \int_0^\infty f(t, x^*(t, x_0), x_1^*(t, x_0))dt,$$

which follows from the relations

$$\left\| \frac{h'(t)}{T} \int_0^\infty f(t, x^*(t, x_0), x_1^*(t, x_0))dt \right.$$

$$\left. - \frac{h'(t)}{T} \int_0^\infty f(t, x_m(t, x_0), \frac{dx_m(t, x_0)}{dt})dt \right\|$$

$$\leq \overline{h'} \max\{K_h; K_{1h}\} \|Q_0\|^m \|(E_2 - Q_0)^{-1}\| \|Z_1^0\| = \|Q_0\|^m \gamma_1' \xrightarrow[m \to \infty]{} 0,$$

where $\gamma_1' = const > 0$ is a constant depending on x_0.

The obtained estimates give possibility to pass coordinatewise in (4.57) and (4.62) to the limit as $m \to \infty$ and to obtain the equalities

$$x^*(t, x_0)$$

$$= x_0 + \int_0^t [f(\tau, x^*(\tau, x_0), x_1^*(\tau, x_0)) - \frac{h'(\tau)}{T} \int_0^\infty f(s, x^*(s, x_0),$$

$$x_1^*(s, x_0))ds]d\tau + \frac{h(t)}{T} H_h\{\varphi(x^*(0); x^*(t_1), x^*(t_2), ...) - \sum_{i=0}^\infty A_i x_0$$

$$- \sum_{i=1}^\infty A_i \int_0^{t_i} [f(\tau, x^*(\tau, x_0), x_1^*(\tau, x_0))$$

$$- \frac{h'(\tau)}{T} \int_0^\infty f(s, x^*(s, x_0), \; x_1^*(s, x_0))ds]d\tau\},$$

$$x_1^*(t, x_0)$$

$$= f(t, x^*(t, x_0), x_1^*(t, x_0)) - \frac{h'(t)}{T} \int_0^\infty f(s, x^*(s, x_0), x_1^*(s, x_0))ds$$

$$+ \frac{h'(t)}{T} H_h\{\varphi(x^*(0); x^*(t_1), x^*(t_2), ...) - \sum_{i=0}^\infty A_i x_0$$

$$-\sum_{i=1}^{\infty} A_i \int_0^{t_i} [f(\tau, x^*(\tau, x_0), x_1^*(\tau, x_0))$$

$$-\frac{h'(\tau)}{T} \int_0^{\infty} f(s, x^*(s, x_0), x_1^*(s, x_0)) ds] d\tau \}.$$

The uniform in t coordinatewise convergence of the functional sequences $\{x_m(t, x_0)\}$ and $\{\frac{dx_m(t,x_0)}{dt}\}$ yields

$$x_1^*(t, x_0) \equiv \frac{dx^*(t, x_0)}{dt} \quad \forall x_0 \in D_{\beta\gamma\varphi h}, \quad t \in [0, +\infty).$$

The last inequality and relation (4.66) yield the validity of the first assertion of Theorem 4.7. Finally, like the proof of Theorem 4.1, it is easy to verify that the limiting transition as $m \to \infty$ can be realized in the recurrence conditions (4.10). This shows that the function $x^*(t, x_0)$ satisfies the boundary condition (4.2). This completes the proof of Theorem 4.7. $\square$

By Q_{01}, we denote the matrix

$$\begin{bmatrix} \dfrac{TK_h(1+\sum\limits_{i=1}^{\infty}\|H_h A_i\|)}{1-K_\varphi\|H_h\|} & \dfrac{TK_{1h}(1+\sum\limits_{i=1}^{\infty}\|H_h A_i\|)}{1-K_\varphi\|H_h\|} \\[2em] \dfrac{\overline{h'}K_h(1+\sum\limits_{i=1}^{\infty}\|H_h A_i\|)}{1-K_\varphi\|H_h\|} & \dfrac{\overline{h'}K_{1h}(1+\sum\limits_{i=1}^{\infty}\|H_h A_i\|)}{1-K_\varphi\|H_h\|} \end{bmatrix},$$

whose elements are positive numbers, since $\|Q_0\| < 1$, and

$$\|Q_{01}\| = \frac{(K_h + K_{1h})\max\{T, \overline{h'}\}}{1 - K_\varphi\|H_h\|}\left(1 + \sum_{i=1}^{\infty}\|H_h A_i\|\right).$$

Conditions $(\mathbf{B_1^0})$ are called conditions $(\mathbf{B_1})$, where the inequality $\|Q_0\| < 1$ is replaced by the estimate $\max\{\|Q_0\|, \|Q_{01}\|\} < 1$.

Corollary 4.7. *If the function $h(t)$ is chosen, then, under conditions $(\mathbf{A_1})$ and $(\mathbf{B_1^0})$, there exists no other value of μ such that the solution of Eq. (4.59) with the initial condition $x(0) = x_0 \in D_{\beta\gamma\varphi h}$ would satisfy the boundary condition (4.2).*

Proof. Let us assume that there exist $\{\mu_1, \mu_2\} \subset \mathfrak{M}$ and the functions $x_1(t, x_0)$ and $x_2(t, x_0)$ such that $x_1(0, x_0) = x_2(0, x_0) = x_0 \in D_{\beta\gamma\varphi h}$,

$$\frac{dx_i(t, x_0)}{dt} = f(t, x_i(t, x_0), \frac{dx_i(t, x_0)}{dt}) + \mu_i h'(t), \quad i \in \{1, 2\},$$

and these functions satisfy relation (4.2).

Then, for $j \in \{1, 2\}$, the following equalities hold:

$$x_j(t, x_0) = x_0 + \int_0^t f\left(\tau, x_j(\tau, x_0), \frac{dx_j(\tau, x_0)}{d\tau}\right)d\tau + \mu_j h(t),$$

$$A_0 x(0) + \sum_{i=1}^{\infty} A_i \left\{ x_0 + \int_0^{t_i} f\left(\tau, x_j(\tau, x_0), \frac{dx_j(\tau, x_0)}{d\tau}\right)d\tau + \mu_j h(t_i) \right\} = \tilde{\varphi}_j,$$

where

$$\tilde{\varphi}_j = \varphi\left(x_0; x_0 + \int_0^{t_1} f\left(\tau, x_j(\tau, x_0), \frac{dx_j(\tau, x_0)}{d\tau}\right)d\tau + \mu_j h(t_1),\right.$$

$$\left. x_0 + \int_0^{t_2} f\left(\tau, x_j(\tau, x_0), \frac{dx_j(\tau, x_0)}{d\tau}\right)d\tau + \mu_j h(t_2), ...\right).$$

Since both functions $x_1(t, x_0)$, $x_2(t, x_0)$ are bounded on the interval $[0, +\infty)$ in the norm, we introduce the notation

$$r = \sup_{t \in [0, +\infty)} \|x_1(t, x_0) - x_2(t, x_0)\|,$$

$$r_1 = \sup_{t \in [0, +\infty)} \left\| \frac{dx_1(t, x_0)}{dt} - \frac{dx_2(t, x_0)}{dt} \right\|.$$

The above-presented equalities yield the relations

$$\|x_1(t, x_0) - x_2(t, x_0)\| \leq \int_0^t [K_h r + K_{1h} r_1] h'(\tau) d\tau + \|\mu_1 - \mu_2\| h(t);$$

$$r \leq [K_h r + K_{1h} r_1 + \|\mu_1 - \mu_2\|] T; \tag{4.67}$$

$$\left\| \frac{dx_1(t, x_0)}{dt} - \frac{dx_2(t, x_0)}{dt} \right\| \leq (K_h r + K_{1h} r_1) \overline{h'} + \|\mu_1 - \mu_2\| h'(t);$$

$$r_1 \leq \{ K_h r + K_{1h} r_1 + \|\mu_1 - \mu_2\| \} \overline{h'}; \tag{4.68}$$

$$\sum_{i=1}^{\infty} A_i \left\{ \int_0^{t_i} \left(f\left(\tau, x_1(\tau), \frac{dx_1(\tau)}{d\tau}\right) - f\left(\tau, x_2(\tau), \frac{dx_2(\tau)}{d\tau}\right) \right) d\tau \right.$$

$$\left. + h(t_i)(\mu_1 - \mu_2) \right\} = \tilde{\varphi}_1 - \tilde{\varphi}_2,$$

which yield the inequalities

$$\|\mu_1 - \mu_2\| \le \frac{1}{T}\{\|H_h\| \cdot \|\tilde{\varphi}_1 - \tilde{\varphi}_2\|$$

$$+ \sum_{i=1}^{\infty} \|H_h A_i\| \int_0^{t_i} (K_h r + K_{1h} r_1) h'(\tau) d\tau\}$$

$$\le \{\frac{1}{T}\|H_h\| K_\varphi \sup_{i \in N}\{\int_0^{t_i} (K_h r + K_{1h} r_1) h'(\tau) d\tau$$

$$+ \|\mu_1 - \mu_2\| h(t_i)\} + \sum_{i=1}^{\infty} \|H_h A_i\|(K_h r + K_{1h} r_1) T\}$$

$$\le \|H_h\| K_\varphi (K_h r + K_{1h} r_1) + \|H_h\| K_\varphi \|\mu_1 - \mu_2\|$$

$$+ \sum_{i=1}^{\infty} \|H_h A_i\|(K_h r + K_{1h} r_1). \quad (4.69)$$

From $(4.67) - (4.69)$, we obtain the estimates

$$\|\mu_1 - \mu_2\| \le \frac{(K_\varphi \|H_h\| + \sum\limits_{i=1}^{\infty} \|H_h A_i\|)K_h r + K_{1h} r_1)}{1 - K_\varphi \|H_h\|};$$

$$r \le \Lambda T, \; r_1 \le \Lambda \overline{h'}, \quad (4.70)$$

where

$$\Lambda = K_h r + K_{1h} r_1 + \frac{K_h(\sum\limits_{i=1}^{\infty} \|H_h A_i\| + K_\varphi \|H_h\|)}{1 - K_\varphi \|H_h\|} r$$

$$+ \frac{K_{1h}(\sum\limits_{i=1}^{\infty} \|H_h A_i\| + K_\varphi \|H_h\|)}{1 - K_\varphi \|H_h\|} r_1.$$

Relations (4.70) yield the coordinatewise inequality

$$\begin{bmatrix} r \\ r_1 \end{bmatrix} \le Q_{01} \begin{bmatrix} r \\ r_1 \end{bmatrix}$$

and the estimate

$$\left\| \begin{bmatrix} r \\ r_1 \end{bmatrix} \right\| \le \|Q_{01}\| \left\| \begin{bmatrix} r \\ r_1 \end{bmatrix} \right\|.$$

Since $\|Q_{01}\| < 1$, we have $r = r_1 = 0$. Hence, $x_1(t, x_0) \equiv x_2(t, x_0)$ on $[0, +\infty)$, and $\mu_1 = \mu_2$. Corollary 4.7 is proved. $\qquad\square$

Like above, a function $\Delta^0_{\varphi h}(x_0)$ of the form (4.60) is called the exact defining function, its value μ at a fixed x_0 is a controlling parameter or the control, and Eq. (4.61) is called the exact defining equation. In addition to Eq. (4.61), we consider the approximate defining equation

$$\Delta^0_{\varphi h m}(x_0) = 0, \qquad (4.71)$$

where

$$\Delta^0_{\varphi h m}(x_0) = \frac{1}{T} H_h \{\varphi(x_m(0); x_m(t_1), x_m(t_2), ...) - \sum_{i=0}^{\infty} A_i x_0$$

$$- \sum_{i=1}^{\infty} A_i \int_0^{t_i} [f(\tau, x_m(\tau, x_0), \frac{dx_m(\tau, x_0)}{d\tau})$$

$$- \frac{h'(\tau)}{T} \int_0^{\infty} f(s, x_m(s, x_0), \frac{dx_m(s, x_0)}{ds}) ds] d\tau \}$$

$$- \frac{1}{T} \int_0^{\infty} f(\tau, x_m(\tau, x_0), \frac{dx_m(\tau, x_0)}{d\tau}) d\tau \qquad (4.72)$$

is the approximate defining function.

Corollary 4.8. *The following propositions hold:*

1) If, under conditions $(\mathbf{A_1})$ *and* $(\mathbf{B_1})$, *there exists a closed subset* $\mathbf{W} \subset D_{\beta\gamma\varphi h}$ *such that, for some* $m \in Z^+$, *the function* $\Delta^0_{\varphi h m}$ *topologically maps* $\mathbf{W}$ *onto* $\Delta^0_{\varphi h m} \mathbf{W}$, *Eq. (4.71) has the unique solution* x^0 *in* $\mathbf{W}$, *and, on the boundary* $\Gamma_{\mathbf{W}}$ *of the set* $\mathbf{W}$, *the inequality*

$$\inf_{x \in \Gamma_W} \|\Delta^0_{\varphi h m}(x)\| \geq (\frac{1}{T} K_\varphi \|H_h\| + (K_h + K_{1h})[1 + \frac{1}{2} \sum_{i=1}^{\infty} \|H_h A_i\|])$$

$$\times \|Q_0^m (E_2 - Q_0)^{-1} Z_1^0\| = \sigma(m), \qquad (4.73)$$

is satisfied, then the boundary-value problem (4.55), (4.2) has a solution $x = x^*(t)$ *with the initial condition* $x^*(0) = x_0^* \in \mathbf{W}$;

2) for any function $h(t)$ *such that conditions* $(\mathbf{A_1})$ *and* $(\mathbf{B_1^0})$ *are satisfied, in order that some subset* $D_2 \subset D_{\beta\gamma\varphi h}$ *contain the initial value* $x^*(0) = x_0^*$ *of the solution of this boundary-value problem, it is necessary that the inequality*

$$\|\Delta^0_{\varphi h m}(\bar{x}_0)\|$$

$$\leq \sup_{x_0 \in D_2} \{\frac{1}{T} \|R_h\| + (\frac{1}{T} K_\varphi \|H_h\| + (K_h + K_{1h})(\frac{1}{2} \sum_{i=1}^{\infty} \|H_h A_i\| + 1))$$

$$\times \max\{1 + \|R_h\|; \frac{\overline{h'}}{T}\|R_h\|\}\|(E_2 - Q_0)^{-1}\|\}\|\bar{x}_0 - x_0\| + \sigma(m, \bar{x}_0), \quad (4.74)$$

be satisfied $\forall m \in Z^+$ *and* $\forall \bar{x}_0 \in D_2$, *where* R_h *stands for* $H_h \sum_{i=0}^{\infty} A_i$, *and* $\sigma(m, \bar{x}_0)$ *denotes the expression*

$$\{\frac{1}{T}K_\varphi\|H_h\| + (K_h + K_{1h})[\frac{1}{2}K_h \sum_{i=1}^{\infty} \|H_h A_i\| + 1]\}\|Q_0^m (E_2 - Q_0)^{-1} Z_1^0\|.$$

Proof. With regard for conditions (4.3) and (4.56) and inequality (4.58), we obtain from (4.60) and (4.72) that the following inequalities hold $\forall m \in Z^+$:

$$\|\Delta_{\varphi h}^0(x_0) - \Delta_{\varphi h m}^0(x_0)\| \le \frac{1}{T}\|H_h\|\|\varphi(x^*(0); x^*(t_1), x^*(t_2), ...)$$

$$- \varphi(x_m(0); x_m(t_1), x_m(t_2), ...)\| + \frac{1}{T}\sum_{i=1}^{\infty}\|H_h A_i\|$$

$$\times \{(1 - \frac{h(t_i)}{T})\int_0^{t_i}\|f(\tau, x^*(\tau, x_0), \frac{dx^*(\tau, x_0)}{d\tau})$$

$$- f(\tau, x_m(\tau, x_0), \frac{dx_m(\tau, x_0)}{d\tau})\|d\tau$$

$$+ \frac{h(t_i)}{T}\int_{t_i}^{\infty}\|f(\tau, x^*(\tau, x_0), \frac{dx^*(\tau, x_0)}{d\tau}) - f(\tau, x_m(\tau, x_0), \frac{dx_m(\tau, x_0)}{d\tau})\|d\tau\}$$

$$+ \frac{1}{T}\int_0^{\infty}\|f(\tau, x^*(\tau, x_0), \frac{dx^*(\tau, x_0)}{d\tau}) - f(\tau, x_m(\tau, x_0), \frac{dx_m(\tau, x_0)}{d\tau})\|d\tau$$

$$\le \frac{1}{T}\|H_h\|K_\varphi\|Q_0^m(E_2 - Q_0)^{-1}Z_1^0\|$$

$$+ \frac{1}{T}\sum_{i=1}^{\infty}\|H_h A_i\|(K_h + K_{1h})\alpha_{1h}(t_i)\|Q_0^m(E_2 - Q_0)^{-1}Z_1^0\|$$

$$+ \frac{1}{T}(K_h + K_{1h})\int_0^{\infty} h'(\tau)d\tau\|Q_0^m(E_2 - Q_0)^{-1}Z_1^0\|$$

$$\le \{\frac{1}{T}K_\varphi\|H_h\| + (K_h + K_{1h})[\frac{1}{2}K_h \sum_{i=1}^{\infty}\|H A_i\| + 1]\}$$

$$\times \|Q_0^m(E_2 - Q_0)^{-1}Z_1^0\| = \sigma(m, \bar{x}_0), \quad (4.75)$$

where $\sigma(m, x_0) \to 0$ as $m \to \infty$, and $x_0 \in D_{\beta\gamma\varphi h}$.

In view of (4.60), we write the following estimate $\forall\{x_0', x_0''\} \subset D_{\beta\gamma\varphi h}$:

$$\|\Delta_{\varphi h}^0(x_0') - \Delta_{\varphi h}^0(x_0'')\| \leq \{\frac{1}{T}\|R_h\| + [\frac{1}{T}K_\varphi\|H_h\|$$

$$+ (K_h + K_{1h})(\frac{1}{2}\sum_{i=0}^{\infty}\|H_hA_i\| + 1)]$$

$$\times \max\{1 + \|R_h\|; \frac{\overline{h'}}{T}\|R_h\|\}\|(E_2 - Q_0)^{-1}\|\}\|x_0' - x_0''\|, \quad (4.76)$$

which yields the equicontinuity of the mapping $\Delta_{\varphi h}^0$ on the set $\mathbf{W}$.

Relations (4.75) and (4.76) for all $m \in Z^+$ and $\bar{x}_0 \in D_2$ yield the inequality

$$\|\Delta_{\varphi hm}^0(\bar{x}_0)\|$$

$$\leq \{\frac{1}{T}\|R_h\| + [\frac{1}{T}K_\varphi\|H_h\| + (K_h + K_{1h})(\frac{1}{2}\sum_{i=0}^{\infty}\|HA_i\| + 1)]$$

$$\times \max\{1 + \|R_h\|; \frac{\overline{h'}}{T}\|R_h\|\}\|(E_2 - Q_0)^{-1}\|\}\|\bar{x}_0 - x_0^*\| + \sigma(m, \bar{x}_0),$$

which ensures the validity of estimate (4.74). Taking (4.73) into account, we complete the proof of Corollary 4.8 analogously to that of Corollary 4.2.

$\square$

We now consider the boundary-value problem for Eq. (4.55) with the boundary condition (4.23). Retaining the designations introduced in Subsection 4.2 for the matrices A_i, C, and the function $\varphi(\psi)$, we consider that the function $f(t, x, x_1) : [0, T] \times D \times D_1 = D_{10} \to \mathfrak{M}$ is continuous in the totality of variables on D_{10}, and, $\forall\{x, x'\} \subset D$, $\{x_1, x_1'\} \subset D_1$, the inequalities

$$\|f(t, x, x_1)\| \leq M,$$

$$\|f(t, x, x_1) - f(t, x', x_1')\| \leq K\|x - x'\| + K_1\|x_1 - x_1'\|, \quad (4.77)$$

hold. Here, M, K, and K_1 are positive constants.

Retaining the content of the designations H, $\beta_{1\varphi}$, β_φ, and $D_{\beta\varphi}$ from Subsection 4.2, we denote the subset of elements of the set D_1, which belong to this set together with their γ_φ-neighborhoods, where $\gamma_\varphi(x) = 2M + \frac{1}{T}\beta_{1\varphi}(x)$, by $D_{1\gamma\varphi}$.

Let us assume that condition a_1 from Subsection 4.2 is satisfied together with the following conditions:

b_1^1) the intersection $D_{\beta\gamma\varphi}$ of the sets $D_{\beta\varphi}$ and $D_{1\gamma\varphi}$ is nonempty;
c_1^1) the matrix norm

$$Q_0^1 = \begin{bmatrix} \frac{KT}{2}(1 + \|H\| \sum\limits_{i=1}^{\infty} \|A_i\|) + K_\varphi\|H\| & \frac{K_1T}{2}(1 + \|H\| \sum\limits_{i=1}^{\infty} \|A_i\|) \\ K(2 + \frac{\|H\|}{2} \sum\limits_{i=1}^{\infty} \|A_i\|) + \frac{1}{T}K_\varphi\|H\| & K_1(2 + \frac{\|H\|}{2} \sum\limits_{i=1}^{\infty} \|A_i\|) \end{bmatrix}$$

is less than 1.

We formally define the sequence of functions $\{x_m(t, x_0)\}_{m=0}^{\infty}$ by the following recurrence relations:

$$x_m(t, x_0) = x_0 + \int\limits_0^t [f(\tau, x_{m-1}(\tau, x_0), \frac{dx_{m-1}(\tau, x_0)}{d\tau})$$

$$- \frac{1}{T} \int\limits_0^T f(s, x_{m-1}(s, x_0), \frac{dx_{m-1}(s, x_0)}{ds})ds]d\tau$$

$$+ \frac{t}{T}H\{\varphi(x_{m-1}(0), x_{m-1}(T); x_{m-1}(t_1), x_{m-1}(t_2), ...)$$

$$- (\sum\limits_{i=0}^{\infty} A_i + C)x_0 - \sum\limits_{i=1}^{\infty} A_i \int\limits_0^{t_i} [f(\tau, x_{m-1}(\tau, x_0), \frac{dx_{m-1}(\tau, x_0)}{d\tau})$$

$$- \frac{1}{T} \int\limits_0^T f(s, x_{m-1}(s, x_0), \frac{dx_{m-1}(\tau, x_0)}{d\tau})ds]d\tau\}, \quad m = 1, 2, ...,$$

$$x_0(t, x_0) = (x_{01}, x_{02}, ...) \equiv x_0, \quad x_0 \in D_{\beta\gamma\varphi}. \quad (4.78)$$

The conditions of existence of sequence (4.78) and its convergence to the function $x^*(t, x_0)$ satisfying the equality

$$x^*(t, x_0) = x_0 + \int\limits_0^t [f(\tau, x^*(\tau, x_0), \frac{dx^*(\tau, x_0)}{d\tau})$$

$$- \frac{1}{T} \int\limits_0^T f(s, x^*(s, x_0), \frac{dx^*(s, x_0)}{ds})ds]d\tau$$

$$+ \frac{t}{T}H\{\varphi(x^*(0), x^*(T); x^*(t_1), x^*(t_2), ...) - (\sum\limits_{i=0}^{\infty} A_i + C)x_0$$

$$-\sum_{i=1}^{\infty} A_i \int_0^{t_i} [f(\tau, x^*(\tau, x_0), \frac{dx^*(\tau, x_0)}{d\tau})$$

$$-\frac{1}{T} \int_0^T f(s, x^*(s, x_0), \frac{dx^*(s, x_0)}{ds}) ds] d\tau\},$$

are considered in the following proposition.

Theorem 4.8. *Let conditions (4.3), (4.77), a $_1$, b_1^1, and c_1^1 be satisfied. Then, as $m \to \infty$, the sequence $\{x_m(t, x_0)\}_{m=0}^{\infty}$ defined by equalities (4.78) converges uniformly in $(t, x_0) \in [0, T] \times D_{\beta\gamma\varphi}$ to the function $x^*(t, x_0)$, and, for all $m \in Z^+$,*

$$\sup_{t \in [0,T]} \{\|x_m(t, x_0) - x^*(t, x_0)\| , \|\frac{dx_m(t, x_0)}{dt} - \frac{dx^*(t, x_0)}{dt}\|\}$$

$$\leq \|Q_0^{1m}(E_2 - Q_0^1)^{-1} Z_2^0\|,$$

where $Z_2^0 = \begin{bmatrix} \beta_\varphi(x_0) \\ \gamma_\varphi(x_0) \end{bmatrix}$. The function $x^*(t, x_0)$ satisfies the boundary condition (4.23) and is a solution of the perturbed equation

$$\frac{dx}{dt} = f(t, x, \frac{dx}{dt}) + \mu,$$

where

$$\mu = \Delta_\varphi^0(x_0) = \frac{1}{T} H\{\varphi(x^*(0), x^*(T); x^*(t_1), x^*(t_2), ...) - (\sum_{i=0}^{\infty} A_i + C)x_0$$

$$-\sum_{i=1}^{\infty} A_i \int_0^{t_i} [f(\tau, x^*(t, x_0), \frac{dx^*(t, x_0)}{dt})$$

$$-\frac{1}{T} \int_0^T f(\tau, x^*(t, x_0), \frac{dx^*(t, x_0)}{dt}) ds] d\tau\}$$

$$-\frac{1}{T} \int_0^T f(\tau, x^*(t, x_0), \frac{dx^*(t, x_0)}{dt}) d\tau;$$

moreover, the condition $\Delta_\varphi^0(x_0) = 0$ is sufficient in order that the function $x^*(t, x_0)$ be a solution of the boundary-value problem (4.55), (4.23).

The proof of this theorem has the same structure as that of Theorem 4.7.

Remark 4.5. If, under assumptions of Theorem 4.8, condition c_1^1 is replaced by the condition

$$c_1^{01}) \qquad\qquad \max\{\|Q_0^1\|,\ \|Q_{01}^1\|\} < 1,$$

where

$$Q_{01}^1 = \begin{bmatrix} \dfrac{TK(1+\|H\|\sum\limits_{i=1}^{\infty}\|A_i\|)}{1-K_\varphi\|H\|} & \dfrac{TK_1(1+\|H\|\sum\limits_{i=1}^{\infty}\|A_i\|)}{1-K_\varphi\|H\|} \\[4mm] \dfrac{K(1+\|H\|\sum\limits_{i=1}^{\infty}\|A_i\|)}{1-K_\varphi\|H\|} & \dfrac{K_1(1+\|H\|\sum\limits_{i=1}^{\infty}\|A_i\|)}{1-K_\varphi\|H\|} \end{bmatrix},$$

then there exists no other value of μ, for which the solution of the equation $\frac{dx}{dt} = f(t, x, \frac{dx}{dt}) + \mu$ with the initial condition $x(0) = x_0$ would satisfy the boundary condition (4.23).

We now formulate analogous results for Eq. (4.55), whose solutions satisfy the multipoint boundary condition (4.26).

We retain the content of the designations $H_p, \beta_{1\varphi p}, \beta_{\varphi p}$, and $D_{\beta\varphi p}$ from Subsection 4.2 and, by $D_{1\gamma\varphi p}$, denote the subset of elements of the set D_1, which belong to this set together with their $\gamma_{\varphi p}$-neighborhoods, where $\gamma_{\varphi p}(x) = 2M + \frac{1}{T}\beta_{1\varphi p}(x)$.

Let us assume that condition a_2 from Subsection 4.2 is satisfied together with the following conditions:

b_2^1) the intersection $D_{\beta\gamma\varphi p}$ of the sets $D_{\beta\varphi p}$ and $D_{1\gamma\varphi p}$ is nonempty;

c_2^1) the matrix norm

$$Q_{0p}^1$$

$$= \begin{bmatrix} \frac{KT}{2}(1+\|H_p\|\sum\limits_{i=1}^{p}\|A_i\|) + K_{\varphi p}\|H_p\| & \frac{K_1 T}{2}(1+\|H_p\|\sum\limits_{i=1}^{p}\|A_i\|) \\[4mm] K(2+\frac{1}{2}\|H_p\|\sum\limits_{i=1}^{p}\|A_i\|) + \frac{1}{T}K_{\varphi p}\|H_p\| & K_1(2+\frac{1}{2}\|H_p\|\sum\limits_{i=1}^{p}\|A_i\|) \end{bmatrix}$$

is less than 1.

Under these conditions, there exists the sequence of functions

$$x_{pm}(t, x_0) = x_0 + \int_0^t [f(\tau, x_{pm-1}(\tau, x_0), \frac{dx_{pm-1}(\tau, x_0)}{d\tau})$$

$$- \frac{1}{T}\int_0^T f(s, x_{pm-1}(s, x_0), \frac{dx_{pm-1}(s, x_0)}{ds})ds]d\tau$$

$$+ \frac{t}{T} H\{\varphi_p(x_{p_{m-1}}(0), x_{p_{m-1}}(T); x_{p_{m-1}}(t_1), ..., x_{p_{m-1}}(t_p))$$

$$- \left(\sum_{i=0}^{p} A_i + C\right) x_0 - \sum_{i=1}^{p} A_i \int_0^{t_i} [f(\tau, x_{p_{m-1}}(\tau, x_0), \frac{dx_{p_{m-1}}(\tau, x_0)}{d\tau}) -$$

$$- \frac{1}{T} \int_0^T f(s, x_{p_{m-1}}(s, x_0), \frac{dx_{p_{m-1}}(\tau, x_0)}{d\tau}) ds] d\tau\}, \quad m = 1, 2, ...,$$

$$x_0(t, x_0) = (x_{01}, x_{02}, ...) \equiv x_0, \quad x_0 \in D_{\beta\gamma\varphi p}, \quad (4.79)$$

which satisfy the recurrence boundary condition

$$A_0 x_{p_m}(0) + \sum_{i=1}^{p} A_i x_{p_m}(t_i) + C x_{p_m}(T)$$

$$= \varphi(x_{p_{m-1}}(0), x_{p_{m-1}}(T); x_{p_{m-1}}(t_1), ..., x_{p_{m-1}}(t_p))$$

for any $x_0 \in D_{\beta\gamma\varphi p}$.

The conditions of convergence of sequence (4.79) to the function $x_p(t, x_0)$ satisfying the equality

$$x_p(t, x_0) = x_0 + \int_0^t [f(\tau, x_p(\tau, x_0), \frac{dx_p(\tau, x_0)}{d\tau})$$

$$- \frac{1}{T} \int_0^T f(s, x_p(s, x_0), \frac{dx_p(s, x_0)}{ds}) ds] d\tau$$

$$+ \frac{t}{T} H_p\{\varphi_p(x_p(0), x_p(T); x_p(t_1), ..., x_p(t_p)) - \left(\sum_{i=0}^{p} A_i + C\right) x_0$$

$$- \sum_{i=1}^{p} A_i \int_0^{t_i} [f(\tau, x_p(\tau, x_0), \frac{dx_p(\tau, x_0)}{d\tau})$$

$$- \frac{1}{T} \int_0^T f(s, x_p(s, x_0), \frac{dx_p(s, x_0)}{ds}) ds] d\tau\}$$

are given in the following proposition.

Corollary 4.9. *Let us assume that conditions (4.77), (4.27) and a_2, b_2^1, and c_2^1 are satisfied. Then:*

1) the sequence of functions $\{x_{pm}(t, x_0)\}_{m=0}^{\infty}$ defined by the relations (4.79) is convergent uniformly in $(t, x_0) \in [0, T] \times D_{\beta\gamma\varphi p}$ as $m \to \infty$ to the limiting function $x_p(t, x_0)$, and

$$\sup_{t \in [0,T]} \{\|x_{pm}(t, x_0) - x_p(t, x_0)\|, \ \|\frac{dx_{pm}(t, x_0)}{dt} - \frac{dx_p(t, x_0)}{dt}\|\}$$

$$\leq \|(Q_{0p}^1)^m (E_2 - Q_{0p}^1)^{-1} Z_{2p}^0\|,$$

where $Z_{2p}^0 = \begin{bmatrix} \beta_{\varphi p}(x_0) \\ \gamma_{\varphi p}(x_0) \end{bmatrix}$;

2) the function $x_p(t, x_0)$ is a solution of the equation

$$\frac{dx}{dt} = f(t, x, \frac{dx}{dt}) + \mu_p \tag{4.80}$$

with the boundary condition (4.26), where

$$\mu_p = \Delta_{\varphi p}^0(x_0) = \frac{1}{T} H_p \{\varphi_p(x_p(0), x_p(T); x_p(t_1), ..., x_p(t_p))$$

$$- (\sum_{i=0}^{p} A_i + C)x_0 - \sum_{i=1}^{p} A_i \int_0^{t_i} [f(\tau, x_p(t, x_0), \frac{dx_p(t, x_0)}{dt})$$

$$- \frac{1}{T} \int_0^{T} f(\tau, x_p(t, x_0), \frac{dx_p(t, x_0)}{dt})ds]d\tau\}$$

$$- \frac{1}{T} \int_0^{T} f(\tau, x_p(t, x_0), \frac{dx_p(t, x_0)}{dt})d\tau;$$

moreover, the condition $\Delta_{\varphi p}^0(x_0) = 0$ is sufficient in order that the function $x_p(t, x_0)$ be a solution of the boundary-value problem (4.55), (4.26).

Remark 4.6. If, under assumptions of Corollary 4.9, condition c_2^1 is replaced by the condition

$$c_2^{01}) \qquad \max\{\|Q_{0p}^1\|, \ \|Q_{01p}^1\|\} < 1,$$

where

$$Q_{01p}^1 = \begin{bmatrix} \dfrac{TK(1+\|H_p\| \sum\limits_{i=1}^{p} \|A_i\|)}{1-K_{\varphi p}\|H_p\|} & \dfrac{TK_1(1+\|H_p\| \sum\limits_{i=1}^{p} \|A_i\|)}{1-K_{\varphi p}\|H_p\|} \\ \dfrac{K(1+\|H_p\| \sum\limits_{i=1}^{p} \|A_i\|)}{1-K_{\varphi p}\|H_p\|} & \dfrac{K_1(1+\|H_p\| \sum\limits_{i=1}^{p} \|A_i\|)}{1-K_{\varphi p}\|H_p\|} \end{bmatrix},$$

then there exists no other value of μ_p such that the solution of the equation $\frac{dx}{dt} = f(t, x, \frac{dx}{dt}) + \mu_p$ with the initial condition $x(0) = x_0$ would satisfy the boundary condition (4.26).

It is clear that, for the boundary-value problems (4.55), (4.23) and (4.55), (4.26), the propositions analogous to Corollary 4.8 can be easily formulated and proved.

4.6   Reduction to a finite-dimensional multipoint problem

Introducing the notation

$$\overset{(n)}{f}(t, \overset{(n)}{x}, \overset{(n)}{x_1}) = (f_1(t, x_1, ..., x_n, 0, 0, ..., x_{11}, ..., x_{1n}, 0, 0, ...), ...$$
$$..., f_n(t, x_1, ..., x_n, 0, 0, ..., x_{11}, ..., x_{1n}, 0, 0, ...))$$

and retaining the content of the designations introduced at the beginning of section 4.3, we consider, in addition to the boundary-value problem (4.55), (4.26), the boundary-value problem for the equation

$$\frac{d \overset{(n)}{x}}{dt} = \overset{(n)}{f}(t, \overset{(n)}{x}, \frac{d \overset{(n)}{x}}{dt}) \tag{4.81}$$

with the boundary condition (4.30). It is the multipoint boundary-value problem in a finite-dimensional space. We will find the conditions, under which the boundary-value problem (4.55), (4.23) is reduced to it.

As above, we consider that the functions $f(t, x, x_1)$ and $\varphi_p(\psi)$ are subordinated to conditions (4.77) and (4.27), respectively. We also consider that $y \in R^n$ belongs to the set $\tilde{D}^{(n)}$ (set $\tilde{D}_1^{(n)}$), if $(y, 0, 0, ...) \in D$ $((y, 0, 0, ...) \in D_1)$. It is obvious that, for any $\{\overset{(n)}{x}, \overset{(n)}{x'}\} \subset \tilde{D}^{(n)}$, $\{\overset{(n)}{x_1}, \overset{(n)}{x'_1}\} \subset \tilde{D}_1^{(n)}$ and $t \in [0, T]$, the following inequalities hold:

$$\| \overset{(n)}{f}(t, \overset{(n)}{x}, \overset{(n)}{x_1}) \| \leq M,$$

$$\| \overset{(n)}{f}(t, \overset{(n)}{x}, \overset{(n)}{x_1}) - \overset{(n)}{f}(t, \overset{(n)}{x'}, \overset{(n)}{x'_1}) \| \leq K \| \overset{(n)}{x} - \overset{(n)}{x'} \| + K_1 \| \overset{(n)}{x_1} - \overset{(n)}{x'_1} \|. \tag{4.82}$$

If $y_i \in \tilde{D}^{(n)}$ $(i = \overline{1, p+2})$, then $Y = (y_1, y_2, ... y_{p+2}) \in \tilde{D}^{(n)^{p+2}}$, and inequalities (4.32) hold $\forall \{Y_1, Y_2\} \subset \tilde{D}^{(n)^{p+2}}$.

Let us assume that condition $_3$ from section 4.3 is satisfied together with the conditions:

b$_3^1$) the intersection $\tilde{D}^{(n)}_{\beta\gamma\varphi p}$ of the sets $\tilde{D}^{(n)}_{(n) \atop \beta_{\varphi p}}$ and $\tilde{D}_1^{(n)}{}_{(n) \atop \gamma_{\varphi p}}$ is nonempty;

here, $\tilde{D}^{(n)}_{(n) \atop \beta_{\varphi p}}$ and $\tilde{D}_1^{(n)}{}_{(n) \atop \gamma_{\varphi p}}$ are the sets of points $\overset{(n)}{x_0} = (x_{01}, x_{02}, ..., x_{0n}) \in R^n$

such that the points $(\overset{(n)}{x_0},0,0,...)$ belong to the domain D together with their $\overset{(n)}{\beta}{}_{\varphi p}$-neighborhoods and to the domain D_1 together with their $\overset{(n)}{\gamma}{}_{\varphi p}$-neighborhoods,

$$\overset{(n)}{\beta}{}_{\varphi p}(x_0) = \frac{T}{2}M + \overset{(n)}{\beta}{}_{1\varphi p}(x_0), \quad \overset{(n)}{\gamma}{}_{\varphi p}(x_0) = 2M + \frac{1}{T}\overset{(n)}{\beta}{}_{1\varphi p}(x_0),$$

and $\overset{(n)}{\beta}{}_{1\varphi p}(x_0)$ is defined in the same way as in condition b_3 from section 4.3;

c_3^1) the matrix norm

$$\overset{(n)^1}{Q}{}_{0p}$$

$$= \begin{bmatrix} \frac{KT}{2}(1 + \|\overset{(n)}{H}{}_p\| \sum_{i=1}^{p} \|\overset{(n)}{A}{}_i\|) + K_{\varphi p}\|\overset{(n)}{H}{}_p\| & \frac{K_1T}{2}(1 + \|\overset{(n)}{H}{}_p\| \sum_{i=1}^{p} \|\overset{(n)}{A}{}_i\|) \\[2ex] K(2 + \frac{1}{2}\|\overset{(n)}{H}{}_p\| \sum_{i=1}^{p} \|\overset{(n)}{A}{}_i\|) + \frac{1}{T}K_{\varphi p}\|\overset{(n)}{H}{}_p\| & K_1(2 + \frac{1}{2}\|\overset{(n)}{H}{}_p\| \sum_{i=1}^{p} \|\overset{(n)}{A}{}_i\|) \end{bmatrix}$$

is less than 1.

With regard for these conditions and inequalities (4.77), (4.27), we can show that the sequence of functions $\{\overset{(n)}{x}{}_{pm}(t, \overset{(n)}{x_0})\}_{m=0}^{\infty}$ defined by the recurrence relation

$$\overset{(n)}{x}{}_{pm}(t, \overset{(n)}{x_0}) = \overset{(n)}{x_0} + \int_0^t [\overset{(n)}{f}(\tau, \overset{(n)}{x}{}_{pm-1}(\tau, \overset{(n)}{x_0}), \frac{d\,\overset{(n)}{x}{}_{pm-1}(\tau, \overset{(n)}{x_0})}{d\tau})$$

$$- \frac{1}{T}\int_0^T \overset{(n)}{f}(s, \overset{(n)}{x}{}_{pm-1}(s, \overset{(n)}{x_0}), \frac{d\,\overset{(n)}{x}{}_{pm-1}(s, \overset{(n)}{x_0})}{ds})ds]d\tau$$

$$+ \frac{t}{T}\overset{(n)}{H}{}_p\{\overset{(n)}{\varphi}{}_p(\overset{(n)}{x}{}_{pm-1}(0, \overset{(n)}{x}{}_0, \overset{(n)}{x}{}_{pm-1}(T, \overset{(n)}{x_0}));$$

$$\overset{(n)}{x}{}_{pm-1}(t_1, \overset{(n)}{x_0}), ..., \overset{(n)}{x}{}_{pm-1}(t_p, \overset{(n)}{x_0}))$$

$$- (\sum_{i=0}^{p}\overset{(n)}{A}{}_i + \overset{(n)}{C})\overset{(n)}{x_0} - \sum_{i=1}^{p}\overset{(n)}{A}{}_i[\overset{(n)}{f}(\tau, \overset{(n)}{x}{}_{pm-1}(\tau, \overset{(n)}{x_0}), \frac{d\,\overset{(n)}{x}{}_{pm-1}(\tau, \overset{(n)}{x_0})}{d\tau})$$

$$- \frac{1}{T}\int_0^T \overset{(n)}{f}(s, \overset{(n)}{x}{}_{pm-1}(s, \overset{(n)}{x_0}), \frac{d\,\overset{(n)}{x}{}_{pm-1}(s, \overset{(n)}{x_0})}{ds})ds]d\tau\}, \quad m = 1, 2, ...,$$

$$\overset{(n)}{x}{}_{p0}(t, \overset{(n)}{x}{}_0) \equiv \overset{(n)}{x_0},$$

is convergent as $m \to \infty$ to the function $\overset{(n)}{x}_p(t, \overset{(n)}{x}_0)$, and the sequence $\{\frac{d \overset{(n)}{x}_{pm}(t, \overset{(n)}{x}_0)}{dt}\}_{m=0}^{\infty}$ converges to the function $\frac{d \overset{(n)}{x}_p(t, \overset{(n)}{x}_0)}{dt}$ uniformly in $(t, \overset{(n)}{x}_0) \in [0, T] \times \tilde{D}_{\beta\gamma\varphi p}$. In this case, the limiting functions satisfy the equality

$$\overset{(n)}{x}_p(t, \overset{(n)}{x}_0) = \overset{(n)}{x}_0 + \int_0^t [\overset{(n)}{f}(\tau, \overset{(n)}{x}_p(\tau, \overset{(n)}{x}_0), \frac{d \overset{(n)}{x}_p(\tau, \overset{(n)}{x}_0)}{d\tau})$$

$$- \frac{1}{T} \int_0^T \overset{(n)}{f}(s, \overset{(n)}{x}_p(s, \overset{(n)}{x}_0), \frac{d \overset{(n)}{x}_p(s, \overset{(n)}{x}_0)}{ds}) ds] d\tau$$

$$+ \frac{t}{T} \overset{(n)}{H}_p\{\overset{(n)}{\varphi}_p(\overset{(n)}{x}_p(0, \overset{(n)}{x}_0), \overset{(n)}{x}_p(T, \overset{(n)}{x}_0); \overset{(n)}{x}_p(t_1, \overset{(n)}{x}_0), ..., \overset{(n)}{x}_p(t_p, \overset{(n)}{x}_0))$$

$$- (\sum_{i=0}^{p} \overset{(n)}{A}_i + \overset{(n)}{C}) \overset{(n)}{x}_0$$

$$- \sum_{i=1}^{p} \overset{(n)}{A}_i [\overset{(n)}{f}(\tau, \overset{(n)}{x}_p(\tau, \overset{(n)}{x}_0), \frac{d \overset{(n)}{x}_p(\tau, \overset{(n)}{x}_0)}{d\tau})$$

$$- \frac{1}{T} \int_0^T \overset{(n)}{f}(s, \overset{(n)}{x}_p(s, \overset{(n)}{x}_0), \frac{d \overset{(n)}{x}_p(s, \overset{(n)}{x}_0)}{ds}) ds] d\tau\}, \quad (4.83)$$

and the function $\overset{(n)}{x}_p(t, \overset{(n)}{x}_0)$ is a solution of the perturbed equation

$$\frac{d \overset{(n)}{x}}{dt} = \overset{(n)}{f}(t, \overset{(n)}{x}, \frac{d \overset{(n)}{x}}{dt}) + \overset{(n)}{\mu}_p \qquad (4.84)$$

with the boundary condition (4.30), where

$$\overset{(n)}{\mu}_p = \overset{(n)}{\Delta}_{\varphi p}(\overset{(n)}{x}_0)$$

$$= \frac{1}{T} \overset{(n)}{H}_p\{\overset{(n)}{\varphi}_p(\overset{(n)}{x}_p(0, \overset{(n)}{x}_0), \overset{(n)}{x}_p(T, \overset{(n)}{x}_0); \overset{(n)}{x}_p(t_1, \overset{(n)}{x}_0), ..., \overset{(n)}{x}_p(t_p, \overset{(n)}{x}_0))$$

$$- (\sum_{i=0}^{p} \overset{(n)}{A}_i + \overset{(n)}{C}) \overset{(n)}{x}_0 - \sum_{i=1}^{p} \overset{(n)}{A}_i [\overset{(n)}{f}(\tau, \overset{(n)}{x}_p(\tau, \overset{(n)}{x}_0), \frac{d \overset{(n)}{x}_p(\tau, \overset{(n)}{x}_0)}{d\tau})$$

$$- \frac{1}{T} \int_0^T \overset{(n)}{f}(s, \overset{(n)}{x}_p(s, \overset{(n)}{x}_0), \frac{d \overset{(n)}{x}_p(s, \overset{(n)}{x}_0)}{ds}) ds] d\tau\}$$

$$- \frac{1}{T} \int_0^T \overset{(n)}{f}(\tau, \overset{(n)}{x}_p(\tau, \overset{(n)}{x}_0), \frac{d \overset{(n)}{x}_p(\tau, \overset{(n)}{x}_0)}{d\tau}) d\tau.$$

The equality $\overset{(n)}{\Delta}_{\varphi p}(\overset{(n)}{x}_0) = 0$ is the sufficient condition for the function $\overset{(n)}{x}_p(t, \overset{(n)}{x}_0)$ to define a solution of the boundary-value problem (4.81), (4.30). By analogy with the proof of Corollary 4.7, we can show that the condition

$$c_3^{01})\qquad \max\{\|\overset{(n)^1}{Q}_{0p}\|, \ \|\overset{(n)^1}{Q}_{01p}\|\} < 1,$$

where

$$\overset{(n)^1}{Q}_{01p} = \begin{bmatrix} \dfrac{TK(1+\|\overset{(n)}{H}_p\|\sum\limits_{i=1}^{p}\|\overset{(n)}{A}_i\|)}{1-K_{\varphi p}\|\overset{(n)}{H}_p\|} & \dfrac{TK_1(1+\|\overset{(n)}{H}_p\|\sum\limits_{i=1}^{p}\|\overset{(n)}{A}_i\|)}{1-K_{\varphi p}\|\overset{(n)}{H}_p\|} \\[2em] \dfrac{K(1+\|\overset{(n)}{H}_p\|\sum\limits_{i=1}^{p}\|\overset{(n)}{A}_i\|)}{1-K_{\varphi p}\|\overset{(n)}{H}_p\|} & \dfrac{K_1(1+\|\overset{(n)}{H}_p\|\sum\limits_{i=1}^{p}\|\overset{(n)}{A}_i\|)}{1-K_{\varphi p}\|\overset{(n)}{H}_p\|} \end{bmatrix},$$

ensures the uniqueness of the control $\overset{(n)}{\mu}_p = \overset{(n)}{\Delta}_{\varphi p}(\overset{(n)}{x}_0)$ on the right-hand side of Eq. (4.84), whose solution satisfies the boundary condition (4.30).

We agree to consider that the function $f(t, x, x_1) \in \hat{C}_{Lip}(x, x_1)$, if it is continuous in the domain D_{10}, bounded in this domain by a constant M, and satisfies the sharpened Cauchy–Lipschitz condition with respect to $\{x, x_1\}$:

$$\|f(t, x', x_1') - f(t, x'', x_1'')\| \le \alpha(t)\varepsilon(m)(\|x' - x''\| + \|x_1' - x_1''\|),$$

where x', x'' are any points of the domain D, whose m first corresponding coordinates coincide, x_1', x_1'' are any points of the domain D_1, whose m first corresponding coordinates coincide, $\alpha(t) \ge 0$ is a function continuous on $[0, T]$, and $\varepsilon(m) \underset{m\to\infty}{\longrightarrow} 0$.

We set $K^* = \max\limits_{t\in[0,T]} \alpha(t) \cdot \varepsilon(0)$ and, in what follows, assume that $K_1 = K = K^*$ in the second condition in (4.77).

We introduce the notation

$$\beta_{\varphi p}^*(x, n_0) = \dfrac{\frac{2}{K^*T} - 1}{\sum\limits_{i=1}^{p}\|\overset{(n_0)}{A}_i\| + \frac{2K_{\varphi p}}{K^*T}}\left(M_\varphi + (\sum_{i=0}^{p}\|A_i\| + \|C\|)\|x\|\right) + \dfrac{M}{K^*};$$

$$\gamma_{\varphi p}^*(x, n_0) = \dfrac{\frac{2}{K^*} - 4}{T\sum\limits_{i=1}^{p}\|\overset{(n_0)}{A}_i\| + \frac{2K_{\varphi p}}{K^*}}\left(M_\varphi + (\sum_{i=0}^{p}\|A_i\| + \|C\|)\|x\|\right) + \dfrac{M}{K^*},$$

$$n_0 \in Z^0;$$

$D^*_{\beta\varphi p}$ is a set, all points of which belong to the set D together with their $\beta^*_{\varphi p}(x)$-neighborhoods;

$D^*_{1\gamma\varphi p}$ is a set, all points of which belong to the set D_1 together with their $\gamma^*_{\varphi p}(x)$-neighborhoods.

Assuming that $\varphi(\psi) \in \hat{C}_{Lip}(\psi)$, we set $\delta(0) = \delta_0(0) = K_\varphi$ in the right-hand sides of inequalities (4.35) and (4.36). One says that $f(t, x, x_1) \in C_{Lip}(t)$, if

$$\|f(t', x, x_1) - f(t'', x, x_1)\| \leq \tilde{K}\|t' - t''\|, \quad \tilde{K} = const > 0,$$

for any $x \in D$, $x_1 \in D_1, \{t', t''\} \subset [0, T]$.

The following proposition is valid.

Lemma 4.3. *Let $f(t, x, x_1) \in \hat{C}_{Lip}(x, x_1) \cap C_{Lip}(t)$, let the intersection $D^*_{\beta\gamma\varphi p}$ of the sets $D^*_{\beta\varphi p}$ and $D^*_{1\gamma\varphi p}$ be nonempty, and let conditions a_2, c_2^{01}, (4.3) and (4.36) be satisfied. If conditions a_3 and c_3^1 are satisfied $\forall n \geq n_0$, then, $\forall x_0 \in D^*_{\beta\gamma\varphi p}$ in the meaning of the coordinatewise limiting transition, equalities (4.37) and the relation*

$$\lim_{n \to \infty} \frac{d\overset{(n)}{x_p}(t, \overset{(n)}{x_0})}{dt} = \frac{dx_p(t, x_0)}{dt} \tag{4.85}$$

hold. In this relation, the functions $x_p(t, x_0)$ and $\overset{(n)}{x_p}(t, \overset{(n)}{x_0})$ are solutions of the boundary-value problems (4.80), (4.26) and (4.84), (4.30), respectively.

Proof. As was shown in the proof of Lemma 4.1, condition (4.3) yields condition (4.27) and inequalities (4.32) $\forall n \in Z^+$.

The inclusion $f(t, x, x_1) \in \hat{C}_{Lip}(x, x_1)$ yields inequalities (4.77) and (4.82) for all $n \in Z^+$. In this case, the constants $M, K = K_1 = K^*$, $M_{\varphi p}$, and $K_{\varphi p}$ are independent of n.

We note that conditions c_2^{01} and c_3^1 yield the estimates

$$\frac{KT}{2}[1 + \|H_p\| \sum_{i=1}^{p} \|A_1\|] \leq \|Q^1_{0p}\| < 1,$$

$$\frac{KT}{2}[1 + \|\overset{(n)}{H_p}\| \sum_{i=1}^{p} \|\overset{(n)}{A_1}\|] \leq \|\overset{(n)}{Q}{}^1_{0p}\| < 1.$$

Therefore, like the proof of Lemma 4.1, we can show that, for $n \geq n_0$, $\overset{(n)}{\beta}_{\varphi p} < \beta^*_{\varphi p}$ and $\beta_{\varphi p} < \beta^*_{\varphi p}$ $\forall x_0 \in D^*_{\beta\gamma\varphi p}$. This means that the inclusion $x_0 \in D^*_{\beta\varphi p}$ yields the relations: $x_0 \in D_{\beta\varphi p}$ and $\overset{(n)}{x_0} \in \tilde{D}^{(n)}_{\overset{(n)}{\beta}_{\varphi p}}$ $\forall n \geq n_0$.

Analogously, the inclusion $x_0 \in D^*_{1\gamma\varphi p}$ yields the inclusions $x_0 \in D_{1\gamma\varphi p}$ and $\overset{(n)}{x_0} \in \tilde{D}^{(n)}_{1\underset{(n)}{\gamma}\varphi p}$. For this purpose, it is sufficient to show that, $\forall n \geq n_0$,

$$\gamma_{\varphi p} < \gamma^*_{\varphi p}, \qquad \overset{(n)}{\gamma}_{\varphi p} < \gamma^*_{\varphi p}.$$

We prove only the first of these inequalities, since the second one can be proved analogously. For $K = K_1 = K^*$, condition c_2^1 yields the inequalities

$$K^*(2 + \frac{\|H_p\|}{2} \sum_{i=1}^{p} \|A_i\|) < 1,$$

$$\|H_p\| < \frac{\frac{2}{K^*} - 4}{\sum_{i=1}^{p} \|A_i\| + \frac{2K_{\varphi p}}{K^*T}} \leq \frac{\frac{2}{K^*} - 4}{\sum_{i=1}^{p} \|\overset{(n_0)}{A_i}\| + \frac{2K_{\varphi p}}{K^*T}}.$$

With regard for condition b_3^1 and the last inequalities, we obtain

$$\gamma_{\varphi p}(x) \leq 2M + \frac{\|H_p\|}{T}(\|d\| + (\sum_{i=0}^{p} \|A_i\| + \|C\|)\|x\|)$$

$$+ \frac{M}{2} \sum_{i=1}^{p} \|H_p A_i\| \leq M(2 + \frac{\|H_p\|}{2} \sum_{i=1}^{p} \|A_i\|)$$

$$+ \frac{1}{T} \frac{\frac{2}{K^*} - 4}{\sum_{i=1}^{p} \|\overset{(n_0)}{A_i}\| + \frac{2K_{\varphi p}}{K^*T}} (M_\varphi + (\sum_{i=0}^{p} \|A_i\| + \|C\|)\|x\|)$$

$$< \frac{\frac{2}{K^*} - 4}{T \sum_{i=1}^{p} \|\overset{(n_0)}{A_i}\| + \frac{2K_{\varphi p}}{K^*}}$$

$$\times (M_\varphi + (\sum_{i=0}^{p} \|A_i\| + \|C\|)\|x\|) + \frac{M}{K^*} = \gamma^*_{\varphi p}(x, n_0),$$

which was to be proved.

It is easy to verify that $0 \in D^*_{\beta\gamma\varphi p}$. Therefore, there exists the control $\overset{(n)}{\mu_p} = \overset{(n)}{\Delta}_{\varphi p}(\overset{(n)}{x_0}) \; \forall n \in Z^+$, and condition c_3^1 yields the inequalities

$$\|\overset{(n)}{\Delta}_{\varphi p}(\overset{(n)}{x_0})\| \leq \frac{\frac{2}{K^*T} - 1}{T \sum_{i=1}^{p} \|\overset{(n_0)}{A_i}\| + \frac{2K_\varphi}{K^*}} (M_\varphi + (\sum_{i=0}^{p} \|A_i\| + \|C\|)\|x_0\|$$

$$+ \frac{MT}{2} \sum_{i=1}^{p} \|A_i\|) + M \leq M' = const < \infty,$$

where $n \geq n_0$, $x_0 \in D^*_{\beta\gamma\varphi p}$. Therefore, the sequence $\{\overset{(n)}{\mu}_p\}^{\infty}_{n=n_0}$ is uniformly bounded in the norm of the space $\mathfrak{M}$. From it, we can separate, by using the method of diagonalization, a subsequence $\{\overset{(s_i)}{\mu}_p\}^{\infty}_{i=1}$ coordinatewise convergent as $i \to \infty$. We now write an appropriate sequence of equations of the form (4.84), by replacing the index n by the index s_i:

$$\frac{d\overset{(s_i)}{x}}{dt} = \overset{(s_i)}{f}(t, \overset{(s_i)}{x}, \frac{d\overset{(s_i)}{x}}{dt}) + \overset{(s_i)}{\mu}_p, \ i \in Z^+. \tag{4.86}$$

Each equation in (4.86) corresponds to the boundary condition (4.30), where the above replacement of indices is carried out. In this case, the solution $\overset{(s_i)}{x}_p(t, \overset{(s_i)}{x}_0)$ of the boundary-value problem (4.86), (4.30) satisfies equality (4.83), where we set $n = s_i$.

Since $\|\overset{(s_i)}{x}_p(t, \overset{(s_i)}{x}_0)\| \leq M_0 \ \forall i \in Z^+, t \in [0, T]$, the sequence of functions $\{\overset{(s_i)}{x}_p(t, \overset{(s_i)}{x}_0)\}^{\infty}_{i=1}$ is uniformly bounded on this interval. We will show that it is equicontinuous there.

Indeed, denoting s_i by k, $\forall\{t^{(1)}, t^{(2)}\} \subset [0, T]$, $t^{(1)} < t^{(2)}$, we have

$$\overset{(k)}{x}_p(t^{(2)}, \overset{(k)}{x}_0) - \overset{(k)}{x}_p(t^{(1)}, \overset{(k)}{x}_0) = \int\limits_{t^{(1)}}^{t^{(2)}} [\overset{(k)}{f}(\tau, \overset{(k)}{x}_p(\tau, \overset{(k)}{x}_0), \frac{d\overset{(k)}{x}_p(\tau, \overset{(k)}{x}_0)}{d\tau})$$

$$-\frac{1}{T}\int\limits_0^T \overset{(k)}{f}(s, \overset{(k)}{x}_p(s, \overset{(k)}{x}_0), \frac{d\overset{(k)}{x}_p(s, \overset{(k)}{x}_0)}{ds})ds]d\tau + \frac{(t^{(2)} - t^{(1)})}{T}\overset{(k)}{H}_p$$

$$\times \{\overset{(k)}{\varphi}_p(\overset{(k)}{x}_p(0, \overset{(k)}{x}_0), \overset{(k)}{x}_p(T, \overset{(k)}{x}_0); \overset{(k)}{x}_p(t_1, \overset{(k)}{x}_0), ..., \overset{(k)}{x}_p(t_p, \overset{(k)}{x}_0))$$

$$-(\sum_{i=0}^p \overset{(k)}{A}_i + \overset{(k)}{C})\overset{(k)}{x}_0 - \sum_{i=1}^p \overset{(k)}{A}_i \int\limits_0^{t_i} [\overset{(k)}{f}(\tau, \overset{(k)}{x}_p(\tau, \overset{(k)}{x}_0), \frac{d\overset{(k)}{x}_p(\tau, \overset{(k)}{x}_0)}{d\tau})$$

$$-\frac{1}{T}\int\limits_0^T \overset{(k)}{f}(s, \overset{(k)}{x}_p(s, \overset{(k)}{x}_0), \frac{d\overset{(k)}{x}_p(s, \overset{(k)}{x}_0)}{ds})ds]d\tau\}.$$

Whence, by analogy with the proof of Lemma 4.1, we obtain the inequality

$$\|\overset{(k)}{x}_p(t^{(2)}, \overset{(k)}{x}_0) - \overset{(k)}{x}_p(t^{(1)}, \overset{(k)}{x}_0)\| \leq (t^{(2)} - t^{(1)})M'', \ M'' = const < \infty.$$

The constant M'' is independent of s_i. Therefore, the last inequality ensures the equicontinuity of the sequence $\{\overset{(s_i)}{x}_p(t, \overset{(s_i)}{x}_0)\}^{\infty}_{i=1}$ on the segment

$[0, T]$. The application of the Arzela theorem and the method of diagonalization allows us to choose a subsequence uniformly convergent in the coordinatewise meaning in $t \in [0, T]$ from this sequence. For convenience, we consider that the sequence itself $\{ \overset{(s_i)}{x}_p(t, \overset{(s_i)}{x_0}) \}_{i=1}^{\infty}$ is such one. We denote again s_i by k. Then, for all $\{ t^{(1)}, t^{(2)} \} \subset [0, T]$ such that $t^{(1)} < t^{(2)}$, relation (4.84) yields

$$\left\| \frac{d \overset{(k)}{x}_p(t^{(2)}, \overset{(k)}{x_0})}{dt} - \frac{d \overset{(k)}{x}_p(t^{(1)}, \overset{(k)}{x_0})}{dt} \right\|$$

$$\leq \left\| \overset{(k)}{f}(t^{(2)}, \overset{(k)}{x}_p(t^{(2)}, \overset{(k)}{x_0}), \frac{d \overset{(k)}{x}_p(t^{(2)}, \overset{(k)}{x_0})}{dt}) \right.$$

$$\left. - \overset{(k)}{f}(t^{(2)}, \overset{(k)}{x}_p(t^{(1)}, \overset{(k)}{x_0}), \frac{d \overset{(k)}{x}_p(t^{(1)}, \overset{(k)}{x_0})}{dt}) \right\|$$

$$+ \left\| \overset{(k)}{f}(t^{(2)}, \overset{(k)}{x}_p(t^{(1)}, \overset{(k)}{x_0}), \frac{d \overset{(k)}{x}_p(t^{(1)}, \overset{(k)}{x_0})}{dt}) \right.$$

$$\left. - \overset{(k)}{f}(t^{(1)}, \overset{(k)}{x}_p(t^{(1)}, \overset{(k)}{x_0}), \frac{d \overset{(k)}{x}_p(t^{(1)}, \overset{(k)}{x_0})}{dt}) \right\|.$$

In view of the estimate $K^* < 1$ and the Lipschitz property of the function $f(t, x, x_1)$ in (x, x_1) and in t, we obtain from the last inequality that

$$\left\| \frac{d \overset{(k)}{x}_p(t^{(2)}, \overset{(k)}{x_0})}{dt} - \frac{d \overset{(k)}{x}_p(t^{(1)}, \overset{(k)}{x_0})}{dt} \right\| \leq K^* \| \overset{(k)}{x}_p(t^{(2)}, \overset{(k)}{x_0}) - \overset{(k)}{x}_p(t^{(1)}, \overset{(k)}{x_0}) \|$$

$$+ K^* \left\| \frac{d \overset{(k)}{x}_p(t^{(2)}, \overset{(k)}{x_0})}{dt} - \frac{d \overset{(k)}{x}_p(t^{(1)}, \overset{(k)}{x_0})}{dt} \right\| + \tilde{K}(t^{(2)} - t^{(1)})$$

$$\leq \frac{1}{1 - K^*} [K^* \| \overset{(k)}{x}_p(t^{(2)}, \overset{(k)}{x_0}) - \overset{(k)}{x}_p(t^{(1)}, \overset{(k)}{x_0}) \| + \tilde{K}(t^{(2)} - t^{(1)})]$$

$$\leq \frac{1}{1 - K^*} [K^*(t^{(2)} - t^{(1)}) M'' + \tilde{K}(t^{(2)} - t^{(1)})]$$

$$= \frac{1}{1 - K^*} [\tilde{K} + K^* M''](t^{(2)} - t^{(1)}).$$

Hence, the sequence $\{ \frac{d \overset{(s_i)}{x}_p(t, \overset{(s_i)}{x_0})}{dt} \}_{i=1}^{\infty}$ is equicontinuous on the segment $[0, T]$. By applying the Arzela theorem and the method of diagonalization one more, it is possible to choose a subsequence $\{ \frac{d \overset{(k_i)}{x}_p(t, \overset{(k_i)}{x_0})}{dt} \}_{i=1}^{\infty}$ from it, which is uniformly convergent in $t \in [0, T]$ in the coordinatewise meaning simultaneously with the sequence $\{ \overset{(k_i)}{x}_p(t, \overset{(k_i)}{x_0}) \}_{i=1}^{\infty}$.

Let

$$\frac{d\,\overset{(k_i)}{x}}{dt} = \overset{(k_i)}{f}(t, \overset{(k_i)}{x}, \frac{d\,\overset{(k_i)}{x}}{dt}) + \overset{(k_i)}{\mu_p}, \; i \in Z^+,$$

be an appropriate subsequence of the sequence of equations (4.86), and let its each equation correspond to the boundary condition obtained from (4.30) by the replacement of the index n by k_i. Retaining the notation introduced in Subsection 4.3 ($\bar\mu_p = \lim\limits_{i\to\infty} \overset{(s_i)}{\mu_p}$, $\bar x_p(t, x_0) = \lim\limits_{i\to\infty} \overset{(k_i)}{x}_p(t, \overset{(k_i)}{x_0})$, where the limiting transitions are realized in the coordinatewise meaning), we will show that equalities (4.39) hold in the case under consideration.

We introduce the following notation:

$$\overset{(k_i)}{x_p}(t, \overset{(k_i)}{x_0}) = \overset{(k_i)}{x_p} = (\overset{(k_i)}{x_{1p}}, ..., \overset{(k_i)}{x_{k_i p}}),$$

$$\frac{d\,\overset{(k_i)}{x_p}(t, \overset{(k_i)}{x_0})}{dt} = \frac{d\,\overset{(k_i)}{x_p}}{dt} = (\frac{d\,\overset{(k_i)}{x_{1p}}}{dt}, ..., \frac{d\,\overset{(k_i)}{x_{k_i p}}}{dt}),$$

$$\overset{(k_i)}{f}(t, \overset{(k_i)}{x_p}) = \overset{(k_i)}{f} = (\overset{(k_i)}{f_1}, \overset{(k_i)}{f_2}, ..., \overset{(k_i)}{f_{k_i}}), \; i \in Z^+.$$

For a fixed natural ℓ, let us consider the sequence $\{\overset{(k_i)}{f_\ell}\}_{i=1}^{\infty}$, for which a finite number of elements corresponding to the values $i \in \{1, 2, ..., m\}$ are zeros, if $k_m < \ell \le k_{m+1}$.

Then, $\forall i \ge m + 1$,

$$\overset{(k_i)}{f_\ell} = f_\ell(t, \overset{(k_i)}{x_{1p}}, ..., \overset{(k_i)}{x_{k_i p}}, 0, 0, ..., \frac{d\,\overset{(k_i)}{x_{1p}}}{dt}, ..., \frac{d\,\overset{(k_i)}{x_{k_i p}}}{dt}, 0, 0, ...),$$

i.e., the indicated sequence takes the form

$$\underbrace{0, 0, ..., 0}_{m}, f_\ell(t, \overset{(k_{m+1})}{x_{1p}}, \overset{(k_{m+1})}{x_{2p}}, ..., x_{k_{m+1}p}, 0,$$

$$0, ..., \frac{d\,\overset{(k_{m+1})}{x_{1p}}}{dt}, \frac{d\,\overset{(k_{m+1})}{x_{2p}}}{dt}, ..., \frac{d\,x_{k_{m+1}p}}{dt}, 0, 0, ...),$$

$$f_\ell(t, \overset{(k_{m+2})}{x_{1p}}, \overset{(k_{m+2})}{x_{2p}}, ..., x_{k_{m+2}p}, 0,$$

$$0, ..., \frac{d\,\overset{(k_{m+2})}{x_{1p}}}{dt}, \frac{d\,\overset{(k_{m+2})}{x_{2p}}}{dt}, ..., \frac{d\,x_{k_{m+2}p}}{dt}, 0, 0, ...), ... \; .$$

We will show that

$$\{ \overset{(k_i)}{f_\ell} \}_{i=1}^\infty \to f_\ell(t, \bar{x}_p(t, x_0), \frac{d\bar{x}_p(t, x_0)}{dt})$$

as $i \to \infty$. For this purpose, we estimate the modulus of the difference

$$I_\ell^{(k_i)} = |f_\ell(t, \overset{(k_i)}{x_{1p}}, \overset{(k_i)}{x_{2p}}, ..., \frac{d\,\overset{(k_i)}{x_{1p}}}{dt}, \frac{d\,\overset{(k_i)}{x_{2p}}}{dt}, ...)$$

$$- f_\ell(t, \bar{x}_{1p}, \bar{x}_{2p}, ..., \frac{d\bar{x}_{1p}}{dt}, \frac{d\bar{x}_{2p}}{dt}, ...)|,$$

where $\overset{(k_i)}{x}_{k_i+1} = \overset{(k_i)}{x}_{k_i+2} = ... = 0$, $i \geq m+1$.

It is easy to verify that

$$I_\ell^{(k_i)} \leq A^0(\ell, g) + B^0(\ell, g),$$

where

$$A^0(\ell, g) = |f_\ell(t, \overset{(k_i)}{x_{1p}}(t), \overset{(k_i)}{x_{2p}}(t), ..., \frac{d\,\overset{(k_i)}{x_{1p}}(t)}{dt}, \frac{d\,\overset{(k_i)}{x_{2p}}(t)}{dt}, ...)$$

$$- f_\ell(t, \bar{x}_{1p}(t), ..., \bar{x}_{gp}(t), \overset{(k_i)}{x}_{(g+1)p}(t), \overset{(k_i)}{x}_{(g+2)p}(t), ..., \frac{d\bar{x}_{1p}(t)}{dt}, ...$$

$$..., \frac{d\bar{x}_{gp}(t)}{dt}, \frac{d\,\overset{(k_i)}{x}_{(g+1)p}(t)}{dt}, \frac{d\,\overset{(k_i)}{x}_{(g+2)p}(t)}{dt}, ...)|,$$

$$B^0(\ell, g) = |f_\ell(t, \bar{x}_{1p}(t), ..., \bar{x}_{gp}(t), \overset{(k_i)}{x}_{(g+1)p}(t), \overset{(k_i)}{x}_{(g+2)p}(t), ...$$

$$..., \frac{d\bar{x}_{1p}(t)}{dt}, ..., \frac{d\bar{x}_{gp}(t)}{dt}, \frac{d\,\overset{(k_i)}{x}_{(g+1)p}(t)}{dt}, \frac{d\,\overset{(k_i)}{x}_{(g+2)p}(t)}{dt}, ...)$$

$$- f_\ell(t, \bar{x}_{1p}, \bar{x}_{2p}, ..., \frac{d\bar{x}_{1p}}{dt}, \frac{d\bar{x}_{2p}}{dt}, ...)|.$$

Since $f(t, x, x_1) \in \hat{C}_{Lip}(x, x_1)$, the estimate

$$B^0(\ell, g) \leq \alpha(t)2(M_0 + M_1)\varepsilon(g) \leq \max_{t \in [0,T]} \alpha(t)2(M_0 + M_1)\varepsilon(g),$$

where $\varepsilon(g) \xrightarrow[g \to \infty]{} 0$, is valid. Then, choosing the index g^0 for an arbitrarily small number $\nu > 0$ so that the inequality $\varepsilon(g^0) < \nu$ is satisfied, we have

$$B^0(\ell, g^0) < 2(M_0 + M_1)\nu \max_{t \in [0,T]} \alpha(t).$$

For $A^0(\ell, g^0)$, the following inequality holds:

$$A(\ell, g^0) \leq \alpha(t)\varepsilon(0)[\sup\{\| \overset{(k_i)}{x}_{1p} - \bar{x}_{1p}| , ..., | \overset{(k_i)}{x}_{g^0 p} - \bar{x}_{g^0 p}|\}$$

$$+ \sup\{|\frac{d\,\overset{(k_i)}{x}_{1p}}{dt} - \frac{d\bar{x}_{1p}}{dt}| , ..., |\frac{d\,\overset{(k_i)}{x}_{g^0 p}}{dt} - \frac{d\bar{x}_{g^0 p}}{dt}|\}].$$

We note that, coordinatewise, $\overset{(k_i)}{x}_p \underset{i\to\infty}{\longrightarrow} \bar{x}_p$ and $\dfrac{d\,\overset{(k_i)}{x}_p}{dt} \underset{i\to\infty}{\longrightarrow} \dfrac{d\bar{x}_p}{dt}$ uniformly in $t \in [0,T]$. Therefore, there exists a number $N(\ell,\nu)$ such that, $\forall\, k_i \geq N(\ell,\nu)$,

$$\sup\{|\,\overset{(k_i)}{x}_{1p} - \bar{x}_{1p}|, ..., |\,\overset{(k_i)}{x}_{g^0 p} - \bar{x}_{g^0 p}|\}$$

$$+ \sup\{|\dfrac{d\,\overset{(k_i)}{x}_{1p}}{dt} - \dfrac{d\bar{x}_{1p}}{dt}|, ..., |\dfrac{d\,\overset{(k_i)}{x}_{g^0 p}}{dt} - \dfrac{d\bar{x}_{g^0 p}}{dt}|\} < \nu.$$

It is obvious that the inequality $k_i \geq N(\ell,\nu)$ is satisfied $\forall\, i \geq i_0 \in Z^+$. Hence, $\forall\, i \geq i_0$, $\quad I_\ell^{(k_i)} \leq \nu(2(M_0 + M_1) + \varepsilon(0)) \max_{t\in[0,T]} \alpha(t)$. Therefore,

$\{\overset{(k_i)}{f_\ell}\}_{i=1}^\infty \underset{i\to\infty}{\longrightarrow} f_\ell(t, \bar{x}_p, \dfrac{d\bar{x}_p}{dt})$ uniformly in $t \in [0,T]$. This means that, in the coordinatewise meaning,

$$\{\overset{(k_i)}{f}(t, \overset{(k_i)}{x}_p, \dfrac{d\,\overset{(k_i)}{x}_p}{dt})\}_{i=1}^\infty \underset{i\to\infty}{\longrightarrow} f(t, \bar{x}_p, \dfrac{d\bar{x}_p}{dt})$$

uniformly in $t \in [0,T]$. Then, executing the coordinatewise limiting transition as $i \to \infty$ in the equalities

$$\dfrac{d\,\overset{(k_i)}{x}_p(t, \overset{(k_i)}{x}_0)}{dt} = \overset{(k_i)}{f}(t, \overset{(k_i)}{x}_p(t, \overset{(k_i)}{x}_0), \dfrac{d\,\overset{(k_i)}{x}_p(t, \overset{(k_i)}{x}_0)}{dt}) + \overset{(k_i)}{\mu}_p, \quad \overset{(k_i)}{x}_p(0, \overset{(k_i)}{x}_0) = \overset{(k_i)}{x}_0,$$

we obtain the relations

$$\dfrac{d\bar{x}_p(t, x_0)}{dt} = f(t, \bar{x}_p(t, x_0), \dfrac{d\bar{x}_p(t, x_0)}{dt}) + \bar{\mu}_p, \quad \bar{x}_p(0, x_0) = x_0. \tag{4.87}$$

In addition, equality (4.41) holds $\forall\, i \in Z^+$ for the function $\overset{(k_i)}{x}_p(t, \overset{(k_i)}{x}_0)$. Let us pass coordinatewise in equality (4.41) to the limit as $i \to \infty$. Like the proof of Lemma 4.1, we can conclude that the function $\bar{x}_p(t, x_0)$ satisfies equality (4.26).

Hence, by virtue of the validity of equalities (4.87), the function $\bar{x}_p(t, x_0)$ is a solution of the boundary-value problem (4.55), (4.26). The uniqueness of the control μ_p for a fixed x_0 yields equalities (4.39). The first of these equalities indicates that $\dfrac{d\bar{x}_p(t,x_0)}{dt} = \dfrac{dx_p(t,x_0)}{dt}$.

The proof of Lemma 4.3 is completed analogously to that of Lemma 4.1. $\qquad\square$

Retaining the content of the designations $\beta_\varphi^*(x, n_0, p_0)$ and $D_{\beta\varphi}^*$ from section 4.3, where we set $K = K^*$, we introduce additionally such notation:

$$\gamma_\varphi^*(x, n_0, p_0) = \dfrac{\frac{2}{K^*} - 4}{T \overset{(n_0)}{\underset{i=1}{\sum}} \|A_i\| + \frac{2K_\varphi}{K^*}}(M_\varphi + (\sum_{i=0}^\infty \|A_i\| + \|C\|)\|x\|) + \dfrac{M}{K^*};$$

$D_{1\gamma\varphi}^*$ is a set, every point of which belongs to the set D_1 together with its $\gamma_\varphi^*(x)$-neighborhood.

Theorem 4.9. *Let $f(t, x, x_1) \in \hat{C}_{Lip}(x, x_1) \cap C_{Lip}(t), \varphi(\psi) \in \hat{C}_{Lip}(\psi)$, let the intersection of the set $D_{\beta\varphi}^*$ and $D_{1\gamma\varphi}^*$ be nonempty, and let conditions a_1^1, a_2^1, and a_3 from Subsection 4.3 and conditions c_1^1, c_2^{01}, and c_3^1, where $K_{\varphi p} = K_\varphi$, be satisfied $\forall\, n \geq n_0,\ p \geq p_0$. Then the limiting relations (4.42) and (4.43) and the relations*

$$\frac{dx^*(t, x_0)}{dt} = \lim_{p\to\infty}\Big(\lim_{n\to\infty} \frac{d\,\overset{(n)}{\tilde{x}}_p(t, \overset{(n)}{x_0})}{dt}\Big), \qquad (4.88)$$

hold $\forall\, x_0 \in D_{\beta\varphi}^* \cap D_{1\gamma\varphi}^*$. Here, the convergences in n and in p are coordinatewise and in the norm of the space $\mathfrak{M}$, respectively, and $x^*(t, x_0)$ and μ are defined in Theorem 4.8.

If condition c_2^{01} is replaced by the condition

$$c_2^{01*}) \qquad \max\{\ \|Q_{0p}^1\|,\ \|Q_{01p}^1\|\ \} \leq q = const < 1,$$

then there exists no other value of $\mu \in \mathfrak{M}$ such that the solution of the equation $\frac{dx}{dt} = f(t, x, \frac{dx}{dt}) + \mu$ with the initial condition $x(0) = x_0$ would satisfy the boundary condition (4.23).

Proof. We show firstly that the inclusion $x_0 \in D_{\beta\varphi}^* \cap D_{1\gamma\varphi}^*$ yields the inclusions

$$x_0 \in D_{\beta\varphi} \cap D_{1\gamma\varphi}, \ x_0 \in D_{\beta\varphi p} \cap D_{1\gamma\varphi p}$$

and

$$\overset{(n)}{x}_0 \in \tilde{D}^{(n)}_{\underset{\beta\ \varphi p}{(n)}} \cap \tilde{D}^{(n)}_{1\underset{\gamma\ \varphi p}{(n)}} \quad \forall\, n \geq n_0,\ p \geq p_0.$$

Then, like the proof of Theorem 4.5, we verify the validity of estimate (4.44).

First of all, we will prove that the implication

$$(x_0 \in D_{\beta\varphi}^*) \quad \Rightarrow \quad (x_0 \in D_{\beta\varphi}) \wedge (x_0 \in D_{\beta\varphi p}) \wedge (\overset{(n)}{x}_0 \in \tilde{D}^{(n)}_{\underset{\beta\ \varphi p}{(n)}}) \quad (4.89)$$

is proper $\forall\, n \geq n_0,\ p \geq p_0$. It was established in the proof of Lemma 4.3 that

$$(x_0 \in D_{\beta\varphi p}^*) \quad \Rightarrow \quad (x_0 \in D_{\beta\varphi p}) \wedge (\overset{(n)}{x}_0 \in \tilde{D}^{(n)}_{\underset{\beta\ \varphi p}{(n)}}) \quad \forall n \geq n_0. \quad (4.90)$$

Comparing (4.89) and (4.90), we see that, in order to prove the validity of implication (4.89), it remains to substantiate the validity of the implications

$$(x_0 \in D^*_{\beta\varphi}) \quad \Rightarrow \quad (x_0 \in D^*_{\beta\varphi p}) \ \forall p \geq p_0,$$

$$(x_0 \in D^*_{\beta\varphi}) \quad \Rightarrow \quad (x_0 \in D_{\beta\varphi}). \tag{4.91}$$

Setting $K_{\varphi p} = K_\varphi$, we have, for all $p \geq p_0$,

$$\beta^*_{\varphi p}(x, n_0) = \frac{\frac{2}{K^* T} - 1}{\sum\limits_{i=1}^{p} \| \overset{(n_0)}{A_i} \| + \frac{2K_\varphi}{K^* T}} (M_\varphi + (\sum_{i=0}^{p} \|A_i\| + \|C\|)\|x\|) + \frac{M}{K^*}$$

$$\leq \frac{\frac{2}{K^* T} - 1}{\sum\limits_{i=1}^{p_0} \| \overset{(n_0)}{A_i} \| + \frac{2K_\varphi}{K^* T}} (M_\varphi + (\sum_{i=0}^{\infty} \|A_i\| + \|C\|)\|x\|) + \frac{M}{K^*} = \beta^*_{\varphi p}(x, n_0, p_0),$$

which yields the first implication in (4.91). The proof of the second implication or, what is the same, the proof of the inequality $\beta_\varphi < \beta^*_\varphi$ is realized by the scheme of the proof of the inequality $\beta_{\varphi p} < \beta^*_{\varphi p}$ in Lemma 4.1.

Implication (4.89) is proved. Analogously, we can show that, $\forall\, n \geq n_0,\ p \geq p_0,$

$$(x_0 \in D^*_{1\gamma\varphi}) \quad \Rightarrow \quad (x_0 \in D_{1\gamma\varphi}) \wedge (x_0 \in D_{1\gamma\varphi p}) \wedge (\overset{(n)}{x}_0 \in \tilde{D}^{(n)}_{1\ \overset{(n)}{\gamma}_{\varphi p}}).$$

The inclusion $f(t, x, x_1) \in \hat{C}_{Lip}(x, x_1)$ implies that we can set, as above, $K = K_1 = \max\limits_{t \in [0,T]} \alpha(t)\varepsilon(0) = K^*$ in the second condition in (4.77). The functions $x_p(t, x_0)$ and $x^*(t, x_0)$ and differentiable and satisfy the equalities written prior to the statements of Corollary 4.9 and Theorem 4.8, respectively. Therefore, it is easy to verify the validity of the relations

$$\|x_p(t, x_0) - x^*(t, x_0)\| \leq \bar{\Gamma}_1 + \bar{\Gamma}_2 + \bar{\Gamma}_3,$$

$$\|\frac{dx_p(t, x_0)}{dt} - \frac{dx^*(t, x_0)}{dt}\| \leq \Gamma_1^1 + \frac{1}{T}(\bar{\Gamma}_2 + \bar{\Gamma}_3),$$

where

$$\bar{\Gamma}_1 = \| \int_0^t [f(\tau, x^*(\tau, x_0), \frac{dx^*(\tau, x_0)}{d\tau}) - f(\tau, x_p(\tau, x_0), \frac{dx_p(\tau, x_0)}{d\tau})$$

$$-\frac{1}{T}\int_0^T (f(s, x^*(s, x_0), \frac{dx^*(s, x_0)}{ds}) - f(s, x_p(s, x_0), \frac{dx_p(s, x_0)}{ds}))ds]d\tau\|$$

$$\le \alpha_1(t)K^*[\sup_{t\in[0,T]} \|x^*(t, x_0) - x_p(t, x_0)\|$$

$$+ \sup_{t\in[0,T]} \|\frac{dx^*(t, x_0)}{dt} - \frac{dx_p(t, x_0)}{dt}\|];$$

$$\Gamma_1^1 = \|f(\tau, x^*(\tau, x_0), \frac{dx^*(\tau, x_0)}{d\tau}) - f(\tau, x_p(\tau, x_0), \frac{dx_p(\tau, x_0)}{d\tau}) -$$

$$-\frac{1}{T}\int_0^T (f(s, x^*(s, x_0), \frac{dx^*(s, x_0)}{ds}) - f(s, x_p(s, x_0), \frac{dx_p(s, x_0)}{ds}))ds\|$$

$$\le 2K^*[\sup_{t\in[0,T]} \|x^*(t, x_0) - x_p(t, x_0)\| + \sup_{t\in[0,T]} \|\frac{dx^*(t, x_0)}{dt} - \frac{dx_p(t, x_0)}{dt}\|];$$

$$\bar{\Gamma}_2 \le K^{(1)}\|H - H_p\| + \|H\|K_\varphi \sup_{t\in[0,T]} \|x^*(t, x_0) - x_p(t, x_0)\|$$

$$+ \|H\|M_0\{\sum_{i=p+1}^\infty \|A_i\| + \delta_0(p+2)\},$$

since $\bar{\Gamma}_2$ coincides with Γ_2 (see the proof of Theorem 4.5);

$$\bar{\Gamma}_3 = \|H\sum_{i=1}^\infty A_i \int_0^{t_i} [f(\tau, x^*(\tau, x_0), \frac{dx^*(\tau, x_0)}{d\tau})$$

$$-\frac{1}{T}\int_0^T f(s, x^*(s, x_0), \frac{dx^*(s, x_0)}{ds})ds]d\tau$$

$$- H_p\sum_{i=1}^p A_i \int_0^{t_i} [f(\tau, x_p(\tau, x_0), \frac{dx_p(\tau, x_0)}{d\tau})$$

$$-\frac{1}{T}\int_0^T f(s, x_p(s, x_0), \frac{dx_p(s, x_0)}{ds})ds]d\tau\|$$

$$\leq \|H\|\|\sum_{i=1}^{p} A_i \int_0^{t_i} [f(\tau, x^*(\tau, x_0), \frac{dx^*(\tau, x_0)}{d\tau}) - f(\tau, x_p(\tau, x_0), \frac{dx_p(\tau, x_0)}{d\tau})$$

$$- \frac{1}{T} \int_0^{T} (f(s, x^*(s, x_0), \frac{dx^*(s, x_0)}{ds}) - f(s, x_p(s, x_0), \frac{dx_p(s, x_0)}{ds}))ds]d\tau\|$$

$$+ \|H\|\|\sum_{i=p+1}^{\infty} A_i \int_0^{t_i} [f(\tau, x^*(\tau, x_0), \frac{dx^*(\tau, x_0)}{d\tau})$$

$$- \frac{1}{T} \int_0^{T} f(s, x^*(s, x_0), \frac{dx^*(s, x_0)}{ds})ds]d\tau\|$$

$$+ \|H - H_p\|\|\sum_{i=1}^{p} A_i \int_0^{t_i} [f(\tau, x_p(\tau, x_0), \frac{dx_p(\tau, x_0)}{d\tau})$$

$$- \frac{1}{T} \int_0^{T} f(s, x_p(s, x_0), \frac{dx_p(s, x_0)}{ds})ds]d\tau\|$$

$$\leq \|H\|\frac{K^*T}{2}\sum_{i=1}^{\infty} \|A_i\|[\sup_{t\in[0,T]} \|x^*(t, x_0) - x_p(t, x_0)\|$$

$$+ \sup_{t\in[0,T]} \|\frac{dx^*(t, x_0)}{dt} - \frac{dx_p(t, x_0)}{dt}\|]$$

$$+ \|H\|\frac{TM}{2}\sum_{i=p+1}^{\infty} \|A_i\| + \|H - H_p\|\frac{TM}{2}\sum_{i=1}^{\infty} \|A_i\|.$$

Setting

$$\sup_{t\in[0,T]} \|x^*(t, x_0) - x_p(t, x_0)\| = \xi,$$

$$\sup_{t\in[0,T]} \|\frac{dx^*(t, x_0)}{dt} - \frac{dx_p(t, x_0)}{dt}\| = \xi_1,$$

we obtain the estimates

$$\xi \leq \frac{K^*T}{2}(\xi + \xi_1) + K^{(1)}\|H - H_p\| + \|H\|K_\varphi\xi$$

$$+ \|H\|M_0\{\sum_{i=p+1}^{\infty} \|A_i\| + \delta_0(p + 2)\}$$

$$+ \frac{\|H\|K^*T}{2}(\xi + \xi_1) \sum_{i=1}^{\infty} \|A_i\|$$

$$+ \|H\|\frac{TM}{2} \sum_{i=p+1}^{\infty} \|A_i\| + \|H - H_p\|\frac{TM}{2} \sum_{i=1}^{\infty} \|A_i\|,$$

$$\xi_1 \le 2K^*(\xi + \xi_1) + K^{(1)}\|H - H_p\| + \frac{\|H\|}{T}K_\varphi \xi$$

$$+ \frac{\|H\|M_0}{T}\{ \sum_{i=p+1}^{\infty} \|A_i\| + \delta_0(p + 2)\}$$

$$+ \frac{K^*\|H\|}{2}(\xi + \xi_1) \sum_{i=1}^{\infty} \|A_i\|$$

$$+ \|H\|\frac{M}{2} \sum_{i=p+1}^{\infty} \|A_i\| + \|H - H_p\|\frac{M}{2} \sum_{i=1}^{\infty} \|A_i\|,$$

which give jointly the componentwise inequality

$$\begin{bmatrix} \xi \\ \xi_1 \end{bmatrix} \le Q_0^1 \begin{bmatrix} \xi \\ \xi_1 \end{bmatrix} + \|H - H_p\|K^{(1)} \begin{bmatrix} 1 \\ 1 \end{bmatrix} + (\|H - H_p\|\frac{TM}{2} \sum_{i=1}^{\infty} \|A_i\|$$

$$+ \|H\|(M_0 + \frac{TM}{2}) \sum_{i=p+1}^{\infty} \|A_i\| + \|H\|M_0\delta_0(p + 2)) \begin{bmatrix} 1 \\ 1/T \end{bmatrix},$$

where we set $K_1 = K = K^*$ in the matrix Q_0^1.

We note that $K^{(1)}$ is a constant independent of p, the series $\sum_{i=1}^{\infty} \|A_i\|$ is convergent, and $\delta_0(p)$ tends to as $p \to \infty$. Therefore, with regard for condition c_1^1, relation (4.44), and the last inequality, we have $\max\{\xi, \xi_1\} \le \eta(p)\|(E_2 - Q_0^1)^{-1}\|$, where $\eta(p) \xrightarrow[p \to \infty]{} 0$. This means that, for all $x_0 \in D_{\beta\varphi}^* \cap D_{1\gamma\varphi}^*$, the sequences $\{x_p(t, x_0)\}_{p=p_0}^{\infty}$ and $\{\frac{dx_p(t,x_0)}{dt}\}_{p=p_0}^{\infty}$ converge in the norm of the space $\mathfrak{M}$ as $p \to \infty$ to $x^*(t, x_0)$ and $\frac{dx^*(t,x_0)}{dt}$, respectively. Using (4.37) and (4.85), we obtain (4.42) and (4.88). Since relation (4.42) yields (4.43), and condition c_1^{01} yields condition c_1^1, Remark 4.5 is valid, which completes the proof. $\qquad\square$

Corollary 4.10. *Under conditions of Theorem 4.9, the inequality*

$$\max_{t \in [0,T]} \{\|x^*(t, x_0) - \overset{(n)}{x_p}(t, \overset{(n)}{x_0})\|, \|\frac{dx^*(t, x_0)}{dt} - \frac{d\overset{(n)}{x_p}(t, \overset{(n)}{x_0})}{dt}\|\}$$

$$\le \bar{L}_1(p) + \bar{L}_2(n), \quad (4.92)$$

where the sum $\bar{L}_1(p) + \bar{L}_2(n)$ is uniformly bounded in $\{p \geq p_0, n \geq n_0\} \subset Z^+$, and $L_1(p) \to 0$ as $p \to \infty$, is satisfied.

Proof.   Using the estimate of the uniform convergence of an approximate solution to the exact one found in the proof of Theorem 4.8, we obtain, for all $t \in [0, T]$, $x_0 \in D^*_{\beta\varphi}$, the inequalities:

$$\sup_{t\in[0,T]} \{\|x^*(t, x_0) - \overset{(n)}{x_p}(t, \overset{(n)}{x_0})\|, \|\frac{dx^*(t, x_0)}{dt} - \frac{d\overset{(n)}{x_p}(t, \overset{(n)}{x_0})}{dt}\|\}$$

$$\leq \sup_{t\in[0,T]} \{\|x^*(t, x_0) - x_p(t, x_0)\|, \|\frac{dx^*(t, x_0)}{dt} - \frac{dx_p(t, x_0)}{dt}\|\}$$

$$+ \sup_{t\in[0,T]} \{\|x_p(t, x_0) - \overset{(n)}{x_p}(t, \overset{(n)}{x_0})\|, \|\frac{dx_p(t, x_0)}{dt} - \frac{d\overset{(n)}{x_p}(t, \overset{(n)}{x_0})}{dt}\|\}$$

$$\leq \|Q_0^{1p}(E_2 - Q_0^1)^{-1}Z_2^0\|$$

$$+ \sup_{t\in[0,T]} \{\|x_p(t, x_0) - \overset{(n)}{x_p}(t, \overset{(n)}{x_0})\|, \|\frac{dx_p(t, x_0)}{dt} - \frac{d\overset{(n)}{x_p}(t, \overset{(n)}{x_0})}{dt}\|\}.$$

We now write the estimates

$$\|x_p(t, x_0) - \overset{(n)}{x_p}(t, \overset{(n)}{x_0})\| \leq \bar{\Gamma}_1^0 + \bar{\Gamma}_2^0 + \bar{\Gamma}_3^0,$$

$$\|\frac{dx_p(t, x_0)}{dt} - \frac{d\overset{(n)}{x_p}(t, \overset{(n)}{x_0})}{dt}\| \leq \Gamma_1^{01} + \frac{1}{T}(\bar{\Gamma}_2^0 + \bar{\Gamma}_3^0),$$

where

$$\bar{\Gamma}_1^0 = \| \int_0^t [f(\tau, x_p(\tau, x_0), \frac{dx_p(\tau, x_0)}{d\tau}) - \overset{(n)}{f}(\tau, \overset{(n)}{x_p}(\tau, \overset{(n)}{x_0}), \frac{d\overset{(n)}{x_p}(\tau, \overset{(n)}{x_0})}{d\tau})$$

$$- \frac{1}{T} \int_0^T (f(s, x_p(s, x_0), \frac{dx_p(s, x_0)}{ds}) - \overset{(n)}{f}(s, \overset{(n)}{x_p}(s, \overset{(n)}{x_0}), \frac{d\overset{(n)}{x_p}(s, \overset{(n)}{x_0})}{ds}))ds]d\tau\|$$

$$\leq \alpha_1(t) \sup_{t\in[0,T]} \|f(t, x_p(t, x_0), \frac{dx_p(t, x_0)}{dt}) - \overset{(n)}{f}(t, \overset{(n)}{x_p}(t, \overset{(n)}{x_0}), \frac{d\overset{(n)}{x_p}(t, \overset{(n)}{x_0})}{dt})\|;$$

$$\Gamma_1^{01} = \|f(\tau, x_p(\tau, x_0), \frac{dx_p(\tau, x_0)}{d\tau}) - \overset{(n)}{f}(\tau, \overset{(n)}{x_p}(\tau, \overset{(n)}{x_0}), \frac{d\overset{(n)}{x_p}(\tau, \overset{(n)}{x_0})}{d\tau})$$

$$- \frac{1}{T} \int_0^T (f(s, x_p(s, x_0), \frac{dx_p(s, x_0)}{ds}) - \overset{(n)}{f}(s, \overset{(n)}{x_p}(s, \overset{(n)}{x_0}), \frac{d\overset{(n)}{x_p}(s, \overset{(n)}{x_0})}{ds}))ds\|$$

$$\leq 2 \sup_{t\in[0,T]} \|f(t, x_p(t, x_0), \frac{dx_p(t, x_0)}{dt}) - \overset{(n)}{f}(t, \overset{(n)}{x_p}(t, \overset{(n)}{x_0}), \frac{d\overset{(n)}{x_p}(t, \overset{(n)}{x_0})}{dt})\|;$$

$$\Gamma_3^0 = \|H_p \sum_{i=1}^{p} A_i \int_0^{t_i} [f(\tau, x_p(\tau, x_0), \frac{dx_p(\tau, x_0)}{d\tau})$$

$$-\frac{1}{T} \int_0^{T} f(s, x_p(s, x_0), \frac{dx_p(s, x_0)}{ds})ds]d\tau$$

$$-\overset{(n)}{H_p} \sum_{i=1}^{p} \overset{(n)}{A_i} \int_0^{t_i} [\overset{(n)}{f}(\tau, \overset{(n)}{x_p}(\tau, \overset{(n)}{x_0}), \frac{d\overset{(n)}{x_p}(\tau, \overset{(n)}{x_0})}{d\tau})$$

$$-\frac{1}{T} \int_0^{T} \overset{(n)}{f}(s, \overset{(n)}{x_p}(s, \overset{(n)}{x_0}), \frac{d\overset{(n)}{x_p}(s, \overset{(n)}{x_0})}{ds})ds]d\tau\|;$$

$\bar{\Gamma}_2^0$ has the same form as $\bar{\Gamma}_2$ in Corollary 4.4, $n \geq n_0$, $p \geq p_0$.

The following inequalities hold:

$$\|f(t, x_p(t, x_0), \frac{dx_p(t, x_0)}{dt}) - \overset{(n)}{f}(t, \overset{(n)}{x_p}(t, \overset{(n)}{x_0}), \frac{d\overset{(n)}{x_p}(t, \overset{(n)}{x_0})}{dt})\|$$

$$\leq \|f(t, x_p(t, x_0), \frac{dx_p(t, x_0)}{dt}) - f(t, \overset{(n)}{x_p}(t, \overset{(n)}{x_0}), \frac{d\overset{(n)}{x_p}(t, \overset{(n)}{x_0})}{dt})\|$$

$$+ \|f(t, \overset{(n)}{x_p}(t, \overset{(n)}{x_0}), \frac{d\overset{(n)}{x_p}(t, \overset{(n)}{x_0})}{dt}) - \overset{(n)}{f}(t, \overset{(n)}{x_p}(t, \overset{(n)}{x_0}), \frac{d\overset{(n)}{x_p}(t, \overset{(n)}{x_0})}{dt})\|$$

$$\leq K^*[\|x_p(t, x_0) - \overset{(n)}{x_p}(t, \overset{(n)}{x_0})\| + \|\frac{dx_p(t, x_0)}{dt} - \frac{d\overset{(n)}{x_p}(t, \overset{(n)}{x_0})}{dt}\|]$$

$$+\sup\{|f_{n+1}(t, \overset{(n)}{x_p}(t, \overset{(n)}{x_0}), \frac{d\overset{(n)}{x_p}(t, \overset{(n)}{x_0})}{dt})|, |f_{n+2}(t, \overset{(n)}{x_p}(t, \overset{(n)}{x_0}), \frac{d\overset{(n)}{x_p}(t, \overset{(n)}{x_0})}{dt})|, ...\};$$

$$\bar{\Gamma}_2^0 \leq \|H_p\|K_{\varphi p} \sup_{t\in[0,T]} \|x_p(t, x_0) - \overset{(n)}{x_p}(t, \overset{(n)}{x_0})\|$$

$$+ \|H_p\| \sup_{\psi_i \in D} \{|\varphi_{n+1}(\psi_1, \psi_2, ..., \psi_{p+2}, 0, 0, ...)|,$$

$$|\varphi_{n+2}(\psi_1, \psi_2, ..., \psi_{p+2}, 0, 0, ...)|, ...\}$$

$$+ \|G_p\|\|x_0 - \overset{(n)}{x_0}\| + M_0\|G_p - \overset{(n)}{G_p}\|,$$

where

$$G_p = H_p(\sum_{i=0}^{p} A_i + C) \ , \ \overset{(n)}{G_p} = \overset{(n)}{H_p}(\sum_{i=0}^{p} \overset{(n)}{A_i} + \overset{(n)}{C});$$

$$\bar{\Gamma}_3^0 \le \|H_p\| \sum_{i=1}^{p} \{\|A_i\|\| \int_0^{t_i} [f(\tau, x_p(\tau, x_0), \frac{dx_p(\tau, x_0)}{d\tau})$$

$$- \overset{(n)}{f}(\tau, \overset{(n)}{x_p}(\tau, \overset{(n)}{x_0}), \frac{d \overset{(n)}{x_p}(\tau, \overset{(n)}{x_0})}{d\tau}) - \frac{1}{T} \int_0^{T} (f(s, x_p(s, x_0), \frac{dx_p(s, x_0)}{ds})$$

$$- \overset{(n)}{f}(s, \overset{(n)}{x_p}(s, \overset{(n)}{x_0}), \frac{d \overset{(n)}{x_p}(s, \overset{(n)}{x_0})}{ds}))ds]d\tau\| + \|A_i - \overset{(n)}{A_i}\|$$

$$\times \| \int_0^{t_i} [\overset{(n)}{f}(\tau, \overset{(n)}{x_p}(\tau, \overset{(n)}{x_0}), \frac{d \overset{(n)}{x_p}(\tau, \overset{(n)}{x_0})}{d\tau})$$

$$- \frac{1}{T} \int_0^{T} \overset{(n)}{f}(s, \overset{(n)}{x_p}(s, \overset{(n)}{x_0}), \frac{d \overset{(n)}{x_p}(s, \overset{(n)}{x_0})}{ds})ds]d\tau\|\}$$

$$+ \|H_p - \overset{(n)}{H_p}\| \sum_{i=1}^{p} \| \overset{(n)}{A_i} \|\| \int_0^{t_i} [\overset{(n)}{f}(\tau, \overset{(n)}{x_p}(\tau, \overset{(n)}{x_0}), \frac{d \overset{(n)}{x_p}(\tau, \overset{(n)}{x_0})}{d\tau})$$

$$- \frac{1}{T} \int_0^{T} \overset{(n)}{f}(s, \overset{(n)}{x_p}(s, \overset{(n)}{x_0}), \frac{d \overset{(n)}{x_p}(s, \overset{(n)}{x_0})}{ds})ds]d\tau\|$$

$$\le \|H_p\| \sum_{i=1}^{p} \{\|A_i\| \frac{T}{2} \sup_{t \in [0,T]} \|f(t, x_p(t, x_0), \frac{dx_p(t, x_0)}{dt})$$

$$- \overset{(n)}{f}(t, \overset{(n)}{x_p}(t, \overset{(n)}{x_0}), \frac{d \overset{(n)}{x_p}(t, \overset{(n)}{x_0})}{dt})\|$$

$$+ \frac{TM}{2}\|A_i - \overset{(n)}{A_i}\|\} + \|H_p - \overset{(n)}{H_p}\| \sum_{i=1}^{p} \| \overset{(n)}{A_i} \| \frac{TM}{2};$$

$$\|G_p - \overset{(n)}{G_p}\| \le \|H_p\|(\sum_{i=0}^{p} \|A_i - \overset{(n)}{A_i}\| + \|C - \overset{(n)}{C}\|)$$

$$+ \|H_p - \overset{(n)}{H_p}\|(\sum_{i=0}^{p} \| \overset{(n)}{A_i} \| + \| \overset{(n)}{C} \|).$$

Retaining the content of the notation $\xi_*(p,n)$, $\xi^0(p,n)$, $\eta^0(n)$, $\mu(n,t,x)$, $\mu^0(n)$, $\overline{A_i}^n$, $\overline{H}_p^n$, $\overline{C}^n$, we introduce one more notation

$$\xi^1(p,n) = \sup_{t\in[0,T]} \left\| \frac{dx_p(t,x_0)}{dt} - \frac{d\overset{(n)}{x_p}(t,\overset{(n)}{x_0})}{dt} \right\|.$$

Then, with regard for the previous inequalities, we obtain the estimates

$$\xi^0(p,n) \le \frac{T}{2}(K^*\xi^0(p,n) + K^*\xi^1(p,n) + \mu^0(n)) + \|H_p\|K_{\varphi p}\xi^0(p,n)$$

$$+ \|G_p\|\eta^0(n) + \|H_p\|\xi_*(p,n) + M_0\|H_p\|(\sum_{i=0}^{p}\overline{A_i}^n + \overline{C}^n)$$

$$+ M_0\overline{H}_p^n(\sum_{i=0}^{p}\|\overset{(n)}{A_i}\| + \|\overset{(n)}{C}\|)$$

$$+ \|H_p\|\sum_{i=1}^{p}\|A_i\|\frac{T}{2}(K^*\xi^0(p,n) + K^*\xi^1(p,n) + \mu^0(n))$$

$$+ \|H_p\|\frac{TM}{2}\overline{A_i}^n + \overline{H}_p^n\frac{TM}{2}\sum_{i=1}^{p}\|A_i\|;$$

$$\xi^1(p,n) \le 2(K^*\xi^0(p,n) + K^*\xi^1(p,n) + \mu^0(n)) + \frac{1}{T}[\|H_p\|K_{\varphi p}\xi^0(p,n)$$

$$+ \|G_p\|\eta^0(n) + \|H_p\|\xi_*(p,n) + M_0\|H_p\|(\sum_{i=0}^{p}\overline{A_i}^n + \overline{C}^n) + M_0\overline{H}_p^n(\sum_{i=0}^{p}\|\overset{(n)}{A_i}\|$$

$$+ \|\overset{(n)}{C}\|) + \|H_p\|\sum_{i=1}^{p}\|A_i\|\frac{T}{2}(K\xi^0(p,n) + K_1\xi^1(p,n) + \mu^0(n))$$

$$+ \|H_p\|\frac{TM}{2}\overline{A_i}^n + \overline{H}_p^n\frac{TM}{2}\sum_{i=1}^{p}\|A_i\|]$$

or, by joining them in a vector inequality,

$$\begin{bmatrix} \xi^0(p,n) \\ \xi^1(p,n) \end{bmatrix} \le Q_{0p}^1 \begin{bmatrix} \xi^0(p,n) \\ \xi^1(p,n) \end{bmatrix} + \begin{bmatrix} \frac{T\mu^0(n)}{2} + \bar{\Phi}(p,n) \\ 2\mu^0(n) + \frac{1}{T}\bar{\Phi}(p,n) \end{bmatrix},$$

where

$$\bar{\Phi}(p,n) = \|G_p\|\eta^0(n) + \|H_p\|(M_0(\sum_{i=0}^{p}\overline{A_i}^n + \overline{C}^n) + \xi_*(p,n))$$

$$+ M_0\overline{H}_p^n(\sum_{i=0}^{p}\|\overset{(n)}{A_i}\| + \|\overset{(n)}{C}\|) + \frac{\|H_p\|T\mu^0(n)}{2}\sum_{i=1}^{p}\|A_i\|$$

$$+\|H_p\|\frac{TM}{2}\overline{A_i}^{\,n} + \overline{H}_p^{\,n}\frac{TM}{2}\sum_{i=1}^{p}\|A_i\|.$$

By virtue of condition c_2^1, the last inequality yields

$$\max_{t\in[0,T]}\{\|x_p(t,x_0) - \overset{(n)}{x_p}(t,\overset{(n)}{x_0})\|, \|\frac{dx_p(t,x_0)}{dt} - \frac{d\,\overset{(n)}{x_p}(t,\overset{(n)}{x_0})}{dt}\|\}$$

$$\leq \|(E_2 - Q_{0p}^1)^{-1}\|[\mu^0(n)\max\{2,\frac{T}{2}\} + \max\{1,\frac{1}{T}\}\bar{\Phi}(p,n)].$$

Let us replace the signs of sums $\sum\limits_{i=0}^{p}$, $\sum\limits_{i=1}^{p}$ in the formula for $\bar{\Phi}(p,n)$ by

$\sum\limits_{i=0}^{\infty}$, $\sum\limits_{i=1}^{\infty}$ respectively, $\|H_p\|$ by the constant

$$\frac{\frac{2}{KT}-1}{\sum\limits_{i=0}^{p_0}\|\overset{(n_0)}{A_i}\| + \frac{2K_\varphi}{KT}},$$

the constants $K_{\varphi p}$ by K_φ, Q_{0p}^1 by Q_0^1, and the quantity $\xi_*(p,n)$ by

$$\xi_*^0(n) = \sup_{\psi_i\in D}\{|\varphi_{n+1}(\psi_1,\psi_2,...)|, |\varphi_{n+2}(\psi_1,\psi_2,...)|, ...\}.$$

We obtain the estimate

$$\max_{t\in[0,T]}\{\|x_p(t,x_0) - \overset{(n)}{x_p}(t,\overset{(n)}{x_0})\|, \|\frac{dx_p(t,x_0)}{dt} - \frac{d\,\overset{(n)}{x_p}(t,\overset{(n)}{x_0})}{dt}\|\}$$

$$\leq \|(E_2 - Q_0^1)^{-1}\|[\mu^0(n)\max\{2,\frac{T}{2}\} + \max\{1,\frac{1}{T}\}\bar{\Phi}^*(n)],$$

i.e., $\forall\, t\in[0,T]$, $p\geq p_0$, $n\geq n_0$,

$$\max_{t\in[0,T]}\{\|x^*(t,x_0) - \overset{(n)}{x_p}(t,\overset{(n)}{x_0})\|, \|\frac{dx^*(t,x_0)}{dt} - \frac{d\,\overset{(n)}{x_p}(t,\overset{(n)}{x_0})}{dt}\|\} \leq \|Q_0^{1p}(E_2$$

$$- Q_0^1)^{-1}Z_2^0\| + \|(E_2 - Q_0^1)^{-1}\|[\mu^0(n)\max\{2,\frac{T}{2}\} + \max\{1,\frac{1}{T}\}\bar{\Phi}^*(n)].$$

This proves inequality (4.92) and, hence, the corollary, since

$$\bar{L}_1(p) = \|Q_0^{1p}(E_2 - Q_0^1)^{-1}Z_2^0\| \leq \|Q_0^{1p}\|\|(E_2 - Q_0^1)^{-1}Z_2^0\|$$

$$\leq \|Q_0^1\|^p\|(E_2 - Q_0^1)^{-1}Z_2^0\| \xrightarrow[p\to\infty]{} 0,$$

$$\bar{L}_2(n) = \|(E_2 - Q_0^1)^{-1}\|[\mu^0(n)\max\{2,\frac{T}{2}\} + \max\{1,\frac{1}{T}\}\bar{\Phi}^*(n)]$$

$$\leq \|(E_2 - Q_0^1)^{-1}\|\cdot 2\max\{2, \frac{T}{2}, \frac{1}{T}\}\max\{\mu^0(n),\ \bar{\Phi}^*(n)\}$$

is the quantity uniformly bounded with respect to $n\geq n_0$. $\qquad\square$

Bibliography

[1] Ateiwi, A. M. (1996). On the numerical-analytic method for second-order difference equations, *Ukr. Mat. Zh.* **48**, 7, pp. 920–924.

[2] Ateiwi, A. M. (1997). To the problem on periodic solutions of one class of systems of difference equations, *Ukr. Mat. Zh.* **49**, 2, pp. 309–314.

[3] Avdeyuk, P. I. (1991). *Invariant Tori of Countable Systems of Differential Equations.* Author's Abstract of the Candidate Degree Thesis (Phys.-Math. Sci.) (Kiev: Inst. of Math. of the NAS of Ukraine) [in Russian].

[4] Bogolyubov, N. N. (1945). *On Some Statistical Methods in Mathematical Physics* (Kiev: AN UkrSSR) [in Russian].

[5] Bogolyubov, N. N. and Mitropol'skii, Yu. A. (1974). *Asymptotic Methods in the Theory of Nonlinear Oscillations* (Moscow: Nauka) [in Russian].

[6] Bogolyubov, N. N. and Mitropol'skii, Yu. A. (1963). *The method of integral manifolds in nonlinear mechanics.* Proceed. of Intern. Symposium on Nonlinear Oscillations. Vol. 1: Analytic Methods (Kiev: Inst. of Math. of the NAS of Ukraine), pp. 93–154 [in Russian].

[7] Bogolyubov, N. N., Mitropol'skii, Yu. A. and Samoilenko, A. M. (1969). *The Method of Fast Convergence in Nonlinear Mechanics* (Kiev: Naukova Dumka) [in Russian].

[8] Boichuk, A. A. (1990). *Constructive Methods of Analysis of Boundary-Value Problems* (Kiev: Naukova Dumka) [in Russian].

[9] Bonsall, F. F. (1960). Positive operators compact in an auxiliary topology, *Pacific J. Math.* **10**, 4, pp. 1131–1138.

[10] Busenberg, S., Fisher, D., and Martelli, M. (1986). Better bounds for periodic orbits of differential equations in Banach spaces, *Proc. Amer. Math. Soc.* **86**, pp. 376–378.

[11] Busenberg, S., Fisher, D. and Martelli, M. (1989). Minimal periods of discrete and smooth orbits, *Amer. Math. Monthly.* **96**, pp. 5–17.

[12] Busenberg, S. and Martelli, M. (1987). Bounds for the period of periodic orbits of dynamical systems, *J. Diff.Equa.* **67**, pp. 359–371.

[13] Bykov, Ya. V. and Linenko, V. L. (1968). *On Some Questions of the Qualitative Theory of the Systems of Difference Equations* (Frunze: Ilim) [in Russian].

[14] Cherevko, I. M. (2004). *Integral Manifolds and Approximation Methods of Study of Differential Functional Equations.* Author's Abstract of the Doctoral Degree Thesis (Phys.-Math. Sci.) (Kiev: Taras Shevchenko National University) [in Ukrainian].

[15] Daletskii, Yu. L. and Krein, M. G. (1974). *Stability of Solutions of Differential Equations in Banach Space* (Providence, RI: Amer. Math. Soc.).

[16] Daletskii, Yu. L. and Krein, M. G. (1950). On the differential equations in a Hilbert space, *Ukr. Mat. Zh.* **4**, 4, pp. 71–74.

[17] Danilov, V. Ya. (1983). *Study of Quasiperiodic Solutions of the Nonlinear Systems of Difference Equations* (Kiev: Preprint/Inst. of Math. of the NAS of Ukraine, 81.8) [in Russian].

[18] Danilov, V. Ya. and Martynyuk, D. I. (1981). Invariant toroidal manifolds of the systems of difference equations with slowly varying phase, *Dokl. AN UkrSSR* Ser. A., 11, pp. 8–11.

[19] Diliberto, S. P. (1956). An application of periodic surfaces. Contributions to the Theory of Nonlinear Oscillations, (Princeton: Princeton Univ. Press) pp. 257–261.

[20] Diliberto, S. P. (1961). Perturbation theorems for periodic surfaces I, II, (Rend. del Circolo Math. Palermo) Ser. 9, pp. 265–299; Ser. 10, pp. 111–161.

[21] Diliberto, S. P. (1970). *An application of periodic surfaces to dynamical systems.* Proceed. of the V-th Intern. Conference on Nonlinear Oscillations. – Vol. : Analytic Methods, (Kiev: Inst. of Math. of the NAS of Ukraine) pp. 257–264.

[22] Dorogovtsev, A. Ya. (1992). *Periodic and Stationary Modes of Infinite-Dimensional Deterministic and Stochastic Dynamical Systems* (Kiev: Vyshcha Shkola) [in Russian].

[23] El'nazarov, A. A. (1998). *Some Questions of the Theory of Countable Systems and Asymptotic Methods.* Author's Abstract of the Candidate Degree Thesis (Phys.-Math. Sci.) (Kiev: Inst. of Math. of the NAS of Ukraine) [in Ukrainian].

[24] Ermolaev, L. A. (1952). On the uniform stability by the first approximation of countable almost linear and nonlinear systems of differential equations, *Izv. AN KazSSR. Ser. Astr. Fiz. Mat. Mekh.* Iss. 1, pp. 88–114.

[25] Evhuta, N. A. and Zabreiko, P. P. (1985). On the method by A. M. Samoilenko for seeking the periodic solutions of quasilinear differential equations in a Banach space, *Ukr. Mat. Zh.* **37**, 2, pp. 162–168.

[26] Evhuta, N. A. and Zabreiko, P. P. (1985). On the convergence of the method of successive approximations by A.M. Samoilenko for seeking the periodic solutions, *Dokl. AN BelSSR* **29**, 1, pp. 15–18.

[27] Evhuta, N. A. and Zabreiko, P. P. (1991). The Poincaré method and Samojlenko method for the construction of periodic solutions to ordinary differential equations, *Math. Nachricht.* **153**. pp. 85–99.

[28] Filippov, M. G. (1990). On the reducibility of the systems of differential equations with almost periodic perturbation given on an infinite-dimensional torus, *Dokl. AN UkrSSR. Ser.* 3, pp. 30–33.

[29] Gorodnii, M. F. (2004). *Properties of Solutions of Difference and Differential Equations and Their Stochastic Analogs in a Banach Space.* Author's Abstract of the Doctoral Degree Thesis (Phys.-Math. Sci.) (Kiev: Taras Shevchenko National University) [in Ukrainian].

[30] Gorshin, S. I. (1949). On the stability of solutions of a countable system of differential equations with permanently acting perturbations, *Izv. AN KazSSR, Ser. Mat. Mekh.* Iss. 3, pp. 32–38.

[31] Gulov, Kh. M. and Perestyuk N. A. (1993). Integral sets of the systems of difference equations, *Ukr. Mat. Zh.* **45**, 12, pp. 1613–1621.

[32] Halanay, A. and Wexler, D. (1952). *Qualitative Theory of Pulse Systems* (Bucharest: Edit. Acad.) [in Romanian].

[33] Hale, J. K. (1961). Integral manifolds of perturbed differential systems, *Annals of Math.* **73**, 3, pp. 946–531.

[34] Hale, J. K. (1963). *Oscillations in Nonlinear Systems* (New York: McGraw-Hill).

[35] Hirsch, M. W. (1976). *Differential Topology* (Berlin: Springer).

[36] Ilyukhin, A. G. (1962). An approximate method of solution of a mixed problem for a nonlinear partial differential equation of the hyperbolic type with a small parameter, *Ukr. Mat. Zh.* **14**, 3, pp. 250–259.

[37] Yorke, J. A. (1969). Periods of periodic solutions and the Lipschitz constant, *Proc. Amer. Math. Soc.* **22**, pp. 509–512.

[38] Kantorovich, L. V. and Akilov, G. P. (1982). *Functional Analysis* (Oxford: Pergamon Press).

[39] Kato, T. (1995). *Perturbation Theory for Linear Operators* (Berlin: Springer).

[40] Khalilov, Z. I. (1952). The Cauchy problem for an infinite system of partial differential equations, *Dokl. AN SSSR* **84**, pp. 229–232.

[41] Khalilov, Z. I. (1961). On the stability of solutions of equations in a Banach space, *Dokl. AN SSSR* **137**, 4, pp. 797–799.

[42] Kharasakhal, V. Kh. (1950). On the fundamental solutions of a countable system of differential equations, *Izv. AN Kaz. SSR. Ser. Mat. Mekh.* Iss. 4, pp. 98–108.

[43] Kolmogorov, A. N. and Fomin, S. V. (1999). *Elements of the Theory of Functions and Functional Analysis* (New York: Dover.).

[44] Kyner, J. H. (1961). Invariant manifolds, *Rend. del Circolo Math. Palermo* Ser.11, 9, pp. 98–110.

[45] Krein, M. G. (1964). *Lectures on the Theory of Stability of Solutions of Differential Equations in a Banach Space* (Kiev: AN UkrSSR) [in Russian].

[46] Krein, S. G. (1972). *Linear Differential Equations in a Banach Space* (Providence, RI: Amer. Math. Soc.).

[47] Krylov, N. M. and Bogolyubov, N. N. (1937). *Introduction to Nonlinear Mechanics* (Kiev: AN UkrSSR) [in Russian].

[48] Kupka, I. (1964). Stabilite des varietes invariantes d'un champ de vecteurs pour les petites perturbations, *Comp. Rend. Acad. Sc. Paris* **258**, 17, pp. 4197–4200.

[49] Kuratowski, K. (1966). *Topology, vol.1* (New York, Warszawa: Academic

Press, Polish Sci. Publ.).

[50] Kurzweil, J. (1968). Invariant manifolds of differential systems, *Differ. Uravn.* **53**, 5, pp. 785–797.

[51] Kwapisz, M. (1992). On modification of the integral equation of Samoilenko's numerical-analytic method, *Math. Nachr.* **157**, pp. 125–135.

[52] Lasota, A. and Yorke, J. (1971). Bounds for periodic solutions of differential equations in Banach spaces, *J. Diff. Equa.* **10**, pp. 83–91.

[53] Lichtenberg, A. J. and Lieberman, M. A. (1992). *Regular and Chaotic Dynamics* (New York: Springer).

[54] Martynyuk, D. I. (1972). *Lectures on the Qualitative Theory of Difference Equations* (Kiev: Naukova Dumka) [in Russian].

[55] Martynyuk, D. I. (1975) Study of the vicinity of a smooth invariant toroidal manifold of a system of difference equations, *Differ. Uravn.* **11**, 8, pp. 1474–1484.

[56] Martynyuk, D. I., Kravets, V. I. and Zhanbusinova, B. Kh. (1989). *On the invariant torus of a countable system of differential equations with delay.* Asymptotic Methods in Problems of Mathematical Physics, (Kiev: Inst. of Math. of the NAS of Ukraine) pp. 77–86.

[57] Martynyuk, D. I. and Perestyuk, N. A. (1975). On the reducibility of difference equations on a torus, *Vych. Prikl. Mat.* Iss. 26, pp. 42–48.

[58] Martynyuk, D. I. and Perestyuk, N. A. (1974). On the reducibility of the linear systems of difference equations with quasiperiodic coefficients, *Vych. Prikl. Mat.* Iss. 23, pp. 116–127.

[59] Martynyuk, D. I. and Perestyuk, N. A. (1975). The reducibility of the linear systems of difference equations with smooth right-hand side, *Vych. Prikl. Mat.* Iss. 27, pp. 34–40.

[60] Martynyuk, D. I. and Ver'ovkina, G. V. (1997). Invariant sets of countable systems of difference equations, *Visn. Kyiv. Univ. Ser. Fiz.-Mat. Nauk.* Iss. 1, pp. 117–127.

[61] Martynyuk, O. M. (1992). *On the solution of a two-point boundary-value problem for countable systems.* Nonlinear Boundary-Value Problems of Mathematical Physics and Their Applications, (Kiev: Inst. of Math. of the NAS of Ukraine) pp. 68–70 [in Russian].

[62] Martynyuk, O. M. (1992). *On the interconnection of solutions of countable and truncated systems of differential equations for an initial problem.* Analytic Methods of Studies of Nonlinear Differential Systems, (Kiev: Inst. of Math. of the NAS of Ukraine) pp. 45–52 [in Russian].

[63] Martynyuk, O. M. (1993). *Studies of Solutions of Boundary-Value Problems for Countable Sys-tems of Nonlinear Differential Equations.* Author's Abstract of the Candidate Degree Thesis (Phys.-Math. Sci.) (Kiev: Inst. of Math. of the NAS of Ukraine) [in Ukrainian].

[64] Martynyuk, O. M. and Martynyuk S. V. (1991). *Studies of periodic solutions of countable systems of second-order differential equations.* Nonlinear Evolutionary Equations in Applied Problems, (Kiev: Inst. of Math. of the NAS of Ukraine.) pp. 88–90 [in Russian].

[65] Massera, J. L. and Schäffer J. J. (1966). *Linear Differential Equations and*

Function Spaces (London: Academic Press).

[66] Medved, M. (1991). On minimal periods of functional differential equations and difference inclusions, *Ann. Pol. Math.* **3**, pp. 263–270.

[67] Milnor, J. W. and Wallace, A. H. (1970). *Differential Topology: First Steps* (New York: Benjamin).

[68] Mitropol'skii, Yu. A. (1964). On the study of the integral manifold for a system of nonlinear equations close to the equations with variable coefficients in a Hilbert space, *Ukr. Mat. Zh.* **V**, 3, pp. 334–46.

[69] Mitropol'skii, Yu. A. and Lykova, O. B. (1973). *Integral Manifolds in Nonlinear Mechanics* (Moscow: Nauka) [in Russian].

[70] Mitropol'skii, Yu. A. and Martynyuk, D. I. (1979). *Periodic and Quasiperiodic Oscillations of Systems with Delay.* (Kiev: Vyshcha Shkola) [in Russian].

[71] Mitropol'skii, Yu. A., Martynyuk, D. I. and Tynnyi, V. I. (1993). The reducibility of linear systems of difference equations with almost periodic coefficients, *Ukr. Mat. Zh.* **45**, 12, pp. 1661–1667.

[72] Mitropol'skii, Yu. A. and Mykhailivs'ka, N. O. (1972). The periodic solutions of discrete difference equations of the second order, *Ukr. Mat. Zh.* **24**, 4, pp. 543–546.

[73] Mitropol'skii, Yu. A., Samoilenko, A. M. and Kulik, V. L. (1990). *Studies of the Dichotomy of Linear Systems of Differential Equations with the Help of Lyapunov Functions* (Kiev: Naukova Dumka) [in Russian].

[74] Mitropol'skii, Yu. A., Samoilenko, A. M. and Martynyuk, D. I. (1984). *Systems of Evolutionary Equations with Periodic and Conditionally Periodic Coefficients* (Kiev: Naukova Dumka) [in Russian].

[75] Moser, J. (1968). The rapidly convergent method of iterations and nonlinear differential equations, *Usp. Mat. Nauk.* **23**, 4, pp. 179–238.

[76] Moser, J. (1962). On invariant curves of area-preserving mappings of an annulus, *Nachr. Akad. Wiss. Göttingen. Math.-Phys. K1.* **11a**, 1, pp. 1–20.

[77] Moser, J. (1961). A new technique for the construction of solutions of nonlinear differential equations, *Proc. Nat. Acad. Sci.* **47**, 11, pp. 1824–1831.

[78] Moser, J. (1969). *On the construction of almost periodic solutions for ordinary differential equations.* Proceed. of Intern. Conf. on Functional Analysis and related Topics, (Tokyo: Tokyo Univ. Press) pp. 60–67.

[79] Moser, J. (1967). Convergent series expansions for quasi-periodic motions, *Math. Ann.* **169**, pp. 136–176.

[80] Pelyukh, G. P. (1994). On the existence of periodic solutions of discrete difference equations and their properties, *Ukr. Mat. Zh.* **46**, 10, pp. 1382–1387.

[81] Pelyukh, G. P. (1998). On the periodic solutions of nonlinear difference equations in the critical case, *Ukr. Mat. Zh.* **50**, 2, pp. 304–308.

[82] Pelyukh, G. P. (2002). On the existence of periodic solutions of nonlinear difference equations, *Ukr.Mat. Zh.* **54**, 12, pp. 1626–1633.

[83] Perestyuk, N. A. and Ronto, A. M. (1995). On a method of construction of successive approximations for the study of multipoint boundary-value problems, *Ukr. Mat. Zh.* **47**, 9, pp. 1243–1253.

[84] Persidskii, K. P. (1976). *Infinite Systems of Differential Equations. Differential Equations in Nonlinear Spaces* (Alma-Ata: Nauka) [in Russian].

[85] Reshetov, M. T. (1950). On the boundedness of solutions and the characteristic numbers of a countable system of linear differential equations of the triangular form, *Izv. AN KazSSR. Ser. Mat. Mekh.* Iss. 4, pp. 109–114.

[86] Ronto, A. (1995). On the boundary value problems with linear multipoint restrictions, *Publ. Univ. of Miskolc. Ser. D. Natur. Sci. Math.* **36**, 1, pp. 81–89.

[87] Ronto, A. M. (1997). *Numerical-Analytic Methods of Studies of Multipoint Boundary-Value Problems.* Author's Abstract of the Candidate Degree Thesis (Phys.-Math. Sci.) (Kiev: Taras Shevchenko National University) [in Ukrainian].

[88] Ronto, A. M. (2001). To the question about the periods of periodic motions in autonomous systems, *Ukr. Mat. Zh.* **53**, 1, pp. 94–112.

[89] Ronto, A. M. (2005). *Initial and Periodic Problems for Functionally Differential Equations.* Author's Abstract of the Doctoral Degree Thesis (Phys.-Math. Sci.) (Kiev: Inst. of Math. of the NAS of Ukraine) [in Ukrainian].

[90] Ronto, M. I. and Martynyuk, O. M. (1991). The study of periodic solutions of countable systems of the second order, *Ukr.Mat. Zh.* **44**, 1, pp. 83–93.

[91] Ronto, M., Ronto, A. and Trofimchuk, S. I. (1996). *Numerical-analytic method for differential and difference equations in partially ordered Banach spaces, and some applications* (Miskolc: Preprint / Univ. Miskolc, Inst. Math.; 96-02).

[92] Ronto, M. I., Samoilenko, A. M. and Trofimchuk, S. I. (1998). The theory of the numerical-analytic method: achievements and new directions of the development I, *Ukr. Mat. Zh.* **50**, 1, pp. 93–108.

[93] Ronto, M. I., Samoilenko, A .M. and Trofimchuk, S. I. (1998). The theory of the numerical-analytic method: achievements and new directions of the development II, *Ukr. Mat. Zh.* **50**, 2, pp. 225–243.

[94] Ronto, M. I., Samoilenko, A .M. and Trofimchuk, S. I. (1998). The theory of the numerical-analytic method: achievements and new directions of the development III, *Ukr. Mat. Zh.* **50**, 7, pp. 960–979.

[95] Ronto, M. I., Samoilenko, A .M. and Trofimchuk, S. I. (1998). The theory of the numerical-analytic method: achievements and new directions of the development IV, *Ukr. Mat. Zh.* **50**, 12, pp. 1656–1672.

[96] Ronto, M. I., Samoilenko, A .M. and Trofimchuk, S. I. (1999). The theory of the numerical-analytic method: achievements and new directions of the development V, *Ukr. Mat. Zh.* **51**, 5, pp. 663–673.

[97] Sacker, R. J. (1965). A new approach to the perturbation theory of invariant surfaces, *Comm. Pure Appl. Math.* **18**, 4, pp. 717–732.

[98] Sacker, R. J. (1969). A perturbation theorem for invariant manifolds and Hölder continuity, *J. Math. and Mech.* **18**, 8, pp. 705–761.

[99] Samoilenko, A. M. (1965). The numerical-analytic method of study of periodic systems of ordinary differential equations. I, *Ukr. Mat. Zh.* **17**, 4, pp. 16–23.

[100] Samoilenko, A. M. (1966). The numerical-analytic method of study of pe-

riodic systems of ordinary differential equations. 2, *Ukr. Mat. Zh.* **18**, 2, pp. 9–18.

[101] Samoilenko, A. M. (1966). The numerical-analytic method of study of countable systems of periodic differential equations, *Mat. Fiz. Iss.* 2, pp. 115–132.

[102] Samoilenko, A. M. (1990). *The Study of a Dynamical System in the Vicinity of a Quasiperiodic Trajectory* (Kiev: Preprint / Inst. of Math. of the NAS of Ukraine; 90.35) [in Russian].

[103] Samoilenko, A. M. (1970). *To the perturbation theory for invariant manifolds of dynamical systems.* Proceed. of the V-th Intern. Conference on Nonlinear Oscillations. Vol. : Analytic Methods, (Kiev: Inst. of Math. of the NAS of Ukraine) pp. 495–499 [in Russian].

[104] Samoilenko, A. M. (1992). The study of a discrete c system in the vicinity of a quasiperiodic trajectory, *Ukr. Mat. Zh.* **44**, 12, pp. 1702–1711.

[105] Samoilenko, A. M. (1987). *Elements of the Mathematical Theory of Multifrequency Oscillations* (Moscow: Nauka) [in Russian].

[106] Samoilenko, A. M. (1991). Dynamical systems in $\mathcal{T}_m \times E^n$, *Ukr. Mat. Zh.* **43**, 10, pp. 1283–1298.

[107] Samoilenko, A. M. (1991). The study of a dynamical system in the vicinity of an nvariant toroidal manifold, *Ukr. Mat. Zh.* **43**, 4, pp. 530–537.

[108] Samoilenko, A. M. and Laptinskii, V. N. (1982). On the estimates of periodic solutions of differential equations, *Dokl. AN UkrSSR Ser. A* 1, pp. 30–32.

[109] Samoilenko, A. M., Martynyuk, D. I. and Perestyuk, N. A. (1973). The existence of invariant tori of the systems of difference equations, emphDiffer. Uravn. **9**, 10, pp. 1904–1910.

[110] Samoilenko, A. M., Martynyuk, D. I. and Perestyuk, N. A. (1994). The reducibility of nonlinear almost-periodic systems of difference equations set on an infinite-dimensional torus, *Ukr. Mat. Zh.* **46**, 9, pp. 1216–1223.

[111] Samoilenko, A. M. and Perestyuk, N. A. (1987). *Differential Impulsive Equations* (Kiev: Vyshcha Shkola) [in Russian].

[112] Samoilenko, A. M. and Petryshyn, R. I. (1998). *Multifrequency Oscillations of Nonlinear Systems* (Kiev: Inst. of Math. of the NAS of Ukraine) [in Ukrainian].

[113] Samoilenko, A. M. and Petryshyn, R. I. (2004). *Mathematical Aspects of the Theory of Nonlinear Oscillations* (Kiev: Naukova Dumka) [in Ukrainian].

[114] Samoilenko, A. M. and Ronto, M. I. (1976). *Numerical-Analytic Methods of Study of Periodic Solutions* (Kiev: Vyshcha Shkola) [in Russian].

[115] Samoilenko, A. M. and Ronto, M. I. (1986). *Numerical-Analytic Methods of Study of Solutions of Boundary-Value Problems* (Kiev: Naukova Dumka) [in Russian].

[116] Samoilenko, A. M. and Ronto, M. I. (1992). *Numerical-Analytic Methods in the Theory of Boundary-Value Problems of Ordinary Differential Equations* (Kiev: Naukova Dumka) [in Russian].

[117] Samoilenko, A. M. and Ronto, M. I. (1980). *Numerical-Analytic Methods of Investigating Periodic Solutions* (Moscow: Mir).

[118] Samoilenko, A. M., Slyusarchuk, V. E. and Slyusarchuk, V. V. (1997). The study of a nonlinear difference equation in a Banach space in the vicinity of a quasiperiodic solution, *Ukr. Mat. Zh.* **49**, 12, pp. 1661–1676.

[119] Samoilenko, A. M. and Slyusarchuk, V. V. (1999). Discrete dynamical systems with invariant asympto-tically stable toroidal manifold, *Ukr. Mat. Zh.* **51**, 4, pp. 466–471.

[120] Samoilenko, A. M. and Teplinsky, Yu. V. (1993). *Countable Systems of Differential Equations* (Kiev: Inst. of Math. of the NAS of Ukraine) [in Russian].

[121] Samoilenko, A. M. and Teplinsky, Yu. V. (2003). *Countable Systems of Differential Equations* (Utrecht-Boston: VSP).

[122] Samoilenko, A. M. and Teplinsky, Yu. V. (1994). The method of truncation in the construction of invariant tori of the countable systems of differential equations, *Differ. Uravn.* **30**, 2, pp. 204–212.

[123] Samoilenko, A. M. and Teplinsky, Yu. V. (1985). On the invariant tori of the impulsive differential systems in the spaces of bounded number sequences, *Differ. Uravn.* **21**, 8, pp. 1353–1361.

[124] Samoilenko, A. M. and Teplinsky, Yu. V. (1994). On the smoothness of an invariant torus of a countable linear extension of a dynamical system on an m-dimensional torus, *Differ. Uravn.* **30**, 5, pp. 781–790.

[125] Samoilenko, A. M. and Teplinsky, Yu. V. (1995). On the reducibility of the countable systems of linear difference equations, *Ukr. Mat. Zh.* **47**, 11, pp. 1533–1541.

[126] Samoilenko, A. M. and Teplinsky, Yu. V. (1998). Invariant tori of the linear countable systems of discrete equations given on an infinite-dimensional torus, *Ukr. Mat. Zh.* **50**, 2, pp. 244–251.

[127] Samoilenko, A. M. and Teplinsky, Yu. V. (1998). *Limit Theorems in the Theory of Systems of Difference Equations* (Kiev: Preprint / Inst. of Math. of the NAS of Ukraine; 98.3) [in Russian].

[128] Samoilenko, A. M., Teplinsky, Yu. V. and Nedokis, V. A. (2004). The method of truncation for counta-bly point boundary-value problems in the space of c number sequences, *Ukr. Mat. Zh.* **56**, 9, pp. 1203–1230.

[129] Samoilenko, A. M., Teplinsky, Yu. V. and Nedokis, V. A. (2007). Boundary-value problems for differential equations, which are not solved with respect to the derivative, in the space of bounded number sequences, *Nonlin. Oscill.* **10**, 3, pp. 391–415.

[130] Samoilenko, A. M., Teplinsky, Yu. V. and Pasyuk, K. V. (2009). On the existence of invariant tori for countable systems of differential-difference equations defined on infinite-dimensional tori, *Nonlin. Oscill.* **12**, 3, pp. 347–367.

[131] Samoilenko, A. M., Teplinsky, Yu. V. and Pasyuk, K. V. (2010). On the existence of infinite-dimensional invariant tori for nonlinear countable systems of differential-difference equations defined on tori, *Nonlin. Oscill.* **13**, 2, pp. 253–271.

[132] Samoilenko, A. M., Teplinsky, Yu. V. and Pasyuk, K. V. (2011). Construction of an infinite invariant torus for a countable linear system of differential-difference equations by the method of reduction in its angular

variable, *Nonlin. Oscill.* **14**, 2, pp. 267–280.

[133] Samoilenko, A. M., Teplinsky, Yu. V. and Semenishina, I. V. (2003). On the existence of a smooth bounded semi-invariant manifold for a degenerate nonlinear system of difference equations in the space m, *Nonlin. Oscill.* **6**, 3, pp. 378–400.

[134] Samoilenko, M. B. (1995). The study of a discrete dynamical system in the neighborhood of an invariant torus, *Ukr. Mat. Zh.* **47**, 12, pp. 1676–1685.

[135] Schaefer, H. H. (1966). *Topological Vector Spaces* (New York: Macmillan).

[136] Schwartz, L. (1967). *Analysis Mathematique. I* (Paris: Hermann).

[137] Sharkovskii, O. N., Maistrenko, Yu. L. and Romanenko, E. Yu. (1967). *Difference Equations and Their Applications* (Kiev: Naukova Dumka) [in Russian].

[138] Slyusarchuk, V. Yu. (2001). *The necessary and sufficient conditions of existence of periodic solutions of the nonlinear equation $x(n+1) - f(x(n)) = h(n)$, $n \in Z$.* Boundary-Value Problems for Differential Equations, (Chernivtsi: Prut) pp. 237–243 [in Ukrainian].

[139] Slyusarchuk, V. Yu. (2002). *The necessary and sufficient conditions of invertibility of the operator $(Dx)(n) = x(n+1) - f(x(n))$, $n \in Z$, in a space of almost periodic number sequences.* Boundary-Value Problems for Differential Equations, (Chernivtsi: Prut) pp. 151–161 [in Ukrainian].

[140] Slyusarchuk, V. Yu. (2000). The necessary and sufficient conditions of the invertibility of nonlinear difference in the spaces $l_p(Z, R), 1 \leq p \leq \infty$, *Matem. Zam.* **68**, 3, pp. 448–454.

[141] Slyusarchuk, V. Yu. (2003). *The Stability of Solutions of Difference Equations in a Banach Space* (Rivne: Ukrainian State Univ. of Water Economy and Nature Management) [in Ukrainian].

[142] Slyusarchuk, V. Yu. (2002). *The Invertibility of Nonlinear Operators* (Rivne: Ukrainian State Univ. of Water Economy and Nature Management) [in Ukrainian].

[143] Teplinsky, Yu. V. and Marchuk, N. A. (2001). On the smoothness of the invariant torus of a countable system of difference equations with parameters, *Ukr. Mat. Zh.* **53**, 9, pp. 1241–1250.

[144] Teplinsky, Yu. V. and Marchuk, N. A. (2000). The method of truncation in the study of the smoothness of the invariant torus of a countable system of difference equations with parameters, *Zb. Nauk. Prats. Kam'yan.-Podil. Derzh. Ped. Univ. Ser. Fiz.-Mat.* Iss. 5, pp. 117–126.

[145] Teplinsky, Yu. V. and Marchuk, N. A. (2002). On the smoothness of the invariant torus of a countable system of difference equations, *Dop. NAN Ukr. Ser. A* 2, pp. 33–37.

[146] Teplinsky, Yu. V. and Marchuk, N. A. (2002). On the C^p-smoothness of the invariant torus of a countable system of difference equations defined on an m-dimensional torus, *Nonlin. Oscill.* **5**, 2, pp. 251–265.

[147] Teplinsky, Yu. V. and Marchuk, N. A. (2003). On the Fréchet-differentiability of the invariant tori of countable systems of difference equations defined on infinite-dimensional tori, *Ukr. Mat. Zh.* **55**, 1, pp. 75–90.

[148] Teplinsky, Yu. V. and Marchuk, N. A. (2003). On the Fréchet-

differentiability of the invariant torus of a nonlinear countable system of difference equations, which is defined on an infinite-dimensional torus and has deviations of the discrete argument, *Nonlin. Oscill.* **6**, 2, pp. 260–278.

[149] Teplinsky, Yu. V. and Nedokis, V. A. (1999). Limit theorems in the theory of countably point boundary-value problems, *Ukr. Mat. Zh.* **51**, 4, pp. 519–531.

[150] Teplinsky, Yu. V. and Nedokis, V. A. (1999). On the countably point boundary-value problems for countable systems of ordinary differential equations, *Nonlin. Oscill.* **2**, 2, pp. 252–266.

[151] Teplinsky, Yu. V. and Nedokis, V. A. (2003). On the countably point nonlinear boundary-value problem on the semiaxis for ordinary differential equations in a space of bounded number sequences, *Nonlin. Oscill.* **6**, 2, pp. 252–256.

[152] Teplinsky, Yu. V. and Samoilenko, M. V. (1996). On the periodic solutions of the countable systems of linear and quasilinear difference equations with periodic coefficients, *Ukr. Mat. Zh.* **48**, 8, pp. 1144–1152.

[153] Teplinsky, Yu. V. and Semenishina, I. V. (2000). On the periodic solutions of difference equations in infinite-dimensional spaces, *Nonlin. Oscill.* **3**, 3, pp. 414–430.

[154] Teplinsky, Yu. V. and Semenishina, I. V. (2003). On the Cauchy problem for degenerate difference equations of the m-th order in the Banach space, *Ukr. Mat. Zh.* **55**, 8, pp. 1127–1137.

[155] Teplinsky, Yu. V. and Semenishina, I. V. (2001). *On the Cauchy problem for degenerate difference equations in the Banach space.* Boundary-Value Problems for Differential equations, (Chernivtsi: Prut) pp. 322–333 [in Ukrainian].

[156] Teplinsky, Yu. V. and Teplinsky, A. Yu. (1996). On the periodic solutions of countable systems of quasilinear difference equations with periodic coefficients in the resonance case, *Dop. NAN Ukr. Ser. A* 1, pp. 13–15.

[157] Teplinsky, Yu. V. and Teplinsky, A. Yu. (1996). On the Erugin and Floquet–Lyapunov theorems for the countable systems of difference equations, *Ukr. Mat. Zh.* **48**, 2, pp. 278–284.

[158] Teplinsky, Yu. V. and Semenishina, I. V. (2001). On finding periodic solutions of second order difference equations in a Banach space, *Nonlin. Oscill.* **4**, 3, pp. 405–421.

[159] Teplinskiy, Yu. V. and Pasyuk, K. V. (2011). *Infinite Dimensional Invariant Tori for Countable Systems of Differential-Difference Equations.* Mathematical Analysis, Differential Equations and their Applications, (Sofia: Academic publishing house "Prof. Marin Drinov") pp. 217–228.

[160] Teplinskiy, Yu. V. and Pasyuk, K. V. (2009). *On existence of invariant tori for countable systems of differential-difference equations.* V International Scientific Conference "The Problems of Differential Equations, Analysis and Algebra", (Actobe: K. Zhubanov State University) pp. 135–137.

[161] Teplinsky, Yu. V. (2012). *On reducibility of difference equations in the space of bounded number sequences.* Proceed. of Intern. Sci. Conf. "Problems of differential equations, analysis and algebra ", (Aktobe: K. Zhubanov State

University) pp. 147–154.

[162] Teplinsky. Yu. V. and Pasyuk, K. V. (2009). On existence of invariant tori for countable linear systems of differential-difference equations, *Sci. Bull. Chernivtsi Univ.* Iss. 454, pp. 108–115.

[163] Teplinsky, Yu. V. and Pasyuk, K. V. (2010). On continuous dependence of invariant tori for countable systems of differential-difference equations on parameters, *Sci. Bull. Chernivtsi Univ.* Iss. 501, pp. 93–103.

[164] Tikhonov, A. N. (1934). On infinite systems of differential equations, *Matem. Sb.* **41**, Iss. 4, pp. 551–560.

[165] Tkachenko, V. I. (1996). On the exponential dichotomy of linear difference equations, *Ukr. Mat. Zh.* **48**, 10, pp. 1409–1416.

[166] Tomilov, Yu. V. (1994). On the asymptotic behavior of sequences given by recurrence relations in a Banach space, *Ukr. Mat. Zh.* **46**, 5, pp. 633–641.

[167] Trofimchuk, S. I. (1985). *Linear Extensions of Dynamical Systems on Compact Spaces. Some Questions of Reducibility* (Kiev: Preprint / Inst. of Math. of the NAS of Ukraine; 85.66) [in Russian].

[168] Trofimchuk, S. I. (1990). Integral operators of the method of successive periodic approximations, *Mat. Fiz. Nelin. Mekh.* Iss. 13(47), pp. 31–36.

[169] Valeev, K. G. and Zhautykov, O. A. (1974). *Infinite Systems of Differential Equations* (Alma-Ata: Nauka) [in Russian].

[170] Ver'ovkina, G. V. (1997). On the introduction of local coordinates for a countable discrete system in a neighborhood of the invariant torus, *Visn. Kyiv. Univ., Ser. Fiz.-Mat. Nauk* Iss. 4. pp. 23–29.

[171] Ver'ovkina, G. V. (1996). The periodic solutions of countable systems of difference equations, *Visn. Kyiv. Univ., Ser. Fiz.-Mat. Nauk* Iss. 1. pp. 17–20.

[172] Ver'ovkina, G. V. (1999). *Invariant Sets of Countable Systems of Differential and Difference Equations.* Author's Abstract of the Candidate Degree Thesis (Phys.-Math. Sci.) (Kiev: Taras Shevchenko National University) [in Ukrainian].

[173] Vulikh, B. Z. (1967). *Introduction to the Theory of Partially Ordered Spaces* (New York: Gordon and Breach).

[174] Zakhlivnaya, O. V. (1996). Introduction of local coordinates for a countable system of differential equations in the vicinity of an invariant manifold, *Ukr. Mat. Zh.* **48**, 4, pp. 443–451.

[175] Zhanbusinova, B. Kh. (1991). *Quasiperiodic Solutions of the Countable Systems of Differential Difference Equations.* Author's Abstract of the Candidate Degree Thesis (Phys.-Math. Sci.) (Kiev: Taras Shevchenko National University) [in Russian].

[176] Zhautykov, O. A. (1965). The principle of averaging in nonlinear mechanics for the countable systems of equations, *Ukr. Mat. Zh.* **17**, 1, pp. 39–46.

[177] Zhautykov, O. A. (1957). On the construction of the integral of a linear partial first-order differential equation with a countable set of independent variables, *Uch. Zapis. Kazpedinst. Ser. Fiz. Mat.* Iss. 8. pp. 32–41.

[178] Zhautykov, O. A. (1959). On the countable system of differential equations with variable parame-ters, *Matem. Sb.* **49**, Iss. 3, pp. 317–330.

Index

asymptotic periodicity of solutions,
253
attractor, 58

boundary-value problem on an
interval, 310
boundary-value problem on the
semiaxis, 290

coordinatewise differentiability, 89
countable-point boundary-value
problems, 289

degenerate case, 207

existence of the Green–Samoilenko
function, 166
extension of solutions, 260, 276

Green–Samoilenko function (GSF),
61, 63

invariant manifold, 173
invariant torus, 63, 138

Lyapunov-stable solution, 37, 59

method of accelerated convergence,
10
method of truncation of the system of
equations, 13
monodromy matrix, 194

multipliers of equation, 194

Newton–Kantorovich method, 263
nondegenerate nonresonance case, 194
nondegenerate resonance case, 198

periodic solutions of
nonlinear equations of the second
order, 229
linear and quasilinear equations,
193
nonlinear difference equations of
the first order, 216

reducibility problems, 4
reduction to a finite-dimensional case,
315, 364

Samoilenko method, 193
semi-invariant manifold, 173, 175
sequence of matrices is
proper, 7
strongly proper, 7
sharpened Cauchy–Lipschitz
conditions, 17
splitting of system, 49
system is degenerate, 173

theorem of
Erugin, 2, 4
Floquet-Lyapunov, 2, 6
truncation method, 89